HEURISTIC SCHEDULING SYSTEMS

ETM WILEY SERIES IN ENGINEERING & TECHNOLOGY MANAGEMENT

Series Editor: Dundar F. Kocaoglu

HEURISTIC SCHEDULING SYSTEMS

With Applications to Production Systems and Project Management

THOMAS E. MORTON AND DAVID W. PENTICO

A WILEY-INTERSCIENCE PUBLICATION

JOHN WILEY & SONS, INC.

NEW YORK   CHICHESTER   BRISBANE   TORONTO   SINGAPORE

Library of Congress Cataloging in Publication Data:
Morton, Thomas E., 1937–
    Heuristic scheduling systems : with applications to production
systems and project management / Thomas E. Morton, David W. Pentico.
        p.   cm. — (Wiley series in engineering & technology
management)
    Includes bibliographical references and index.
    ISBN 0-471-57819-3
    1. Production control.   2. Scheduling (Management)   3. Heuristic
programming.   I. Pentico, David W.   II. Title.   III. Series: Wiley
series in engineering and technology management.
TS157.5.M67   1993
658.5'3—dc20   92-37936

Printed in the United States of America

10 9 8 7 6 5 4 3 2 1

This book is dedicated to Professor Ken Baker
who loves to teach but does not sacrifice meaning

CONTENTS

PART II   ONE-MACHINE PROBLEMS

PREFACE

Heuristic scheduling systems that deal with the complexities of manufacturing under global competition are currently much sought after. It is increasingly recognized that all the strengths of traditional operations research, knowledge based systems, and sophisticated user interfaces will be necessary to build the needed systems of the future. Such successful integration of approaches has not really occurred yet, although there are many systems being built by users, software houses, and academics that attempt some integration.

What seems to be needed are builders who understand all three approaches well. This book is intended to provide a solid foundation in OR approaches, both exact and heuristic, for those who need to design, build, or critique systems, while minimizing excess mathematical baggage. Discussion of np-completeness and worst-case-bound analysis will not be found here, because these are available elsewhere. Brief introductions *are* provided to the problems and opportunities inherent in combining OR, AI, and MIS approaches. While the book most completely develops new and powerful heuristic methods for building scheduling systems, it presents a thorough foundation for exact methods, so that the book is suitable as an introductory textbook in scheduling and project management.

This book should be useful and accessible to practitioners with backgrounds in operations research, artificial intelligence, industrial engineering, computer science, information systems, MIS, and other technical manufacturing backgrounds. As a textbook it is intended to be flexible enough for advanced undergraduate, masters, and Ph.D. curricula. For example, no prior exposure to integer programming or dynamic programming is assumed in the text. A chapter in the text provides enough intuitive understanding of such techniques to master the text and use the software provided. Undergraduates and less technical MBAs, for example, might easily make a good semester course by omitting integer programming, dynamic programming, and other search methods, concentrating on the full development of dispatch heuristics and learning to develop heuristic systems via the software provided. Somewhat more technical students without OR backgrounds could use the complete text and more advanced software to make a more challenging semester course. Of course, students with good OR backgrounds and reasonable programming skills would be in the position to move quickly through the text and the software and to develop and test their own ideas, using the software as a research tool.

A bit of advice is probably in order for selecting from this text to teach a semester course. Except for very advanced students, it is probably not reasonable to use the entire text. Therefore we have marked with (A) advanced materials that can easily be cut without much loss of continuity or scope of the text (whether a chapter, a section, a subsection, a proposition/heuristic, or a proof). Similarly, we mark with (S) additional topics that can easily be cut, but at the expense of a loss of scope. (There is other capstone material that could be deleted, but at a loss of unity of the material, such as Chapter 21.)

The authors cut their teeth on *Introduction to Sequencing and Scheduling* by Kenneth R. Baker, published in 1974, and now out of print. With that author's kind permission, a fair amount of material from that book that is still very relevant has been adapted for use here; for example, certain discussions of classical mathematical techniques, many examples, and exercises.

The authors would like to acknowledge the significant contributions of Venkatesh Narayan, Prasad Ramnath, Susan Slotnick, and Ramakrishnan Srinivasan to the development of this book and the accompanying software.

THOMAS E. MORTON
DAVID W. PENTICO

HEURISTIC SCHEDULING SYSTEMS

PART I
PRELIMINARIES

___1
INTRODUCTION

1.1 SCHEDULING IS STRATEGIC!

The Japanese long-range concerns for just-in-time production (JIT), raising product quality, reducing shop throughput times, and increasing system flexibility now permeate the consciousness of American industry as well. This in turn has dramatically increased the perceived importance of scheduling. Twenty to 30 years ago automatic scheduling for complex shop systems was largely an academic exercise; today there are a number of successful commercial projects for both shop and project scheduling, and new software development is proceeding at a rapid pace, both for customized and off-the-shelf systems.

In the old days, before JIT, it was considered reasonable to have large work-in-process (WIP) buffer stocks and finished goods stocks to absorb scheduling errors and to decouple complex systems. The resulting simpler modules could be managed adequately based on the practical experience of a single human expert. Today it is widely accepted that these comfortable buffers need to be progressively eliminated for a number of reasons:

(a) The increasing complexity and rapid obsolescence of products.

(b) More rapid learning to eliminate process flaws.

(c) Customer desires for much shorter lead times with easier order revision during that lead time.

(d) More flexible reaction to desired changes in product mix.

(e) More flexible reaction to emergency floor problems.

(f) More rapid awareness of variation in product quality and its exact source.

Without large WIP inventories to decouple the stages of the system, the human expert must struggle to adapt to the larger system directly. At best, this leads to a lessened ability to deal with the system in as much detail; at worst, the broad understanding of the problem may be lost as well. This situation is made worse by the fact that the system itself is now typically changing rapidly and becoming more complex.

By the way, the Japanese have not yet solved the general problem of scheduling with low buffers, although they are working hard on it. JIT and Kanban were developed for rather simple flow systems. While Japanese (and world) effort continues (properly) to redesign complex systems into simpler systems, technological forces toward greater complexity are large.

Breakthroughs in scheduling methodology, practice, and software are sorely needed. All useful approaches should be pursued. The following major approaches are complementary to each other:

(a) Improved methods for training human experts.
(b) "Expert systems" to imitate human experts.
(c) Mathematical scheduling systems (exact or approximate).
(d) Hybrid systems combining the strengths of the other approaches.

Let us very briefly give an overview of the inherent strengths and weaknesses of each of these approaches. We have already explained that the human scheduler will have much greater difficulty due to reduced buffers, rapid system change, and increases in complexity. A related issue is that it is becoming much more difficult for a retiring expert to adequately train replacements. But we must not forget the great strengths of human expertise as well. Humans have (a) a very broad base of "common" sense, (b) intuitive, poorly understood thinking abilities, and (c) the ability to strategically summarize knowledge. (In particular, "common" sense is not common at all; it is the ability to apply broadly expertise learned in one area to other areas. Only humans have this ability.) These abilities taken together often work to allow experts to react quickly and accurately to glitches and problems.

Highly touted "pure" expert systems (such as the Digital Equipment Company computer configurer) can codify more routine historical human expertise and, with some difficulty, can mechanize and mass produce it. However, these "pure" systems cannot (a) codify subtle and/or controversial expertise, (b) create new expertise in new situations, or (c) utilize broad based commonsense knowledge. In particular, in rapidly changing environments, these systems must be revised carefully and frequently with the assistance of an adequate expert who must determine the changes and then help implement them. (It is often said in this regard that expert systems are "brittle" or that they model "surface expertise." Much of the leading edge research in such systems concentrates on how to model "deeper expertise" so that the system will be more adaptable to change.)

One advantage of mathematical scheduling systems is that, theoretically at least, a real shop floor could be modeled very accurately and very completely. Also, it may be possible to write the software with generic modules, so that changes in the floor could be modeled simply by choosing a few modules and reconnecting existing ones. Properties of "deep structure" are inherently provided by the optimization procedure itself.

A major difficulty with this approach is that, if the system is modeled in detail, it is much too large to be solved in a timely way, even by the most powerful computers. If the system is modeled to be solvable, so much detail will be lost as to make the solution unusable. A related problem is that designing and implementing new modules into the software are technically intensive and time consuming; there are an almost unlimited variety of modules that would be needed, since scheduling systems are so diverse. The bottleneck dynamics approach to be developed in this book mitigates some of these problems. Accurate approximate procedures are developed, which can solve very large problems in a reasonable time. Schedules can be corrected for small changes in the input without resolving the model. Very general economic principles allow building new modules relatively easily. (See Chapter 2, especially Section 2.4.)

Nevertheless, in the long run it seems necessary for viable scheduling systems to combine the best of historical human expertise, theoretical or mathematical knowledge, and the common sense of the current user. Such a cooperative multi-expertise system would be difficult to achieve but would pay handsome dividends. While this topic is largely outside the scope of this book, a brief development is given in Chapter 21.

Our purpose here is primarily to discuss procedures for designing mathematical scheduling systems. We interpret the term "mathematical" broadly, as models having a more or less complex representation of the system, an objective to be optimized, and a broad spectrum of possible methods for making decisions from the most primitive to the most complex. In Section 2.2 we review the strengths and weaknesses of various scheduling methods used in the past. In Sections 2.3 and 2.4 we give an overview of some of the newer methods being developed, including bottleneck dynamics. (The OPT system designed by Creative Output, which enjoyed great commercial success in the early to mid-1980s, is an early example of a system using bottleneck methods.)

An entirely different reason why scheduling is strategic is that, broadly interpreted, it pervades all economic activity. Stated most generally, scheduling is the process of organizing, choosing, and timing resource usage to carry out all the activities necessary to produce the desired outputs at the desired times, while satisfying a large number of time and relationship constraints among the activities and the resources. Thus scheduling systems include (among others) projects of all types, logistics systems, material requirements planning systems, job shops, process shops, assembly shops, transfer lines, research teams, management planning groups, high-volume repetitive lines, and dispatch and control stations. If underlying principles can be found for job shops, which are sufficiently general, then it can be hoped they will have application in these related areas. Section 1.3 discusses classification of scheduling problems at different aggregation levels in a little more detail. Section 1.4 presents a classification of different types of systems at each level of aggregation.

In Chapter 3, Section 3.1 presents a more detailed classification of the types of job shop problems in the order of development employed in this book. (This classification is somewhat richer than for the classical one-machine problem.) Section 3.2 gives a brief summary of each of the remaining chapters of the book. Finally, Section 3.3 gives an overview of bottleneck dynamics extensions, which are beyond the scope of this book.

1.2 WHAT IS SCHEDULING?

1.2.1 The Broad Idea

Scheduling occurs in a very wide range of economic activities. It always involves accomplishing a number of things that tie up various resources for periods of time. The resources are in limited supply. The things to be accomplished may be called "jobs" or "projects" or "assignments" and are composed of elementary parts called "activities" or "operations" and "delays." Each activity requires certain amounts of specified resources for a specified time called the "process time." Resources also have elementary parts, called "machines," "cells," "transport," "delays," and so on.

Scheduling problems are often complicated by large numbers of constraints relating activities to each other, resources to activities and to each other, and either resources or activities to events external to the system. For example, there may be *precedence constraints* connecting activities that specify which activities must precede other activities, and by how much of a delay, or by how much allowed overlap. Or two particular activities may interfere with each other and be unable to use the same required resource simultaneously. Or it may not be possible to use two resources simultaneously during certain parts of the day or on the same activity. Or a resource may be unavailable during specified intervals due to planned maintenance or planned use outside the system. Since these complex interrelationships can make exact or even approximate solutions of large scheduling problems very difficult, it is natural to attempt to solve simpler versions of a problem first to gain insight. Then one can test how sensitive the solution is to this complexity and find approximate solutions to difficult problems where the complexity proves central.

1.2.2 What Is a Good Schedule?

Finding a good objective to maximize or minimize can be difficult for a scheduling problem for several reasons. First, such important objectives as customer satisfaction with quality or promptness are difficult to quantify and do not appear among the accounting numbers. Second, a shop is usually dealing with three types of objectives, which are quite different:

(a) maximize shop throughput over some time period;

(b) satisfy customer desires for quality and promptness; and

(c) minimize current out-of-pocket costs.

Several approaches are possible:

(a) solve problems with one objective at a time;

(b) solve for trade-off curves between objectives; and

(c) combine objectives by assigning costs to customer desires and lack of utilization.

We consider all these approaches in this book, although our ultimate approach is to work toward assigning dollar values to service and utilization—that is, choice (c). We return to this approach later.

1.2.3 Grouping Activities

It is natural to inquire: When is a group of activities one job or project and when is it several? Similarly, when is a set of machines a single resource and when is it several? To answer the first question, we note that there are some particular activities of special interest: those whose completion allows a completed product to leave the shop and thus bring in an economic benefit. (Or at least their completion signals some progress payments or something of that sort.) We might term these "product completion" or "project completion" or "project milestone" activities. Other activities are of indirect interest: their completion allows other activities to start, which eventually allow a completed product. We call these "intermediate" activities. A product completion, together with all its necessary supporting intermediate activities, is called a "job." Similarly, a project completion and its supports are called a "project" or, if part of a larger project, a "milestone."

1.2.4 Grouping Resources

Defining single and multiple resources is a little more complicated. The stylized single (fixed) resource is defined by there being no internal decisions left to be made and a single queue of activities waiting for its use. In particular, at any time there will be activities in process, activities waiting in the queue, and activities yet to arrive. In order to be considered a single resource, the only decisions to be made over time are

(a) the next activity to be started,

(b) the start (release) time for the activity on the resource, and

(c) the completion (release) time for the activity to release the resource.

To illustrate the idea here, a single milling machine is one resource by this definition. Given a fixed set of waiting activities, the next activity to be started is a sequencing decision. Typically, the completion time is simply the finish time of the last activity, after which the material being worked on is removed. Thus there is only a single decision here: "the next activity to be started."

Next, consider two identical milling machines. If there are two queues, and each machine is scheduled separately, then there are two resources. But if there is a single queue with the discipline that the next job goes on the first available machine, it reduces the situation again to a single resource problem, with the only decision to be made being to sequence the waiting jobs.

If several resources are involved in completing a given activity more or less simultaneously, then, for that activity, we must determine its start time and stop time on each resource and its resulting completion time. This can become quite complicated, especially if an activity requires different amounts of and times on the different resources or if different activities are quite varied in their mix of resource needs.

1.2.5 Multiple Resources: Routing

In the more general multi-resource system, an activity or set of activities may have several choices of resources to accomplish their completion. For example, a shaft may be milled in a new FMS cell, or it could be turned out more slowly on a general purpose manual machine. Three different choices for an activity are called three "routes"; finding the best route is called the "routing problem." The routing problem may also exist for a group of activities as a whole: for example, a roll of paper could be made on an old line or a new line.

1.2.6 Scheduling Reconfiguration

To date we have given a very general description of the scheduling problem: predefined activities and predefined resources, with perhaps time windows of availability for each activity and resource, and an objective involving timely completion and reasonable operating costs. This is the problem most treated by scheduling research and books in the past; we may call it the "classical scheduling problem."

But there are a number of other decisions that interface with the classical problem in practice. If we add some of these interface decisions to our system model we may speak of the "extended scheduling problem." One such decision is to change the amount and the internal and/or external configuration of the resources while the scheduling problem progresses in order to compensate for the changing load on the system. For example, a machine might be able to speed up during an extremely busy period to level the load somewhat, at the expense of increased tool wear. Or a machine might be reconnected physically (via changing transport connections) to add capacity to a different part of the system. Or extra personnel might be added dynamically to different parts of the system over time. Or extra resources might be leased, or an obsolete line started up temporarily. It is also possible that resources are shared between two systems—with some statement of time intervals of availability to each or windows that might be due to scheduled maintenance. One decision could be to increase the window for a critical resource. We summarize these types of decisions under "resource reconfiguration" or simply "configuration" for short.

In a similar fashion we can reconfigure activities, jobs, and projects. For example, there may be several different choices of technology for performing an activity even in the short run, requiring different amounts of the resources, and requiring more or less processing time as a result. (We call this the resource/time trade-off). This is an internal reconfiguration. An external reconfiguration would be to negotiate due dates and penalty costs with the customer based on our projection of the shop load and competing customers. Or it may be possible to increase the window (arrival time, due date) alternatively by expediting processing in a previous system or transport, or by planning to expedite processing in a successor system or transport.

1.2.7 Information System Issues

So far, we have talked about the primary classical scheduling decisions, which are of three types: sequencing, timing, and routing. We have also talked about somewhat larger related decisions that must often be considered, such as reconfiguring resources or reconfiguring activities. There are also a number of management information system (MIS) decisions that need to be made to support the primary decisions discussed above. These include *forecasting, labeling, grouping, aggregating,* and *disaggregating.* We discuss each of these briefly in the following paragraphs.

We think of three types of *forecasting* for scheduling: *external, internal, and distributed.* A typical *external* forecast would be to attempt to estimate customer requirements over the next three months. Such a forecast is typically a mixture of known orders, orders in negotiation, and statistical estimation of orders as yet unknown. This is a rich topic that cannot be considered adequately in this book. We typically assume a known nonstationary set of demands and due dates, with the understanding that it will be revised frequently. (An exception is due date setting models, which we also consider.)

Internal forecasts, however, are typically developed by the scheduling heuristic itself as part of the solution process. For example, the program might wish to make a rough estimate of the lead time, or time to completion, for various jobs in the shop. The known due dates out of the shop then allow estimating the times the jobs are due off the current operation. A simple heuristic would then be to give the job with the earliest operation due date the highest priority. Bottleneck dynamics makes extensive use of such forecasts, including current lead times, current job delay costs, current resource criticality, future resource criticality, and future arrival times.

Large scheduling systems may be *distributed.* For example, there may be one scheduling subsystem for each department in the shop, with a planning system controlling them. Then an upstream subsystem might supply projected arrival times to the current system, while the downstream subsystem might supply needed due dates off the current system. Similarly, the planning system might send down an overall view of resource criticalities, while the subsystem of interest might send corrections back based on its detailed knowledge of its own area.

Keeping track of where a customer's order is on the floor currently is clearly important and sounds straightforward. However, if a given piece of WIP can have multiple final uses, the dynamic *labeling* issue arises. There is usually need for considerable flexibility in this regard. For example, the scheduling program may label a particular bit of work in process or inventory as intended for fulfilling one job that is deemed to be of the highest priority. If a new need arrives with higher priority, or if priorities change, then the system may automatically relabel this WIP for a different purpose. If this happens too often, customers may complain. We discuss various ways of dealing with this "nervousness" phenomenon. In particular, one way of dealing with final inventory safety stocks is to assign these stocks to projected customers and do the safety stocks in terms of safety time rather than safety stock.

The term *grouping* can have two slightly different meanings. If our detailed scheduling model tells us to run a set of activities together, this is a detailed *timing* or *lot-sizing* problem. Although we may say we have "grouped" the items in casual speech, we do not use group in that sense here. Grouping here always means kind of heuristic aggregation, which may cause some loss of accuracy in return for convenience and easy computation. For example, if we decide that a set of items have similar characteristics and always keep them together throughout the shop with fixed internal priority, we have in fact aggregated these activities into a group. It is also possible to disaggregate the group at some machines. The whole issue of aggregation of activities, resources, and models is highly complex. We deal with it briefly in Chapter 21.

1.2.8 Summary

So then, to summarize, what is scheduling? A scheduling system dynamically makes decisions about matching activities and resources in order to finish jobs and projects needing these activities in a timely and high-quality fashion while simultaneously maximizing throughput and minimizing direct operating costs. (The scheduling plan may be at different levels of detail/realism depending on the need.) Types of basic scheduling decisions to be made (classical scheduling) include

(a) sequencing,

(b) timing/release, and

(c) routing.

Added decisions for extended scheduling models include

(d) resource reconfiguration and

(e) activity reconfiguration.

In addition, a number of supporting MIS decisions are usually needed, including forecasting, labeling, grouping, aggregating, and disaggregating.

(Roughly speaking, the topics above have historically received decreasing attention as one goes down the list, due to their increasing mathematical difficulty, especially in terms of exact mathematical solution. Because we are oriented toward heuristics in this book, we redress the balance somewhat. However, problems that can be solved mathematically provide insight into the more difficult problems and hence still receive a somewhat larger emphasis than their real world importance would dictate.)

To date we have been talking on a fairly abstract plane. In the next two sections things become somewhat more concrete.

1.3 LEVELS OF SCHEDULING

1.3.1 Overview

It should be clear from the discussion in the previous section that whether a group of assets is considered a single resource or not for scheduling purposes depends on whether internal decisions are being considered on our current level of abstraction or are simply given by some fixed "rules" that are transparent to us. Thus a given job shop might be considered to be a number of resources when modeled in detail; one level up, however, we may only model the decision of what to release to the shop next and when. That is, at this next level of abstraction the shop is being considered a single resource. This can be repeated several times, leading to a classification of manufacturing problems at several levels. Most such schemes have four or five levels. We present one with five levels in Table 1.1.

While all these levels can be considered scheduling in that they all have issues of sequencing, timing, routing, reconfiguration, forecasting, labeling, grouping, aggregation, and disaggregation, this formal similarity should not be pushed too far. There are a wealth of scheduling models at Level 4; these constitute the principal topic of this text. In attempting to model a different level, it is useful to recognize the similarities, but the differences are also crucial. Careless transfer of models intact to a different level is usually disastrous. We discuss modeling issues at each

TABLE 1.1   Classification of Scheduling Levels

Level	Examples of Problems	Horizon
1. Long-range planning	Plant expansion, plant layout, plant design	2–5 years
2. Middle-range planning	Production smoothing, logistics	1–2 years
3. Short-range planning	Requirements plan, shop bidding, due date setting	3–6 months
4. Scheduling	Job shop routing, assembly line balancing, process batch sizing	2–6 weeks
5. Reactive scheduling/control	Hot jobs, down machines, late material	1–3 days

level in a little detail, paying special attention to similarities and differences with Level 4 scheduling. These levels differ, especially in forecasting issues. We turn next to discussing each level separately.

1.3.2 Level 1: Long-Range Planning

At the long-range planning level, types of "jobs" or "projects" include location, sizing, and design of plants, warehouses, departments, transfer lines, and FMS cells, expansion of plants and warehouses, layout, and design. Activities are various smaller pieces of these projects, while resources include financing capability, engineering capacity, and management capacity. With several projects and many concurrent activities, sequencing and timing issues are clearly relevant. Routing is concerned with such issues as alternate sources of financing and alternate machinery vendors.

However, these formal similarities with scheduling are often swamped in practice by the overwhelming forecasting issues and related needs for extensive activity and resource reconfiguration. What will demand be five years from now? What kinds of scenarios can be envisioned? What technology will be available? What is the risk of buying a high-tech assembly line that may be obsolete in three years? What type of factory is safest? Should production plants be dispersed geographically?

Not only do forecasting problems dominate Level 1 problems, but current forecasting methods are very weak. A common method is to calculate a number of possible scenarios, each based perhaps on a deterministic forecast with or without safety stocks. Good stochastic models are clearly needed, but it would seem that current methods are not adequate, since models with known means and variances uncorrelated from period to period simply do not capture the situation.

As an example of the dangers of applying models across levels, consider the issue of timing of successive plant expansions to balance the economics of scale in expansion and costs of idle cash invested in excess capacity. These problems have often been attacked by lot sizing models originally developed for Level 4 scheduling. Unfortunately, those models abstract away reconfiguration and forecasting issues and have not been very successful.

There has also been a lot of work applying project scheduling models to these types of very large and long-range projects. These applications are somewhat more successful, although the forecasting issue remains extremely difficult. In particular, reactive correction and redesign of such a schedule are not well understood.

1.3.3 Level 2: Middle-Range Planning

At the middle-range planning level, scheduling methods have been more successful. Production smoothing deals largely with resource reconfiguration over time to provide balanced capacity facing seasonal demand loads. External resource changes include hiring, firing, layoff, and subcontracting; internal changes include overtime, opening lines, and shifting resources between shops. A production schedule is definitely required, so sequencing, timing, and routing issues are directly relevant. In the past, highly aggregated versions of production activities and available resources took away much of the scheduling flavor. With advances in computational power, and more accurate heuristics, it is now practical, for example, to schedule eight product lines on 20 resources monthly over an 18-month period.

Forecasting issues remain very difficult. Production smoothing models and logistics models are often handled in practice as being deterministic with adjustments for safety stocks (as in Levels 3, 4, and 5, although this is somewhat less successful here). In logistics models, methods that incorporate congestion effects deterministically into network models seem to hold promise. We discuss this issue briefly in Chapter 21.

Both stochastic and deterministic models have been employed successfully at Level 2. Although forecasting issues are important and not fully resolved, issues of grouping, aggregation, and disaggregation probably dominate.

1.3.4 Level 3: Short-Range Planning

For short-range planning, models of interest include material requirements planning (MRP), shop bidding, and due date setting. MRP accepts as input the production leveling plan from Level 2, known orders, orders in negotiation, and statistical forecasts of remaining orders over the 3–6-month horizon to produce a "master schedule," which is a time phased forecast of what needs to be produced by product type, quantity, and time period. Each piece of final forecast need at a point in time is then "exploded," that is, broken down into components that are offset backward by a standard lead time for final assembly. Each component is then broken down and offset backward again, and so on. When components become raw materials due in at a certain date, this is translated into an order from a vendor. In this first pass, resource capacity is not taken into account. Timing is solved in a very simple way, by offsetting back from the due date by historical lead times. Then the shop basically is scheduled on a first-come first-serve basis, obviating both sequencing and timing. Routing choices are not considered. Such a simple procedure often produces an erratic shop load. MRP estimates a crude profile of resource usage, adjusts the master schedule, and tries again. This may be repeated.

More sophisticated MRP methods to be discussed in this book actually do an aggregated form of scheduling. Resource constraints are taken into account automatically, and lead times are actual lead times for the process rather than historical. Such a Level 3 system can also serve as a planning module for Level 4 scheduling systems in an automatic mode.

Shop bidding and due date setting models allow balancing the shop in a different fashion than repeatedly readjusting the master schedule. Higher prices and slower delivery can be negotiated for the overworked shop; lower prices and fast delivery can be offered when the shop schedule has slack. Thus these approaches are primarily activity reconfigurations in nature. In the past, shop bidding and due date setting models have been done largely as a "seat of the pants" exercise by an experienced negotiator. We discuss methods in this book which use a dynamic scheduling representation of the current and near-future shop to suggest which types of prices and delivery promises are appropriate.

With a 3–6-month horizon, the master schedule is a mixture of known orders, orders in negotiation, and unknown orders. The latter are estimated by statistical forecasting techniques such as exponential smoothing. Theoretical work on such forecasting mixtures is sparse; however, see Morton et al. (1990). In practice, deterministic systems with safety stocks are usually employed for material requirements planning, while shop bidding or due date setting models may be either deterministic or probabilistic.

1.3.5 Level 4: Scheduling

With scheduling, we are down to a 2–6-week horizon, depending on whether activity processing times are more on the order of a day or a week. The upper level MRP (or similar) system lays out a forecast of arrivals to the scheduling system, giving arrival times, due dates, and quantities of each job, together with the activities necessary to complete it. The full schedule for, say, a 6-week period might be developed once a week, using currently updated input and shop status. This is typically a major activity even for heuristics and might take perhaps 8 hours on a mainframe. In a week, the schedule would be solved again so that only the first week of each schedule is actually used. The remainder is just to make sure the current week's plans fit properly into future weeks. There will frequently be corrections to the schedule needed between the weekly rescheduling. But a major 8-hour run is not practical many times a week, not only due to the cost but because the resulting system would not be real time. This problem is dealt with at Level 5 below.

The Level 4 scheduling operates using a fairly accurate master schedule of upcoming jobs, priorities, and due dates for the next few weeks. The standard rolling horizon model mentioned earlier, with reactive correction for glitches and emergencies, seems to be preferred in practice to fancier stochastic modeling.

We do not spend any more time discussing scheduling issues here, since they occupy much of the rest of the book. Dominant issues at the scheduling level are sequencing, timing, and routing, although forecasting and other MIS issues need more attention than they have received classically.

1.3.6   Level 5: Reactive Scheduling/Control

At the most detailed level of reactive scheduling and control, central issues are emergencies and glitches, such as machine breakdown, material failure, or late arrivals. Detailed constraints for activities and resources that might be ignored at Level 4 must be treated. Much activity is involved around correcting the Level 4 schedule for these emergencies, glitches, and details. (Much of this detail has been handled manually, although it seems appropriate for a simple expert system cooperating with a Level 4 scheduling system.) As far as reactively scheduling in midweek to correct for breakdowns or late arrivals, old manual scheduling systems were robust. Since they used simple priority dispatch rules, which were just executed "on the fly" during the week, changes could usually be incorporated by applying the same priorities to the changed shop. Or else one or more priorities could be changed, and then scheduling could proceed. Successor mathematical scheduling systems could not usually allow scheduling modification in such an easy fashion. The impracticality of making a complete multi-hour run after every minor break-down or change in arrival times led in practice to complete manual control of the system throughout the week. This in turn tended to lower confidence in the automatic system, and in many cases to its demise.

The bottleneck dynamics approach attempts to incorporate the best of both worlds. A multi-hour run is still made once a week. This run establishes dynamic priorities, which in turn allow the system to be scheduled during the week, incorporating the changes by running a low-cost simulation (say, 15 minutes). A large number of such updating runs can be made without losing the necessary real-time feature of the system.

Forecasting issues for Level 5 have largely been discussed. Methods for more accurately predicting emergencies and glitches are important but are not considered here.

1.4 TYPES OF LEVEL 4 SCHEDULING ENVIRONMENTS

In the last section we looked at scheduling at different levels of time horizon and aggregation. Here we fix our attention on Level 4 scheduling environments, which range from small, complex, one-of-a-kind custom job shops to high-speed, low product variety transfer lines, from discrete parts manufacturing to continuous process flows. Table 1.2 provides a summary of these environments.

1.4.1 Classic Job Shop

A typical example of a classic job shop is a research machine tool milling company. Each order is unique and has a unique routing. Operations are performed sequentially on a single lot of parts, which travel together through the shop. There are no floor inventories that are not identified with a single activity. Scheduling is highly complicated and does not repeat in any simple way. The classic job shop is also called a "closed" shop because orders are distinct and WIP cannot typically be borrowed from one job to another.

However, while it may not be readily apparent, many very different production environments can be identified as having many of the characteristics of a closed job shop. For example, any customized one-time project, from designing and building a fancy home to research and development on the prototype space shuttle to nonstandard paperwork that flows across a desk, shares many of the features of a job shop. Knowledge of how to schedule a machine tool milling shop gives many insights into these other problems. However, there are also large differences that should not be ignored. Many of these center around time scale. With longer time horizons and concomitant changes in technology, a number of new issues come to the fore. How

TABLE 1.2   Scheduling Environments

Type	Characteristics
1. Classic job shop	Discrete, complex flow, unique jobs, no multi-use parts
2. Open job shop	Discrete, complex flow, some repetitive jobs and/or multi-use parts
3. Batch shop	Discrete or continuous, less complex flow, many repetitive and multi-use parts, grouping and lotting important
4. Flow shop	Discrete or continuous, linear flow, jobs all highly similar, grouping and lotting important
5. Batch/flow shop	First half, large continuous batch process; second half, typical flow shop
6. Manufacturing cell	Discrete, automated grouped version of open job shop or batch shop
7. Assembly shop	Assembly version of open job shop or batch shop
8. Assembly line	High-volume, low-variety, transfer line version of assembly shop
9. Transfer line	Very high-volume and low-variety linear production facility with automated operations
10. Flexible transfer line	Modern versions of cells and transfer lines intended to bring some of the advantages of high-volume production to job shop items

do frequent engineering change orders and other redesigns affect the scheduling effort? How does one estimate activity duration and progress? These questions are not approached in this book; we only desire to alert the reader to the advantages and perils of drawing scheduling analogies.

1.4.2 Open Job Shop

Traditionally, a shop that produces to final inventory rather than directly to orders is called an "open shop." We use the term here in a similar but more general fashion where there may be several customers with demand for the same (or nearly the same) products so that it makes sense to maintain final (or nearly finished) inventories, or larger WIP, or to divert activities or jobs meant for one customer to another customer of higher priority. Scheduling issues are similar to those of the closed shop except that the labeling of partially or fully completed jobs according to customer and due date becomes more complex and dynamic. A careful procedure for relabeling becomes necessary. A common example might be a job shop that produces machine tool replacement parts for several automotive companies.

The production run for the space shuttle would perhaps be somewhere between a closed and an open job shop. Production of mainframe computers is often a confusing mixture of custom orders and repeat orders. A tract house builder is an excellent example of an open job shop. Extremely long time horizons and high uncertainty and revision are less of a problem here but are still important.

1.4.3 Batch Shop

A batch shop is basically an open shop for which the duplication in WIP and final production between customers becomes so large that large batch processing is typical to take advantage of economics of scale in processing similar parts (lot sizing). Flow through the shop is not completely linear, but it is usually less complex than for closed or open job shops. One example of a discrete batch shop would be a garment factory. Another example might be a manufacturer who supplies various small parts to other manufacturers.

   An example of a continuous batch shop might be an oil refinery. Another might be a chemical process factory. A batch process with a longer horizon and corresponding engineering changes during production might be undergraduate or graduate education. Ph.D. student production at some schools might be a job shop, and at others a batch process!

1.4.4 Flow Shop

A flow shop is basically a batch shop with linear flow. Flow can be discrete, continuous, or semicontinuous (bottling). Grouping and lotting tend to be important. In the simplest case each job consists of the same set of activities to be performed in the same sequence on the same set of machines for every job. That is, in the simplest case, there is a machine that does the first activity in each job, another that does the second, and so on.

   So far, we have described a simple flow shop. In a compound flow shop, each machine in the series may be replaced by a set of parallel machines (or a batch machine), which may be identical or very different. Each job goes to one machine in the first cluster, then one in the second, and so on. Another variant, called "reentrant flow," involves using the same machine or cluster more than once. Suppose, for example, a pipe is repeatedly extruded to ever smaller diameters. After each extrusion the pipe might return to the same furnace for annealing. We might call this a "reentrant" flow shop. In other variants some machines might be skipped by some jobs. If rework is important, some activities might be repeated when inspection reveals flaws.

   There are few absolutely pure flow shops, but many many compound flow shops with minor variations. Bottling companies, certain printing companies, and steel mills are some examples. From this point on in our classification it is more difficult to find long-range, highly uncertain versions of these scheduling situations, since flow shops are typically only feasible after the production process has become quite standardized.

1.4.5 Batch/Flow Shops

A surprisingly large number of production processes divide rather naturally into two halves: the first half is a large continuous batch process, where the raw and early intermediate materials are "cooked." The second half is a more typical compound flow shop. A food processing company, for example, firsts literally cooks the edible product in large kettles connected by piping and so on; this system is usually not fully linear. The second half involves bottling/canning/boxing the food on several parallel high-speed lines. For example, three lines might be old and identical, one line old and suitable for larger bottles only, one line new and fast and capable of handling new products as well as the standard products.

A paper mill is another example. It first shreds and "cooks" logs into pulp in vats and lines. The second process involves making paper in long rolls on several parallel lines. It is only semicontinuous, since, when one batch is finished, a new one must be started on the line with a rather conventional setup time. Many simple things that are packaged are produced in a similar fashion.

1.4.6 Manufacturing Cell

The manufacturing cell attempts to combine the flexibility of the job shop with the low cost and low confusion of the flow shop. Basically, items to be produced in the job shop are grouped into several similar families. Each family is served by one cell, which is typically an automated handler surrounded by a cluster of machines.

The machines are also electronically scheduled and synchronized. The machines are usually NC controlled and quite flexible in the activities they can perform. The fact that all jobs in a group are similar makes it feasible for seven or eight machines to perform all necessary activities. The fact that machines may be accessed in any order with nearly the speed and efficiency of a flow shop allows many former job shop items to be produced with flow shop efficiency.

Cells are still relatively experimental, but most large metal-working companies have at least pilot projects.

1.4.7 Assembly Shop

An assembly shop is just an open shop or batch shop in which activities join to sub-subassembly activities, which in turn join to subassembly activities, which in turn join to assembly activities forming the final product (or nearly final product waiting for options). Labeling and cannibalization issues, which are important in open shops, remain important here. A particular WIP item may be inventoried at several levels of the process simultaneously.

Whether or not to consider the assembly shop as distinct from the open shop is somewhat ambiguous. But it leads rather naturally into the assembly line.

1.4.8 Assembly Line

An assembly line is a medium- to high-volume, low product variety version of the assembly shop. A conveyor belt or other similar device moves the WIP along the line to a number of machines or worker stations. There is no buffer inventory except at the beginning of the line. Thus production is very efficient, but lockstep and somewhat rigid. If a station takes too long, the job must be sent along anyway to be "fixed" later at a repair station at the end of the line. An assembly line need not be linear. It can have several feeder lines at certain points. The feeder lines can themselves have feeder lines.

Newer versions of assembly lines for automobiles reintroduce inventories at certain points and use AGVs (automatic guided vehicles) for transport. Thus flexibility and higher quality are restored at the cost of making the system look more like an open job shop.

Assembly lines are used for manufacturing commercial aircraft, automobiles, appliances, and farm equipment.

1.4.9 Transfer Line

A transfer line is a very high volume and very low variety production line, which is usually rather linear. Whereas assembly lines use a large amount of human labor, the transfer line itself is often completely automated. Again, there are no buffers on the line, so that rejects must be dealt with afterward. The flow side of a batch/flow shop may be composed of several transfer lines. Transfer lines are used in manufacturing light bulbs, hardware, toys, games, and paint, in addition to several products mentioned under batch/flow shops.

1.4.10 Flexible Transfer Line

A flexible transfer line bears the same resemblance to a transfer line that the manufacturing cell does to the open shop, only in the reverse direction. More sophisticated buffers, electronics, and transport devices allow items to be processed in different orders on the line, thus allowing a much greater diversity of product to be accommodated at smaller volumes for each. A number of larger companies are experimenting with the concept.

___2
APPROACHES TO
SCHEDULING

2.1 OVERVIEW

In this chapter, we give a general overview of approaches for scheduling. Initially, it is useful to give a glossary of some terms. Many of these are fairly standard, some are modified, and some are new.

2.1.1 Glossary of Terms

Interval schedule. A formal schedule is given in advance; most actual processing is expected to conform to that schedule, even if unanticipated emergencies and other happenings force some changes. A good example is an executive's appointment book. Interval scheduling is useful when the use of several critical resources must be coordinated. Interval scheduling can cause large gaps and inefficiencies in the schedule. In practice, schedules often become quite patchwork.

Dispatch schedule. A formal schedule may or may not be given in advance, but simple practical changes may be handled just by adjusting/slipping the whole schedule in a flexible way. The emphasis is on scheduling resource by resource, keeping each resource busy with the most important activity available. When a resource becomes free, the "highest priority" activity among those currently available is performed next. This tends to produce a compact, efficient schedule, but any necessary complicated resource matching may be more difficult. While some users may object to a schedule with constant minor changes, the entire schedule remains logical and tightly knit in terms of the known priorities.

Simple dispatch schedule. In a "simple" dispatch schedule, resources are never held idle in anticipation of the arrival of "hot" jobs. (A synonym would be a "nondelay" dispatch schedule.) We usually use just "dispatch" as a synonym for "simple dispatch schedule" and call the other the "inserted idleness" case. (Some shops allow inserted idleness; some do not. Analytically, the simple case is often used to approximate the inserted idleness case.)

Critical job schedule (lot for lot). All the activities of the "most critical" job are scheduled at once throughout the shop, then the second most critical, and so on. The idea is to take care of the most important jobs first. Less important jobs may face a haphazard availability of resources.

Critical resource schedule (bottleneck schedule). The "bottleneck" or most important, overused resource is scheduled first, then the other resources are scheduled around it. The idea is that efficient use of the central resource will help many critical jobs.

Critical operation schedule. Looking inside jobs, the activity/resource pair with the highest priority is scheduled first. This method tries to combine the advantages of the previous two methods.

Forward scheduling. Schedule forward in time. Simple forward scheduling produces current feasibility and compactness, at the possible expense of not meeting the most critical due dates. Forward scheduling is usually dispatch, although it can be interval/critical job.

Backward scheduling. Schedule backward from due dates. Backward procedures satisfy due dates at the possible expense of current infeasibility. Backward scheduling is usually interval/critical job, although it can be dispatch.

Heuristic dispatch scheduling. A forward dispatch method where at each choice point (sequencing, timing, routing, etc.) priority indices are calculated by some rule (static, dynamic, iterative) and the highest priority choice is chosen in simulation mode.

Advanced dispatch scheduling. Heuristic dispatch scheduling that forecasts due date problems and critical resources dynamically. (Bottleneck dynamics is such a method; see Chapter 10.)

Combinatorial scheduling. Some subset of all possible schedules is evaluated via a search tree. This can be forward, backward, or neither. Specific methods here include integer programming, approximate integer programming, beam search, and dynamic programming. The main advantage is that combinatorial scheduling seeks perfect or "optimal" answers. Disadvantages include prohibitive computation time and storage space for large problems and, just as important, lack of any ordered intuition as to how to change the solution for emergency changes, slippages, and so on.

2.1.2 Categorization of Approaches

We turn next to broadly categorizing scheduling approaches. These may roughly be given as:

- Manual-interval
- Manual-dispatch
- Simulation-dispatch
- Mathematical-exact
- Mathematical-heuristic
- Pure expert system
- Mixed AI/OR/DSS systems

Manual-interval scheduling arises when precise matching of resources and activities is essential, such as an executive's schedule book, launching a test rocket, concerts, or lecturing. Manual-dispatch scheduling arises when the exact timing of activities can be changed but the overall priorities should remain coherent. Much of more complicated manufacturing is of this variety. Simulation-dispatch essentially computerizes a very simple version of manual-dispatch scheduling.

Exact mathematical approaches choose an objective to be optimized (tardiness, utilization, etc.), formalize the resources and constraints, and solve the problem by mathematical programming. Heuristic mathematical approaches try only to solve approximately the mathematical problem thus formulated. A pure expert system computerizes a more complex version of manual-dispatch scheduling, attempting to incorporate realism unavailable to a mathematical model. Mixed AI/OR/DSS systems attempt to combine all the advantages of pure expert systems, mathematical systems, and decision support systems.

2.2 PAST APPROACHES

2.2.1 Manual Approaches

Manual approaches to scheduling have evolved over many centuries and, as any person trying to build an expert system can attest, they can be deep, subtle, and highly varied across industries. Two broad varieties have evolved: interval and dispatch. Interval methods often work backward from due dates. The due dates may themselves be derived in order to efficiently schedule some bottleneck resource. While the interval method is important in many cases, it is not discussed much in this book. See, for example, Fox and Smith (1984) and Smith (1987). Current methods are quite patchwork and do not lend themselves well to schedule changes. The bottleneck dynamics methods developed in this book can (we believe) be extended to interval scheduling situations, but this is beyond our scope here.

Most manual-dispatch procedures tend to be forward in nature, making heuristic priority choices at decision points, subject to numerous judgmental exceptions. Primary, secondary, and tertiary operation ordering and routing for a given type of job are determined by manufacturing job engineers in advance. Jobs are released to the floor by some simple rule, such as least slack (due date minus historical average lead time). They are prioritized for sequencing on each machine by another simple rule such as "choose the available operation for this machine with the highest ratio of (value)/(local machine processing time)." Lot sizes are usually fixed throughout the shop and over time, using perhaps an economic lot size model with accounting setup and holding costs. Exceptional problems such as lot splitting, enforced idleness (holding a machine for a coming hot job), rerouting due to breakdowns, expediting important late orders, and the like are usually handled by human common sense and expertise on the spot.

Although human strengths and weaknesses in performing manual scheduling should be studied very thoroughly before software systems are employed, space does not permit expanding on this rich subject here. See Baker (1974). See also descriptions of the process of studying the shop scheduling process in Prietula et al. (1991) and McKay et al. (1992).

Strengths of manual scheduling include

(a) quick and accurate mixing of human expertise and formal priorities and
(b) broad common sense for adaptability to crises.

Weaknesses of manual scheduling include

(a) the inability to test a large number of different formal priorities or to accurately estimate the effects of various actions,
(b) difficulties in mixing expertise (committee problem),
(c) difficulties in responding to rapid change (future shock), and
(d) increasingly complex systems (data overload).

2.2.2 Computer Simulation Approaches

In the 1950s and 1960s the rise of computers made it possible to represent the structure of the shop, activities, jobs, and other constraints in some detail so that, given appropriate input data and simple (heuristic) dispatch rules at decision points, the computer could extrapolate a given schedule into the future at relatively low cost. Investigators could generate various types of artificial or historical input data and simulate the effect of using different types of simple heuristics under different conditions.

The literature on simulation in scheduling is vast; see especially Baker (1974), Baker and Dzielinski (1960), Moore and Wilson (1967), Pai and McRoberts (1971), Glaser and Hottenstein (1982), and Barret and Barman (1986).

Simulation can represent quite realistic systems at modest computational cost and high-level simulation languages make the programming relatively modest. Simulation also has the advantage of providing a more natural approach for interfacing with human expertise. (Input and design remain difficult). However, the disadvantage is that results obtained are not even approximately optimal, and there are likely to be no good benchmarks to establish how good current solutions are or how to efficiently get to better ones.

Almost all newer approaches are more computationally intensive than simple dispatch simulation. Mathematical programming methods often do not use the simulation idea at all. Most other methods, such as advanced dispatch and AI/OR/DSS systems, keep simulation as their core but add more expensive and accurate decision-making procedures to it.

2.2.3 Mathematical Approaches

In the 1960s, Balas (1965, 1967) and Gomory (1965, 1967) exploited the growing power of the computer to develop modern integer programming, which allows rather realistic job shop scheduling problems to be formulated in a way that theoretically permits them to be solved exactly. A large number of scheduling integer programming formulations and procedures followed. See, for example, Balas (1969, 1970), Brown and Lomnicki (1966), Florian, et al. (1971), Greenberg (1968), Ignall and Schrage (1965), Lomnicki (1965), McMahon and Burton (1967), and Schwimer (1972).

Although we deal with branch-and-bound in some detail in Section 5.5, it may be useful intuitively to sketch how it would work for a simple problem. Suppose 30 jobs are to be sequenced on one machine, and we have a function to calculate how good a given schedule is. Conceptually, create a decision tree with 30 branches for the possible choices for first job (first level). Each first level branch then branches again 29 times for the choices for second job, leaving $30 \times 29 = 870$ second level branches. Then the tree branches again and again. There are, finally, $30 \times 29 \times 28 \cdots \times 1 = 30!$ branches at the 30th level, a huge number (2×10^{32}). Rather than look at all possible solutions and evaluate them, we find ways to prove that certain parts of the tree can simply be chopped off or pruned, since they can be shown to have only nonoptimal solutions. We keep looking and pruning, looking and pruning, until only one possible solution is left (or the computer runs out of time).

It has been suggested that the method be called "branch-and-prune" since that is really what is going on. However, exact pruning is only possible by obtaining bounds on how good a branch could be. "Branch-and-prune" would also describe approximate procedures such as beam search. (One of the authors once, in a waggish mood, called the procedure "branch-and-branch" since in large problems there may be a great deal more branching than pruning.)

In the 1960s and 1970s Srinivasan (1971) and others used dynamic programming for sequencing problems, using all possible subsets of jobs as elements of the state space. Such methods have been competitive with branch-and-bound mostly for a restricted class of problems. Again, we discuss dynamic programming in more detail in Section 5.4. However, it may again be worthwhile to give a simple intuitive sketch of the method. Suppose, as before, we need to sequence 30 jobs on one machine. Suppose for some reason we were asked to give a table such that, given any six of the jobs, the table would specify the optimal cost and ordering if those six jobs were the first six in the schedule. (This would be a very large table—part of the problem.) How could we construct such a table? For each entry, one of these six must come last. The optimal solution for the other five could be looked up in a similar table of five elements, which would, in turn, depend on a table of four elements, and so on! So the answer is: build a table of single elements, then work up to two elements, and so on, and so on, until finally we have a table for 30 elements, which solves the original problem.

For both integer programming and dynamic programming, however, while very small problems can be solved to find "perfect" priorities, routings, and so on, large problems have remained intractable, and really realistic problems can be expected to remain so, since these problems are usually known to be "*np*-complete," which means roughly that no method can be discovered that doesn't grow exponentially in problem size (e.g., see Garey et al., 1976). For example, no computer is available that could check out all permutations of 50 jobs on one machine. Even if such an extraordinary machine *were* developed, to solve a problem with 55 jobs would require a computer about 300,000,000 times faster. Of course, exact mathematical methods are more clever than this, but running times do still go up exponentially in problem size in the same way.

For this reason, various heuristic methods began to be developed for approximately solving these large mathematical programs. Neighborhood search is a rather general technique, used for scheduling by Wilkerson and Irwin (1971) and many others. First, pick a feasible starting solution, using any method. (Usually the best bet is to start with the best other heuristic known, since the method is better at fine-tuning than coarse search.) Next, try all possible ways of changing the schedule slightly (i.e., staying in the "neighborhood" of the current solution) and evaluate each resulting schedule. If there is no improvement, the method is finished. Otherwise, take the largest improvement and begin looking for small changes from that, and so on. For the 30-job sequencing method above, the small change might be interchanging any two adjacent jobs. Then there are $(n - 1)$ such changes to start with at any current solution. Neighborhood search is a special example of a general nonlinear programming method called "hill-climbing." In our example pairwise interchange proves very fast and, for many problems, very accurate. Note, however, that it arrives at only a local optimum, and there is little intuition about how changing the problem changes the solution. Nevertheless, it is very useful, especially if used in conjunction with good starter heuristic methods, as mentioned. Neighborhood search is discussed more thoroughly in Sections 5.2 and 5.3. It also forms the basis for many modern methods such as tabu search and simulated annealing, discussed below.

Lagrangian relaxation (Shapiro, 1979; Fisher, 1981, 1985; Karmarkar and Kekre, 1985; Potts and Van Wassenhove, 1985; Dobson et al., 1987; Van De Velde, 1991) solves a simpler integer programming problem by dropping some of the constraints and paying a penalty proportional to the amounts by which they are violated. (If shop labor is exceeded, assume some can be rented from outside at some price and solve the resulting easier integer program.) This is then combined with some kind of a search procedure. The method is useful when the simpler integer programs can be solved easily, which tends to rule out more complex shops. Lagrangian relaxation is a cousin to the bottleneck dynamics heuristics developed in this book. While often powerful, it is complex to use and is not yet a general purpose method.

Neighborhood search is an example of what Glover (1990) calls "intensification strategies," procedures for "sticking with a winner." Another early heuristic method—random sampling (discussed by Baker, 1974)—Glover would call a "diversification strategy": more like "it's time for a change"—anything to get out of the current rut. A problem with no structure at all lends itself to diversification strategies; one with a simple grand structure lends itself to intensification strategies. Most problems are in between. In its simplest form, random sampling is just neighborhood search except that many random starting solutions are generated. Thus a number of different neighborhood searches are performed, hopefully in different parts of the search space, and the best solution can be chosen. It is not necessary to carry out diversification at the start, it can occur during the hill-climbing or only to try to move beyond the local optimum. Randomization is one way to do diversification, but there are a number of others. Modern techniques exploiting mixed strategies of intensification and diversification include tabu search, simulated annealing, genetic algorithms, beam search, and other approximate branch-and-bound methods. We turn next to discussing these more recent approaches.

2.3 NEWER APPROACHES

There are a number of heuristic search techniques that, although not yet fully mature, have already made noticeable contributions to scheduling. They fall roughly into three categories:

(a) Intensification/diversification methods.
(b) Bottleneck methods.
(c) Expert systems, mixed AI/OR/DSS systems.

We discuss all three in this section, except that while bottleneck dynamics is one of the bottleneck methods, it has been given a separate section.

2.3.1 Intensification/Diversification Methods

Tabu Search. Scheduling heuristics employing tabu search have been reported by Glover and Laguna (1989), Laguna et al. (1989), and Widmer and Hertz (1989). Glover (1990) also has an excellent tutorial piece on tabu search. In its simplest form, tabu search is neighborhood search with a list of recent search positions. "Tabu" comes from the fact that these positions (Glover likes seven, but there are advocates of much longer lists) may not be repeated while in the active list. Other than that, one still takes the move improving the objective most, or at least hurting it the least. The best solution to date is also always saved in case no better solution is ever found. Thus, when a local optimum is reached, the procedure will move on to a worse position in the next move and will be prevented from returning immediately.

The moves to the local optimum are intensifying, while those moves out of the local optimum into hopefully new territory are diversifying. (Remember we always save our best so far!)

Simulated Annealing. Scheduling heuristics employing simulated annealing are described by Kirkpatrick et al. (1983), Vakharia and Chang, (1990), Coroyer and Liu (1991), Ishibuchi et al. (1991), and Van Laarhoven et al. (1992). Simulated annealing also adds diversification to a neighborhood search procedure, but in a somewhat different way. Again, in its simplest form, a random amount is added to each possible move's evaluation. That is, as before, the objective function for a potential move is calculated and compared with the current objective. A random amount is added to this difference, producing an adjusted difference. (The amount is small with high probability, with smaller and smaller probabilities of larger amounts.) Then the move with the highest positive adjusted difference is chosen. Following the original difference would be the intensification strategy; following the random amount would be a diversification strategy. Combining the two produces a mixed strategy. An overwhelmingly improving move tends to dominate any diversification; a very large perturbation tends to dominate most improving moves. Basically, instead of waiting to diversify until a stalemate is reached, the method balances diversification and intensification at every move. Parameters can be set to vary the balance of power.

Genetic Algorithms. Genetic algorithms (Holland, 1975; Dorigo, 1989; Davis, 1991; Falkenauer and Buffoix, 1991; Nakano and Yamada, 1991; Della Croce et al., 1992) can refer to any search process simulating the natural evolutionary process. There is a current population of possible solutions to the problem. In each generation, the best solutions (most fit individuals) are allowed to produce new solutions (children) by mixing features of the parents (or by mutation); the worst children die off to keep the population stable, and on to the next generation. It is possible to consider a broadened concept of genetic algorithms as a very general class of techniques that can include neighborhood search, tabu search, simulated annealing, and beam search (see below) as special cases.

From a slightly different point of view, tabu search, simulated annealing, and genetic algorithms might *all* be termed "extended neighborhood search." We discuss these methods in Section 6.3.

2.3.2   Beam Search (Partial Enumeration)

Beam search is one of a number of methods developed by artificial intelligence software engineers for partially searching decision trees (Lowerre, 1976; Rubin, 1978; Ow and Morton, 1988). These methods have been responsible for powerful computer chess playing programs (Anthony and Schaffer, 1990). They all employ carefully crafted mixtures of intensification and diversification strategies. Beam search is rather like a type of branch-and-bound procedure, but, instead of waiting to throw away a part of the tree that is *guaranteed* useless, we may throw away parts of the tree that are *likely* to be useless. One essential is to have a good measure of what "likely" means; another is to throw away parts that save a lot of effort without taking much risk.

Let's go back to the same problem: sequencing 30 jobs on a machine. Set the beam width, say, four. We must also choose an evaluation function, that is, a way of choosing the best four paths to save at any point. One way is to use a heuristic that assigns a priority to a job to be sequenced next. All possible jobs give 30 branches at the start. Choose the four branches with the highest priorities, throwing away the rest. Branch on these, giving $4 \times 29 = 116$ branches. Now calculate a composite priority for each of the 116, and choose the four with the highest priority and throw the rest away. Now the number of nodes to be expanded is just four at each level, irrespective of the number of jobs. A 60-job problem has 60 levels, and a 30-job problem has 30 levels. Thus the number of nodes to be expanded for 60 jobs is just twice that for 30 jobs. Finding a decent composite priority or "evaluation function" is a rich subject that we get into a little more in Section 6.2. There, we also discuss more sophisticated beam search methods, approximate dynamic programming methods, and other partial enumeration methods.

Basically, neighborhood search allows no diversification; therefore it is very fast but gives limited quality solutions. Branch-and-bound and dynamic programming insist on complete diversification, and thus give perfect solutions with typically unacceptable running times. Beam search takes a step from complete branch-and-bound diversification to a middle ground. Tabu search takes a step from no diversification toward a middle ground. It thus becomes possible for the two methods to end up exploring quite similar territory.

2.3.3 Bottleneck Methods

OPT, from Creative Output (Lundrigan, 1986; Meleton, 1986; Vollman, 1986; Fry et al., 1992), is representative of many other current finite scheduling systems (NUMETRIX, Q-Control), which make clever approximate solutions to mathematical programming scheduling models. In a preliminary pass, OPT identifies the bottleneck resource (most heavily loaded in some sense). Priorities are then set in the rest of the system in an attempt to "baby the bottleneck." That is, jobs that require extensive processing in noncritical parts of the system but use only small amounts of bottleneck time may be expedited.

OPT's basic advantage is that, by focusing on a single bottleneck resource, the problem is sufficiently simplified that it can be solved well and with intuitive support to the user. OPT also has a number of disadvantages that keep it from being an ideal system: (a) it cannot accommodate multiple and shifting bottlenecks, (b) the user interface is not strong, (c) reactive correction to the schedule requires full rerunning, and (d) it is proprietary, so not only is the software rigid, but also simple judgmental modifications of schedules are difficult. Nevertheless, OPT remains the standard against which any new scheduling system must be judged. OPT is discussed in more detail in Section 10.3.

The shifting bottleneck method for job shop scheduling was first developed by

Adams et al. (1988) and has recently been extended by Balas and Vazacopoulos (work in progress). Related work is being carried out by Ovacik and Uzsoy (1992). The method may be considered a high computational cost, high accuracy version of OPT. While the method is currently limited to the makespan objective and to moderately complex shops, it has been applied successfully to scheduling steel mills and clearly has good potential. The method depends on being able to accurately solve a simplified version of the full problem, in which all resources except one have fixed decision rules (such as sequence) specified in advance. A special version of this is that all resources but one have plenty of excess capacity. We call the first the "embedded one-resource problem," and the latter the "embedded relaxed one-resource problem" (EORP and ERORP).

The ERORP with the highest objective function cost defines the bottleneck resource (a sophisticated definition). Then, holding the bottleneck solution constant, the procedure is repeated on the remaining resources to define the bottleneck among the remaining, and so on. As this procedure is repeated, the solution eventually becomes feasible for the entire problem, and the cost decreases, until finally a local optimum is achieved. A version of partial enumeration can be added to produce even more accurate solutions.

The shifting bottleneck method is discussed in some detail in Section 10.4.

2.3.4   Expert Systems, Neural Networks, and Mixed AI/OR/DSS Systems

Expert Systems. ISIS (Fox and Smith, 1984) was one of the first major artificial intelligence scheduling prototypes to be built. It uses backward interval scheduling and beam search, with an evaluation function defined by estimation of constraint violations. OPAL (Bensana et al., 1988) is similar. CALLISTO (Sathi et al., 1985, 1986), an outgrowth of ISIS to do project scheduling, adds some bottleneck dynamics heuristics. There have also been a number of similar applications to flexible manufacturing systems (Shaw and Whinston, 1986; Lecocq et al., 1988, Shaw, 1988).

PATRIARCH (Morton et al., 1986, 1988) is a prototype two-level planning and scheduling system. The upper level utilizes bottleneck dynamics; the lower level is a simple multiple-expert system. OPIS (Smith et al., 1986) builds on PATRIARCH to produce a more sophisticated multiple-expert system. MERLE (Prietula and Thompson, work in progress) is a multi-expertise scheduling system that can allow negotiation between the human user and the internal AI/OR system. Other recent work in multi-expert systems is due to Keng et al. (1988), and Ow et al. (1988).

Neural Networks. Neural networks are an AI development that try to simulate the learning process of the human brain (Marquez et al., 1992). Consider a neural network procedure to be a "black box." For any particular set of inputs (particular scheduling problem), the black box will give a set of outputs that are suggested actions to solve the problem. These can be simulated to give an objective function value, which can be compared against the optimum obtained by some other very expensive conventional method. The black box inside consists of several layers of very complicated networks, allowing inputs to be added together, strengthened, stopped, nonlinearized, and so on. The black box also has a great number of knobs on the outside, which can be fiddled with to adjust the output.

For the given input problem (training problem), we fiddle with the knobs until the output solution is close to optimum as measured by closeness to the optimum (target). (Actually, the procedure itself adjusts the knobs.) Unfortunately, training the network to do this one problem does not guarantee that the same settings will do well with other problems. So next, we input maybe 10 rather different problems simultaneously, and the machine tries to find knob settings to produce close to optimal scheduling decisions for each of the 10 problems simultaneously.

The hope is then that, if a new somewhat similar problem is introduced, the black box as currently trained, and with no new targets to allow new training, will perform well. (Training on the 10 problems is like doing a regression. The line may fit interpolated points well, but to what degree is it safe to *extrapolate* the solution?) There is a fair amount of current work being done to apply neural networks to scheduling (Zhou, et al., 1990, 1991; Lo and Bavarian, 1991). However, while this direction shows promise, there are not too many results to point to at this time.

Other references for this section include Fox (1987), Goldratt and Cox (1984, 1986), Jacobs (1984), Byrd and Hauser (1991), Shaw et al. (1992), and Vaithyanathan and Ignizio (1992).

2.4 BOTTLENECK DYNAMICS

2.4.1 The Big Picture

Basically, bottleneck dynamics estimates prices (delay costs) for delaying each possible activity within a shop and, similarly, prices (delay costs) for delaying each resource in the shop. Trading off the costs of delaying the activities versus the costs of using the resources allows calculating, for example, which activity has the highest benefit/cost ratio for being scheduled next at a resource (sequencing choice), or which resource should be chosen to do an activity to minimize the sum of resource and activity delay costs (routing choice). Thus these prices allow making local rather than global decisions with easy and intuitive cost calculations, allowing solution by advanced dispatch simulation. Decision types that can be handled include sequencing, routing, release and other timing, lot sizing, and batching.

If the scheduling problem were convex (basically, if neighborhood search could be guaranteed to always achieve the global optimum), the method would be easy to implement and guaranteed to work well. But the problem is not convex, and thus no consistent set of prices exists to make this procedure optimal. However, the intuition is that in many large shops where no individual activities dominate, the structure is "close to convex" in some sense. But now it requires a lot of work to validate this intuition:

(a) Look for a number of different ways to estimate prices easily.

(b) Demonstrate by extensive computer testing and other validation that some of the pricing procedures work well.

Fortunately, research to date has been encouraging. In this book we talk a great deal about different scheduling problems, how to schedule given prices for them, how to estimate prices, computer testing experience, and further research needed.

2.4.2 Estimating Activity Prices (Delay Costs)

To simplify the discussion, suppose that each job in the shop has a due date and that there is a customer dissatisfaction that increases with the lateness and inherent importance of the job. Again to keep it simple, suppose our objective is to minimize the sum of dissatisfaction of the various customers for jobs. Then, if a certain activity is estimated to have a given lead time to finish, this yields a current expected lateness. If this activity is expedited by, say, a day, then the reduction in lateness, and thus the reduction in customer dissatisfaction, can be estimated. *This gives an estimate of the activity price or delay cost.* The difficult issue here is to estimate the lead time since, obviously, the lead time depends on shop sequencing decisions that have not been made yet.

If the marginal cost of customer lateness is constant (important cases are minimizing the weighted flowtime or job lateness), then the lead time is not relevant, and the price is constant, giving a particularly easy and important case. For the more general case, such as if the customer is unhappy about tardiness but doesn't care if the job is early, the current lead time of each activity must be estimated by one of a number of methods such as:

(a) Historical fixed multiple of total remaining process time.

(b) Lead time iteration (running the entire simulation several times to improve the lead time estimates).

(c) Fitting historical (or simulation) data of lead times versus shop load measures.

(d) Human judgment based on historical experience (input via user interface).

There is especially a great deal of experience from MRP users on (a) and (d). There is also quite a lot of computer testbed experience with (b) obtained since about 1980. Good experience is just beginning to accumulate on (c). These issues are explored more intensely in Sections 10.2 and 10.5 and also in later chapters that discuss bottleneck dynamics solutions to particular scheduling models.

2.4.3 Estimating Resource Prices

All bottleneck methods for estimating resource prices (same as resource delay costs and resource usage charges) are based ultimately on the same busy period analysis.

Look at a given scheduling problem with a certain current approximate solution. Suppose we wish to estimate the price of a given resource at a given point in time. First, there will be a given queue of activities waiting and known activities to arrive at various points of time. Now we shut down the resource for a day and ask: "What will be the extra cost to the system?" All jobs in the queue will be delayed by one day also, so that the resource delay cost or price will be at least the sum of all the prices of activities in the queue, times one day. But each job that arrives before the queue will be empty will be slowed down by one day also. That is, all jobs in the current *busy period* will be slowed down. Thus the price of the resource is just the sum of the prices of the activities in the resource's current busy period.

Thus we need estimates of the activity prices, which we have discussed, and estimates of the activities in the current busy period. Unfortunately, we must estimate the busy period as well. There are several different methods that have been tested and are continuing to be tested.

Fortunately, experimental results show that the accuracy of bottleneck dynamics heuristics is not unduly sensitive to how accurate the prices are. Thus a number of different methods sometimes turn out to give good results and have been tested and continue to be tested. Pricing methods in wide use historically include myopic methods and OPT-like methods. Methods based on the line length and historical utilization include static and dynamic queuing approximations. More sophisticated and computationally intensive methods include exhaustive search and iterative methods.

These issues are discussed in Chapter 10, along with bottleneck dynamics approaches to scheduling situations later in the book.

We discuss next how to use prices to make decisions in various types of problems.

2.4.4   Bottleneck Dynamics for Sequencing Decisions

Myopic decisions for sequencing were developed in the 1960s, OPT-like rules came in the late 1970s and early 1980s, while bottleneck dynamics started in the early 1980s but has matured in the late 1980s and into the 1990s. Versions of all three procedures can be constructed that make priority estimates and decide sequencing at resources one at a time. These versions would be special cases of the following rule for sequencing at one resource:

(a) Prioritize in order of decreasing benefit/cost ratio.
(b) The benefit of choosing an activity is the importance of the job times a slack factor, which measures whether the job is close to due or not.
(c) The cost of choosing an activity is the remaining resource expediting cost for the job of which the activity is a part.
(d) The remaining resource expediting cost is the expediting rate times the remaining resource cost.

(e) The remaining resource cost is the process time of each remaining activity times the price(s) of the resource(s) it uses, summed over all remaining activities.

The expediting rate is basically the interest rate to be charged for using the resources early. Fortunately, in this simple benefit/cost model, the expediting factor appears the same way in every priority and hence can be ignored.

Note that, for the myopic rule, this simply reduces to delay cost divided by current activity processing time. For OPT-like rules, this reduces instead to delay cost divided by future activity time on the bottleneck machine downstream. We discuss the relative performances of myopic, OPT-like, and bottleneck dynamics in the chapters on the appropriate models.

2.4.5 Bottleneck Dynamics for Other Scheduling Decisions

We sketch briefly here the use of bottleneck dynamics for other types of decisions. In *job release* the question is when to release the job to the floor, where it begins to accumulate inventory and waiting costs. One good rule is: release the job when the sequencing benefit/cost ratio first exceeds 1.0. (The shop release problem also has other issues, such as floor congestion, but these require different kinds of heuristics.)

An approach to the *routing* problem, where each job has the choice of routes through the shop, is simply to calculate the total resource cost for each route (sum of resource price × activity duration) and add the expected lateness cost if sent by that route. Finally, simply choose the route with the lowest cost. There are some difficulties here that will be made apparent later. A special case of this routing problem is where all routes are the same except for a given activity, which may be done on one of several machines. This is often called the "parallel" machine problem. Bottleneck dynamics is especially good at handling the case where different machines have comparative advantages for different activities.

In considering *lot sizing* and/or *sequence-dependent* decisions, the primary issue is how to convert setup times into setup costs so that traditional OR models and heuristics for lot sizing may be used.

For a given activity and resource (with or without specifying the preceding activity on the resource), look up the setup time. Multiply the setup time by the current resource price to obtain the heuristic setup cost.

In the *enforced idleness* case, it may be worthwhile to keep a resource idle to await a soon-to-arrive high priority (hot) job. To decide whether this is worthwhile, first find the net benefit of switching the jobs. Subtract from this the lost waiting time times the resource price. If the difference is positive, wait for the hot job.

For *simple preemption* (*preempt resume*), start with the base nonpreemptive case; that is, at each completion, choose the available activity with the highest priority. Then, in addition, when a high-priority activity arrives while another is processing, compute a revised priority for the in-process activity, using the remaining processing time. (It will thus be harder to displace the longer it has waited.) Given a displacement, let it compete in the line with its shorter process time.

In the *general preemption* case, there may be a setup cost for taking the job off. The arriving job must be able to pay this additional cost before preempting. *Preempt repeat* is an example of this, where the setup cost is a new setup cost, complete repeating of processing to date, and possible loss of material as well.

Bottleneck dynamics also has the potential to handle:

- Assembly
- Inventory of subassemblies
- Furnaces (batch resources)
- Aggregation
- Hierarchical systems
- Multiple projects
- Project resource/time trade-offs
- NPV projects
- Tool wear
- Quality/speed trade-offs
- Real-time scheduling revision
- Interval scheduling

Many of these issues are discussed in detail in later parts of the book.

2.4.6 Empirical/Theoretical Results for Bottleneck Dynamics

Rachamadugu and Morton (1982) developed myopic heuristics for the one-machine weighted tardiness problem and the similar parallel machine problem with heuristics that were the forerunner of the bottleneck dynamics benefit/cost approach. Vepsalainen and Morton (1987) extended these procedures to flow shops and simple job shops by developing the first version of lead time iteration (Vepsalainen and Morton, 1988). As mentioned in Section 2.3, several AI systems appeared about this time, using very early bottleneck dynamics methods. A more advanced version of bottlenecks dynamics for job shops was incorporated into SCHED-STAR (Morton et al., 1988), using iterative pricing and lead time methods. A paper developing theoretical busy period results to support bottleneck dynamics followed (Morton and Singh, 1988). Recent activity has included multi-project scheduling with a weighted tardiness criterion and multiple resources (Lawrence and Morton, 1991), and explicit comparison of various bottleneck dynamics pricing approaches with myopic and OPT-like rules (Morton and Pentico, 1993). Work on bottleneck dynamics on parallel machines with the objective to maximize utilization is being carried out by Kumar and Morton (1991).

Other research, which is either in the conceptual stage or in earlier phases of completion, includes:

(a) Dynamic resource pricing and job routing in a heavily loaded shop.

(b) Myopic versus OPT-like versus bottleneck dynamics in a dynamic job shop.

(c) Cost accounting for timeliness, quality, and flexibility in a global manufacturing environment.

(d) Tardiness/cost trade-offs in multi-project scheduling.

(e) Combining bottleneck dynamics and shifting bottleneck methods in scheduling.

All these research efforts, whether complete, in late stages, in early stages, or conceptual, are discussed much more thoroughly in appropriate parts of the book.

____3
BOOK SUMMARY

3.1 PRELIMINARIES

It is perhaps time to take stock, to review the broad picture of scheduling problems and of approaches for solving them, to summarize the remainder of the book chapter by chapter, and to take a glimpse at sketches of possible future research. In the process, perhaps, you may discover whether this area of heuristic scheduling interests you enough to take the plunge into mastering it by a mixture of a dab of mathematics, a dibble of examples, a splash of experimenting with software, all bonded with a mixture of blood, sweat, and common sense.

Other scheduling books include Conway et al. (1967), Baker (1974), French (1982), and Lawler et al. (1985). Another reference for this section is Johnson and Montgomery (1974).

3.2 BOOK SUMMARY

3.2.1 Part I: Preliminaries

In Chapter 1 we wrestled with the questions of how important scheduling is, exactly *what* it is, and how to recognize and classify scheduling problems at different levels and types of scheduling environments. We decided that the older conventional wisdom of scheduling as a low importance, "seat of the pants" activity does not jibe with the fact that JIT is forcing ever decreasing work-in-process buffer stocks, which in turn dramatically increases the difficulty and importance of scheduling. Scheduling can be defined as a matching of needed activities with limited resources

36

to maximize customer satisfaction, maximize shop utilization, and minimize operating costs. Scheduling decisions include sequencing, release and timing, routing, and reconfiguration, as well as related MIS issues.

Scheduling levels can be organized by decreasing horizons and scope of planning. Level 1, long-range planning (e.g., plant expansion), has a 2–5-year horizon; Level 2, middle-range planning (e.g., logistics), has a 1–2-year horizon; Level 3, short-range planning (e.g., MRP), has a 3–6-month horizon; Level 4, scheduling (e.g., job shop scheduling), has a 2–6-week horizon; Level 5, reactive scheduling (e.g., shop floor), has a 1–3-day horizon. Higher levels have a much greater difficulty with forecasting and the adequacy of deterministic rolling horizon models. Lower levels are more complex and difficult to solve mathematically.

Types of scheduling environments, roughly in order of similarity and volume of related jobs, include the following: classic job shop, open job shop, batch shop, flow shop, batch/flow shop, manufacturing cell, assembly shop, assembly line, transfer line, and flexible transfer line. Advanced automation and computer techniques have blurred the boundaries between them somewhat.

In Chapter 2 we classified broad approaches to scheduling and explained their elements intuitively. Most manual approaches, for example, are forward in nature and make simple heuristic priority choices at decision points, with frequent exceptions. Strengths include human flexibility, adaptability, and common sense. Weaknesses include increasing future shock and data overload. Computer simulation expands the power of the simple forward dispatch approach but loses much adaptability and common sense.

Exact mathematical scheduling approaches include integer programming (branch-and-bound) and dynamic programming. Complex problems can be formulated with these methods, but it is not usually practical to solve large problems with them. This leads us to consider heuristic methods.

Neighborhood search is a type of "hill-climbing" successively trying to improve a given solution, usually converging to a local optimum. Results are often quite good. Other mature methods include Lagrangian relaxation and random sampling.

Many newer heuristics use intensification/diversification strategies in that they combine continual local improvement with occasional expansion of the search area. These include extended neighborhood search methods such as tabu search, simulated annealing, and genetic algorithms. Beam search is an approximate form of branch-and-bound that does not bother to search unpromising areas of the tree. Bottleneck methods include OPT, which chooses a bottleneck and schedules to suit it, and a related method called the shifting bottleneck method. Expert systems, mixed AI/OR/DSS systems, and neural network methods attempt to obtain the advantages of both mathematical systems and human common sense. These have great long-range potential, but really successful systems have not yet appeared.

Bottleneck dynamics estimates prices (delay costs) for delaying each possible activity within a shop and, similarly, estimates prices (delay costs) for delaying the use of each resource in the shop. By using these prices, local decisions can be costed out and the cheapest ones (e.g., sequencing choices with the greatest benefit/cost ratio) chosen. Most research efforts in bottleneck dynamics focus on the best way to estimate the prices and the associated activity lead times cheaply, accurately, and robustly.

3.2.2   Part II: One-Machine Problems

Remember that we have defined a *resource* as a grouping of (possibly complex) productive capabilities with a single input queue. The only decisions to be made are sequencing and release timing of jobs or operations from the input queue to the resource. (There may be internal queues, which are "slaves" in that sequencing and release are automatic by simple disciplines such as "first-come first-serve" but no internal current active decisions need to be made.) For example, a single machine is one resource, but a set of parallel machines served by one line on a "highest priority activity goes to first available machine" is a single-resource problem also. A flexible manufacturing cell with adequate tooling may be a single resource; however, if there is internal contention for tooling, and thus implicit queues for the tools, it becomes a multi-resource problem.

Remember that we have defined a single *processor* as a productive capability able to handle exactly one job or operation. This is traditionally called a "machine"; however, this causes some confusion. For example, a batch machine such as an annealing furnace handles several jobs simultaneously and hence is a multi-processor. A one-processor, one-resource problem may equivalently be called a "one-machine" problem in the traditional fashion, if we label multi-processor machines carefully to avoid confusion. We follow this convention.

Many complex scheduling systems can be decomposed into single resource blocks, each of which handles one or more operations of a job. Complex single-resource blocks may usually be aggregated into nearly equivalent one-machine problems in which jobs arrive and leave over time. The output from one machine provides the input to other machines. These dynamic one-machine problems are very difficult in themselves to solve exactly. A commonly studied *static* approximation in which all operations arrive at the same time is not very useful as a building block for more complicated problems. However, the static approximation is much easier to solve and turns out to provide strong intuition for developing heuristic approximations for the desired dynamic building blocks.

In Part II we develop these fundamental building blocks for scheduling by exploring exact and heuristic results for single-machine (one-processor) problems. In Part III, we study multi-processor, one-resource problems such as batch machines, equal and proportional parallel machines, interleaved jobs, certain lines, and flexible manufacturing cells. We develop procedures to aggregate these apparently complex subsystems to allow using single-machine results to model them. Thus the rather strange definition of "single resource" pays large dividends.

Chapter 4 lays foundations for the basic building block dynamic single-machine problem. Next, it presents a number of elementary results for the special static case (all jobs arrive simultaneously) and a number of common objective functions, such as "minimize flowtime" or "minimize lateness": that is, one-machine problems with relatively tractable objective functions and all activities available initially. While regular objectives are quite common in practice, practical static problems are relatively rare. These static results form the intuitive foundation for heuristics for more realistic dynamic problems. (They also help lay the foundation for the bottleneck dynamics methodology.) First, the dynamic single-machine problem is carefully defined. Second, some broad properties of optimal scheduling solutions are derived. Third, exact priority rules are derived for regular static problems with such objectives as minimizing makespan, minimizing weighted flow, minimizing maximum lateness, and minimizing the number of tardy jobs. Finally, exact but partial results are given for regular static problems with more difficult objectives such as minimizing weighted tardiness and minimizing the weighted number of tardy jobs.

By this point we will have developed most of the exact results that can be derived for regular static problems without more advanced methods. Thus in Chapters 5, 6, and 10 we give an introduction to more general purpose mathematical solution methods. In Chapter 5 we introduce classic methods, such as neighborhood search, dynamic programming, and branch-and-bound, initially in the context of one-resource, regular static problems. In Chapter 6 we similarly take a look at newer heuristic methods such as beam search, approximate dynamic programming, tabu search, simulated annealing, genetic algorithms, neural networks, and bottleneck dynamics, again all initially for one-resource, regular static problems. (Later, in Chapter 10, we discuss a class of newer heuristic approaches that might be termed "bottleneck" approaches, including the forecast myopic dispatch approach, OPT, the shifting bottleneck algorithm, and bottleneck dynamics. These are delayed until later because they require somewhat more complex models to be understandable.) Chapter 5 therefore gives an introduction to mathematical solution methods such as special structure, neighborhood search, intensification and diversification, branch-and-bound, dynamic programming, and beam search. The discussions in Chapters 5, 6, and 10 are much more comprehensive than the brief discussion in Chapter 2 and, in particular, work through a number of numerical examples in detail. Thus the reader will be able to employ these methods by hand or by using the software provided, or even be able to program simple versions of these methods.

So far, we have been discussing mostly the static case: all activities are present at the start, and the resource gradually works through a single busy period to become idle. (Remember that this case is highly unrealistic but allows us to lay a foundation.) In the *dynamic* case, the resource has a queue at the start, but activities arrive over time. In Chapter 7 we first look at harder static problems, those that are regular (earlier completion always preferred for any activity) and nonregular (e.g., activities may prefer not to be early or tardy). We next turn to simpler dynamic problems where exact priority rules or partial results can be derived, and also to harder dynamic problems that can still be reduced to sequencing problems, both for the case in which the resource will not deliberately wait for hot jobs (no inserted idleness) and for the case in which it will.

Chapter 8 considers timing issues, that is, static or dynamic one-machine problems where specifying the sequence is not enough, since the next activity cannot be or should not be processed immediately on completion of the last activity. These include machines with a setup time for the next activity that depends on the activity processed last, so-called sequence-dependent setups. In an important special case, there are groups of jobs with little or no setup between them but larger setups when the group type changes. How many of the jobs in one of these groups (lot size) should be run before the larger setup time is incurred? These are lot sizing issues. A different situation arises when some activities cannot be processed until other predecessor activities have finished, for example, when the activities together form a project. Ideas to be developed here are central for Part V on project scheduling and management. Finally, we discuss the preemptive case, assuming that an activity can be interrupted to allow a high-priority job to proceed and then the activity be restarted. In the simple preemption case, the activity can be preempted and restarted at no penalty. In the complex preemption case, a higher than usual setup time may be incurred by the interrupting job and/or another extra setup may be required to reinitiate the partially finished job, and/or part or all of the partial processing time may be wasted, and/or the interrupted workpiece may be scrapped.

3.2.3 Part III: Multi-machine Problems

Part III turns to an introduction to multi-machine problems. In some cases there may still only be one queue, yielding only one complex resource, as for parallel machines. In this case, the only decisions to be made are the sequencing and release timing of the external queue. Internal processing either requires no decision-making or depends on heuristics prespecified before the scheduling problem is considered.

In Chapter 9 we consider embedded simple one resource problems. Consider a multi-resource complex shop. Focus on one critical resource. Pretend the other resources are in large enough supply that queues will not develop for them, so that there is only one queue and hence, by our definition, only one effective resource. The resulting problem is called the "simple embedded one resource problem." It is much simpler to deal with (exactly or approximately) than the full problem. The principle used in OPT-like procedures suggests that the solution obtained to the simple embedded problem may be a good approximation to the full solution.

An extension to this idea, called the "complex embedded one-resource problem," is to model some of the other resources to be in limited supply but to specify exact sequencing rules for them in advance. The resulting problem again has only one decision queue and is easier to analyze. The shifting bottleneck method is a sophistication of OPT-like procedures, which solves simple and complex embedded problems in an iterative "neighborhood search" way. Thus we will be especially interested (to use the method effectively) in developing ways to solve simple/complex embedded problems very quickly and easily (either exactly or approximately). The simplest case of interest is that where each job is a set of activities to be performed in series, the "classic job shop." However, we also consider the cases of scheduling a project or multiple projects, relaxing all but one resource in the same way.

Chapter 10 completes the development of mathematical solution methods for bottleneck approaches, as mentioned previously. Chapter 11 considers parallel machines and batch machines. Parallel machines involve a single queue serving several resources (such as bank tellers). Each machine behaves as a single resource except for the common queue. For regular measures of performance, it can be proved optimal to correctly prioritize the queue and then allocate the next job to the machine that can finish it first. If the machines are equal, this reduces to the machine that can start it first. Thus sequencing rules for the one-machine problem can often be adapted here. Proportional parallel machines differ only from identical machines in that some machines are faster by a constant speed factor than other machines. Discussion of the general case (neither equal nor proportional machines) is deferred to Chapter 12 (routing). A batch machine, such as an annealing furnace, simply processes a group of activities simultaneously. There may be rules as to which activities may be processed together; those requiring the same temperature and cooking time, for example. There may be individual setup times and/or a joint setup time that is dependent on the group that was processed last.

Chapter 12 considers basic routing questions. We first consider the general parallel machine problem, where different machines may have different comparative advantages for processing different activities. Now there are simultaneous questions of how to prioritize the activities and which machine to choose to do the processing. Next, we consider routing through the shop. The simplest question is where to route a single activity, which is analogous to the general parallel machine problem. A more complex problem is to simultaneously choose the routing for a related set of activities (so-called process routing). It may also be possible to try several different choices of resource combinations for activities and to develop cost/resource trade-offs for a job. Finally, problems for activities that can be done in more than one way, so that there are resource/cost trade-offs for each activity, may be studied as routing problems.

3.2.4 Part IV: Flow Shops and Job Shops

Next, we move on to considering various flow shop and job shop models. Chapters 13 and 14 discuss scheduling flow shops. A simple flow shop is a series of machines that process jobs whose activities have the same serial structure. There are some variations where the same machine might appear more than once (reentrant flow) or some jobs might skip some machines. A compound flow shop is a simple series of machine groups; each group may be a complex set of resources. Examples include a series of groups of parallel machines or a batch line followed by parallel machines. Compound flow shops are very common but not always recognized due to the physical complexity. They will be more common in the future as automation causes more subsystems to take on the characteristics of a single resource. These chapters present exact and heuristic results for a number of criteria including makespan, economic makespan, weighted flow, and weighted tardiness. Results from a number of flow shop testbed experiments are presented. Finally, the case of a nuclear fuel tube shop is presented, both to illustrate a real-world complex flow shop and to sketch how to model it and develop an appropriate heuristic scheduling system.

Chapters 15 and 16 introduce scheduling in job shops. Flow shops have some analytical simplicity since there are no routing decisions. All jobs have their operations in series; the same operation in the sequence on each job uses the same machine. Classic job shops preserve two out of three of these properties: there are no routing decisions, and all jobs have their operations in series. Thus a classic job shop is simply a flow shop with a different routing assigned to each job. We may also talk of simple classic job shops and compound classic job shops. If we add a fuller freedom of routing choices, the compound job shop becomes quite a rich model. (An interesting fact is that once heuristic rules are specified for scheduling the job shop as a function of its input queue and arrival times, it can be considered as an aggregate single resource to be scheduled in a higher level aggregate system with several such aggregate resources.) In Chapter 16, computational results are presented for a number of objective functions and classic and modern heuristics for the classic job shop. In addition, a number of model extensions are briefly discussed, as well as a more realistic job shop form of the nuclear fuel tube shop.

3.2.5   Part V: Project Scheduling and Management

Although we have discussed projects before (jobs whose activities may have preceding required activities or precedence constraints), it is time for a fuller development. In Chapter 17 we discuss classic network projects (no resource constraints). First, we introduce simple network ideas for such projects. Second, we develop the classic deterministic critical path analysis, giving the minimal project duration, critical path, and early and late start times allowable for each activity. Third, we develop simple time/cost trade-offs by allowing activities to be shortened in duration at higher costs. (The early versions of these models were termed CPM for "critical path method.") Fourth, we develop several probabilistic versions of these ideas. (The most popular model in which durations are probabilistic and independent is called PERT for "program evaluation and review technique.") Finally, we discuss simple resource leveling as a prelude to the more sophisticated resource-constrained methods in the next chapter.

In Chapter 18 we work with the more realistic resource-constrained project scheduling problem. First, we discuss project job shops, which are basically classic job shops in which each job has more complicated precedence constraints. Second, we move to the case of scheduling multiple large projects, where the emphasis is more aggregate, and an activity is simply taken to use a fraction of the overall available resource. Third, we turn to cases in which activities can be done in differing ways, producing resource/cost trade-off models. Fourth, we turn to a brief discussion of a number of model extensions, including consumable resources, resource replanning, net present value formulations, and milestones/dependent projects. Finally, we give a real world case example, both to illustrate a realistic multi-project situation and to sketch an appropriate heuristic scheduling system for it. The case example is producing custom optimal equipment, a typical (but not classic) project shop.

3.2.6 Part VI: Other Issues

Chapter 19 discusses complex resource problems and/or complex activity problems. We first consider the multiple machines per resource case. One example is interleaved jobs: more than one activity is being processed in an interleaved series. That is, one activity may be finishing while another is starting. A simple case is a machine with a revolving turret, so that one activity may be set up while another is running. Another is an automatic serial processor that is fed by a single queue. More complex resources such as lines, sectors, and flexible manufacturing cells are more complex versions of parallel machines or serial processors. They usually have inner queues and apparent inner decision choices and hence do not seem to be single resources. But often these decisions are made automatically by simple rules, so that only the external queue need be optimized. Thus they become very complicated one-resource problems. At the end of this chapter we develop a number of complex single-resource problems that are mixtures of the cases discussed above.

Next, individual activities requiring complex resources are discussed. For example, an activity may require both a certain milling machine and a particular milling tool out of the automatic tool magazine. Some of the questions raised here are quite difficult and bring up issues related to interval scheduling. The chapter closes with a case involving a computer board line.

Chapter 20 considers a number of types of model extensions. First we consider open shops, assembly shops, and continuous shops. The issues here are that when shops progress to lower variety and higher volume, there will often be work in process inventories that could be used for more than one final customer order, leading to the need for flexible job labeling methods. In the next section we briefly discuss advantages, disadvantages, and methodologies for push systems versus pull systems. Push systems, such as classical scheduling systems, release material into the shop and push it through. Pull systems, such as Kanban, JIT (traditional version), and Conwip, operate so that when finished material is pulled out of the system, new production is initiated. Modifications for this situation are sketched. Finally, we look at a number of models where the machine can be speeded up to help relieve a bottleneck. Trade-offs of speed versus tool wear versus product quality versus capacity will be investigated.

In Chapter 21 we discuss a number of broader planning, scheduling, and control issues such as reactive scheduling, decision support systems, aggregate level systems, and integration of planning and scheduling. Reactive scheduling involves ways of making changes to a schedule for emergency events happening between regular weekly scheduling runs. Next we discuss systems that are analogous to scheduling but occur at more aggregate levels of this system, such as MRP for the manufacturing planning function and logistics systems for distribution. We also discuss revisions needed in cost (management) accounting to use the information available from the heuristic part of a bottleneck dynamics scheduling system. Then we discuss integration of planning and scheduling systems, with a case of a printed circuit board line. Next we discuss decision support systems that allow human and machine to talk back and forth interactively, mostly under manual control, to facilitate reactive scheduling, among other things. Finally, we discuss expert systems, which have had some success in codifying human schedulers' procedures and imitating them.

3.2.7 Appendices

Appendix A is a comprehensive chapter-by-chapter bibliography for the book. Appendix B contains an acronym glossary and symbol glossary. Appendix C is a general overview of Parsifal℠, a software package for solving sequencing and scheduling problems by using the optimization and heuristic methods discussed in this book. Appendix D is a Parsival user guide.

3.3 FUTURE DIRECTIONS

A major need for bottleneck dynamics, or any other heuristic scheduling framework for that matter, is the writing of *shells*. A shell is a general piece of software that would allow software houses or even sophisticated users to craft a scheduling system easily by inputting a description of the system structure and general parameters. This is very difficult to do for general shops; it might be better to start with more specific shells, such as classic job shop and complex flow shop.

Another related basic need for any heuristic scheduling framework is a general testbed for comparing heuristics and optimal methods, and for improving the heuristics. Again, it is very difficult to write a completely general testbed. The software with this book will address this issue to some extent.

A number of heuristic methods besides bottleneck dynamics are quite promising and need further testing and development. This is true especially of neighborhood search, tabu search, simulated annealing, genetic algorithms, beam search, the shifting bottleneck method, and partial enumeration methods. In particular, not only a testbed but a systematic way to test and improve these methods would help.

Modeling of shops would be much improved if there were better constructed case studies of real-world shops. These preferably need to be done by someone with a sophisticated understanding of underlying shop structure.

PART II
ONE-MACHINE PROBLEMS

A great many complex scheduling systems and project management systems have the following special form:

(a) There are a number of jobs or projects to perform.
(b) Each has importance and timeliness of completion issues.
(c) Each job consists of a number of activities.
(d) When ready, each activity queues for a unique resource.
(e) The system is much too large to be solved perfectly.

One of the great functions of human intelligence is to decompose such a problem into building blocks that can be solved perfectly or almost perfectly as a function of the unknown inputs and outputs of each block. A second great function of intelligence is to reassemble these building blocks back into the full problem and then search for compatible inputs and outputs between blocks to provide a nearly perfect (or at least very good) solution to the original problem.

In the problem we have described, the most obvious building block is the multi-process one-resource (single-input queue) problem (such as an FMS cell) with known dynamic input. But this building block is, in itself, too complicated to solve directly. Instead, we shall demonstrate that in most cases the dynamic one-resource problem can adequately be approximated as an equivalent *dynamic single-machine problem*.

Even the dynamic single-machine problem is too complex to solve directly in most cases. At this point, it becomes useful to remove some of the constraints in this problem, to simplify it further. This is called a *relaxation*. One type of relaxation, the *static* problem, allows all activities to arrive at the beginning of the problem. A second type of relaxation, the *preemptive* problem, allows partially completed activities to be stopped, removed from one machine, and finished later. These relaxations are relatively easy to solve and to obtain insight about. They are not all that useful in and of themselves, but they provide a great deal of insight into building good solutions for the dynamic single-machine building block.

Chapter 4 lays the foundation for the dynamic single-machine building block and develops a number of relaxed solutions for it. Chapters 5 and 6 introduce mathematical tools for exactly or approximately solving the dynamic single-machine problem. Chapters 7 and 8 develop solutions for the general single-machine problem.

In Part III, we begin our development of multi-machine problems. Chapters 9 and 10 develop newer mathematical methods for multi-machine situations. Chapters 11 and 12 develop parallel machines and routing.

Finally, we buttress our arguments about the importance of this building block by the following detailed observations:

(a) Many full problems have only one resource (e.g., architect, trucker, super-computer).

(b) Conventional dispatch methods repeatedly solve single-machine problems to give heuristics for full system problems.

(c) OPT-like methods solve the bottleneck resource problem and extend the answer as a heuristic for the full system problem.

(d) Shifting bottleneck methods use an iterative procedure based on successively improving solutions to embedded one-resource problems.

(e) Bottleneck dynamics methods are dispatch methods combined with improvement procedures and hence utilize accurate approximations to the one-resource problem.

____4
SINGLE MACHINE: FOUNDATIONS

4.1 INTRODUCTION

In this chapter we build a foundation for the dynamic single-machine problem. We then develop many exact and approximate solutions for static and/or preemptive relaxations of the problem, to provide a foundation for addressing the full problem.

In Section 4.2.1 we discuss the meaning of the dynamic problem more precisely and also define the static and preemptive relaxations in more detail. In Section 4.2.2 we define the dynamic one-machine model in more detail, including typical objective functions. In Section 4.3 we discuss a number of basic propositions about single-machine problems. Section 4.4 develops exact results for a number of objective functions under the static relaxation. Section 4.5 develops a number of partial results for more difficult objective functions.

4.2 PRELIMINARIES

4.2.1 Overview

In this section we define the dynamic single-machine building block more carefully in terms of (a) the fundamental model, (b) types of relaxation, and (c) types of objective functions. We do this informally in this section and more formally in Section 4.2.2.

Informal Statement of the Fundamental Model. The building block subproblem is defined for a time interval between a given starting time and a given stopping time. Each job (operation) for this subproblem has a known arrival time, processing time, and due date (if any). (In practice these are all forecasts, but we treat them as deterministic.) The machine may have a time after the start of the problem at which it is first available (e.g., due to finishing a partially complete job). Some of the jobs may be on hand or arrive at the beginning of the time interval, giving an initial queue.

How does one choose "good" beginning and end points for a dynamic single-machine problem? There is a natural beginning point: *now* in the shop is the time at which decisions must be made. Choosing the end point is more subtle. This is called choosing the horizon. Choosing a very short horizon will require much less computation time but may involve ignoring near-term important arrivals, which could change our current decisions. Choosing a very long horizon will be computationally expensive but will minimize chances that our current decisions will be poor ones for the longer problem of which this subproblem is the start. (We only need accurate current decisions, since when we get to the nonaccurate part of the schedule, we can solve a new subproblem starting from that point.) If possible, we would prefer to make the horizon long enough so that the machine will eventually have a slack period, that is, out to the end of the current busy period. This will guarantee that future arrivals will not affect current decisions. Lacking this, we will choose as long a horizon as we can afford and ignore operations (jobs) that do not finish by this point in estimating the performance of our solution period.

In this chapter and in Chapter 7 we assume any setup for a job is independent of the last job on the machine, so that we may incorporate the setup into the processing time and schedule jobs back-to-back with no time loss. We also assume jobs (operations) are independent and always ready to schedule when they arrive. More general assumptions will be considered in Chapter 8.

Types of Relaxations of the Dynamic Single-Machine Model.
As we have said, the dynamic single-machine problem is too difficult to solve exactly in most cases. In mathematics, the most common way in which difficult problems are simplified and studied is by *relaxation*: remove part of the constraints and analyze the easier resulting problem.

[To illustrate relaxation, consider the problem of minimizing the function $f(x) = 12 - x + 6x^2$, where x must be an integer. Relax the problem to remove the constraint that x must be an integer, and solve the resulting problem by calculus:

$$f'(x) = -1 + 12x; \quad f''(x) = +12$$

with minimum at $x = \frac{1}{12}$, $f(\frac{1}{12}) = 12 - \frac{1}{12} + 6(\frac{1}{12})^2 = 11.96$. Due to convexity, the relaxed and original problems decrease smoothly on both sides around $x = \frac{1}{12}$. Thus we get the insight that the solution to the constrained problem must come either at $x = 0$ or $x = 1$. Since $f(0) = 12$ and $f(1) = 17$, the solution to the original problem is seen to be $x = 0$.]

Clearly, the cost of an optimal solution for the relaxation of a cost minimization problem is less than or equal to the cost of the solution for the original problem. We say the relaxation solution provides a *lower bound* for the original problem. If the optimal solution is not known, this lower bound is useful in evaluating how good a heuristic is. For example, if a heuristic solution gives a cost of 8.7, and a lower bound of 8.5 is known, the heuristic cost is no more than 0.2 from the cost of the optimal solution.

Static Versus Dynamic. An important relaxation of the dynamic single-machine problem is moving the arrivals of all activities to the beginning of the interval. The resulting problem is called the *static* single-machine problem. Occasionally, a dynamic problem actually involves all activities arriving simultaneously. Otherwise, this relaxation usually changes the problem too much to provide good bounds on the original problem. However, it is extremely useful in providing insight to solution methods and approximations for the dynamic problem.

Preemptive Versus Nonpreemptive. Another important relaxation is allowing activities to be interrupted (one or more times) without cost in order to process other important activities more quickly. Occasionally, a dynamic problem actually allows interruption without cost. Fairly often the preemptive relaxation gives good bounds on the original problem. But once again, the preemptive relaxation allows insight into the solution for the original nonpreemptive case.

A slight variation that is often computationally easier rounds arrival times, process times, and due dates to integers and allows processing only in unit blocks. We term this type of relaxation the *unit-preemptive* case. Note that by making units very small (say, 0.1 hour) the two cases can be made as similar as desired. Thus we shall freely focus on solving whichever case is the most convenient mathematically. (There are three different methods for spreading an activity's costs of timeliness/production among these unit pieces. This will be discussed in the next subsection.)

Types of Problem Objectives. Problems may also be categorized by whether their objective functions are *regular* or *nonregular*. For regular objectives, it is always considered preferable (other things equal) to finish an activity earlier, rather than later; in nonregular objectives, this may not be so. (For example, in a just-in-time environment, finishing jobs too early may represent excess work in process.) Regular objectives, in turn, may be categorized into (a) maximize some measure of resource *utilization*, (b) minimize some measure of job *flowtime*, and (c) minimize some measure of job *tardiness*.

We discuss problem objectives in more detail in the next subsection.

Classification of Single-Machine Problems. We thus arrive at a classification of single-machine problems:

(a) Static versus dynamic.
(b) Preemptive versus nonpreemptive.

(c) Regular versus nonregular.

(d) Utilization versus flowtime versus tardiness.

In this chapter we look primarily at regular static problems, for both the preemptive and nonpreemptive cases, with independent activities (jobs). In Chapter 7 we also consider dynamic single-machine problems with independent activities. We discuss single-machine problems with dependent jobs in Chapter 8.

We turn in Section 4.2.2 to a more formal specification of these issues.

4.2.2 Formal Dynamic Single-Machine Model

In dealing with job or activity attributes for the dynamic one-machine model, we need to distinguish between information that is known (or forecasted) in advance and information that is generated as the result of scheduling decisions. Information that is known in advance serves as input to the scheduling procedure. (We try to consistently use lowercase letters to denote this type of data.) We rather informally present the model structure and inputs.

Structure and Model Input (I1 = "Input Assumption 1")

I1   The machine processes activities one at a time serially.

I2   The resource is available over the *scheduling interval* t_s *to* t_e (without loss of generality we henceforth assume $t_s = 0$).

I3   *n single-operation jobs* (activities) arrive over the interval (early enough to all finish by t_e; constrained to do so).

I4   Activity input: *processing time* p_j.

I5   Activity input: *ready time* r_j (= 0 for static case).

I6   Activity input: *due date* d_j.

I7   Preemption model choice: (a) nonpreemptive, activities finish without interruption; (b) preemptive, activities can be interrupted without penalty.

I8   Objective function choice: (a) regular, all activities always prefer finishing earlier to finishing later; (b) nonregular, remaining types of objective functions.

(To complete the model, I7 and I8 must be spelled out in more detail. This is discussed below.)

A few points about the model and its inputs.

(a) The machine is assumed to be continuously available over the interval and must finish all the jobs by its end.

(b) Inputs are assumed to be deterministic, or at least to have been forecasted satisfactorily, and do not depend on the processing sequence. (Activities can have setups, as long as they don't depend on the previously processed activity. The stated processing time is understood to include both direct processing time and facility setup time.)

(c) Due dates may be absolute, or due dates may be violated by paying some penalty. The absence of pertinent due dates is easily handled; for example, the due dates may all simply be moved to the left of the interval or to the right of the interval as appropriate.

Now that we have the problem structure and inputs, we turn to the actual mechanics of scheduling the problem, and then to simple output measures to help evaluate the schedule.

General Scheduling Procedure. All possible schedules for the dynamic problem with or without preemption and with or without a regular objective can be formed by repeating the following steps:

S1 Choose the time to remove the activity currently being processed from the resource.
S2 Choose the next activity for processing.
S3 Choose the time to put that activity on the resource.
S4 Repeat steps S1 to S3 until all activities are completed.

Following these steps produces a complete schedule built forward in time. Of course, certain constraints must be satisfied:

C1 An activity cannot be processed before it arrives.
C2 An activity cannot be processed after it is completed.
C3 Two activities cannot be processed at once.

Note that the static relaxation of this problem eliminates the need to consider C1. Similarly, the nonpreemptive case removes the choice S1; removal of activity j is always p_j units of time after starting it.

To evaluate the goodness of a given schedule, we need to know the time at which each activity finishes. For a general preemptive dynamic schedule, finding these times will require a fair amount of bookkeeping. For the nonpreemptive case with new activities started as soon as possible, we may calculate the times with a formula in terms of only the order in which the activities are processed. This formula is particularly simple for the static nonpreemptive case. We come back to this issue later in the chapter.

Preemption Versus Nonpreemption. Remember that in Section 4.2.1 we said that the preemption assumption was useful for three reasons:

(a) Sometimes it is actually appropriate to the problem.
(b) Sometimes this relaxation gives a good lower bound.
(c) The preemptive solution gives insight about the nonpreemptive problem.

We also said that unit preemption in which activities can only be preempted in unitary pieces can always be substituted for ordinary preemption, since units can be made as small as desired, and approximates normal preemption as accurately as desired. However, there are several types of unit preemptions that can arise depending on costing assumptions. These are a little complicated but important. We discuss them here.

A central problem is: How do we spread the original objective value for the activity among the unit pieces ("mini-jobs")? There are a number of possibilities, but we shall have occasion to use three:

(a) *Cost-at-End* (*CAE*). All mini-jobs of an activity have the original due date; the last mini-job to be completed incurs the full original objective function.

(b) *Spread-Cost* (*SC*). All mini-jobs of an activity have the original due date; each of the mini-jobs (p_j in number) has an objective function that is $1/p_j$ of the original.

(c) *Tight-Spread-Cost* (*TSC*). Mini-jobs of an activity have due dates d_j, $(d_j - 1), (d_j - 2), \ldots$; each of the p_j mini-jobs has an objective function $1/p_j$ of the original.

CAE preemption has value as a reasonable scheduling discipline, but it has little or no value as a relaxation. The reason is that the cost is attached to only one mini-job, chosen in advance. Therefore, at least for regular objectives, the scheduling procedure will finish this mini-job first and violate our condition that the last one finished incurs the cost. Thus constraints must be added to force the final mini-job to actually finish last. This makes the solution as complicated mathematically as the nonpreemptive case. In fact, we shall see that for static regular problems, CAE preemption cannot provide lower costs than the ordinary nonpreemptive case.

SC and TSC preemption, on the other hand, do simplify the mathematics considerably. Often these two methods will provide the same schedule. SC is sometimes easier to use in proofs; on the other hand, TSC gives a better lower bound and, in fact, duplicates the nonpreemptive case when possible.

Next, we wish to discuss different types of measures of how "good" a schedule is (objective function). We first consider major groupings of objectives, which we call primary.

Primary Output Measures

P1   *Completion time C_j*—time at which activity j completes processing.

P2   *Flowtime F_j*—amount of time activity j spends in the system:

$$F_j = C_j - r_j$$

P3   *Lateness L_j*—amount of time (positive or negative) by which the completion of activity j exceeds its due date:

$$L_j = C_j - d_j$$

P4   *Tardiness T_j*—the lateness of activity j if it is positive; if the lateness of activity j is not positive, the tardiness is zero:

$$T_j = \max\{0, L_j\}$$

P5   *Earliness E_j*—the negative of the lateness of activity j if the lateness is negative; if the lateness is positive, the earliness is zero:

$$E_j = \max\{0, -L_j\}$$

What is the significance of these primary output measures? P1, completion time, is in itself the only *necessary* output from the schedule since all the other primary output measures are determined from it.

P2, flowtime, measures the response of the system to individual demands for service. The flowtime is the amount of time between the job's arrival into and departure from the system. Clearly the flowtime of an activity is closely related to its work-in-process (WIP) cost. A number of other names are used for flowtime, such as customer lead time and customer response time.

P3, lateness, measures the conformity of the schedule to a given due date. Lateness rewards jobs for being early equally much as it punishes them for being tardy. We will see that flowtime and lateness are really two disguises for the same shop output measure and lead to the same best scheduling policy.

P4, tardiness, reflects the fact that, in many situations, distinct penalties and other costs will be associated with positive lateness, but no penalties or benefits will be associated with negative lateness. Hence many objectives measure the amount and frequency of tardiness.

P5, earliness, reflects the fact that in some situations there may be a penalty for finishing activities before they are due. (The customer may not accept the order until the due date, and the finished inventory cost may be higher than the WIP cost.) Penalties for earliness lead to nonregular objectives, which are discussed in Chapter 7.

Schedules are generally evaluated by aggregate quantities that summarize the scheduling performance for all jobs. Three types of aggregation are most common:

(a) A weighted average of a primary output measure.

(b) The maximum or minimum of some primary output measure.

(c) Some mixture of (a) and (b).

The weighted average will be weighted by the relative activity importance: a *weight* w_j for job j implies a weight of w_j in the objective function. We are ready to look at common objective functions, which use the primary outputs above.

Common Objective Functions (Minimize)

F1	Makespan	$C_{max} = \max_j\{C_j\}$
F2	Weighted flowtime	$F_{wt} = \Sigma_j\ w_j F_j$
F3	Weighted lateness	$L_{wt} = \Sigma_j\ w_j L_j$
F4	Weighted tardiness	$T_{wt} = \Sigma_j\ w_{Tj} T_j$
F5	Maximum flowtime	$F_{max} = \max_j\{F_j\}$
F6	Maximum lateness	$L_{max} = \max_j\{L_j\}$
F7	Maximum tardiness	$T_{max} = \max_j\{T_j\}$
F8	Weighted number of tardy jobs	$N_{wt} = \Sigma_j\ w_{Nj}\delta(T_j)$

$$\text{where } \delta(x) = 1 \quad \text{if } x > 0$$
$$\delta(x) = 0 \quad \text{otherwise}$$

| F9 | Weighted earliness plus weighted tardiness | $ET_{wt} = \Sigma_j\ (w_{Ej} E_j + w_{Tj} T_j)$ |

We need to say a few words about some of these objectives.

Makespan. The schedule with the smallest *makespan* is often taken as a surrogate for the schedule with the highest utilization. The intuitive idea is that finishing the given set of activities earlier will allow new activities to be started earlier. This approach is an excellent approximation for the single-machine case and a useful first cut for more general problems. However, in multi-processor/multi-resource problems, makespan gives no credit for resources that finish well before the last one finishes. A bottleneck dynamics utilization criterion called *economic makespan* has been developed, which addresses this issue (see Chapter 11).

Weighted Flowtime and Weighted Lateness. *Weighted flowtime* and *weighted lateness* are both very popular objectives for a number of reasons. First, we shall see that they are both very easy to use; their optimal solutions are identical and very intuitive. (This is why we have not distinguished their weighting factors, using w_j for both.) Second, they are robust in the sense that schedules that are optimal for them often produce good schedules for problems with somewhat different objectives.

Weighted Tardiness/Other Tardiness Measures. In many situations *weighted tardiness* is a good objective, but problems using this objective are very difficult to solve exactly. The bottleneck dynamics approach provides strong heuristics for this objective. The *weighted early–tardy* objective is also very important when customers do not want jobs to be tardy but will not take early delivery. This objective is the only nonregular objective we have presented so far. The *weighted number of tardy jobs* objective reflects the situation when customers simply refuse to accept tardy jobs, so that the order is lost.

Note that the weights for flowtime (w_j), tardiness (w_{Tj}), earliness (w_{Ej}) and number tardy (w_{N_j}) are all distinct. Heuristics for these objectives are discussed in Chapter 7.

Maximum Flowtime, Lateness, Tardiness. Minimizing *maximum tardiness* is important when customers tolerate smaller tardinesses but become rapidly and progressively more upset for larger ones. Minimizing *maximum lateness* is primarily important because it is a relatively easy problem and can be used as an aid for solving other problems. For example, minimizing *maximum flowtime* can be accomplished by setting the due dates equal to the arrival times and minimizing maximum lateness. Minimizing maximum tardiness can similarly be accomplished by minimizing maximum lateness and truncating to zero if that value is negative. Finally, there is a technical use for the maximum lateness objective in solving complex multi-resource problems with a makespan objective. If we set an arbitrary common due date for all jobs in the problem, then minimizing maximum lateness will minimize the makespan.

Weights (w_j). Sometimes, for lack of good data, all weights are simply assumed to be $1/n$. To indicate this case, we shall use, for example, F_{av} instead of F_{wt}. Unfortunately, this practice, while occasionally necessary, tends to obscure the fundamental economic rewards and penalties inherent in the shop situation. Difficulty in estimating w_j does not detract from the urgency to estimate it as well as possible. Using F_{av} assumes that all penalties are equal, which is a very strong statement. Some authors require that $\Sigma_j\ w_j = 1.0$, and then speak of "weighted mean flowtime", for example. We prefer to keep the direct economic interpretation that the w_j are implicit dollar penalties per day of flow for the job. This idea is at the foundation of bottleneck dynamics. While normalization does not affect policy form or proof, it obscures the intuition of our approach.

The weight w_j represents the cost of delaying activity j for, say, one day. This weight should be related to the value of the activity (or value added); it should also be related to the importance or "antsiness" of the customer. (The authors find practitioners can estimate w_j more easily if it is written as $w_j = D_j A_j$, where D_j is the dollar value of the activity and A_j is a customer priority factor.)

Weights (w_{E_j}, w_{T_j}). Note that there are two weights for every job in the weighted early/tardy problem: one for the cost per day of finishing early and the other for the cost per day of finishing tardy. Some simplifying assumptions may be useful, such as assuming that the ratio between early weight and tardy weight is the same for every job.

4.2A Numerical Exercises

4.2A.1   Consider the following six-job problem with all arrival times 0, with processing times, job weights, and due dates for each one:

Job j:	1	2	3	4	5	6
p_j	8	6	4	2	6	8
w_j	1	1	1	2	2	2
d_j	10	16	20	12	12	25

Create any schedule with no preemption and no idle time. Evaluate your schedule by each of the following objectives: makespan, weighted flow-time, weighted lateness, weighted tardiness, maximum flowtime, maximum lateness, maximum tardiness, and weighted number tardy. Use the same weights for all objectives.

4.2A.2   Have someone demonstrate for you the "Interactive Manual Interchange" routine in the software.

    (a) Enter the data from 4.2A.1. Produce four quite different schedules, and record the various objectives.

    (b) Try to produce a schedule with low maximum tardiness. How does it behave with respect to weighted number tardy?

4.2A.3   Now create your own input for a larger problem. (Make sure the machine is too heavily loaded to get zero tardiness.) Repeat 4.2A.2 for this larger problem.

4.2A.4   Redo 4.2A.3, now allowing preemption and idle time, as well as nonzero arrival times, for a smaller problem.

4.2A.5   Redo 4.2A.3 as follows: for each objective function, create a schedule that will do better than the original one; repeat to get the best schedule you can. (The software shows you instantly when an objective improves.)

4.2B Software/Computer Exercises

4.2B.1   Write a computer program that accepts as input a scheduling problem and one or more objective functions. It then queries the user for a schedule and evaluates it by the objective(s). The query is then repeated until the user ends the session.

4.2C Thinkers

4.2C.1   Suppose each job has two due dates. After d_{1j} it is tardy; after d_{2j} the customer will cancel the order. (This is a common real-world situation. d_{2j} is called the "drop dead" date.) Design an objective function to handle this case.

4.3 FUNDAMENTAL PROPOSITIONS

Our philosophy throughout this book is to try to keep all ideas and propositions as clear as possible, while using a minimum of mathematical notation and detail, as befits our priorities to allow a broad class of users access to methods for developing powerful heuristics. Often proofs (even simplified and rough) may be passed over on the first reading, for readers primarily interested in building software customized to an application. Numerical exercises and software exercises should be of broad interest, although more complicated software exercises involving extensive programming are probably not of interest for all users. More mathematical students are likely to be interested in the "thinker exercises", try to understand proofs more clearly and more carefully master the discussions of mathematical approaches in Chapters 5, 6, and 10.

4.3.1 General Case Results

General Mathematical Forms of Objective Functions

Definition. A *completion-based* objective function is denoted by $Z = f(C_1, C_2, \ldots, C_n)$.

Definition. Z is *additive* if $Z = \Sigma_j f_j(C_j)$.

Definition. Z is *maxi* if $Z = \max_j\{f_j(C_j)\}$.

Definition. Z is *regular* if Z will increase only if at least one C_j increases.

Proposition 1. If Z is additive, it will be regular if and only if each function $f_j(C_j)$ is monotonically increasing in C_j.

Proof. The "if" part of the proposition is clear. To prove the "only if" part, choose a set of C_j where some $f_k(C_k)$ is locally decreasing in C_k. Hold the others constant, and decrease C_k. The function $f_k(C_k)$ increases, which increases Z. Therefore Z was increased by decreasing C_k, which contradicts the original assumption. ∎

Proposition 2. If Z is maxi, it will be regular if each function $f_j(C_j)$ is monotonically increasing; it will be nonregular if at least one function $f_k(C_k)$ is locally decreasing for at least one set of completions $(C_1, C_2, \ldots, C_n)$ for which $f_k(C_k) = \max_j\{f_j(C_j)\}$.

Proof. Similar to that of Proposition 1. ∎

Developing an understanding of why Proposition 2 cannot be as strongly stated as Proposition 1 is left as an exercise.

Proposition 3. Consider the unit-preemptive relaxation (SC or TSC case) of the general dynamic problem.

(a) The relaxation can be considered a sequencing (permutation) problem.

(b) For an additive objective the problem is convex and can be solved by the assignment problem specialization of linear programming.

(c) For a maxi objective the problem is convex and can be solved by the maxi version of the assignment problem.

Proof. Consider first the additive case with the SC variant. ∎

Divide the availability interval $(0, t_e)$ into units. Augment the available activity units by enough dummy "idleness" units (with full availability, no due date, and no cost) so that there is a one-to-one correspondence between activity units and time slots. Activity j is divided into p_j units, labeled $j1, j2, j3, \ldots, jk$, each with due date d_j and with assignment cost $(1/p_j)f_j(C_{jk})$. (C_{jk} is the completion time of unit piece k of job j.) If the assignment chosen violates the availability of the activity, the cost is M (any large cost to prevent the assignment).

The maxi case is similar, except that the assignment cost is simply $f_j(C_{jk})$. ∎

The TSC version is left as an exercise.

Power of Proposition 3. Proposition 3 is powerful for a number of reasons. First, the assignment problem is convex and solvable by a special structure LP algorithm that is very fast. Thus very large assignment problems are easily solved. (However, note that if there are 50 jobs divided into 100 units each, then the matrix is at least 5000 by 5000, so that the assignment problem would itself be very large. There is thus incentive for a cruder discretization: 500 by 500 might be much more manageable.)

Second, convexity suggests that a number of approximate methods for solving the relaxed problem, such as pairwise interchange and/or Vogel's method (basically minimize maximum regret), could be expected to be very accurate and in some cases optimal.

Third, if a solution to the relaxed problem can be found in which all the pieces of each original activity are actually processed nearly without gaps, then a solution to the original nonrelaxed problem has been found. (Later in this chapter, and elsewhere, we shall use this idea repeatedly.)

Fourth, real problems often allow some preemption with rules that are somewhat vague. Thus both the nonpreemptive and preemptive solutions may be considered approximations to the true situation.

Many scheduling books number the activities initially 1 to n; then, to indicate the position in the schedule, the job scheduled first is denoted [1]. Thus if job 7 is scheduled first, we would say [1] = 7. We feel that this careful notation tends to obscure proposition statement and proof unnecessarily in many cases. Where it will cause little confusion, we number the jobs *after sequencing them*; for example, F_1 means the flowtime of the first activity scheduled. (We revert to the other more complicated notation when absolutely necessary.)

Regular Dynamic Problems with No Preemption

Proposition 4. For the regular dynamic problem with no preemption allowed, the schedule is uniquely determined by the activity sequence. The next activity in sequence starts as soon as it arrives and the machine becomes free, with

$$C_j = \max\{C_{j-1}, r_j\} + p_j \text{ or}$$
$$C_j = C_{j-1} + \max\{(r_j - C_{j-1}), 0\} + p_j$$

Proof. While the sequence may force some idle time waiting for the next job in sequence to arrive, regularity ensures that the next activity starts as soon as feasible. The middle term on the right of the second expression for C_j is the inserted idleness mandated by choosing activity j to go next.   ∎

Proposition 5. For the regular dynamic problem with no preemption allowed and no inserted idleness allowed (i.e., for the job in the sequence, we require that $r_j \leq C_{j-1}$), Proposition 4 reduces to $C_j = \Sigma_{i=1,j} p_i$
An important special case is when $r_j = 0$ for all j, that is, the static case.

Proof. In Proposition 4 the second equation simplifies since the inserted idleness terms are zero.   ∎

In general dynamic problems, establishing whether a given sequence requires inserted idleness requires considerable extra effort, so that Proposition 4 may as well be used directly. The static case, however, automatically allows the simpler formula. Propositions 4 and 5 will be useful in Chapter 7 when we discuss sequencing problems for dynamic job shops in some detail. At this point, however, we wish to derive some exact solutions for regular static problems. We first present several propositions for this case.

4.3.2 Propositions for Regular Static Problems

Intuitively, in a static shop with all jobs available at all times and a regular measure of performance (so that all jobs want to get done earlier), there is no particular reason to use a CAE preemptive policy. Once the final mini-job is scheduled, we can shift the other mini-jobs back in the sequence until all the mini-jobs run in series, producing a nonpreemptive solution. Due to CAE, this shift will not increase the objective value.

Proposition 6. If a one-machine problem is regular and static, then CAE preemption is unnecessary.

Proof. Start with an optimal preemptive plan. Find a split activity, and consider the last two pieces of that activity in the schedule. Interchange the order of the next to the last of these pieces and the piece following it. The piece moved ahead can only benefit its activity by the regularity assumption. The piece moved back does not affect that activity's completion time and thus does not hurt its contribution to the objective function.
Now repeat this process to recombine the activity one piece at a time. Finally, the activity will be nonpreempted without costing anything. Repeat the entire process for any other split activities, one at a time.   ∎

(It is interesting to ask what goes wrong with this proof if regularity or staticness is violated, or for non-CAE preemption. We leave this as an exercise.)

Proposition 7. For the one-machine regular static problem, inserted idleness is unnecessary.

Proof. The proof is similar, but easier, and it is left as an exercise. ∎

Proposition 8. For the one-machine regular static problem, the sequence of jobs ($n!$ choices) completely specifies all necessary scheduling solutions, and $C_j = \Sigma_{i=1,j} p_i$.

Proof. Obvious. ∎

4.3A Numerical Exercises

4.3A.1 Consider the following sequencing problem with three jobs and a weighted flow objective function:

Job j:	1	2	3
p_j	4	5	3
w_j	8	5	1

Suppose that CAE unit preemption is allowed, and the unit joblets are named after the original jobs. A pundit claims the following sequence of the 12 joblets is optimal:

$$2, 3, 3, 1, 2, 2, 2, 2, 1, 1, 1, 3.$$

Use the proof idea of Proposition 6 to construct a better sequence that is not preemptive. Note that the Interactive Manual Interchange of the software routine can do most of the work for you. (*Hint*: In CAE preemption all the weight is attached to the last joblet; the others have weight 0.)

4.3A.2 Give an example of an objective function that is not regular. Construct a problem in which this measure of performance is optimized by a schedule that is not a permutation schedule (defined simply by the ordering of the jobs).

4.3A.3 Give an example of a regular objective function that is neither additive nor maxi. Is there a shop situation for which it seems reasonable? (*Hint*: Think combination rules.)

4.3A.4 Create a small problem similar to 4.2A.1; let all times be integers, and be sure the sum of the processing times is not greater than 10. Also form the unit-preemptive relaxation of the problem. For a given schedule, and for each of the cost variants CAE, SC, and TSC, demonstrate numerically that the relaxation has a cost no higher than the original. The Interactive Manual Interchange software may be helpful.

4.3B Software/Computer Exercises

4.3B.1 Write a program that accepts a problem and a schedule as inputs, and constructs the unit-preemptive relaxation, printing out the value of the solution for both the original problem and the relaxation.

4.3C Thinkers

4.3C.1 Prove Proposition 7. Give examples to show Proposition 7 is not true if any part of the assumptions is not true.

4.3C.2 Similarly demonstrate Proposition 6 is not true if any part of the assumptions is not true.

4.3C.3 Prove Proposition 3 for the TSC unit-preemptive case.

4.3C.4 Proposition 2 cannot be stated as strongly as Proposition 1 because it is possible for a cost function for one job to be nonmonotonic but never to be the largest cost function for all the jobs for any set of completions.
(a) Give a simple example of such a case.
(b) Do you think such cases are important practically?

4.4 REGULAR STATIC PROBLEMS: EXACT CASES

We turn now to discuss a number of regular objectives, which, for static problems, give convenient, intuitive, exact priority rules. In the next section we derive some preliminary results for other regular objectives that are more difficult. In Chapter 7 we use all these exact results for the static relaxation as aids in deriving good heuristics for the more general dynamic versions of these problems.

One way to find the nonpreemptive solution (when it works) is to solve the preemptive version and then rearrange the result to find an optimal nonpreemptive solution to it. The unit variation of preemption is particularly useful for this purpose.

We may group most scheduling objectives into four possible categories: (a) utilization based, (b) flow based, (c) due date based, and (d) economic based.

4.4.1 Utilization Based Objectives

Utilization is an important objective for heavily loaded shops when jobs may have to be turned away or expansion is indicated. Makespan, while it has theoretical limitations, is an important objective since it is easy to analyze. This is particularly true for the static problem.

Proposition 9. For a static problem, any nondelay ordering of the activities yields the minimum makespan, which is $C_{max} = \Sigma_{j=1,n}\, p_j$ for one machine (see exercise 4.4A.2).

Proof. For any one-machine problem, static or dynamic, the makespan is clearly the sum of all the processing times (which are fixed) plus the sum of idle time. An equivalent objective is to minimize idle time between activities. Therefore any nondelay (no inserted idleness) schedule is optimal for the static problem. ∎

(We shall see in Chapter 7 that minimizing makespan for dynamic arrivals and independent jobs is almost as easy: when the processor becomes free, schedule any available job. Adding precedence relationships adds little difficulty. But more general timing issues, such as sequence dependent setups, change the problem from being very easy to very hard. Finally, for multi-process and/or multi-resource problems, minimizing makespan is almost always complex, and heuristics are required for large problems.)

Makespan is an excellent surrogate for utilization for one-machine problems, or even much more complicated problems, if the time interval of the problem is carefully chosen so that the resource(s) is naturally idle at the end. (This would occur for the dynamic one-machine problem, for example, if the end of the time interval were chosen as the end of the current busy period for the machine).

On the other hand, makespan is a less useful surrogate for utilization if, as is more typical, the machine is quite busy at the end of the time interval and unfinished jobs are simply allowed to finish while suppressing further new arrivals. This is basically a boundary or edge effect. One solution to this problem is to take a large interval; doubling the interval halves the percentage boundary error. Chapters 11 and 13 outline more sophisticated corrections for this problem.

4.4.2 Flow Based Objectives

Reducing turnaround time through the shop will often be the primary objective for a shop that is not fully loaded and competes primarily on delivery lead times. It will also be of primary importance when there is strong interest in JIT (just-in-time) or reducing WIP in general. Weighted completion time, weighted flowtime, weighted lateness, and weighted shop WIP are all important flow objectives; in fact, all are equivalent.

We first use two propositions to derive the optimal scheduling rule for the weighted flow objective. Next, we show that weighted completion, weighted lateness, and weighted shop WIP give the same scheduling rule. [Some of these simple flow results were first published by Smith (1956).]

Proposition 10. For the static F_{wt} objective, if all $p_j = 1.0$, then an optimal sequence of jobs must satisfy $w_1 \geq w_2 \geq w_3 \geq \cdots \geq w_n$.

Proof. Suppose in some optimal solution $w_j < w_{j+1}$. Interchanging w_j and w_{j+1} produces a schedule with weighted flow decreased by $w_{j+1} - w_j$. Because the new schedule is better than the original, the original schedule is not optimal. ∎

Our purpose in proving Proposition 10 is to apply it to the unit-preemptive relaxation of the original problem.

Proposition 11. For the static F_{wt} problem, an optimal sequence of jobs must satisfy $(w_1/p_1) \geq (w_2/p_2) \geq (w_3/p_3) \geq \cdots \geq (w_n/p_n)$.

Proof. Form the unit-preemptive SC relaxation, for which units have weights (w_j/p_j). Solve this relaxation by Proposition 10. Now, in the relaxation, the solution is not affected by rearranging mini-activities with the same weight. (If the mini-activities from the two activities are interspersed, by Proposition 10 the mini-activities must have the same weights.) Thus we can schedule all units (mini-activities) of the same activity together, giving a nonpreemptive solution with the same cost. Hence the nonpreemptive solution is optimal. ∎

WSPT

In the future we shall denote this rule as *WSPT* (weighted shortest processing time). (This is a classic name and we shall not change it; however, "importance per processing time unit" would be better.)

This scheduling rule makes excellent sense intuitively. Suppose you have three reports due. The first is worth $1000 and will take 50 hours. The second is worth $200 and will take 5 hours. The third is worth $2000 and will take 120 hours. By our rule, the first has a priority of $20/hr; the second has a priority of $40/hr; the third has a priority of $16.67/hr. To maximize dollars per hour flow through your "shop" (desk), complete #2, then #1, then #3.

Proposition 12. In the static or dynamic problem, for any of the objectives of minimizing (a) weighted flowtime, (b) weighted completion time, (c) weighted lateness, or (d) weighted shop WIP, any optimal schedule must be WSPT.

Proof. We first argue (a), (b), and (c) are equivalent problems, and then argue (d) is equivalent to (a).

Now weighted flow for any schedule can be written

$$F_{wt} = \Sigma\, w_j(C_j - r_j) = \Sigma\, w_jC_j - \Sigma\, w_jr_j = C_{wt} - r_{wt}$$

And weighted completion time is just

$$C_{wt} = \Sigma\, w_jC_j = C_{wt}$$

And weighted lateness is just

$$L_{wt} = \Sigma\, w_j(C_j - d_j) = \Sigma\, w_jC_j - \Sigma\, w_jd_j = C_{wt} - d_{wt}$$

Now note that r_{wt} and d_{wt} are weighted combinations of arrival dates and due dates, respectively; hence r_{wt} and d_{wt} are constants that do not depend on the schedule chosen. Thus the three objectives differ only by a constant and must be minimized for the same schedule.

Finally, if the weights are interpreted as dollars per day of inventory charges (instead of dollars per day of customer satisfaction lost), then clearly inventory is a flow measure. (Of course, we might like to use different weights for inventory than for customer desire for promptness; both optimal policies would still be WSPT but could differ due to the differing weights.) ∎

4.4.3 Due Date Based Objectives

Reducing the amount and frequency by which individual flow times exceed promised times (due dates) will often be the primary objective when customers desire reliable time delivery to meet their own tight scheduling purposes. Maximum lateness, weighted tardiness, and weighted number of tardy jobs are all very important objectives. There will also be increasing emphasis on JIT for both the producer shop and the customer shop, which makes early–tardy objectives increasingly important.

Due date objectives tend to be difficult to analyze. There are three exceptions for which we may obtain exact results: minimizing maximum lateness, maximum tardiness, and the unweighted number of tardy jobs, all for the static case. (It might seem that weighted lateness, which refers to due dates, should be included in this list. However, we saw in the last section that due dates only change the weighted lateness function by a constant. Hence we have classified that objective as a flow rule.) In the next section we present partial exact results for the more difficult cases of weighted tardiness, weighted number of tardy jobs, and weighted flow/tardy (which is a mixture of flow and tardiness objectives).

Minimizing Maximum Lateness

Proposition 13. For the static problem with minimizing maximum lateness as the objective, there is an optimal policy satisfying $d_1 \leq d_2 \leq d_3 \leq \cdots \leq d_n$ (this is called EDD, the "earliest due date rule").

Proof. We first show the result for the special case where all process times are 1.0. Suppose two such activities were in fact sequenced so that $d_j > d_{j+1}$. Now activity $j + 1$ has a due date at least one earlier than j and is processed one time unit later. Hence the lateness is at least two more than that of activity j. Interchanging will decrease the lateness of activity $j + 1$ by one, and increase the other by one. This change cannot possibly increase the maximum lateness.

Now return to the original problem and form the unit-preemption SC relaxation, which we have just shown can be optimized by arranging the mini-jobs in due date sequence. Rearranging mini-jobs with the same due date will not affect the objective. Rearrange so that all mini-jobs of an activity are together. We now have a solution to the original problem without increasing the objective function; it must be optimal. ∎

Note that using the EDD rule helps to prevent the embarrassment of very large latenesses, but the rule is often poor with respect to average lateness and/or average tardiness. EDD may be useful in a shop where management only intervenes in exceptional cases. Minimizing maximum lateness is also very important if all activities are going to be needed for the same project somewhere downstream. In such a case one of the inputs being very late would not be good.

Minimizing Maximum Tardiness

Proposition 14. The maximum lateness objective and the maximum tardiness objective are both minimized by the same policy. Thus for the static case EDD minimizes maximum tardiness as well.

Proof. If the policy that minimizes maximum lateness has a positive minimum, then maximum tardiness cannot possibly give a smaller minimum. Therefore the policy gives the same answer for the two objectives.

If the maximum lateness policy has a negative minimum, then zero tardiness (which is the best possible) is achievable with this policy.   ∎

Number of Tardy Jobs

Proposition 15. For the static problem with all $p_j = 1$, an optimal algorithm for minimizing the weighted number of tardy jobs is:

Step 1: Order the activities in EDD order.

Step 2: If there are no tardy jobs, stop; this is the optimal solution.

Step 3: Find the first tardy job, say, k, in the sequence.

Step 4: Move the job with smallest weight w_{N_j}, $1 \leq j \leq k$, to the end of the sequence.

Step 5: Revise the completion times and return to step 2.

Proof. After step 1, since the whole sequence is in EDD order, the subset from job 1 to the first tardy job, k, will also be in EDD order. Since EDD minimizes maximum tardiness for the latter set, and there is one tardy job in the set with the current ordering, then at least one of these jobs will inevitably be tardy for any ordering. Now job $k - 1$ was not tardy, and job k has completion time only one greater, and $d_{k-1} \leq d_k$. Thus k will only be tardy by one time unit. If we remove the job with the smallest weight, k will no longer be tardy.

Now the remaining problem (except for last job) is still in EDD order, so repeat the procedure until there are no tardy jobs remaining except for those removed to the end.   ∎

Note that the removed jobs will all be tardy in the final optimal sequence. Depending on the size of the corresponding penalties, these jobs may simply be scrapped and not performed on the machine.

It is easy to employ preemptive Proposition 15 to solve the nonpreemptive unweighted number of tardy jobs problem, in Proposition 16; but our relaxation method will not work exactly for the general weighted case. (See the next section.)

Proposition 16. (Hodgson's Algorithm). For the static problem with the number of tardy jobs objective (all $w_{N_j} = 1$; that is, min N_{av}), the algorithm of Proposition 15 remains optimal if step 4 is replaced by "move the single job j $(1 \leq j \leq k)$ with the longest process time to the end of the list."

Proof. The proof is the same as that of Proposition 15, except now we have no choice of how much weight to remove. Therefore we might as well choose the activity with the longest process time to ensure less tardiness among later jobs. ∎

Some further results on the number tardy problem were given by Maxwell (1970) and Moore (1968).

As an example, consider the job set in the table below.

Job j	p_j	d_j
1	1	2
2	5	7
3	3	8
4	9	13
5	7	11

A worksheet is given below summarizing Hodgson's procedure:

Step 1: Initialize in EDD order: 1–2–3–5–4.

Job j	d_j	p_j	C_j	T_j
1	2	1	1	0
2	7	5	6	0
3	8	3	9	1
5	11	7	16	5
4	13	9	25	12

Step 2: Job 3 is the first tardy one.
Step 3: Job 2 is removed as longest among 1, 2, and 3: 1–3–5–4.

Job j	d_j	p_j	C_j	T_j
1	2	1	1	0
3	8	3	4	0
5	11	7	11	0
4	13	9	20	7

Step 4: Now job 4 is the first tardy one.
Step 5: Remove job 4 as tardy: 1–3–5.

Job j	d_j	p_j	C_j	T_j
1	2	1	1	0
3	8	3	4	0
5	11	7	11	0

Step: No more jobs are tardy.

Step 7: An optimal sequence is thus 1–3–5–(2–4).

(The parentheses mean the order between 2 and 4 is immaterial.)

The fact that due date objectives are usually very difficult to solve exactly will lead us later to look for good heuristics.

4.4A Numerical Exercises

4.4A.1 Consider the following scheduling problem data:

Job j	p_j	d_j	w_j
1	1	7	2
2	3	18	8
3	1	8	1
4	1	4	1
5	3	13	7
6	1	1	1
7	2	0	4
8	2	6	6
9	3	8	7
10	1	11	2
11	1	11	4
12	1	7	4

Input these data as a problem into the software's Interactive Manual Interchange routine. Input weighted lateness or weighted flowtime as the objective function. Use Proposition 11 and the interchange capability to reorder the jobs into the optimal sequence. Record the optimal sequence and objective value. Check by adjacent pairwise interchange, $(n - 1)$ tries, that the solution cannot be improved easily. (Do this by watching the objective function for weighted flow.) Subsequent problems below can most easily be worked by remaining in the routine with these data.

4.4A.2 For the same data and interchange routine, shift the focus to minimizing maximum lateness or maximum tardiness. (*Note:* It is not necessary to change the objective in the computer; all needed objectives are always listed.) Using Proposition 13, reorder the jobs into the optimal sequence.

Record the optimal sequence and objective value. Check that it is very difficult to improve this solution.

4.4A.3 Now order the sequence to minimize the number of tardy jobs, using Proposition 16. Record the solution, and check that it seems to be optimal.

4.4A.4 Illustrate by an example that Proposition 9 remains true for dynamic arrivals; that is, any nondelay schedule yields optimal makespan. (We prove this formally later.)

4.4A.5 Show by an example that this is not true if the resource can process two jobs simultaneously (e.g., two bank tellers with one line).

4.4A.6 Create three different static problems, or use data sets provided by the software. Schedule each by EDD, WSPT, Hodgson's algorithm, random, and FCFS. Now, for each of the following objective functions, give the objective value for each rule, stated as a percentage above the optimum: weighted lateness, number of tardy jobs, and maximum lateness.

4.4A.7 Create a fairly large problem for the weighted number of tardy jobs objective, with all $p_j = 1$, and solve by Proposition 15. (You may use the software and any provided data sets at your discretion.)

4.4B Software/Computer Exercises

4.4B.1 Using the software, re-solve 4.4A.6, where now there are 20 statistically generated problems.

4.4B.2 Write a computer program to be able to do 4.4A.7 more automatically, with computer output at each step to show the algorithm in action.

4.4C Thinkers

4.4C.1 Prove a version of Proposition 9 for the dynamic case.

4.4C.2 Prove Proposition 11 without using relaxation. (*Hint*: Interchange any two adjacent jobs, work out the net change in the objective function, and find conditions for the interchange to improve the objective function.)

4.4C.3 Similarly prove Proposition 13 without using relaxation. (*Hint*: Similarly show that two adjacent jobs can always be switched into due date order.)

4.4C.4 A rule that is often proposed is the "least slack rule": prioritize so that $d_1 - p_1 \leq d_2 - p_2 \leq d_3 - p_3 \leq \cdots$. Show in exact analogy to due date minimizing maximum lateness, that the least slack rule for all $p_j = 1$ maximizes minimum lateness.

4.4C.5 Show that the result in 4.4C.4 also holds true for arbitrary processing times. (*Hint*: Form the unit-preemptive relaxation, and apply 4.4C.4.)

4.5   REGULAR STATIC PROBLEMS: FURTHER PARTIAL RESULTS

4.5.1   Due Date Based Objectives

While many due date problems (such as weighted number of tardy jobs, weighted tardiness, and the weighted early/tardy job problem) are too complex to solve exactly, our methods will allow us to learn much about the optimal solution. These facts will aid us later when we build heuristics or employ search techniques for these problems.

Weighted Number of Tardy Jobs

Proposition 17. (a) For the static problem with objective to minimize N_{wt}, an optimal algorithm has the following general form (where "optimal removal job" has not been fully specified).

Step 1: Order the activities in EDD order.

Step 2: If there are no tardy jobs left, we are done.

Step 3: Find the first tardy job, say, k, in the sequence.

Step 4: Find an "optimal removal job" between the first and k inclusive.

Step 5: Move that job to the end of the sequence; if the rest of the sequence still has a tardy job return to step 4, otherwise go to step 6.

Step 6: Revise completion times; return to step 2.

(b) The optimal removal job is either job k or some *undominated* job, where a job j is called undominated in this context if there is no other suitable job i with both $w_i \leq w_j$ and $p_i \geq p_j$.

Proof. Left as an exercise.   ∎

Intuitively, we want to either remove k, which removes the tardiness problem immediately, or else remove a job with a small weight (to avoid much weight in tardy jobs) and a large processing time (to help all remaining jobs be less tardy). Since the two desires are in possible conflict, we may have to look at several trade-offs between w_i and p_i. The special treatment of job k is only important if we expect no later tardy jobs. If worse problems follow, then k may be considered to be just another candidate to trade off w_i and p_i. Clearly, if the problem has very many jobs, none with a very large p_i, we should consider removing the job with smallest w_i/p_i.

This problem (weighted number of tardy jobs) is known to be "np-complete," which means roughly that any method solving it exactly becomes computationally too expensive (even on supercomputers) as the number of jobs is increased. Although we shall not discuss mathematical methods for solving such problems (exactly or approximately) until Chapters 5 and 6, perhaps it is worthwhile to use this problem to illustrate how those methods would attack this problem.

(a) Each time a job is to be removed, a job from the current undominated set must be removed. Consider each such choice to be a branch of a decision tree, which will branch and branch again. Use clever methods to decide which branches are not really worth looking at. If all such choices are looked at, this is called "branch-and-bound." This is not practical for large problems.

(b) A relatively high quality heuristic would be: each time a job is to be removed, choose the job with lowest w_i/p_i. This is based on the relaxation results in Proposition 15.

(c) A fancier heuristic would be: each time a job is to be removed, try the jobs with the three lowest w_i/p_i, and prune the tree to maintain three active explorations. This is based on a method that approximates branch-and-bound at much lower computation, called beam search.

Search methodologies suggested here will be developed in Chapters 5 and 6.

Weighted Tardiness. We turn next to providing some preliminary results for a very important and difficult problem—the weighted tardiness problem. [Some early heuristics for tardiness were given by Wilkerson and Irwin (1971) and Baker and Martin (1974).]

Proposition 18. For the static weighted tardiness problem (T_{wt}), if any sequence can produce $T_{wt} = 0$, EDD is optimal.

Proof. A sequence can produce 0 total tardiness only if some sequence can produce 0 maximum tardiness. But EDD minimizes maximum tardiness.   ■

(A) Proposition 19. For static T_{av}, if EDD produces one tardy job, then T_{av} is minimized. (Not true for T_{wt}.)

Proof: Left as an exercise.   ■

(A) Proposition 20. For the static T_{wt} problem, where all jobs have the same due date, T_{wt} is minimized by WSPT if either

(a) all $w_j = 1.0$, or
(b) all $p_j = 1.0$.

Proof. Left as an exercise.   ■

While Proposition 20 is not guaranteed to be true for general w_j and p_j, it is true for arbitrary weights for the unit relaxation of the general case. Therefore it may be useful to treat Proposition 20 as true for arbitrary weights in heuristics.

(A) Proposition 21. For the static T_{wt} problem, if WSPT makes all jobs tardy, then T_{wt} is minimized by WSPT if either

(a) all $w_i = 1.0$, or

(b) all $p_i = 1.0$.

Proof. Left as an exercise.   ■

As for Proposition 20, Proposition 21 holds for arbitrary weights for the unit relaxation. Therefore it may be useful to treat Proposition 21 as true for arbitrary weights in heuristics.

Proposition 22. For the static T_{wt} problem, if no possible sequence can make any job nontardy, then T_{wt} is minimized by WSPT.

Proof. Obvious.   ■

Proposition 23. For the static T_{wt} problem, if a job is not tardy when scheduled second instead of first, then it need not be scheduled first.

Proof. Suppose it were scheduled first, then we could interchange the first and second job.   ■

Proposition 24. For the static T_{wt} problem, if the job with the highest WSPT priority is tardy even if scheduled first, then it should be scheduled first.

Proof. Left as an exercise.   ■

A heuristic principle seems to emerge from these propositions: *Don't bother to schedule a job until it is almost late.* Once all jobs of interest are tardy or nearly tardy, schedule by WSPT. This gives some credence to a popular idea: "schedule lightly loaded shops by EDD to minimize tardiness; schedule heavily loaded shops by WSPT to minimize tardiness." In Chapter 7 we develop a weighted tardiness heuristic that uses a sharper version of this idea.

4.5.2 Economic/Mixed Objectives

Often the motives of a shop are a mixture of those we have discussed up to this point. In addition, jobs can often be expedited in various ways at higher direct costs. These models are handled well by bottleneck dynamics. (See Chapter 7 for further development of these ideas.) Here we illustrate the possibility of deriving partial optimality properties for mixed/more complicated models.

Weighted Flow/Tardiness. For illustrative purposes, we choose a shop in which the objective F_{wt} is desired in order to minimize, say, WIP inventories, but we also desire to penalize tardy jobs by T_{wt}. That is, the objective is given by $FT_{wt} = F_{wt} + T_{wt}$. (Since both simpler objectives are in units of \$/week, they can simply be added together instead of picking a new weight to average them together.)

In particular, jobs finishing before their due date have effective weights of w_j;

jobs finishing after their due dates have weights of $w_j + w_{T_j}$, which we abbreviate by $w_j + w_{T_j} = w_{M_j}$. We also need to distinguish between a WSPT sequence produced by the early weights w_j, and those produced by the tardy weights w_{M_j}. For the purpose of this section only, we shall call the first sequence WSPT1, and the second sequence WSPT2.

Proposition 25. For the static FT_{wt} problem, if WSPT1 produces no tardiness, then WSPT1 is optimal.

Proof. WSPT1 gives a minimum for F_{wt} by Proposition 12, and gives $T_{wt} = 0$, so the sum of these is minimized. ∎

Proposition 26. For the static FT_{wt} problem, if no possible sequence can make any job tardy, then WSPT1 is optimal.

Proposition 27. For the static FT_{wt} problem, if no possible sequence can make any job nontardy, then WSPT2 is optimal.

Proposition 28. For the static FT_{wt} problem, if the job with the highest priority by WSPT2 is tardy even if done first, then the job should be scheduled first.

Proof. Proofs for Propositions 26, 27, and 28 are left as exercises. ∎

A heuristic begins to emerge here also. In a lightly loaded shop with no tardiness problems, schedule by WSPT1. Once all jobs of interest are tardy, schedule by WSPT2. In the middle, jobs that are going to be tardy have WSPT2 type priorities, while those that are not going to be tardy have WSPT1 type priorities. What about those in between? We return to this problem in Chapter 7.

Group Coordinated Job Problems. Sometimes the job set to be run on a machine can be divided into groups G1 = (11, 12), G2 = (21, 22, 23, 24), G3 = (31) , for example, with the general name of job g_j. Within each group, each job has a due date d_{g_j}, set so that all the items in the group should be able to complete some processing after this machine, and arrive at a common due date d_g. (e.g., time at which to pack the group for customer shipment).

Thus, if the appropriate criterion is to minimize the sum of weighted tardiness for each group, we might call this the *grouped weighted tardiness problem*, and write this as min $\Sigma w_g \max_j[(C_{g_j} - d_{g_j})^+]$.

Proposition 29. For the static grouped weighted tardiness problem (or any other grouped regular additive objective), there is an optimal solution for which each group will be ordered in EDD order.

Proposition 30. For the dynamic grouped regular additive objective problem, there is an optimal solution such that when pairwise interchange is feasible within a group, it will be in EDD order.

4.5A Numerical Exercises

4.5A.1 Generate several interesting examples, where the true objective is weighted tardiness. Try as heuristics at least all of the following: random, EDD, WSPT, Hodgson's rule, WLPT (reverse of WSPT), and least slack. Which works better for which problems, and why? [To do this problem with the software, choose (a) the one machine option, (b) file or manual (provided or input data?), (c) static, (d) scheduling heuristics, (e) all desired heuristics, or (f) weighted tardiness objective.]

4.5A.2 Use the examples as in 4.5A.1, but now with the true objective of minimizing the weighted number of tardy jobs. Add the heuristic suggested in the text: modify Hodgson to be WSPT–Hodgson, which is contained in the software. Using the software in the same way, determine how the various heuristics perform for this problem.

4.5A.3 Give a numerical example to prove that Proposition 20 is not true for all w_j and p_j.

4.5A.4 Give a similar numerical example to prove that Proposition 21 is not true for all w_j and p_j.

4.5B Software/Computer Exercises

4.5B.1 Use the appropriate software statistical generation of problems to repeat exercise 4.5A.1 on a much larger problem set, taking averages over the problems.

4.5B.2 Similarly repeat 4.5A.2 on a much larger statistically generated problem set.

4.5B.3 Program an algorithm for the weighted number of tardy jobs problem that basically follows the form of Proposition 17, pausing to let the user select among dominant jobs the one to remove next.

4.5C Thinkers

4.5C.1 Consider the following compound objective: (1) minimize the maximum tardiness. If this is zero, (2) find the 0-tardiness schedule that in addition minimizes weighted flow among these schedules.
Consider *generalized Smith's rule*:

Step 1: First verify that a 0-tardiness schedule exists.

Step 2: Identify the lowest priority (w_i/p_i) job that may be scheduled last without being tardy, and place it last.

Step 3: Repeat for the next to the last job, and so on.
    (a) Show generalized Smith's Rule is optimal if all $p_j = 1$.
    (b) Show it is optimal if all $w_j = 1$.

(c) Show by counterexample it is not true in general.
(*Hint*: Longer processing time jobs "want" to be at the end, as do lower weight jobs; however, for a particular job the two motives may be in conflict.)

(d) Discuss (or test) it as a heuristic in general.

4.5.C.2 Prove Proposition 17 on the general form of the weighted number of tardy jobs algorithm.

4.5C.3 Prove Proposition 19.

4.5C.4 Prove Proposition 20. (*Hint*: Suppose the optimal solution has an adjacent pair not satisfying WSPT. Show that interchanging this pair cannot hurt the objective.)

4.5C.5 Prove Proposition 21 (same hint as for 4.5C.4).

4.5C.6 Prove Proposition 24.

4.5C.7 Prove Propositions 26, 27, and 28.

___5

MATHEMATICAL SOLUTION METHODS: CLASSIC

5.1 INTRODUCTION

In Chapter 4 we developed some exact results for the regular static one-resource problem. However, we found that many problem objectives, such as weighted number of tardy jobs and weighted tardiness, did not yield to these methods. The methods therefore may not be expected to suffice in more complex shop problems. Thus in this chapter and Chapters 6 and 11 we give an introduction to more general purpose mathematical solution methods. In this chapter we introduce classic methods, such as neighborhood search, dynamic programming, and branch-and-bound, initially in the context of regular static single-machine problems. In Chapter 6, we similarly take a look at newer heuristic methods such as tabu search, simulated annealing, genetic algorithms, Lagrangian relaxation, beam search, approximate dynamic programming, and elementary bottleneck dynamics. Again we start by applying these methods to regular static single-machine problems. Chapter 10 develops other newer heuristic approaches, which might be termed "bottleneck" approaches, including OPT, the shifting bottleneck algorithm, and multi-machine bottleneck dynamics. (These are delayed until later because they require somewhat more complex dynamic models.)

In Section 5.2 we discuss special structure methods such as dominance, transitive pairwise interchange, and priorities. In Section 5.3 we give a general discussion of neighborhood search (pairwise interchange is a special case). In Sections 5.4 and 5.5 we discuss the classic methods of dynamic programming and branch-and-bound.

Although all the methods discussed in this chapter can be generalized to the full scheduling problem, we largely introduce and compare them for static problems with the weighted tardiness objective T_{wt} for expositional reasons.

5.2 SPECIAL STRUCTURE: PAIRWISE INTERCHANGE, PRIORITIES

5.2.1 Dominance

A *dominance* property is any property that specifies a subset of the set of all sequences which can be guaranteed to contain an optimal sequence. Dominance reduces the number of possible solutions that need to be considered. (It is quite possible, however, that the property eliminates relatively few candidate sequences and increases software complexity and computational costs for each sequence remaining. Often experimentation is necessary to establish usefulness.)

An example of a simple dominance property is the following.

Proposition 1. Consider the regular static problem with T_{wt} as the objective. If, for any job k,

$$d_k \geq \Sigma p_j$$

then k may be assigned to the last position in the sequence.

Proof. Left as an exercise. ∎

A brute force approach to solving T_{wt} would apply this dominance property repeatedly to the data set until no more can be removed, thus reducing $n!$ choices to, perhaps, $(n - 2)!$ or $(n - 3)!$ choices. However, any decent heuristic will usually ignore such jobs until very late in the action, so the extra bookkeeping may not be worth the effort.

Another example of a dominance property was seen in Proposition 17 of Chapter 4 on minimizing the weighted number of tardy jobs. We found it was unnecessary to remove jobs that had both higher weights and shorter processing times than other choices. Dominance properties will be presented throughout this book, usually in the form of propositions.

5.2.2 Transitive Pairwise Interchange

In the previous chapter, we showed that an adjacent pairwise interchange argument could be used in proving the optimality of certain sequencing rules (e.g., to show that WSPT minimizes F_{wt} and to show that EDD minimizes T_{max}). The intuitive idea behind the adjacent pairwise interchange method is: if we keep interchanging until there is no pairwise interchange that will improve the solution, this may be an optimal solution.

Hill Climbing. In maximization problems, changing the current solution little by little to gradually improve its value is called "hill climbing," for obvious reasons. If we climb bit by bit until all directions are "down," we might say we are at the top of the hill. This works well if there is only one hill. However, if we get to the top of a small hill, with many larger hills around us, we may not have accomplished very much.

In the context of this book, the expression "hill climbing" is perhaps a little unfortunate: we are minimizing costs rather than maximizing profits, so a more correct term might be "valley descending." However, this terminology is very old and very well established. So we will still talk about "hill climbing" to mean in general "improving the objective."

Many times we can prove in advance that there can be only one hill in the problem, so that hill climbing will work perfectly. This is guaranteed, for example, if the problem is known to be *convex*. We do not take time to review this important concept. Suffice it to say that problems solvable by linear programming or the transportation algorithm are convex; thus most simple relaxations we have studied to date are convex and are well suited for pairwise interchange. Problems with integer choices only often are not convex and may have many hills and valleys, and thus pairwise interchange cannot guarantee the optimal solution. If the relaxation, however, is close to the original problem in value, then the relaxation may help us start on the right hill (or almost the right hill) and thus help us toward the true optimum.

Suppose that we have a single machine sequencing problem such as we studied in the last chapter. Suppose we want to minimize an objective function Z (valley descending), and that a solution (sequence) S is found for which all adjacent pairwise interchanges lead to an increase in Z. Does this mean that S is an optimal sequence? If Z is represented by F_{wt} and S corresponds to WSPT sequencing, the answer is "yes"; but the answer is more often "no." We illustrate the problem that may arise with the T_{wt} criterion and the following three-job static set:

Job j	p_j	d_j	w_j
1	1	4	1
2	2	2	1
3	3	3	1

The optimal sequence is 2–1–3 with $T_{wt} = 3$, as we can see by writing down all six sequences and the corresponding objective function values:

Sequence	Objective
1–2–3	4
1–3–2	5
2–1–3	3
2–3–1	4
3–1–2	4
3–2–1	5

You should verify that, for an initial sequence of 1–3–2, adjacent pairwise interchange can lead to 1–2–3 and then to 2–1–3, which is optimal. However, starting at the same 1–3–2, adjacent pairwise interchange can also lead to 3–1–2, which is not optimal but cannot be improved by adjacent pairwise interchange. The adjacent pairwise interchange method is sufficient to prove optimality for a limited class of sequencing rules. Understanding the properties of this class allows us to recognize the problems for which adjacent pairwise interchange always converges to the optimum. This knowledge also allows us to identify those situations where, given a "good enough" initial guess (or "seed"), adjacent pairwise interchange converges to the optimum. Finally, we study the properties of problems that almost have the properties for optimal adjacent pairwise interchange. These insights lead to excellent heuristics and form part of the insights of bottleneck dynamics.

Definition. For two adjacent jobs i before j, j starting at time t, we define the *dynamic ordering relation* $iR(t)j$ to mean "starting at time t, i scheduled before j is preferred to j scheduled before i." (i.e., interchanging i and j would not improve the objective function).

Note that at least one of $iR(t)j$ and $jR(t)i$ must hold.

Definition. The *global* (or *stationary*) *ordering relation* iRj means "$iR(t)j$ whenever i and j are adjacent, for any choice of t."

The intuitive idea is that the global statement iRj means i should come before j whenever adjacent, irrespective of whether they are both scheduled early in the sequence, or both in the middle, or both at the end. Please note in the three-job example above that 2 should be scheduled before 1 if adjacent and early in the sequence, but 1 should be scheduled before 2 if adjacent and late in the sequence. Note, on the other hand, that for "nice" problems such as $T_{\max}$ or F_{wt}, all relations are global. These ordering relations suggest insights about adjacent pairwise interchange.

Definition. R is *globally transitive* for a problem if the ordering relation is global for every pair of jobs when adjacent, and if iRj and jRk always implies iRk.

This definition makes the rather intuitive statement that if i should be scheduled before j when together, and j should be scheduled before k when together, then i should be scheduled before k when together. Note again that transitivity does not hold for our three-job example above, but it does for our nice problems. This leads us to a useful result.

Proposition 2. For a static regular problem, if the interchange relation R is globally transitive, then adjacent pairwise interchange is guaranteed to produce the optimal solution; in fact, the final solution can be ordered by R.

Proof. Look at any i and j at time t. Either $iR(t)j$ or $jR(t)i$. Thus by the global property, either iRj or jRi holds. Then by transitivity the elements can be simply ordered. Let the first job in this ordering be m. We claim that in any optimal solution, m can be at the beginning of the sequence, since it can be moved to the front by pairwise interchange. Similarly for the second most dominant element, and so on.   ∎

This proposition points to a simple way of using adjacent pairwise interchange methods in solving new sequencing problems. Under the hope that the problem is globally transitive, interchange can be analyzed analytically, and a condition that specifies how two jobs should be ordered can be derived. If the condition turns out to be globally transitive, pairwise interchange can be guaranteed optimal. (In fact, if calculating a simple formula quickly establishes whether iRj or jRi, the full effort of pairwise interchange may not be necessary.) This leads us to the topic of priorities.

(A) 5.2.3 Priorities

We may also be more ambitious and analyze the situation somewhat more abstractly. Suppose we have some function $D(i, j, t)$, which tells us how much the objective function changes when i is followed immediately by j, with i starting at t, and then i and j are interchanged.

We confine our interest to the case where the objective function is a sum of convex functions of the individual completion times. (That is, the marginal penalty for increasing lateness for any completion time is increasing. This is a common case and includes the weighted tardiness objective.) Then remember from Section 4.2.1 that the objective function Z is the sum of costs of the form $f_i(t)$, for each job i where t is the completion time of job i. As we have done before, we first simplify the problem by looking at the case where all jobs have process time 1.0.

Definition. The *unit priority* π of a job i at time t is given by the rate of increase of its objective per unit time that its completion increases; that is, $\pi(i, t) = f_i(t + 1) - f_i(t)$.

Note that π has an interpretation as a benefit/cost ratio: the benefit of scheduling i a time unit earlier divided by the cost of using one time unit of the resource one time unit earlier.

Dispatch heuristics calculate (approximate) priorities for all jobs at $t = 0$ and then schedule (all the units of) the highest priority job first. All priorities are recalculated at time p_1 to see what job to schedule second. Then, at time $p_1 + p_2$, all priorities are evaluated again, and so on. Thus $n + (n - 1) + (n - 2) + \cdots = n(n + 1)/2$ priorities must be evaluated to schedule a simple one-machine problem. In special cases, however, the priorities are time invariant, so that only n priorities must be calculated once. WSPT and EDD are time-invariant rules, which helps to explain their popularity. Typical heuristics for most other problems, such as T_{wt}, are not invariant. Any heuristic that uses a priority that depends on the slack of the job will have each job's priority grow with time. Some may grow faster than others, causing switching in the rankings as time passes.

Proposition 3. The effect of a unit interchange is just the difference in the two priorities of i and j; that is, $D(i, j, t) = \pi(i, t + 1) - \pi(j, t + 1)$.

Proof. $D(i, j, t) = f_i(t + 2) - f_i(t + 1) + f_j(t + 1) - f_j(t + 2)$; now apply the definition of π. ∎

Proposition 4. For the unit-preemptive problem, the job i with the highest priority $\pi(i, t + 1)$ will dominate pairwise interchange with any other job at time t.

Proof. Left as an exercise. ∎

Proposition 5. For the case with processing times p_j, if for some t $\pi(i, t + m_1) \geq \pi(j, t + m_2)$ for all j, and for all m_1, m_2 such that $1 \leq m_1, m_2 \leq 1 + 2\max_k(p_k)$, then the highest priority job i will dominate adjacent pairwise interchange with any other job at time t.

Proof. Form the unit-preemptive relaxation of the problem. For any two jobs at time t, use Proposition 4 repeatedly so that the higher priority job's joblets will all move ahead of the other job. The resulting solution is optimal and no longer preemptive. Thus this solution is optimal for the original problem. ∎

When Proposition 5 is combined with Proposition 2, we seem to have very strong results. We have local transitivity at one point in time, and global transitivity guarantees us that adjacent interchange produces the global optimum. We can also predict the outcome of any adjacent pairwise interchange just using priorities. However, we must be careful. First of all, Proposition 5 requires that priorities change quite smoothly. Certainly our three-job example of the last section does not exhibit smooth changes in priorities. A job's priority is zero until it is tardy, and then increases instantly on becoming tardy, to remain at w_j/p_j.

Another difficulty is that local transitivity does not imply global transitivity. Given a fairly smooth objective function, Proposition 5 really assures us that we can take a shortcut over considering all interchanges at any time t and simply choose the job with the highest priority, provided that priorities change smoothly; we are not guaranteed global optimality.

However, if pairwise interchange is preceded by a good heuristic (which places completion times for jobs somewhat near optimal), then the approximate local transitive priorities established by pairwise interchange will in fact be stable from iteration to iteration. Therefore these priorities have a reasonable chance to lead to the optimal solution. Another way of stating this is presented in the following proposition:

Proposition 6. For initial solutions "sufficiently close" to the optimal solution, neighborhood search will produce the optimum solution. (Compare this proposition with the idea of the "radius of convergence" of Newton's method in NLP for some intuition).

Proof. Left as an exercise. ∎

Proposition 5 would suggest that we redefine the effective priority at time t to be some kind of average of $\pi(i, t + m)$ over relatively small values of m. As an example, it might make sense to average over, say, the next $2p_{av}$; that is, $\pi_{adj}(i, t) = (1/2p_{av})\Sigma_m \pi(i, t + m)$, where $0 \leq m \leq 2p_{av}$.

In future chapters when we work out a heuristic for the T_{wt} objective, we will expand on this idea.

Finally, we must decide when to terminate the procedure. Classically, if a single neighborhood is being used, the procedure stops when there are no improved solutions over the full neighborhood of solutions. This idea is actually too simple. In a very large problem one might terminate even though tiny improvements were available, just to save computation. Or one might continue past the local optimum using a larger neighborhood, or other diversification procedures to be discussed later.

Some early results on dominance and the effect of interchanges were given by Rau (1971, 1973) and Srinivasan (1971).

5.2A Numerical Exercises

5.2A.1   Create an eight-job problem (static T_{wt}) in which the dominance property of Proposition 1 may be applied four successive times; see if you can solve the remaining problem by inspection.

5.2A.2   Develop a four-job problem (static T_{wt}) to illustrate that there may be more than one nonoptimal solution that cannot be improved by pairwise interchange.

5.2A.3   For a small three-job problem (static T_{wt}), develop the unit-preemptive version. Demonstrate Proposition 4 and Proposition 5 by example.

5.2B Software/Computer Exercises

5.2B.1   Use the software package for static one-machine sequencing to evaluate adjacent pairwise interchange for some T_{wt} test problem for two different seeds.

5.2C Thinkers

5.2C.1   Prove Proposition 1.

5.2C.2   Prove Proposition 4.

5.2C.3   Prove Proposition 6.

5.2C.4   The sequencing rule R' is called the antithetical rule of sequencing rule R if the job that is assigned position j in sequence under R is assigned position $(n - j + 1)$ under R'. Suppose that a measure of performance Z is minimized by a particular transitive rule R_0. Show that Z is maximized among all possible sequences by R_0'.

5.3 NEIGHBORHOOD SEARCH

5.3.1 Overview

We have already had a fair amount of experience with pairwise interchange methods, primarily for a subclass of special structure problems, where (a) the method could be proved to give optimal answers, and (b) the optimal position of a job in the final optimal sequence could be summarized by an easily calculated priority (such as w_j/p_j for T_{wt}). In this section we wish to develop an understanding of the larger class of neighborhood search techniques.

As we have discussed repeatedly, the computational effort required to exactly solve combinatorial problems grows remarkably fast as problem size increases. In fact, it is difficult for most newcomers to these problems to understand just how fast. There is the old story about the king who offered a wise advisor a large reward for his service. The advisor asked for one grain of rice on the first square of the chess board, two on the second, four on the third, and so on. The king readily agreed. Now it turns out, there would be roughly 0.5 million grains of rice on the 20th square, which the king could certainly manage. But there would be 1 million sets of 0.5 million grains of rice on the 40th square, which would probably wipe out the king's treasury. There would have to be 1 million sets of 500 trillion grains of rice on square 60, and 16 million sets of 500 trillion grains of rice on square 64, which is more than the world's production of rice over all the ages!

Dynamic programming and branch-and-bound are exact combinatorial techniques that have this problem of exponential growth. They might handle sequencing 25 jobs with some difficulty, possibly 40 utilizing great cleverness, but with 70 it is out of the question. (These numbers would be somewhat larger for special structure problems.)

Neighborhood search is a rather general purpose heuristic technique, which, if used thoughtfully, can provide results often very close to optimal at practical computational cost. In its pure form, neighborhood search is very intensive (myopic); it just tries to improve the current solution as much as possible at each step. In Section 6.3 we consider modern forms of extended neighborhood search such as tabu search, simulated annealing, and genetic algorithms, which mix intensification and diversification strategies. (Diversification strategies occasionally experiment with trying less likely choices.)

In the next subsection we explain the fundamental neighborhood search procedure rather carefully; in the following subsection we give some examples and limited computational experience.

5.3.2 The Neighborhood Search Procedure

The basic elements of the neighborhood search procedure are:

(a) A starting solution to the problem of interest—the *original seed.*
(b) All solutions "close to" the original solution—the *neighborhood of the seed.*

(c) A method for selecting the new seed (improved solution)—the *selection criterion*.

(d) A method for terminating the procedure—the *termination criterion*.

The neighborhood search procedure may be used for quite complicated problems where a solution is itself very complex. However, for the purpose of clarity, we restrict ourselves to regular static sequencing problems, so that a solution is simply given by some permutation of the numbers 1, 2, . . ., n.

Initial Seed. There are a number of ways of obtaining the initial seed. A common method is simply to pick the seed at random. However, it might make more sense to pick the seed as the solution to a good heuristic for the problem. By having a "good seed," local improvement methods would have some chance of achieving the optimal value (as mentioned in the special structure discussion in the last section). Or, the heuristic could be asked to generate a number of choices of similar quality, and each could be used as a seed. This is discussed more thoroughly in Section 6.3.

Neighborhood of the Current Solution. By far the most common method of generating a neighborhood of a current solution is the *adjacent pairwise interchange* operation. If the current sequence were 1, 2, 3, . . ., n, then the neighborhood of the seed would be exactly the following $(n - 1)$ sequences:

$$2, 1, 3, 4, \ldots, n - 2, n - 1, n \quad \text{(interchange 1 and 2)}$$
$$1, 3, 2, 4, \ldots, n - 2, n - 1, n \quad \text{(interchange 2 and 3)}$$
$$\vdots$$
$$1, 2, 3, 4, \ldots, n - 1, n - 2, n \quad \text{(interchange } n - 2 \text{ and } n - 1)$$
$$1, 2, 3, 4, \ldots, n - 2, n, n - 1 \quad \text{(interchange } n - 1 \text{ and } n)$$

This neighborhood is relatively small and easy to generate, but it also limits the number of new choices to look at, so there is a trade-off.

A different neighborhood would be to consider all possible jobs k to insert into the first position (which would have the highest "priority"), moving the first $(k - 1)$ jobs each back one position. Then the neighborhood of the seed (perhaps termed *top priority up front*) would be exactly the following $(n - 1)$ solutions:

$$2, 1, 3, 4, \ldots, n - 2, n - 1, n$$
$$3, 1, 2, 4, \ldots, n - 2, n - 1, n$$
$$4, 1, 2, 3, \ldots, n - 2, n - 1, n$$
$$\vdots$$
$$n, 1, 2, 3, \ldots, n - 3, n - 2, n - 1$$

The choice of the generating mechanism can affect the size of the neighborhood markedly. For example, *general pairwise interchange* (which is the second most commonly used neighborhood in practice) considers the neighborhood generated by every possible pairwise interchange (not just adjacent jobs), leading to a neighborhood of a particular sequence containing about $n(n - 1)/2$ new sequences. Another large neighborhood might be called *general top priority to the front*. Try all $(n - 1)$ choices in the front. Fix the best first choice, and try all $(n - 2)$ choices in position number 2, and so on. This also produces a neighborhood with about $n(n - 1)/2$ new possible sequences. (General top priority to the front might have advantages in a rolling horizon situation if only the first few decisions are actually to be implemented.)

An interchange move something like top priority to the front is called k-move. A 1-move considers moving any job one to the left or one to the right, a total of $(n - 1)$ choices, since there is duplication. A k-move considers moving $1, 2, \ldots, k$ moves to the left or right, a total of about $(2k - 1)n$ choices in all. k-move neighborhoods are gaining favor, because they are often about as accurate as general pairwise interchange, with only about $2k/n$ as much cost.

General three-way interchange is defined similarly, with a neighborhood size of $n(n - 1)(n - 2)/6$. k-way interchange has a neighborhood size of the order of n^k. In general, the larger the neighborhood, the smaller the chance of being caught at a local optimum. However, the search cost of going to a larger neighborhood is usually very high. For example, n-way interchange always produces the optimum answer but requires searching the entire $n!$ space of all solutions. In practice even three-way interchange is rarely used. A common procedure would be to do adjacent pairwise interchange until no improvement can be attained, and then switch to general pairwise interchange at much greater cost. To obtain even higher quality solutions at very high cost, three-way interchange is occasionally employed. We will find in Section 6.3 that there are many other (often less costly) ways to break out of a local optimum.

Selection Criterion. So far, we have talked about selecting the seed and defining the neighborhood. We turn next to selecting the next seed (improved solution). (The terminology gets a little confusing here. Sometimes when people talk about the "seed," as here, they mean the next improvement in the same search; sometimes they mean the next original seed to try in a second or third procedure. We will try to keep the distinction clear. Here we are talking about a single procedure and "next seed" means "next solution after pairwise interchange.") There are two important but related issues here: (a) Which solution in the neighborhood would make the best seed for the next iteration of the procedure? (b) How do we recognize a reasonable, if not perfect, next solution at lower computational cost? The classical answer to (a) was: "Evaluate all solutions in the neighborhood, and choose the one giving the best improvement in the objective function. If none exists, either terminate or go to a larger neighborhood." The more recent answer to (b) is "Find the first improved solution and take it as the seed without evaluating the rest of the neighborhood."

There is a trade-off between getting the largest possible improvement in the given iteration and making iterations more quickly. (This is like the controversy in linear programming as to how carefully to choose the entering basis column.) We may also be able to do some rough preliminary calculations to decide which solutions in the neighborhood to evaluate first. For example, in the general top-priority-to-the-front neighborhood, we might estimate priorities for jobs and look at them in the order of decreasing priority. This method could simultaneously save most of the computation and produce a very good new seed.

However, all the selection methods we have mentioned have been intensive, or myopic. There may also be diversification motives for choosing a new seed, as discussed in Section 6.3.

Termination. Finally, we must decide when to terminate the procedure. Classically, if a single neighborhood is being used, the procedure stops when there is no improved solution over the full neighborhood of solutions. This idea is actually too simple. In a very large problem, one might terminate even though tiny improvements were available, just to save computation. Or, one might continue past the local optimum, using a larger neighborhood (or by other diversification procedures to be discussed in Chapter 6). Or one might use a crude procedure to estimate likely improved solutions to actually be computed carefully. (This is called a *screening* procedure.)

We summarize what we have said to date somewhat more formally. (The following procedure is deliberately stated rather generally so that it will be useful in discussing the extended neighborhood procedure later.)

Neighborhood Search Procedure

Step 1: Obtain a seed solution using criteria of (a) proximity to optimum and (b) intensification/diversification.

Step 2: If stopping criterion is met, stop; else continue.
(Criterion would roughly be "likelihood of further improvement at reasonable cost.")

Step 3: Choose a neighborhood, trading off (a) richness, and (b) computational difficulty.

Step 4: Choose a strategy to order the evaluation of solutions, possibly using a subsidiary heuristic.

Step 5: After evaluating some part of the solutions in the neighborhood, choose one or more new seeds based on proximity/intensification/diversification.

Step 6: Return to step 1.

5.3.3 Examples of Neighborhood Search; Computational Experience

To illustrate neighborhood search in its simple classical form, suppose that the objective is to minimize N_{wt} with all weights equal to 1.0. That is, we wish to minimize the number of tardy jobs. Since we know that Hodgson's algorithm gives an optimal answer for this problem, we will be in a position to judge how well neighborhood search does on the same problem. We consider again the same five-job example introduced in Section 4.4, in which all $w_j = 1$.

Job j	p_j	d_j
1	1	2
2	5	7
3	3	8
4	9	13
5	7	11

Suppose that we give the following implementation of neighborhood search for this problem:

(a) The initial seed is given by WSPT sequencing.
(b) The neighborhood is adjacent pairwise interchange.
(c) Evaluation proceeds simply from front interchanges to back interchanges. The best improvement identifies the new seed.
(d) Terminate when the neighborhood provides no improvement.

The worksheet in Table 5.1 shows the implementation of neighborhood search for this problem. The optimum solution is obtained. (It is left as an exercise to show

TABLE 5.1   Worksheet to Illustrate Neighborhood Search

	Stage 1	
Seed:	1–3–2–5–4	$N_{wt} = 3$
Neighborhood:	3–1–2–5–4	$N_{wt} = 4$
	1–2–3–5–4	$N_{wt} = 3$
	1–3–5–2–4	$N_{wt} = 2*$ new seed
	1–3–2–4–5	$N_{wt} = 3$

	Stage 2	
New seed:	1–3–5–2–4	$N_{wt} = 2$
Neighborhood:	3–1–5–2–4	$N_{wt} = 3$
	1–5–3–2–4	$N_{wt} = 3$
	1–3–2–5–4	$N_{wt} = 3$
	1–3–5–4–2	$N_{wt} = 2$

Termination with $N_{wt} = 2$

TABLE 5.2   Neighborhood Search Versus Branch-and-Bound
(16 eight-job problems, objective T_{wt}, weights = 1)

Initial Seed	Neighborhood	Number Optimal	Average Error (%)
Random	Adjacent pairwise	4 of 16	20.00
EDD	Adjacent pairwise	14 of 16	0.02
Random	General pairwise	14 of 16	0.02

Source: Baker (1974).

that if the initial seed were EDD, then adjacent pairwise search would fail to find the optimum. The same exercise investigates what would happen if the neighborhood were expanded to general pairwise search at this point.)

As we have said, neighborhood search terminates with a local optimum. Unfortunately, there is in general no way to know whether the terminal sequence is also a global optimum. However, Baker (1974) tested the performance of various types of neighborhood search on the T_{wt} problem (with weights = 1.0). In particular, he solved 16 problems with eight jobs each, small enough to be solved exactly by branch-and-bound. He obtained the results in Table 5.2.

While eight-job problems are not very difficult for either branch-and-bound or heuristics to deal with and the sample of 16 problems is small, there are several points of interest. First note that random is a poor seed, EDD is a medium quality seed for the T_{wt} problem. No high quality seed was tried, such as heuristics specifically designed for T_{wt} (see Section 7.2). Thus we are tempted to conclude that:

(a) A low-quality seed and adjacent pairwise interchange tend to lead to low-quality results.

(b) A medium-quality seed and adjacent pairwise interchange tend to lead to high-quality results.

(c) A low-quality seed and general pairwise interchange tend to lead to high-quality results.

(Baker gives the 16 test problems completely in his Appendix A. Students may at some point wish to test high-quality seeds and adjacent or nonadjacent interchange for this same data set.) Some other early work on pairwise interchange includes Elmaghraby (1967, 1971) and Randolph et al. (1973).

Very Large Problems. For very large problems, such as a 1,000-job dynamic weighted tardiness problem, the fundamental calculation of the exact effect of a pairwise interchange or of a k-move may become too expensive since many thousands may need to be done. In such cases it may be possible to develop methods to cheaply eliminate considering bad interchanges or to approximately calculate the effect of each interchange. See Morton and Ramnath (1992).

5.3A Numerical Exercises

5.3A.1 Change the five-job example in the text by using the following weights for the jobs: 1,1,1,4,1, producing an N_{wt} objective. Apply the same neighborhood search procedure.

5.3A.2 Repeat 5.3A.1 with different starting seeds and weights.

5.3A.3 Determine the optimum solution for the problem in 5.3A.1 and 5.3A.2 by inspection. What can you say about the apparent accuracy or reliability of the method? Dependence on the seed?

5.3B Software/Computer Exercises

5.3B.1 Use the software package for static one-machine sequencing to evaluate pairwise interchange for T_{wt}, starting from different types of seeds. Design an orderly experiment, and do a fair amount of testing.

5.3B.2 Write a neighborhood search routine for T_{wt}, with options for various types of neighborhoods, options for expanding the neighborhood when a local optimum is reached, and options for single or multiple starting seeds. Test the value of these options on some test problems of your own generation.

5.3C Thinkers

5.3C.1 One crucial trade-off in implementing a neighborhood search involves the quality of the terminal solution versus the time required to reach termination. Discuss the implications of the following options for this trade-off:

(a) Generating neighborhoods by adjacent versus full pairwise interchange.

(b) Choosing a new seed as soon as improvement is obtained versus searching the full neighborhood.

5.3C.2 Discuss the optimizing properties of neighborhood search if the objective is F_{wt} and neighborhoods are adjacent pairwise interchange.

5.3C.3 Sometimes neighborhood results may be improved by running the problem a number of times with randomly chosen seeds and choosing the best result. Suggest and defend other methods than random for making multiple starts. What are the advantages and disadvantages of random versus your methods?

(A) 5.4 DYNAMIC PROGRAMMING

(Dynamic programming material in Sections 5.4 and 6.2 may be omitted if desired in a first reading.)

Remember that a measure of performance is regular and additive if

$$Z = \Sigma_j f_j(C_j)$$

and each $f_j(C_j)$ is monotonically increasing (at least nondecreasing) in C_j.

For the static problem we have been studying, dynamic programming can be used to find the optimal sequence for any regular additive measure of performance. (It can also be used in more complicated problems; we do not address this here.)

For example, if Z is the weighted tardiness penalty then

$$f_j(C_j) = w_j(C_j - d_j)^+$$

(The + means that if the expression is negative, it is replaced by 0.) Suppose we are considering how to solve a 10-job problem of this type. Also suppose we know jobs 6, 2, and 8 will be the first three in the sequence for the optimal solution, but we don't know the order. Since the finish time of 6, 2, and 8 is not dependent on their order, that decision will not affect the rest of the problem. Then the principle of optimality of dynamic programming states: *Jobs 6, 2, and 8 must be scheduled to minimize their own weighted tardiness, with no thought to the characteristics of the remaining jobs.*

We use this principle recursively as follows. We recognize that the 10-job problem must be composed of some 9-job problem plus a 10th job at the end. If we have a table of costs and solutions to all 9-job problems, we could compose 10 such problems depending on what job goes last, add up the cost of each, and choose the job to go last with the minimal cost. However, we don't know the solutions to the 9-job problems either. We could, however, find these in the same way from the solutions to the 8-job problems. We repeat this reasoning until, finally, we work back to the 1-job problems, which we can solve. The job to go last in each 1-job problem is forced!

It may be easier to say all this forward. For each possible 1-job problem, determine the optimal tardiness if it goes first. Next, for each possible subset of two jobs, make two choices for which goes second, and use the fact that a table is available giving the optimal solution for 1-job subsets. Work up to all 3-job subsets, and so on, until, finally, the desired solution to the 10-job problem is obtained.

Let J denote any subset of the n jobs (e.g., $\{2, 3, 8\}$, where the curly brackets $\{\ \}$ mean a set rather than an ordered triple). In particular, let $\{j\}$ represent the subset with the single job j. Also let C_J represent the completion time for all the jobs in J, which, by our assumptions, does not depend on the ordering. That is,

$$C_J = \sum_{j \in J} p_j$$

Now let

$G(J) = $ the minimum cost for the set of jobs J if they are scheduled first, in an optimal order

We recursively develop $G(J)$ for all subsets of the full set N, working up from single-job subsets.

For single-job subsets we have

$$G(\{j\}) = f_j(C_j) = f_j(p_j)$$

In general, we have, recursively,

$$G(J) = \min_{j \in J} [G(J - \{j\}) + f_j(C_J)]$$

Finally, when we have generated $G(N)$, we have the optimal cost for the whole problem. At each stage we mark the job to be placed last. Thus, to find the optimum ordering for N, we first look up the job j to be placed last, and then look up the solution for $N - \{j\}$. For that set we find the job to be placed last, and so on.

We illustrate the method with a 4-job problem with T_{wt} objective and the following data:

Data for Four Jobs and Sum of Tardiness Objective

Job j	p_j	d_j	w_j
1	1	2	1
2	2	7	1
3	3	5	1
4	4	6	1

Dynamic Programming Worksheet

				Stage 1								
J				$\{1\}$		$\{2\}$		$\{3\}$		$\{4\}$		
C_J				1		2		3		4		
$j \in J$				1		2		3		4		
f_j				0		0		0		0		
$G(J - \{j\})$				0		0		0		0		
$G(J)$				0		0		0		0		

						Stage 2							
J	$\{1, 2\}$		$\{1, 3\}$		$\{1, 4\}$		$\{2, 3\}$		$\{2, 4\}$		$\{3, 4\}$	
C_J	3		4		5		5		6		7	
$j \in J$	1	2*	1	3*	1	4*	2*	3*	2*	4*	3	4*
f_j	1	0	2	0	3	0	0	0	0	0	2	1
$G(J - \{j\})$	0	0	0	0	0	0	0	0	0	0	0	0
$G(J)$		0		0		0	0		0	0		1

Stage 3

J	$\{1, 2, 3\}$			$\{1, 2, 4\}$			$\{1, 3, 4\}$			$\{2, 3, 4\}$		
C_J	6			7			8			9		
$j \in J$	1	2*	3	1	2*	4	1	3	4*	2*	3	4*
f_j	4	0	1	5	0	1	6	3	2	2	4	3
$G(J - \{j\})$	0	0	0	0	0	0	1	0	0	1	0	0
$G(J)$	0			0			2			3		

Stage 4

J	$\{1, 2, 3, 4\}$				
C_J	10				
$j \in J$	1	2	3	4*	Optimal sequence
f_j	8	3	5	4	1–3–2–4
$G(J - \{j\})$	3	2	0	0	Total weighted tardiness = 4
$G(J)$				4	

To illustrate these calculations, consider the set $J = \{1, 2, 4\}$ encountered at Stage 3. For this set $C_J = 7$, since $p_1 + p_2 + p_4 = 7$. If job 1 comes last in the set J, then $f_1(7) = 5$, and for the remaining jobs $G(\{2,4\}) = 0$ from Stage 2, so that the total cost when $\{1, 2, 4\}$ comes first and job 1 comes last within it is $5 + 0 = 5$. On the other hand, the next column indicates that if job 2 comes last within it, $f_2(7) = 0$ and, for the remaining jobs, $G(\{1, 4\}) = 0$, so that the total cost when $\{1, 2, 4\}$ comes first and job 2 comes last in that set is $0 + 0 = 0$. The final choice is job 4 comes last in that set, giving a total cost of $1 + 0 = 1$. Thus this numerical solution of one set for Stage 3 demonstrates by calculation that if the set $\{1, 2, 4\}$ were to come first, then 2 should be last in that set, with an optimal cost of 0. This is seen from the table in that $G(\{1, 2, 4\})$ is shown as 0, and job 2 is marked with an asterisk.

To reconstruct the optimal sequence from the example, note that the asterisk at Stage 4 indicates that job 4 should come last in the full sequence. Thus the set $\{1, 2, 3\}$ comes first. Looking in Stage 3 for the solution to $\{1, 2, 3\}$, we find that job 2 should come last in that set. Thus the set $\{1, 3\}$ comes first. Looking in Stage 2, we see that job 3 should come last in that set, leaving job 1 to be first. Thus the optimal sequence ordering is $1–3–2–4$.

Next, we turn to estimating the computational complexity of the problem. For an n-job problem, there are $2^n - 1$ subsets of the n jobs and, since the recursive equation must be solved for each subset, the total computation of dynamic programming for this problem is of the order of 2^n, which is exponential. (Remember our grains of rice example. For our worksheet we had only to evaluate $2^4 = 16$ sets. For sequencing 10 jobs we would need to consider about 1000, for 20 jobs 1,000,000, for 30 jobs 1,000,000,000 and so on. Thus it is not, and is not ever likely to be, feasible to use this method for very large problems, such as, say, 50 jobs. Dynamic programming remains a reasonable approach for middle sized problems for which there are no special structure algorithms (such WSPT for F_{wt}) and exact solutions are desired. Dynamic programming and branch-and-bound are often used to provide optimal answers to middle sized problems to check the accuracy of various types of heuristics. In Section 6.2 we discuss some approximations to dynamic programming that can be useful as heuristics.

Even though the computational demands of dynamic programming grow at an exponential rate with increasing problem size, the approach is much more efficient than the complete enumeration of all feasible sequences, since complete enumeration grows by $n!$, and $2^n << n!$. Because dynamic programming considers most sequences only indirectly, without actually evaluating them explicitly, the technique is often described as an *implicit enumeration* approach.

The dynamic programming approach we have given here is forward oriented, in the sense that the subsets we are optimizing are at the front of the schedule. Many other implementations are backward oriented (see Baker, 1974). One important advantage of the forward approach is that computations and formula complexity are slightly simplified. A more important advantage of the forward approach is related to planning horizons and rolling horizon procedures. In a schedule of, say, 20 jobs we are usually more interested in the first three jobs in the sequence, which will be implemented in the very near future. Before the later jobs are implemented, new information and breakdowns will modify our planning somewhat. Thus it is most critical to know the jobs to be scheduled first.

Suppose now that we had 30 jobs and had worked out the solution for all subsets with less than or equal to five jobs in them, and then we run out of computer time. Suppose we notice that for every subset of five jobs containing job 17 (one in six), job 17 is always scheduled first. This allows us to state a planning horizon result: either job 17 is scheduled first in the full problem or job 17 is scheduled sixth or later in the full problem. Intuitively, the second statement seems unlikely since job 17 dominated every other job in the system many times over. Thus we might take the risk to schedule job 17 first. Note that by running backward DP we would only be able to guess that job 26, perhaps, should be run last, which is not nearly as useful. We use these ideas to develop heuristics for DP in Section 6.2.

Early papers using dynamic programming in scheduling include those by Srinivasan (1971) and Rachamadugu and Morton (1982).

5.4A Numerical Exercises

5.4A.1 Modify the example in this section to have simple weights other than 1.0. Solve this problem by dynamic programming.

5.4A.2 Modify the problem in 5.4A.1 so that the objective is F_{wt}. Solve this problem both by dynamic programming and by known properties of the optimal solution. Compare the effort and insight required for the two approaches.

5.4A.3 Consider the N_{wt} 5-job problem solved heuristically by neighborhood search in 5.3A.1. Solve this problem exactly by dynamic programming.

5.4B Software/Computer Exercises

5.4B.1 Write a simple dynamic programming routine to solve total weighted tardiness problems for four or five jobs optimally. Test your routine on some of the earlier problems in the book.

5.4C Thinkers

5.4C.1 Formulate the problem of minimizing T_{max} as a dynamic programming problem (i.e., write the appropriate recursion relations).

5.4C.2 Indicate how to identify any alternate optima with dynamic programming.

5.4C.3 Derive formulas for (a) the number of additions and (b) the number of comparisons required by a dynamic programming solution to an n-job problem. Compare these with comparable expressions for a procedure consisting of complete enumeration of all feasible sequences. In particular, what numbers are obtained from the formulas for $n = 5$, 10, and 20?

5.4C.4 Prove: If it is never optimal for a job to appear first in any 5-job possible initial subset, then it is never optimal for that job to appear first in the full optimal solution.

5.4C.5 Prove any other propositions similar to 5.4C.4 that you can think of.

5.5 BRANCH-AND-BOUND

5.5.1 Introduction

Branch-and-bound is a rather general methodology for solving many types of combinatorial problems. The basic idea of *branching* is to conceptualize the problem as a decision tree. From each decision choice point, called a *node*, for a partially completed solution there grow a number of new branches, one for each possible decision. These in turn become new nodes for branching again, and so on. Leaf nodes, which cannot branch any further, represent complete solutions. (They might also represent dead ends, but this may be treated as a very "bad" solution.) If we find solution values for all the *leaf* nodes, the cheapest of these will be the optimal solution. Of course, this will be too expensive except for very small problems; this is where the *bounding* procedure (to be developed later) comes in. However, let us define the decision tree problem somewhat more carefully before getting into more clever solution methods.

5.5.2 The Method

As an example of a branching procedure, let P denote a single-machine sequencing problem containing n jobs. The problem P can be solved by solving n related

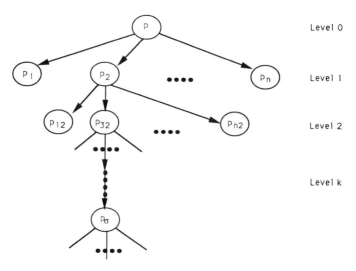

Figure 5.1. The decision tree for simple sequencing.

subproblems, $P_1, P_2, \ldots, P_j, \ldots, P_n$. The subproblem P_j means that job j has been constrained to come last, but all other choices are open. Clearly, these subproblems are somewhat easier than the original problem because only $(n - 1)$ positions remain to be assigned. Note that P_j may be considered a partially solved version of P. In addition, the set of subproblems P_j is a mutually exclusive and exhaustive partition of P in the sense that if each P_j is solved, the best of these n solutions will represent an optimal solution to P. (To be consistent with our DP discussion, we would fix jobs coming first rather than last. However, for simple bounding procedures, it is often easier to work from back to front. We discuss this issue more later.)

Next, each of the subproblems can be partitioned (see Figure 5.1). For instance, P_2 can be partitioned into $P_{12}, P_{32}, P_{42}, \ldots, P_{n2}$. In P_{12} jobs 1 and 2 occupy the last two positions in the sequence in that order; in P_{32} jobs 3 and 2 occupy the last two positions in the sequence in that order. Therefore the second-level partition P_{j2} bears the same relation to P_2 as the first-level partition P_j bears to P. At level k, then, each subproblem contains k fixed positions and can be further partitioned into $(n - k)$ subproblems, which form part of level $(k + 1)$. If this branching procedure were to be carried out completely, there would be $n!$ subproblems at level n, each corresponding to a distinct feasible solution to the original problem. In other words, exhaustive pursuit of the branching tree would be equivalent to complete enumeration of all sequences. The function of the bounding process is to provide a means for dramatically reducing this computation. (σ represents some general set of jobs constrained to be last in order, such as 18532. P_σ represents the resulting constrained subproblem.)

As we have said, evaluating all $n!$ leaf nodes is not practical. We must be more clever. Suppose for an 8-job problem we had available a solution (not necessarily optimal) $\sigma_1 = 18352476$ with a total tardiness of 17.6. Suppose by some other method we knew that any sequence ending in 62 must have a total tardiness of at least 19.5. In such a case 19.5 is called a *lower bound* on subproblem P_{62}. Why bother forming branches from P_{62}, and branches from them, and look at all the leaf nodes? All these subproblems and leaf nodes are known to be inferior. They all end in jobs 62 at the end, and we already know there is a better solution available than any of these. Let us put this a little more formally.

Suppose that at some intermediate stage a complete solution has been obtained that has an associated performance measure Z. Suppose also that a subproblem encountered in the branching process has an associated lower bound $b > Z$. Then that subproblem need not be considered any further in the search for an optimum; we may prune all branches sprouting from this node and their descendants from the tree. When such a subproblem is found its node is said to be *fathomed*. By not branching any further from fathomed nodes, the enumeration process is curtailed. (Solutions shown to be inferior indirectly are said to be enumerated *implicitly*.)

The complete solution used in comparisons that allow branches to be fathomed is called a *trial solution*. Such a solution may be obtained at the very outset by first running a heuristic; or it can be obtained and improved in the course of the tree search. (One way is to pursue the tree directly to be bottom as rapidly as possible.)

There are many types of bounding procedures, some easy to use and some requiring a computer in themselves. Unfortunately, bounding procedures vary a great deal by the type of problem and objective function. We illustrate the idea by developing a simple bounding procedure for the T_{wt} problem.

Let σ be a partial sequence of jobs assumed to be last for a subproblem, and let $j\sigma$ denote the further completed partial sequence in which σ is immediately preceded by job j. Let P_σ represent a subproblem at level k in the branching tree. This subproblem will be the original problem P with the last k positions in sequence assigned, where σ specifies the assigned partial sequence. Associated with P_σ is a lower bound v_σ, which is a known part of the contribution of the jobs assigned to the k end positions to the total v_σ tardiness of any solution; that is,

$$v_\sigma = \sum_{i \in \sigma} w_i T_i$$

The T_i values in this sum can be calculated because the completion time of each $i \epsilon \sigma$ is known even though the complete sequence has not yet been specified.

It is possible to develop much more complicated (and computationally intensive) bounding procedures than this one. It is also possible to develop dominance properties to further reduce the solution procedure. Discussions of these more complicated procedures would be distracting here.

A branch-and-bound algorithm must maintain a knowledge of the remaining unsolved subproblems, either by maintaining a list or through other logical means. These are the subproblems that haven't been shown inferior by a lower bounding procedure (not fathomed) and whose own subproblems have not yet been generated. These are called *active* subproblems. It is sufficient to solve or fathom all active subproblems to determine the optimum; this is not easy since new problems are being generated as old ones are eliminated.

Two basic types of strategy serve to classify branch-and-bound procedures: (a) the lower bounding and dominance procedures employed and (b) an ordering strategy for which problem to tackle next.

The lower bound value for a node or subproblem may be considered a rough estimate of how good the solution may be expected to be. The lower the lower bound, the lower the cost of a solution that may be hoped for. *Best first search* (also called *jumptracking* because it jumps all over the tree) follows this strategy. It tends to encounter good partial solutions all over the tree and hence keeps a good strategic understanding of the overall decision tree at all times. It has two sources of overhead: a large and somewhat unpredictable required storage space for subproblems, and the time spent in identifying the current problem to expand. *Depth first search* (also called *backtracking*) looks for a very quick good solution by following good lower bounds right to the bottom of the tree and then searching in this part of the tree. It has a small list and needs little overhead. The early solutions obtained by backtracking are often relatively poor, but useful in pruning unlikely branches. (Either method is about as easy to use by hand, or with available software. Depth first search is a good beginning technique for students wishing to write branch-and-bound computer programs.)

Best First Search Algorithm for Branch-and-Bound

(*n* jobs, static sequencing, T_{wt} objective)

Step 1: (Initialization) Set $\sigma = \phi$ (empty set). Place this initial P_σ on the active list, with associated values $v_\sigma = 0$; and $C_\sigma = \Sigma_j\, p_j$.

Step 2: Select the current P_σ with the lowest v_σ from the list. If level $k = n$, stop; it is the optimal solution.

Step 3: Replace P_σ on the list by the $(n - k)$ subproblems $P_{j\sigma}$, one for each j not in σ. For $P_{j\sigma}$ let

$$C_{j\sigma} = C_\sigma - p_j$$
$$v_{j\sigma} = v_\sigma + w_j(C_\sigma - p_j)^+$$

Return to step 2.

Note that the above algorithm can be improved if we start with a trial solution, which can be left on the active list. No subproblem with a lower bound higher than the trial solution need be left on the active list. Such a trial solution can always be obtained from a heuristic for the problem.

In order to illustrate the best first search algorithm given, consider the following example (total tardiness, weights = 1.0). Although tie-breaking is a minor issue, we specify a rule just to standardize our results:

Tie-Breaking Rule

Step 1: Choose the subproblem of the greatest depth.

Step 2: For two subproblems of the same depth, choose the one with the largest index for the last job assigned, then next to the last, and so on.

Job j	p_j	d_j
1	4	5
2	3	6
3	7	8
4	2	8
5	2	17

The branching tree for this example is displayed in Figure 5.2. The lower bound v_σ for each subproblem is entered just below the corresponding node. The order of branching is indicated by the number that appears in a box just above the corresponding node.

Initially, the tree consists of P_ϕ, with $v_\phi = 0$ and $C_\phi = 18$. At step 2, the initial problem is removed from the active list and subsequently replaced by P_1, P_2, P_3, P_4, and P_5. As shown in Figure 5.2, $v_1 = 13$, $v_2 = 12$, $v_3 = 10$, $v_4 = 10$, and $v_5 = 1$.

The best first strategy calls for branching next from P_5, since it has the best lower bound. After expanding P_5, P_{35} and P_{45} both have the new best lower bound of 9, and, after tie-breaking, P_{45} is expanded, giving nothing of immediate interest. P_{35} is expanded next giving P_{435} with a lower bound of 10, along with the existing P_3 and P_4, which also have lower bounds of 10. Using tie-breaking we expand P_{435}, giving new subproblems with lower bounds of 12 and 11. We then jump back to a little worked part of the tree. P_4 is expanded, and then P_{54} before this area becomes unattractive. P_3 is expanded and then P_{53}, giving P_{453} with a lower bound of 11, which is competitive. With a tie at 11, we go again for greater depth and expand P_{435} giving P_{2435} at 11, and expand P_{2435} giving the full solution P_{12435} at 11. This is guaranteed optimal because it dominates all active solutions. Note some typical characteristics of the best first search approach:

(a) The total number of steps to the optimal solution is relatively small.

(b) No full solution is generated until late in the process.

(c) We learn a great deal about solutions in all parts of the tree.

Next, we give an informal statement of the depth first version of branch-and-bound. (A more precise statement is left to the exercises.)

Depth First Search Algorithm for Branch and Bound (Informal)

Step 1: Same basic structure as the best first search algorithm, starting with P.

Step 2: New choice hierarchy for picking the next subproblem:

(a) Choose the subproblem at the greatest depth first.

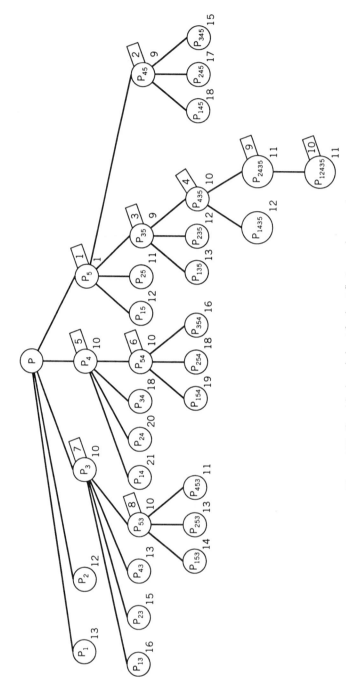

Figure 5.2. Graphical worksheet for best first search example.

(b) At the same depth, choose the one with smallest lower bound.

(c) Remaining tie-breakers as before.

(d) Eliminate subproblems with bounds above the current best solution.

We also leave it as an exercise to verify that the depth first algorithm acts as follows:

Step 1: Proceed myopically directly to the bottom of the tree to obtain a solution.

Step 2: When it is not possible to proceed deeper, backtrack to the first level for which it is possible to proceed; use the best branch remaining at that level.

(This is why the algorithm is often called backtracking.)

For this example the depth first procedure may be followed on the same diagram with a different order of proceeding. We leave it as an exercise to verify that for depth first the order of expanding problems is: P, P_5, P_{35}, P_{435}, P_{2435}, P_{12435}, back up to P, P_4, P_{54}, P_{354}, back up to P, P_3, P_{53}, P_{453}. Note that for this simple example, depth first finds what turns out to be the optimal solution much earlier, but then spends a long time verifying optimality. (Note also that for a much larger problem the optimal solution might be in a different part of the tree entirely than the first solution; then depth first might spend a great deal of time in the wrong part of the tree.)

Remember that in dynamic programming we argued that for rolling horizon purposes, we would rather solve the problem beginning at the start and working to the end, rather than backward. We also found this to be efficient. The same "natural" issues apply for branch-and-bound, except that it is more difficult to get very simple lower bounds for practical computation unless one solves the problem backward. For this reason branch-and-bound is often solved backward.

However, a more sophisticated procedure to find a lower bound for, say, a problem with the sequencing starting with 5–4 would be to solve the unit-preemptive version using the assignment algorithm with the first part of the solution constrained. This would be a lot of work, but would give very good bounds, and would allow solving the problem forward.

Note in these problems that the lower bound really serves two purposes: (a) it suggests a good problem to look at next and (b) it identifies problems that never need be looked at. These purposes are quite different. A lower bound is good for the second purpose, since lower bounds are about being absolutely safe. But a lower bound may not be very good for guessing a *likely* problem to look at next. For this purpose we would like to know, as best as possible, how good a solution will come out of that problem. This leads us (in the next chapter) into beam search and other approximate branch-and-bound methods. A further discussion of early applications of branch-and-bound to scheduling may be found in Agin (1966), Lawler and Wood (1966), Mitten (1970), and Schwimer (1972).

5.5A Numerical Exercises

5.5A.1 Solve the example given in the text by depth first search. Explain exactly how you made your choice of which problem to treat next.

5.5A.2 Solve the example given in the text by best first search, working forward instead of backward. Use the following procedure: the lower bound is at least the tardiness of jobs scheduled to date, plus the maximum tardiness of remaining jobs if scheduled in EDD order. (If the problem is too large, make a serious effort.)

5.5A.3 Try to solve a 5- or 6-job problem you invent by backward best first search, such that the first solution encountered is not optimum.

5.5A.4 Try 5.5A.3 by depth first search.

5.5B Software/Computer Exercises

5.5B.1 Write a simple aid for solving branch-and-bound problems similar to those in the text. It would initially accept as input the due dates and processing times of the jobs. When considering subproblem P_{54}, for example, the user types in 54, and the program types back the three expanded problems P_{154}, P_{254}, and P_{354} and their associated lower bounds. Use this aid to solve some new problems.

5.5B.2 Write a computer program to solve the T_{wt} problem with weights of 1.0 by branch-and-bound. It should have the options of best first, depth first, and possibly other strategies.

5.5C Thinkers

5.5C.1 In the text the algorithm for depth first search is given in table form. In the two paragraphs following that table, a different characterization is given in terms of backtracking. Show that these two apparently different statements of the algorithm are actually the same.

5.5C.2 A stronger lower bound can result by noting that the tardiness of a subproblem, say, P_{54}, must be at least as large as the minimum of the ordinary lower bounds of P_{154}, P_{254}, P_{354}. Develop a formula for a stronger lower bound incorporating this idea.

5.5C.3 Discuss the effect of obtaining a trial solution initially in branch-and-bound, when best first search is used and when depth first is used. What trade-offs are involved in deciding whether to generate an initial trial solution?

5.5C.4 Explain the statement: "Depth first search is intensification oriented, with diversification secondary. Best first search is diversification oriented with intensification secondary."

5.5C.5 Consider the following mixed branch-and-bound strategy. In mode 1 use best first search. In mode 2 do depth first search going deeper in the tree until either a better complete solution than the current trial solution has been found, or the lower bound exceeds the current trial solution, terminating mode 2. Use the following procedure:

Step 1: Choose a new node by mode 1, unless all nodes have been evaluated, then terminate.

Step 2: Switch to mode 2.

Step 3: When mode 2 terminates, go to step 1.

(The idea here is to split energies between the likely parts of the trees and better, faster solutions for pruning the tree.) Show this would have found the optimal solution much earlier in the text example.

6

MATHEMATICAL SOLUTION METHODS: NEW DIRECTIONS

6.1 INTRODUCTION

In Chapter 5 we introduced such classic mathematical solution methods as neighborhood search, dynamic programming, and branch-and-bound, designed for ease of exposition in the context of one-machine regular static problems. Here, we similarly develop a number of newer mathematical methods that are largely modifications of the classical methods: beam search, approximate dynamic programming, tabu search, simulated annealing, genetic algorithms, Lagrangian relaxation, and neural networks. We also discuss those aspects of bottleneck dynamics that can be developed in this restricted context. Further development of bottleneck dynamics and other bottleneck methods, such as OPT and the shifting bottleneck algorithm, will wait until Chapter 10 when we will have richer models with bottlenecks available.

6.2 BEAM SEARCH, APPROXIMATE DP, AND RELATED METHODS

6.2.1 Overview

Branch-and-bound and dynamic programming offer the comfort and safety of exact optimal solutions at the expense of computational effort that grows exponentially with problem size, so that large problems are not practical to solve. Might it be possible to modify them in some way to gain the practical capability of solving large problems while sacrificing only some small part of the safety?

Indeed, in many cases this is quite possible, although the authors' strong statement here may be somewhat controversial. Many investigators do not like to make such statements unless it is possible to mathematically derive a *worst case bound* such as: "no problem solved by the approximate procedure can ever be more than, say, 5% from optimal." The authors' (along with other heuristics "believers") point of view is that the following statement is easier to obtain, and really more useful: "in studying the approximation on 1000 representative problems, the approximate procedure differed on average 0.5% from optimal, with a maximum of 3%."

Approximation procedures for solving decision tree problems have been widely developed in artificial intelligence for many years. We present in detail one such technique—beam search—which has been fairly widely studied. Recently, operations researchers have also been paying more attention to approximate branch-and-bound, under such names as *partial enumeration*.

Approximation techniques for dynamic programming have been less studied, although they also seem important. We present some early ideas on the subject, many related to the rolling horizon ideas we have discussed earlier.

6.2.2   Beam Search

Consider once again our branch-and-bound example from Section 5.5. Several sorts of commonsense interests suggest themselves in our minds.

It would be best to focus our attention on the part of the tree that seems the most worthwhile. The lower bounds give us one way of trying to estimate what is worthwhile. Are they the best way? They do not tell us the solution value to expect on a branch, only the very best we could hope for. Is there a way of guessing what to expect as a "*likely*" solution from a given branch? (The choice of words here is difficult; we would like to speak of "average" or "expected" results, but more in terms of experience than a formal mathematical average.)

If we knew exactly the best solution value to expect from each branch at Stage 2, we could just go down the best one! We would investigate one branch at each level, so our effort would be proportional to the number of levels; that is, computation would be proportional to n rather than $n!$.

If we knew even approximately how good a solution to expect from each branch at Stage 2, we could take a chance and throw away all branches except, say, the three "most likely." Suppose we expand these three giving 12 active branches, and then find an approximate best solution value from each of these 12, and save the best three. We could expand these three giving nine active branches, and so on. Of course, we could save more subproblems for more safety, or fewer for more speed.

Beam Width. The number of subproblems saved is called the *beam width b*. If there are n jobs, then the number of subproblems considered is less than bn^2. This is a very practical number of subproblems even for large n (say, 200).

Of course, there is one catch. How do we give a good estimate of the value of the best trial solution starting from, say, P_{54}? This is really not hard; we simply run any high-quality heuristic with the restriction "jobs 5 and 4 at the end." We could use very fast heuristics such as EDD or WSPT, which are sometimes proposed for the tardiness problem, and use their solution as our trial solution. Or we could be a little more ambitious and use pairwise interchange. Or we could simply use the first part

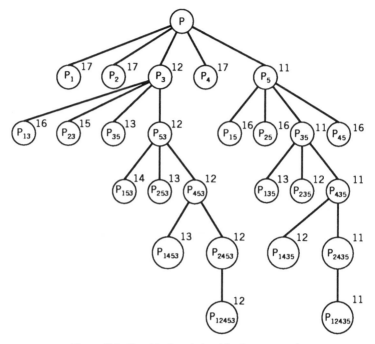

Figure 6.1. Graphical worksheet for beam search.

of depth first search to run right down to a trial solution. We term this *depth first beam search*. This is likely to be a good estimate, with an effort something like kn^2, where k is some factor. Thus the total computation time of depth first beam search is less than $kb(n^4)$. This is a practical, if somewhat expensive, heuristic. We will discuss ways of economizing down to perhaps n^2 or n^3 later.

For small hand problems, depth first beam search may actually be more work than branch-and-bound, but this changes dramatically for large problems. Beam search also has the characteristic of identifying many near optimal solutions, which is often useful to managers who must consider many factors not in the model. Figure 6.1 summarizes depth first beam search for the problem in Section 5.5. We have suppressed lower bounds and only show estimated best solutions at each point. Note that beam search found the optimal solution 1–2–4–3–5 with value of 11, but also found three different solutions with value 12 and four different solutions with value 13. We can also answer questions like: what is the likely extra cost if management insists on scheduling job 1 last? (Answer: 17 - 11 = 6.)

Valuation Function. The method we use for (approximately) evaluating each subproblem is called the *valuation function*. Calculating a trial solution from the problem is a very natural valuation function, but it is rather expensive computationally. It may be worthwhile to use a poorer valuation function that is easier to calculate. For example, in our problem, the lower bound for a subproblem could be used as the valuation. The reader will be asked to verify, as an exercise, for our same problem that the simple lower bound valuation function obtains the optimal solution for a beam width of one, but that, depending on tie-breaking procedures, it may take a beam width of three to obtain all solutions within one of the optimal.

The lower bound valuation function here is called *global* because the valuation function at a given level k is comparable across the tree. P_{53} has a lower bound of 10 and is therefore preferred to P_{25} with a lower bound of 11. It is possible to generate approximate valuation functions that are still global. One possibility is to run our trial heuristic only for some fixed h jobs down the tree, rather than the full n. Evaluate the objective function for just these h jobs, and then add a rough "salvage value" estimate of the remaining jobs. This rough evaluation has a cost only $O(h)$ instead of $O(n)$, and thus for very large problems with very large n will be much much cheaper. Morton and Ramnath (1992) are currently applying this idea to very large job shop problems. (See Chapter 16.)

But now suppose our valuation function was crude and not comparable across the tree. Suppose, for example, we didn't know the lower bounds, but only their rankings (e.g., this might be the output of some priority functions). Then P_{53} might be rank 1 among its siblings from a parent P_3 ranking 2 or 3 among its siblings. P_{25}, on the other hand might be third among its siblings. Which would be preferred? It is difficult to know. *Ranking* is called a local valuation function for this reason. Of course, we can try to crudely estimate a global ranking. Suppose we decided to add the rankings of a subproblem and its parents together, for example. Under this scheme P_{53} has a rank of 1.0, its parent 2.5, giving an aggregate rank of 3.5. P_{25} has a rank of 3.0 and its parent has a rank of 1.0, giving 4.0. Under this scheme P_{53} would be preferred to P_{25}. As it turns out, P_{53} leads to a trial solution of 12, while P_{25} leads to 16, so that it works out in this example. We can see, however, that this approach is risky. Some sort of global evaluation function, whether lower bound or heuristic trial solution, is desirable. An improved procedure similar to adding the rankings will be deferred until Section 15.4.

Miscellaneous Points. It is not necessary that the beam width be rigidly constant. It is disconcerting, with a beam width of 2, to be forced to choose one of two tied contenders arbitrarily. It is perhaps better to temporarily change the beam width to 3, carry the extra computational burden for a level, and work out the issue at the next level. On the other hand, if the beam width is 5, and all but two trial solutions at the level are atrocious, it makes sense to expand only the two nodes, but go back to 5 at the next level. One scheme would be: beam width = min{5, # of trial solutions within 3.5 of min trial solution}.

Other Approximate Branch-and-Bound Methods

Beam Search/Best First Search-1. There are a number of other possibly important approximate branch-and-bound methods. A simple one is *beam search/best first search.* This is roughly a combination of best first branch-and-bound and beam search. Do exact best search branch-and-bound, but limit the table size to, say, 1000 nodes. Whenever a subproblem is expanded that doesn't rank in the first 1000 lower bounds, it is simply eliminated. Calculate both a lower bound and a beam search type trial solution for each problem. (Trial solutions can be improved by neighborhood search.) Sort the problems to look at both by the lower bound, as before, and, in a separate table, by the trial solution, as in beam search. Choose the next problem to expand by the trial solution table, as in beam search. However, when there is an improved trial solution, use the lower bound table to delete dominated problems (out of both tables). When the maximum length of the table is exceeded, set a marker that the solution becomes a heuristic and record upper and lower bounds on the solution at that point. Heuristically delete problems by highest trial solution to maintain table feasibility. Note that this procedure tries to remain an optimal procedure and should find good solutions in relatively few expansions. On the other hand, it requires more bookkeeping and hence would need extensive testing to demonstrate whether it is superior to a simpler procedure. To the authors' knowledge this procedure has never been tested in any serious way.

Beam search/best first search-2. The above procedure could be made faster, at some risk, by utilizing "approximate lower bounds," which are stronger and fast to calculate but occasionally may violate the lower bound requirement slightly. The authors have had good success with the following pseudo-lower bound. Form the unit preemption of the problem at the node. Run a good dispatch heuristic plus neighborhood search, to get a "near-optimal" solution. (This is essentially the "poor-person's LP relaxation" at the point.) One could also subtract some tolerance from these "bounds" to reduce the risk of violation.

Chess Decision Trees. Chess playing decision trees have similarities to both branch-and-bound and beam search. (Chess is a two-person game, so the expanding decision tree alternates between maximizing and minimizing level by level. We ignore this here and discuss the problem as if chess were a one-person game.) Basically, chess playing programs try to do branch-and-bound and eliminate inferior branches by lower bounds. However, there are far too many branches to get to the end of the game and work backward. So the program may work eight to ten levels down, for example, and be left with hundreds of thousands or millions of partially completed subproblems. Experienced chess players have spent years working out sophisticated ways to evaluate these ending positions, that is, to come up with *valuation functions.* A small improvement in the valuation function can make a major improvement in the chess program. Of course, hardware and software to speed up the search and allow searching deeper are equally important. Researchers have recently discovered that if one or two lines of play have evaluations much much better than all the others, it may be safe to expand only those one or two lines. We have discussed a similar idea in beam search.

(A) 6.2.3 Approximate Dynamic Programming and Related Methods

We summarize the principal ideas of beam search:

(a) Look at an exact search procedure.
(b) Form an idea of likely and unlikely search areas.
(c) Ignore the unlikely search areas.

It is easy to apply these same ideas to dynamic programming. First, we choose forward dynamic programming as our exact search procedure. Second, we separately solve the problem by some high-quality heuristic. (We have many to choose from by now; for example, pairwise interchange from a decent seed, or depth first beam search). We will use the *apparent rankings* from the heuristic as a principal input to the approximate procedure. We also need to choose a *ranking tolerance* factor, analogous to the beam width in beam search. Basically, the ranking tolerance is just the error in rank we think will be the largest in our heuristic. If the heuristic ranks a job 7th, and we have a tolerance of 3, we assume the job will appear between positions 4 and 10 in our final answer. As before, the larger the tolerance, the more expensive the procedure is computationally but the more accurate the answer. (If our heuristic gives quantitative benefit/cost ratios for each job, the tolerance could alternatively be for the maximum error in the heuristic benefit/cost ratio. If a job ranked 7th possessed a normalized value of 1.5, with an (error) tolerance of 0.2, and the job ranked 15 had a benefit/cost ratio of 1.3, we would again assume the job would rank no lower than 15th in the final solution.)

Let us give an example. Suppose we want a near-optimal answer for 50 jobs, typically far too large to do optimally by dynamic programming. Suppose we have a good heuristic and are willing to grant it a ranking tolerance of 3. Consider computing optimal subsets of, say, 30 elements out of the 50 to be scheduled first. In the original dynamic program we would need to consider all subsets of 30 out of 50, or $(50)(49) \cdots (21)/30!$, which is a huge number. With a ranking tolerance of 3, however, we would know that jobs ranked 1 to 27 by the heuristic are forced into our 30-element set, and jobs ranked 34 to 50 are forced out. Thus we really only need to pick 3 elements out of jobs 28 to 33, or 3 out of 6. There are $(6)(5)(4)/3! = 20$ ways to do this, a dramatic savings. In general, there will be $(2b + 1)(2b) \cdots (b + 2)/b!$ choices, leading to Proposition 1.

Proposition 1. While forward dynamic programming has computational complexity $O(2^n)$, ranking tolerance DP has computational complexity $O(n(2b)^b)$, where n is the number of jobs to be sequenced and b is the ranking tolerance parameter.

Proof. Left as an exercise. ■

The authors conceived of ranking tolerance DP while writing this book and are not aware of any work on this or similar procedure. It remains an open research question.

Sometimes a different strengthening of the original procedure can be created as follows. Suppose, as before, we have a 50-job sequencing problem to be solved, and we start by running forward DP. Suppose, while computing optimal solutions to subsets, we also keep track of the ranking of jobs in those solutions.

Suppose after running up through, say, all 8-job subsets, we discover that for every subset that job 26 is a part of, job 26 ranks first. Then it is clear we can say something about the optimal solution to the 50-job problem: job 26 will either rank 1 or 9th or lower. We formalize this as a proposition.

Proposition 2. Consider running forward DP for a problem with n jobs to be sequenced. If, for all k-element subsets containing job j, job j always ranks in the first $m < k$, then, in the full optimal solution, job j either ranks in the first m or in position $k + 1$ or below.

Proof. Left as an exercise. [This proof is basically a planning horizon type proof (Morton, 1981).] ∎

We can again combine this result with a ranking tolerance procedure to get a good heuristic. Look at the ranking heuristic. Say it shows job 26 ranks 3, and we choose a ranking tolerance of 6, which means we don't choose to believe the true ranking is below 9. Now we run *all* subsets up through size 9 and find that every time job 26 is part of the subset it is ranked first. *We thus conclude that job 26 may in all practicality be run first.*

Once we have marked job 26 as first, we remove it from the problem and repeat our procedure on the remaining 49 jobs as a new problem.

If it turned out instead that every time job 26 were part of the subset it is ranked first or second, then we could create two subproblems, one in which job 26 is sequenced first and the other in which it is sequenced second. Thus we could obtain a branching tree of dynamic subproblems, and some curtailing of the branches by a beam search type valuation might be necessary.

In general, approximations and heuristics based on the mixture of branch-and-bound and dynamic programming ideas seem quite interesting. However, it is also important for heuristics to be robust and simple; hence small improvements in this direction would probably not be considered worthwhile.

6.2A Numerical Exercises

6.2A.1   Expand the 5-job T_{av} problem for branch-and-bound in Section 5.5 to include a sixth job: processing time 5, due date 10.

   (a) Using the software, evaluate a number of heuristics for this problem, including R&M.

   (b) Using R&M as a starting seed, solve the problem with both adjacent and general pairwise interchange, using the software.

   (c) Find an approximate solution to this problem with beam search by hand, using beam width 2 and trial solutions found by the EDD heuristic.

6.2A.2   Continuing with the problem being studied in 6.2A.1:

(a)   Find an approximate solution to this problem from beam search on the software, using a beam width of 20, and trial solutions found by the R&M heuristic.

(b)   Expand up to 2000 nodes of best first branch-and-bound using the R&M heuristic as the start with the provided software.

(c)   Expand up to 5000 nodes of depth first branch-and-bound using R&M and the provided software.

6.2A.3   Redo 6.2A.1(c) more crudely by using the simple WSPT ranking of the job added as the evaluation function. Add the rankings for the subproblem and its parents to approximate a global evaluation function.

6.2A.4   Try some kind of variable beam width variation for 6.2A.1(c).

6.2A.5   Solve the problem of 6.2A.1 by the ranking tolerance DP approximation, using tolerance of 1.

6.2A.6   Do 6.2A.1 with objective changed to number of tardy jobs.

6.2A.7   Do 6.2A.2 with objective changed to number of tardy jobs.

6.2B Software/Computer Exercises

6.2B.1   Modify your program from 5.5B.1 to become a simple aid for solving either branch-and-bound or beam search problems. Use this aid to solve some new problems.

6.2B.2   Extend your computer program of 5.5B.2 to solve branch-and-bound, beam search, or the version that tries to solve branch-and-bound but changes to a heuristic as necessary.

6.2B.3   Write a computer program to do ranking tolerance DP approximations. Solve a few problems.

6.2C Thinkers

6.2C.1   Why do you think beam search tends to solve all problems on one level of expansion at the same time? (*Hint*: You may get a different answer depending on whether the valuation function is a trial solution or not.)

6.2C.2   Invent variations on beam search and explain when and where they might be useful.

6.2C.3   Read up on chess playing routines (old *Scientific American* issues may be helpful). What similarities and differences with our problem do you find?

6.2C.4   Prove Proposition 1.

6.2C.5   Prove Proposition 2.

6.2C.6   Can you invent other combinations of exact methods and/or heuristics that might be useful? Explain your ideas carefully.

6.3   EXTENDED NEIGHBORHOOD SEARCH: TABU SEARCH, SIMULATED ANNEALING, AND GENETIC ALGORITHMS

6.3.1   Overview

Recall that simple neighborhood search has a number of advantages: it is fast and easy to program, and it seeks a local optimum to the problem, so that many times it is surprisingly accurate, especially if it is started from a good heuristic seed. Its disadvantage is that it is very single-minded: it employs almost entirely intensification rather than diversification procedures. By contrast, DP, branch-and-bound, and various approximations to them take the trouble to explore the solution space more completely: they balance intensification against diversification.

It is rather natural to ask if neighborhood search can be modified to incorporate diversification strategies without losing its primary advantages. The answer is yes; we might term the resulting methods *extended neighborhood search*. (Some might also argue that all these methods could be considered special cases of genetic algorithms.)

In Section 6.3.2 we study branching diversification methods for neighborhood search, such as random sampling and beam neighborhood search. Next, we review some newer methods that diversify without branching, and point out that combinations of techniques should be of interest. In particular, in Section 6.3.3 we examine tabu search, in Section 6.3.4 simulated annealing, and in Section 6.3.5 genetic algorithms. Later, in Section 16.3, we discuss a few recent job shop studies using some of these methods.

6.3.2   Extended Neighborhood Methods: Branching Methods

One obvious way to search out a larger part of the solution space (i.e., to diversify) is to start from, say, m different starting seeds, run the procedure m different times, and choose the best solution.

The simplest possible way to choose m different seeds in this fashion would be by *unbiased random sampling*. Suppose there are 24 jobs to be sequenced. We simply assign each the probability $p = 1/24$, and choose each sequence in this way. Suppose we desire a sample of $m = 200$. Choose 200 different seeds in this fashion. Now run neighborhood search 200 times and choose the best solution.

Of course, we may have strong knowledge that job 17 shouldn't be scheduled first; therefore, giving it a 1/24 chance of being scheduled first seems to waste information. A better way to choose m different seeds in this fashion would be by *biased random sampling.* For this we need a strong *master seed.* This might be, for example, the sequence produced by the output of ordinary neighborhood search with a decent heuristic seed input.

Next, we choose a sequence of probabilities of choosing a given element in the master seed, which goes down, for example, geometrically. That is, we set the probability of obtaining the j ranking element as

$$p_j \text{ prop } \alpha^j$$

We say "prop" for proportional to, rather than "equal" because, as elements are chosen, we have to normalize the remaining probabilities to add to 1.0. For example, when $n = 8$ and $\alpha = 0.6$, then the initial set of probabilities is:

j	1	2	3	4	5	6	7	8
p_j	0.40	0.24	0.15	0.09	0.05	0.03	0.02	0.01

A large value of α distributes the selection of the next choice in a more random fashion, while a small value of a makes the highest rank item in the heuristic almost certain to be selected. Thus a large value of α tends to destroy the help of the master seed and make the searches random over the space, so that there is too much diversification. On the other hand, too small a value of α tends to make all the searches produce the same result as the master seed, so that there will be no diversification at all.

Therefore one of the requirements of the method is to find out by experimentation or other means the best sample size m and the best α for a particular problem structure and desired solution accuracy. One way to attempt this would be to use sequential sampling theory. Unfortunately, to do this theoretically requires a knowledge of the distribution of solution values in the problem space, which is not likely to be available. In practice one could sample in lots of 20 and fit by regression (on log paper?) the apparent cost/benefit ratio of additional sample points. Even this will only work if the objective function is fairly well behaved.

A different way of generating a number of initial seeds might be termed "beam-seeding" since it bears a relationship to beam search. Take an initial master seed, chosen in some good fashion. Form the neighborhood of the seed with respect to pairwise interchange or whatever is being used. Rank the possible selected moves in order of decreasing improvement of the objective. Choose the first m as seeds of the multiple procedure; m might be called the "seed beam width." Note that it may happen that some of the seeds have worse objectives than the original. This is perfectly acceptable, for the purpose of the multiple seeds is diversification. A worse seed at the start may possibly still be improved by neighborhood search into the best final solution.

Note that beam-seeding need not be limited to the initial seed, but can be performed anytime throughout a neighborhood search, based on criteria specified ahead of time. For example, one obvious possibility would be to wait until the rate of improvement per iteration in neighborhood search drops below a certain threshold ("in a rut"), and then diversify to look for new areas. There are at least two issues here:

(a) How to keep all the diversified paths from dropping back to the original "rut".
(b) How, after several diversifications, to prune the branches back to a manageable total effort.

The first issue is addressed by *tabu search, simulated annealing,* and *genetic algorithms* in different ways. We will discuss them in the next three subsections. The second question, deciding which branches to keep and which to terminate, is fairly difficult (see methods for similar issues in genetic algorithms). The answer is certainly not, after the latest beam-seeding, to simply choose the m branches with the lowest m costs and terminate the rest. A given branch, for example, may have the lowest value of the objective function, but it may not have improved over many iterations, nor may previous extensive beam-seeding from it have been fruitful. We may very well decide to save the trial solution, but avoid spending further computational effort in this part of the space. Thus we need some kinds of complex measures of the value of sample information.

One kind of simple rule that avoids exponential explosion in computation would be: follow each path to its local minimum and beam-seed only at the initial master seed and at local minimums. Terminate any path after reaching the third local minimum.

6.3.3   Tabu Search

Tabu search is a rich set of methodologies for building extended neighborhood procedures, with particular emphasis on avoiding being caught in a local optimum. We have space here only to present very simple versions, and a few of the additional ideas useful in larger, more complex problems. The reader is referred to "Tabu Search: A Tutorial" by Glover (1990).

General Concepts. The idea of tabu search is quite simple. Why do we get caught in local optima? Because once we reach the best state within a neighborhood (local optimum), all moves increase the objective function, and hence none will be chosen. How can this be fixed? Simple: always take the best move available, *even if this makes the objective somewhat worse.* This is basically a diversification move, because intensification is momentarily of no advantage.

Now, of course, if we move up out of the local optimum, on the very next move we can probably save the most by moving right back! We must force ourselves to continue diversifying for a few moves. The approach that tabu search employs is to keep a list of our last m moves (Glover mentions $m = 7$ as typical; we favor much larger lists) and not to allow them to be repeated while they remain on the list (they are currently "tabu"). One approach that is often very robust for medium sized problems is to simply include all previous moves on the tabu list so that it grows.

With good sorting and data management, this may remain practical up to 500 or 1000 moves.

Before reaching the local optimum the neighborhood procedure will improve at each step, so that no repetition is possible and the current m moves, which are "tabu," would never be chosen anyway. But after leaving the local optimum and attempting to try to climb over the hill (if we are minimizing) into a different region of solutions, the tabu list hopefully forces diversification until the old solution area is left behind.

Note several things:

(a) The local optimum trial solution can be saved as the best to date, so that nothing is lost by going on (just as in beam search).

(b) The main extra work in the procedure is in keeping an updated list of the last m solution sequences and in checking whether each proposed new step is "tabu."

(c) The procedure will not stop even if, in fact, the global solution has been found; there must be some other termination procedure even if it is only the number of moves or elapsed time.

Dealing with Larger Problems. Since tabu search requires the "best" move at each choice point, rather than simply an "improving" one (which may not exist), the selection of a move may become very expensive for very large problems if the neighborhood (number of possible next choices) of a move is very large and/or if the computation involved in evaluating each interchange is very large. One way out of the neighborhood problem is to prove that a much smaller neighborhood actually contains all new moves of interest. A second way out is to do a preliminary screening or filtering of the moves by some approximation to allow choosing those of interest. On the computation of the effect of interchange, approximations may again be employed to prescreen candidates. Finally, certain objectives, such as makespan, have fast algorithms for determining the effect of interchange. All these methods may be required to make tabu search practical, as we shall see in the tabu search study on the job shop problem with makespan criterion (Dell'Amico and Trubian, 1991) in Section 16.3.4.

Other Issues. There are a number of issues involved with constructing tabu lists besides the length. One question is what to do if all current moves are tabu? One simple answer is to pick one of the tabu moves at random and do it anyway. Another question is how to absolutely avoid cycling. (This is usually not a problem; the second time a solution is visited differing things will be tabu, producing a different trajectory.) There is also the idea of an *aspiration criterion*. If a move satisfies the aspiration criterion (e.g., has best balanced machines), it is exempt from being tabu. That is, we may be very interested in exploring new directions out of this point rather than the point itself.

Tabu search could easily be used in connection with beam-seeding procedures instead of using an aspiration point. We simply beam-seed several branches at such an interesting point. At a diversifying point, the tabu list can force diversification by preventing new branches from returning to the branching point.

6.3.4 Simulated Annealing

Simulated annealing was developed initially in the study of the cooling and re-crystallization process (annealing) of hot materials and hence has quite a different origin than tabu search. Yet they are similar in that both are primarily concerned with avoiding or escaping local optima. Simulated annealing also bears a similarity to the biased sampling neighborhood search method we discussed previously.

In simulated annealing we again are considering extended neighborhood search. After evaluating a new neighborhood, we do not always choose the apparent best move (the one that increases the objective function the most). Instead, we choose that best move with highest probability, the second best move with next highest probability, and so on. Whereas biased random sampling typically goes down geometrically by the same factor for each next move, simulated annealing proba-bilities go down exponentially according to the size of the improvement given by the move. That is,

$$p_j \text{ prop } \exp[\, -k(D_{\text{best}} - D_j)/K\,]$$

In this formula, p_j represents the probability of making move j from among the neighborhood choices. D_{best} represents the improvement to the objective function for the best choice, and D_j represents the improvement for choice j. We also normalize so that probabilities add to 1.0, and normalize with a factor K depending on when a given difference in the two D's should be considered to be large or small. Note that neither D_{best} nor D_j need be positive, but that if D_j is much smaller than D_{best}, it has a very small probability of being chosen, whereas if it is nearly equal, it has nearly a maximum probability of being chosen. Note also the effect of the *temperature* (control factor k). For k nearly equal to zero, every step of the neighbor-hood search is highly random and diversifying, while for large enough k, the procedure behaves more like ordinary neighborhood search or tabu search. When normalized for the particular problem, k/K is called the normalized temperature.

Note that we have the flexibility to vary k dynamically. One typical scenario would be to start with a high k to be intensive, and get maximum improvement, but to gradually reduce k as the search gets more difficult, to allow more diversification. At the local optimum k could be reduced further. It could be raised again once the local optimum has been left behind.

The primary difference between simulated annealing and tabu search is that simulated annealing diversifies by randomizing, while tabu search diversifies by forcing new solutions. It would not be hard to produce mixed procedures that diversify both by constraints and by randomizing; it would be a matter of testing to determine if this could be helpful.

One major difficulty experienced by several researchers trying to experiment with this technique is how to normalize the temperature to fit the problem. A second major difficulty is finding the best way to vary the temperature dynamically as the solution procedure progresses. Much more investigation of these issues is warranted.

A recent job shop study (Van Laarhoven et al. 1992) is discussed in Section 16.3.

6.3.5 Genetic Algorithms

General Concepts. Genetic algorithms can refer to any search process simulating the natural evolutionary process. There is a current population of possible solutions to the problem. In each generation, the best solutions (most fit individuals) are allowed to produce new solutions (children) by taking the best features of each parent and mixing the remaining features (or by mutating features of a single parent). The worst children die off (on average) to keep the process stable, and on to the next generation.

By keeping our definitions of these ideas general enough, almost any of our extended neighborhood search ideas can be considered special cases. For example, if we start with a single parent, consider the mutation to be pairwise interchange of priorities onto a machine, have a child for every possible mutation of the parent, kill off the parent and all but the most fit individual, and stop when the parent is as fit as any child; then we have a description of ordinary neighborhood search.

If we keep a memory about not repeating children under certain conditions, we have tabu search. If the child selected may vary to some extent from the most fit, we have simulated annealing. With beam-seeding we allow larger populations at the end of a generation. If we have more complicated rules for generating the neighborhood under various conditions, as in beam-seeding, then we begin to allow a more complex varying size population.

We turn to a more careful statement of genetic algorithms for sequencing in scheduling. Then we describe as an example the detailed version for a particular problem, such as the static weighted tardiness problem. Next, we extend the example to the multi-machine (job shop) case. (In Section 16.3 we present a genetic algorithm study on minimizing makespan in the classic job shop.)

Genetic Algorithms for Sequencing in Scheduling. One way that genetic algorithms can be applied to sequencing problems (no routing choices allowed) is the following:

Step 1: The terms "individual," "chromosome," and "complete set of instructions for a method to solve the problem" are used interchangeably.

Step 2: The chromosome encodes the solution instructions using a string of characters called "genes."

Step 3: The chromosome is organized into groups of subchromosomes, each encoding instructions for scheduling one of the machines.

Step 4: A subchromosome consists of a preferred permutation sequence for all operations to be performed on a given machine. (Exact use of the preferred sequence will be explained below.)

Step 5: An initial set of solutions is selected by simple random sampling, or beam-seeding around a strong master seed (Section 6.3.2), to form the initial "population" of individuals.

Step 6: The objective function for each individual is calculated by simulating the directions of the chromosomes. This is called the individual's "fitness."

Step 7: A set of individuals is selected to "reproduce" in this "generation." The probability of reproducing is related to some measure of fitness (such as fitness percentage between best and worst).

Step 8: Reproduction, which may be bisexual or monosexual, produces new individuals with somewhat different chromosomes. (This is discussed in more detail below.)

Step 9: Fitnesses are calculated for the new offspring.

Step 10: Enough individuals are caused to die to maintain the desired population. The probability of dying is made proportional to some measure of unfitness. This marks the end of a generation.

Step 11: If the generation cutoff has been reached, terminate. The result of the heuristic is the objective function of the most fit current individual.

Step 12: Otherwise go to step 7.

Scope for Variations in Genetic Algorithms for Scheduling. As we have seen to be the case for simulated annealing, one of the great strengths (and weaknesses) of genetic algorithms is their great capacity for being varied and shaped to individual situations. We will now go through a few of these possibilities to demonstrate.

The chromosome or set of instructions for sequencing the complete problem has unlimited numbers of forms available to general problems. For sequencing problems, it is natural for it to give the preferred ordering of jobs to be performed at each machine. It is not useful to make the preferred ordering absolute, however, since if the next operation preferred on the machine will not arrive to the machine for a long time, this would involve an excess of idle time on the machine (which in perverse cases could even be infinite—gridlock). A common simplification is called "dispatch." When a machine becomes free, the most preferred operation *currently available* is scheduled next. This simplification tends to increase machine utilization. However, very high priority jobs might then arrive while another job is being processed, when it would have been preferable for the machine to wait. An "extended dispatch" would allow waiting for up to some fraction of a processing time of the highest dispatch priority job.

Unless there is some special problem structure, calculating the fitness of each individual amounts to a complete simulation of the cost of using a particular set of priorities. Thus if a genetic algorithm routine creates 9000 individuals, it must perform 9000 complete simulations of the job shop and qualifies as a rather high-quality (and high-cost) heuristic.

When reproduction is bisexual, two parent chromosomes are first selected. They are duplicated, so that the original parents remain unchanged. Genes that the two parents have in common are kept for the children. (Since two rather fit parents agree on what might be otherwise thought to be randomly kept genes, they are more likely to be involved in fitness.) As for the rest of the chromosome, pieces of chromosome are swapped in some way (which may have to be modified to assure a legal resulting complete chromosome), so that each of the two children will have all the known good characteristics and some of the remaining characteristics of each parent (as in real-life reproduction). This is done, for a particular problem type, in some way that will intuitively give a good chance for improvement.

When reproduction is monosexual, one parent chromosome is selected and duplicated. The child chromosome is rearranged in some way or mutated (new material put in or deleted or both at once). Note that for our current problem the subchromosome simply represents a permutation sequence. Hence monosexual reproduction represents some permutation of the existence sequence. In particular, reproduction might represent a random adjacent pairwise interchange, or general pairwise interchange, such as for other neighborhood search methods. The parent is unaffected by having a child. In addition, a parent might occasionally undergo a mutation, such as a general pairwise interchange at random. This helps prevent the whole population remaining at a local minimum.

The dying mechanism is basically inverse to the mechanism of having children. For example, the probability of dying could be proportional to $(1.0 -$ normalized fitness.)

In some special cases measuring the fitness of an offspring may be of relatively low cost, such as for pairwise interchange on policies for a static one-machine problem. Usually, however, the cost will be much more like a full simulation of the problem.

The decision of whether to let the population expand over time is open. A typical decision made in practice is to keep the population size constant.

Example. It is difficult to fully understand a genetic algorithm approach from a small example. Such an example can still make the simpler ideas more concrete, however. We consider again the 5-job static unweighted tardiness problem of Section 5.5, repeating the problem data for convenience. The criterion is weighted tardiness, with all $w_j = 1.0$.

Job j	p_j	d_j
1	4	5
2	3	6
3	7	8
4	2	8
5	2	17

(It may also be useful to refer to the depth first search solution of this same problem in Figure 5.2, which is sequence 12435 with tardiness 11.)

There are only 120 possible members of the population. We choose a permanent population size of 3. In each generation the single most fit individual reproduces using adjacent pairwise interchange chosen simply at random. That is, there are four possible children; each is chosen with probability 1/4. Duplication of children is permitted and children can duplicate other members of the population. (This reproduction is far too simple to convey the full power of bisexual reproduction, but it keeps the example tractable.) Then the least fit member dies to complete the generation.

We choose the initial population as random permutation sequences.

GENERATION 1

Individual	25314	14352	12345
Fitness (cost)	25	17	16

12345 is allowed to reproduce, with offspring 13245 and cost 20.

GENERATION 2

Individual	13245	14352	12345
Fitness	20	17	16

Note that the best fitness has not improved, but that the average fitness has and that diversity has been preserved. Now 12345 is allowed to reproduce again, with offspring 12354 and cost 17.

GENERATION 3

Individual	12354	14352	12345
Fitness	17	17	16

Now 12345 is allowed to reproduce again, with offspring 12435 and cost 11.

GENERATION 4

Individual	14352	12345	12435
Fitness	17	16	11

Note that in fact, although the procedure does not recognize it, the unique optimal policy has now been found. If further generations are formed, 12435 will always be the parent and will never spawn superior children. The average fitness of the population will improve, however, until the best neighbors of 12435 are represented.

While the example is quite helpful in providing intuition, it has several limitations due to the many simplifications employed, besides not using the power of bisexual reproduction. In the first place, since only the most fit member is allowed to reproduce (or be mutated), the same member will continue to reproduce, unless replaced by a superior child. Thus the method here is not very different from neighborhood search, except that members of the neighborhood are generated randomly at cost 1/N of the cost of generating the entire neighborhood. The total computation time is thus about the same as for neighborhood search. Neighborhood search has the advantage of automatically verifying when one arrives at a local optimum.

It is possible to enrich this example slightly with more children and bisexual reproduction. We leave this for the exercises.

Recently, there has been large-scale application of genetic algorithms to job shop makespan problems (Della Croce et al., 1992). Their study is discussed in Section 16.3.

6.3A Numerical Exercises

6.3A.1 Find a provided problem set for T_{wt} on the software where differing starting seeds for adjacent pairwise interchange sometimes give different answers.

(a) Solve this problem by R&M followed by adjacent pairwise interchange.

(b) Solve this problem repeatedly by RANDOM followed by adjacent pairwise interchange.

(c) How many sequences did you evaluate in (b) before attaining or beating the result of part (a)?

6.3A.2 For the problem in 6.3A.1, but by hand, use the EDD rule for the master seed and biased random sampling with a = 0.6 and a random number table, to do biased random sampling seed selection. Input these seeds to the software and evaluate as in 6.3A.1. How many sequences did you evaluate before obtaining or beating the result in 6.3A.1(a)?

6.3A.3 Try the problem in 6.3A.1, starting from a seed that does not lead to the global optimum. Use tabu search from the software with tabu list length of at least 10. Does tabu search find the optimum in less than 500 moves?

6.3A.4 Starting from the same seed as in 6.3A.3, use simulated annealing with k set at 0.01 times the EDD solution cost. Repeat for $k = 1$. Compare the two results and comment.

6.3A.5 Repeat the example given in the text for genetic algorithms, with the change that two individuals are allowed to reproduce with two children each. Children may not repeat.

6.3A.6 Add a reasonable method of bisexual reproduction to 6.3A.5 and solve the problem there again.

6.3B Software/Computer Exercises

6.3.B.1 Develop a simple aid for solving various kinds of neighborhood search problems. It should expand the neighborhood, and rank the next steps by improvement amount. The user specifies a choice and the computer keeps score.

6.3B.2 Develop a tabu search/simulated annealing combination program that also allows various combinations. Try some problems.

6.3B.3 Develop, write, and test your own invention to do extended neighborhood search.

6.3B.4 Write a program to extend the genetic algorithms example to allow variable population size and number of individuals reproducing. Include a mechanism for bisexual reproduction. Given a small weighted tardiness problem as input, the program should output a trace generation by generation.

6.3C Thinkers

6.3C.1 Do some serious thinking on the issue of which branches to drop in beam-seed search. Discuss your conclusions.

6.3C.2 Do some serious thinking on the issue of the best design of a genetic algorithm for the static one-machine weighted tardiness problem, for large problems such as $n = 100$.

6.4 LAGRANGIAN RELAXATION; NEURAL NETWORKS

6.4.1 Introduction

There are other important techniques that should be discussed. Some techniques, such as neural networks, are very new in scheduling and do not have sufficient testing to validate their competitiveness. Other techniques, such as Lagrangian relaxation, are more an art than a science and are difficult to discuss in a general textbook in sufficient detail to adequately explore their full potential. We introduce these methods briefly in this section.

(A) 6.4.2 Lagrangian Relaxation

Overview. Here we give an overview of Lagrangian relaxation, based in large part on the excellent tutorial by Fisher (1985). In the next subsection we develop this technique for a simple example: the one-machine weighted tardiness problem. For the latter see work by Potts and Van Wassenhove (1985); there is work in process in several places.

Lagrangian relaxation has been applied widely in operations research problems over the past 20–25 years. Its strength from an integer programming standpoint is that it usually gives stronger lower bounds to large difficult integer programming problems than does linear programming. These stronger lower bounds can then be used directly in branch-and-bound procedures for the same problem. The stronger bounds vastly reduce the search, which is usually felt to more than justify the high cost of performing the Lagrangian procedure.

From a heuristic point of view Lagrangian relaxation has three strengths:

(a) The tight lower bound provides a benchmark for the adequacy of heuristics being derived independently when it is not feasible to solve the problem exactly to give a perfect benchmark.

(b) Various rounding procedures can be applied to the Lagrangian solution directly (how to do this depends very much on the problem structure) to produce a heuristic.

(c) Lagrangian prices can potentially provide a sophisticated, high cost source for the prices needed in bottleneck dynamics. (See the next section.)

There are difficulties in using Lagrangian relaxation as well:

(a) It is necessary to partition the constraints and relax only some of them. It is a somewhat delicate matter that the resulting problem not be too easy or too hard, since it represents an integer program that must be solved exactly many many times in solving a single Lagrangian relaxation.

(b) There are a number of other judgmental questions as well, such as the best way to search for the "best" prices and the best way to round the final solutions.

Lagrangian relaxation is based on the experience that many harder integer programs can be modeled as an easier integer programming problem complicated by a set of integer side constraints. One way to use this observation in *convex* problems would be to use Lagrange multipliers, which involves finding dual prices that cause the constraints to be exactly satisfied and the optimal solution obtained. We thus spend most of our time solving smaller problems and searching for the optimal dual prices.

The same idea works for integer programs, except that we must be much more careful. Prices will be harder to search for, and there may be no sets or multiple sets that cause the constraints to be satisfied. Furthermore, when we find the solution, it will usually just give a lower bound on the optimal cost and be infeasible to the original problem. Nevertheless, as we indicated above, such a solution can be very useful.

So we create a Lagrangian problem in which the difficult constraints have their amount of violation multiplied by their price; this cost is then added to the objective function. If we have planned correctly, the Lagrangian problem is reasonably easy to solve and provides a lower bound on the cost of the optimal solution to the original problem.

Using Fisher's example, we discuss the Lagrangian relaxation concept in rather abstract terms. Consider some hard integer program (e.g., cost minimization) that we are trying to solve:

$$Z = \min cx$$
$$Ax \leq b \quad \text{(hard constraints)}$$
$$Dx \leq e \quad \text{(easy constraints)}$$
$$x \geq 0, \text{ and integer}$$

(These are vectors and matrices in usual compact notation.)

Think of the hard constraints as resource constraints. Next, we choose any set of positive prices u to reward us for any unused resources, and subtract this reward from the objective function:

$$\min cx - u(b - Ax)$$
$$Ax \leq b$$
$$Dx \leq e$$
$$x \geq 0, \text{ and integer}$$

Note that the reward term is guaranteed to be nonnegative by the constraints, and so the solution to this problem is a lower bound on the original.

Now we don't really need the hard constraints in the problem twice, since we are going to vary the u to keep this constraint in line. Dropping this constraint can only decrease the objective function, so that the resulting Lagrangian problem is also a (smaller) lower bound on the original:

$$L(u) = \min cx - u(b - Ax)$$
$$Dx \leq e$$
$$x \geq 0, \text{ and integer}$$

For a fixed value of the price vector u (dual variables), we are assuming that this resulting problem is relatively easy to solve. This is important because, since every $L(u)$ is a lower bound on the original problem, we would obviously like to vary the u to maximize L. Every time we change u, however, we have a new integer program to solve.

Three important problems arise:

(a) Which constraints should be relaxed?

(b) How do we search effectively for good prices u?

(c) How do we "round" the final solution to get a good heuristic for the original problem?

We cannot afford to take the time to go deeply into these issues here, but we make a couple of intuitive points:

(a) The constraints not relaxed should be those that form a well known problem structure that is easily solved, such as a knapsack problem, or simple network problem, or a permutation sequence problem.

(b) There are two common methods to search for prices. One, the subgradient method, is an adaptation of convex gradient search methods (watching out for places without derivatives). Other procedures, called multiplier adjustment methods, depend on a special structure of the problem.

(c) Rounding methods are highly problem specific. See Erlenkotter (1978) and Fisher et al. (1982).

Lagrangian Relaxation: Scheduling Example. We turn again for a practical example of Lagrangian relaxation to the one-machine static weighted tardiness problem. To apply Lagrangian relaxation, we first need an integer programming statement of the problem.

Let $1 \leq j \leq n$ be the original labeling of job j.

Let p_j be job processing times.

Let d_j be job due dates.

Let w_j be job tardiness weights.

Let $1 \leq j* \leq n$ be the job finally in position j in the sequence.

Let T_{j*} be the actual tardiness (final order) of $j*$.

Let C_{j*} be the actual completion time (final order).

Let w_{j*} be the weight of the job ending in position j.

Let d_{j*} be the due date of the job ending in position j.

Let x_{ij} be 1 if j directly follows i in the final ordering, 0 otherwise.

The integer program for the problem is:

$$\min \sum_{j=1,n} w_{j*} T_{j*} \tag{1}$$

$$\text{s.t.} \quad C_{o*} = 0 \tag{2}$$

$$C_{j*} = C_{(j-1)*} + \sum_{k=1,n} x_{jk} p_k \tag{3}$$

$$w_{j*} = \sum_{k=1,n} x_{jk} w_k \tag{4}$$

$$d_{j*} = \sum_{k=1,n} x_{jk} d_k \tag{5}$$

$$\sum_{i=1,n} x_{ij} = 1 \tag{6}$$

$$\sum_{j=1,n} x_{ij} = 1 \tag{7}$$

$$x_{ij} \in \{0,1\} \tag{8}$$

$$T_{j*} \geq C_{j*} - d_{j*} \tag{9}$$

$$T_{j*} \geq 0 \tag{10}$$

Here (1) says to minimize weighted tardiness of final orderings; (2) and (3) say completion times are built up of processing times of final orderings; (4) and (5) develop weights and due dates for final orderings; (6) and (7) say every final ordering job has a unique predecessor and a unique successor; (8) says the ordering variable is 0–1; (9) says tardiness is at least as large as lateness; (10) says negative latenesses are ignored.

Note that although writing this problem formally as an integer programming problem is fairly complicated, all constraints but the last two really just say: "choose a permutation sequence, and note the resulting completion times, weights, and due dates resulting from the final ordering." The final two constraints and the objective function simply say "minimize the weighted tardiness for that problem over these sequences." In the following, we therefore simply use a shorthand statement of this formulation:

$$\min \sum_{j=1,n} w_j T_j$$

s.t.   [perm. seq. constraints] (these we also suppress below)

$$T_j \geq C_j - d_j$$

$$T_j \geq 0$$

Since we know how to exactly solve many objectives with the permutation sequence constraints, it seems rather likely that we should treat this and the non-negativity of tardiness as the "easy" constraints and relax the tardiness constraint:

$$\max \left\{ L(u) = \min \sum_{j=1,n} [w_j T_j - u_j (T_j - (C_j - d_j))] \right\}$$

s.t. $T_j \geq 0$

Now in order to maximize $L(u)$, to get the tightest possible lowest bound (and to get the most meaningful prices), we cannot allow any $u_j \geq w_j$ or the minimization would be unbounded below ($-$infinity). In this case, the minimization will always occur for all $T_j = 0$. Thus our overall Lagrangian problem simplifies to

$$\max_{0 \leq u \leq w} \left\{ L(u) = \min \sum_{j=1,n} u_j (C_j - d_j) \right\} \tag{LgP}$$

We see that the inner problem has indeed become easy to solve! This is just the static one-machine lateness objective with the u_j as the weights. Thus for a given set u_j, the solution is to sequence by WSPT:

$$u_j/p_j \geq u_{j+1}/p_{j+1} \quad \text{and so on}$$

Also, considering the outside maximization, we see that any u_j for a tardy job in this sequence can be *increased* to the minimum of the point where WSPT would be violated, or to where $u_j = w_j$. Similarly any nontardy job can have its u_j *decreased* to the first point where WSPT would be violated, or to where $u_j = 0$.

The more general subgradient method (Geoffrion, 1974; Fisher, 1981) would involve successive improvement of the u's as follows.

Subgradient Method

Step 1: Pick a "good" initial u estimate.

Step 2: Find the optimal sequence $S(u)$ by WSPT and objective $L(u)$.

Step 3: Estimate changes to improve u by methods analogous to gradient search.

Step 4: Repeat steps 2 and 3 until there is adequate convergence.

The special multiplier adjustment method of Potts and Wassenhove (1985) depends on the special structure of this particular problem. The idea is that if the optimal sequence were known, by using that sequence in (LgP) we could find the best u values to maximize this expression and yet have those u values produce the sequence by WSPT. This should give the solution to the Lagrangian. Failing to do that, we should apply the same procedure to a strong heuristic, say R&M. Thus we would try to solve

$$\max_u \sum_{j=1,n} u_j T_{j \text{ heur}}$$

$$\text{s.t.} \quad u_j/p_j \geq u_{j+1}/p_{j+1} \qquad (\text{LB}_{\text{heur}})$$

$$0 \leq u_j \leq w_j$$

It is intuitively obvious how to solve this problem:

Step 1: Set $u_1 = w_1$.

Step 2: Recursively set $u_{j+1} = \min [w_{j+1}, u_j(p_{j+1}/p_j)]$.

To illustrate these ideas, let us consider the same unweighted tardiness problem we have been working with from Section 5.5. We know the optimal sequence for that problem is 1–2–4–3–5 with optimal total tardiness of 11. Since we are lucky enough to know the optimal solution, we may use it to solve (LB_{heur}) above, to study the prices, and see how far this lower bound estimate is from the optimum.

Job	p_j	C_j	d_j	T_j	w_j	w_j/p_j	u_j	u_j/p_j	$u_j T_j$
1	4	4	5	0	1	0.250	1	0.250	0.000
2	3	7	6	1	1	0.333	0.75	0.250	0.750
4	2	9	8	1	1	0.500	0.50	0.250	0.500
3	7	16	8	8	1	0.143	1	0.143	8.000
5	2	18	17	1	1	0.500	0.29	0.143	<u>0.286</u>
								Lower bound	9.536

Note that the lower bound obtained by using the known optimal solution is 9.536, which is fairly close to 11 but differs by 15%. (It seems possible also that a sequence other than the optimum might give a better lower bound here.)

The difficulty in getting sharper lower bounds in this case follows from the conflict between the WSPT sequence and the desire for large prices. It is left as an exercise to demonstrate that for problems in which all jobs are tardy, the lower bound will be perfect. We consider some of these issues further in the exercises.

6.4.3 Neural Networks

Neural networks have their origins in artificial intelligence (*DARPA Neural Network Study*, 1988; Marquez et al., 1992). Just as genetic algorithms try to emulate the evolutionary process of natural selection, neural networks try to emulate the learning process of the human brain, primarily its high connectivity. It has achieved some success in a diversity of fields, and there has been a considerable amount of "hype" that it is a new supermethod.

The truth is that it is just a fancy nonlinear forecasting method operating with many of the strengths and weaknesses of linear forecasting methods, such as regression, discriminant analysis, logit, probit, and so on. These methods are not useful in solving large integer programs in a precise way, and there is no reason to expect neural networks to do so either. The few reports of neural networks and scheduling that exist today are certainly not very convincing (Zhou et al., 1990, 1991; Lo and Bavarian, 1991).

However, forecasting methods have found some use in planning and scheduling, especially in highly uncertain situations where accuracy is less important and decisions must be made repeatedly. Thus it is to be expected that neural networks will also. Therefore we give a brief introduction.

In Figure 6.2 we give a rather general diagram of the way a typical forecasting system works. We go through this diagram first for conventional linear forecasting models (regression), and then a second time to show the correspondence for neural networks.

First consider a forecasting problem. We have sales and advertising expenditure data over the last 6 months. These come in at ①; the inputs—namely, advertising and time—go in as inputs to the model at ②, while the goal output or actual results for the period ③ go into a procedure to optimize the model to fit the data, as well as current predictions the model would make with its current coefficients ④, the

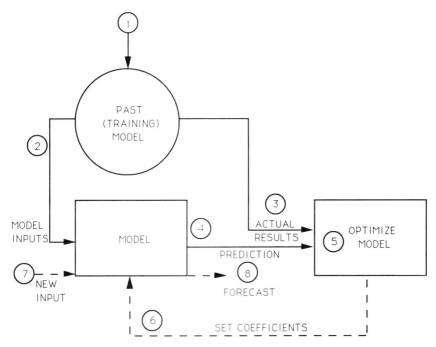

Figure 6.2. Diagram of general forecasting system.

forecast error between ③ and ④ becomes an error signal into ⑤, the model optimization, which corrects the model ⑥ so that it forecasts well for this particular data. At ⑦ new data are input for which a forecast is required, in the hope that the data are similar to the data just used to generate the model.

Neural networks are very similar. The past data ① are called *training data* because they will be used to "train" the model. ③ becomes training goals, and the error signal is again used to "optimize" the model.

The model is like an electrical network in an analog computer. The network is organized into a number of nodes at stage one, stage two, and stage three (there could be more stages.) Each stage is typically connected in every possible way to the next, by arcs that multiply the current flowing through them by a constant (these constants are the controls for "optimizing" the network). Constants are often normalized between 0 and 1. All currents flowing over different arcs into the same node are summed. The current at the node is then often truncated to be between 0 and 1 by an S-shaped sigmoid function. Other network formats and functions can also be used, depending on the purpose.

How do we change the many coefficients to "optimize" (regress) the model from the given input data? There is a simple method called *back propagation*, which transmits the error signal backward through the network, to correct the signal. The comparison between linear regression and neural networks is again demonstrated in Figure 6.3.

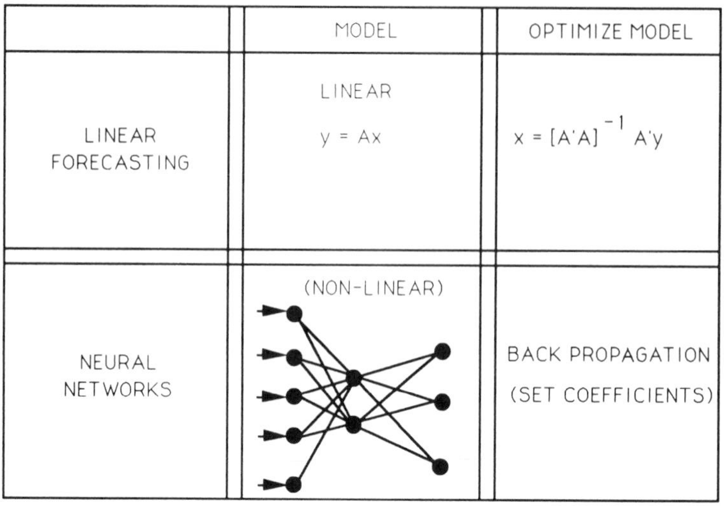

	MODEL	OPTIMIZE MODEL
LINEAR FORECASTING	LINEAR $y = Ax$	$x = [A'A]^{-1} A'y$
NEURAL NETWORKS	(NON-LINEAR)	BACK PROPAGATION (SET COEFFICIENTS)

Figure 6.3. Linear regression versus neural networks.

Unfortunately, just as for linear regression, training the neural network even for a large number of data points does not predict perfectly. It is true that the more broadly we train the network, the more likely it will handle a new fairly similar problem without being retrained. However, we pay a cost in that the broader training will not fit on average as well.

Once again, training on a problem set is something like doing a regression. The line may fit similar problems fairly well (interpolation), but what about the accuracy in solving new problems that are "different"? How different is different?

Neural networks will probably grow in importance in the future, in particular, for more aggregate planning problems. However, more coverage will not be given here.

Other references for this section include Fisher (1973) and Vaithyanathan and Ignizio (1992).

6.4A Numerical Exercises

6.4A.1 For the same 5-job unweighted tardiness problem of Section 5.5, find a good starting u in the search for the best prices by some intuitive method. How good of a lower bound does your starting u provide?

6.4A.2 Search iteratively for improved values of u (say, five iterations) without using either method described in the text.

6.4A.3 Add two or three jobs to this same data set to make the problem harder.

(a) Solve optimally by the branch-and-bound software.

(b) Get the best lower bound you can by intuitive Lagrangian relaxation methods.

(c) Find the optimal u values for the known optimal sequence.

(d) Evaluate the resulting lower bound.

6.4B Software/Computer Exercises

6.4B.1 Devise a method and write the software for doing heuristic Lagrangian relaxation for the weighted tardy problem.

(a) Search for new approximate solutions to the problem using some version of beam search.

(b) For every approximate solution obtained, find its corresponding lower bound to the problem by (LB_{heur}).

(c) Keep a running record of the best LB to date.

(d) Stop after the last M evaluations have not improved the lower bound.

6.4B.2 Test your program in 6.4B.1 extensively.

6.4C Thinkers

6.4C.1 For a weighted tardy problem in which all jobs are tardy for the optimal sequence, prove that the optimal lower bound from Lagrangian relaxation is equal to the optimal value of the original objective function.

6.4C.2 Apply the Lagrangian relaxation method to the weighted early/tardy problem.

(a) Which constraints need to go in the objective function?

(b) Is the resulting problem sufficiently easy to solve analytically?

(c) Develop as many results as you can.

6.5 BOTTLENECK DYNAMICS: SINGLE RESOURCE

6.5.1 Overview

Recall from our general discussion in Section 2.4 that bottleneck dynamics estimates the dynamic value of extra capacity for each resource (resource price) and also the dynamic value of having capacity earlier rather than later for each resource (resource delay cost). (It is easily seen that the resource delay cost is just the interest rate from the firm's cost of capital times the resource price.) Similarly, bottleneck dynamics estimates delay prices (delay costs) for each activity within the shop.

Trading off the resource delay costs for expediting an activity against the activity delay costs saved allows the activity with the highest net savings per unit of resource cost (benefit/cost ratio) to be scheduled first, with no knowledge of the rest of the system except that conveyed by the estimated prices. Inclusion of other types of cost and discounted present value formulations are straightforward. Decision types that can be handled include sequencing, routing, release and other timing, lot sizing, and batching.

The advantages of transfer pricing in terms of decomposing large systems are well known. Now, in general, scheduling systems are not convex and hence no exactly consistent set of prices exists. However, the central intuition here is that in many large shops, where no individual activities dominate, the structure is "close to convex" in some meaningful sense.

In the next subsection we develop our pricing formulation and illustrate decision formulas, without dealing with the issue of where the prices come from. In the final subsection we take a fresh look at the static F_{wt} and T_{wt} problems, in the first case to validate our method against known results and, in the second case, to derive new results. Further discussion of the T_{wt} heuristic, as well as other new models, will be considered in the next chapter.

Heuristic pricing methods for a static single resource problem are relatively easy because:

(a) The unknown resource price scales all priorities equally, and hence need not be determined.

(b) The lead time from the completion of the activity until the completion of the job need not be forecast; it is zero.

In more general shops the estimation and updating of resource and activity prices and activity lead times are more difficult. We defer the discussion of these issues until Chapter 10.

6.5.2 Heuristic Pricing Formulation

We developed a big picture understanding of bottleneck dynamics and its associated use of heuristic prices in Section 2.4. It would be very helpful to review that section at this time. We also give a brief review here. Basically, bottleneck dynamics estimates an approximate dual price (delay cost) for delaying each possible activity in a shop by estimating the corresponding delay of delivery in the final job of which it is a part. This requires estimating the lead time from the current activity to the final job conclusion to estimate the time urgency of the job.

Bottleneck dynamics also estimates an approximate dual price (delay cost) for delaying any possible resource in a shop, by adding up the delay costs for all activities already waiting for (or using) the resource or that will arrive before the resource is next empty. (This is termed "the current busy period.") The resulting price $R(t)$ is not an accounting price for the machine; it is an estimate of the opportunity cost for taking time on the machine in terms of jobs pushed back.

We first develop our heuristic pricing formulation informally. Next, we show the correctness of the approach for some simple models we have already examined. But the principal justification for the approach will come in developing new accurate models for the one-machine problem in Chapters 7 and 8, and later for multiple resource problems in Chapter 10.

Heuristic Pricing Definitions (Single Resource)

(a) Let $R(t)$ be the (implicit) price per unit time of using the resource at time t.

(b) Let I be the interest rate derived from the firm's cost of capital.

(c) Therefore $IR(t)$ is the premium per unit of time that usage of a unit time's worth of the resource is shifted either earlier or later. (That is, I and R both have units of "per unit of time.")

(d) Suppose the completion time of j is the only influence j has on an additive or maxi objective function.

(e) Let the nontime importance w_j of job j be given by $w_j = D_j V_j$, where D_j is the value added, and V_j is the importance of the customer.

(f) Let the slack $S_j(t)$ be given by $S_j(t) = d_j - p_j - t$.

(g) Let $U_j(t) = f_j(S_j(t))$ be the activity time urgency (marginal cost of delay) if the job is currently scheduled first and expects to complete with slack S_j. That is, f_j is the marginal cost of decreasing slack of j in the objective function with completion time $t + p_j$.

(h) The delay price or delay price of delaying job j is given by $w_j U_j(t)$.

As a first example, we apply this framework to the static one-resource case we looked at in Chapter 4. To estimate what priority job j should be given in the line for the single resource, let the resource have price R and let job j have weight w_j and urgency U_j. If we decide to process it a time Δ earlier (or later), then the saving (or cost) for j is $\Delta w_j U_j$; but the resource is also used earlier, with resulting extra cost $\Delta I R p_j$. Thus the net savings for processing the job earlier would be

$$\Delta w_j U_j - \Delta I R p_j$$

Since there is a single resource, the highest priority job would be the one with the highest net savings for unit of resource cost, or the priority π_j would be given by

$$\pi_j = \Delta[(w_j U_j)/(I R p_j) - 1.0]$$

Finally, since Δ, I, and R are the same for every job, a priority with the same ranking can be given simply by the following formula.

Heuristic Formula 1. For the static one-resource problem, sequencing priorities may be estimated by

$$\pi_j = (w_j/p_j)U_j$$

where w_j is the job importance, p_j is the processing time, and U_j is the current time urgency factor. [Remember that $U_j(t)$ changes with time so that priorities must be recalculated whenever time changes.]

This is a very suggestive formula. It says to prioritize by the basic job value per unit of processing time (bang for the buck) scaled up or down by a "slack" factor. The one remaining difficulty here is the estimation of U_j.

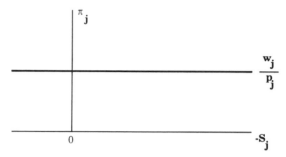

Figure 6.4. Case 1: weighted lateness.

6.5.3 Application to Sequencing Problems

In this subsection we apply the bottleneck ideas to some simple objectives for which we already know the answer, to obtain insight about the procedure.

Case 1: Minimize Weighted Lateness. Since in this case we have $w_j U_j(t) = w_j$ for all t, it follows that the marginal cost of lateness (or slack) is independent of lateness or slack as shown in Figure 6.4. Hence we need not worry about estimating the slack or its variability. $U_j(t) = 1.0$ for all t. The priority function is simply

$$\pi_j = w_j/p_j$$

Note that from our work in Chapter 4, we know that this formula is exact, whether for the unit-preemptive case, or the nonpreemptive case.

Case 2. Minimize Weighted Tardiness—No Correction for $E\{U_j\}$. Now we try to use the same idea, although the urgency is no longer independent of the slackness. If j is scheduled first at time t, its slack will be given by

$$S_j(t) = d_j - (p_j + t)$$

And the U_j function as a function of negative slack is shown in Figure 6.5.
Working through the same analysis as before we would predict the following.

Heuristic 1 for Weighted Tardiness. (No U correction)

$$\pi_j = w_j/p_j \text{ for jobs past due}$$
$$= 0 \text{ otherwise}$$

This is a fairly good heuristic, as you will be asked to verify in the exercises. However, it can be improved in two different ways, both of which are worth understanding.

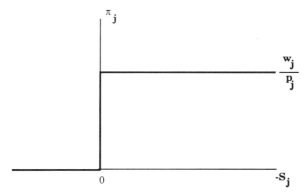

Figure 6.5. Case 2A: weighted tardiness, uncorrected.

The first is a correction for discreteness. A job cannot really be delayed just a small Δ amount if it is delayed. It must be delayed for an amount p_i, the processing time of the job with which it is being switched. Now if job j has initial slackness greater than p_i, there is no effect. Priority is 0 before and after the switch. But suppose that $S_j/p_i = \beta < 1.0$, and conceive of job j being divided into mini jobs. Clearly, a fraction β will still have priority 0, but $1 - \beta$ will have full priority. Then the average priority will be $(1 - \beta)(w_j/p_j)$.

It has been shown by Rachamadugu and Morton (1982) that there is an optimal sequence for the weighted tardiness problem satisfying this priority rule; that is, it is necessary but not sufficient for optimality.

Proposition 1. Let i and j be any two successive jobs in an optimal sequence for the static one-machine T_{wt} problem, i preceding j. Then it is true that

$$(w_i/p_i) \{1 - S_i^+/p_j\}^+ \geq (w_j/p_j) \{1 - S_j^+/p_i\}^+$$

Proof. Left as an exercise. ∎

This is a much better rule, but it has one defect. We really want the priority for j to depend only on j and not on the jobs with which it is being switched. This is both to make the rule fast and to allow priorities to be used in other ways. The simplest approximation to correct for this, which doesn't seem to reduce solution quality very much, is to calculate p_{av}, the average processing times of jobs competing for top priority, and to substitute this for the unknown p_i. This leads us to the following.

Heuristic 2 for Weighted Tardiness. (*U* corrected for discreteness)

$$\pi_j = (w_j/p_j) \{1 - (S_j)^+/p_{\text{av}}\}^+$$

where p_{av} is an average processing time for currently competing jobs.

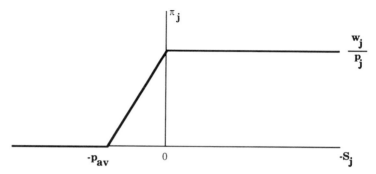

Figure 6.6. Case 2B: weighted tardiness, corrected for discreteness.

A graph of π_j for Heuristic 2 is shown in Figure 6.6.

The last correction on the urgency factor U is a little more subtle. Suppose five jobs all have processing times 1.0. Four of the jobs have weight of 8 and due date $t = 4$. One of the jobs has weight of 1 and a due date of $t = 1$. It is time zero and a decision is to be made. By our last rule, the four jobs are more than a processing time away from being tardy and hence have priority 0. The job with weight 1 has a priority of 1 and is processed first. But one of the jobs with weight 8 will now be tardy, costing 8; whereas if the high-weight jobs were run first, the low-weight job would have had a total cost of 4.

The problem here is that estimating slacks and priorities job by job may be wrong if there are several jobs with similar due dates. If the four jobs were *aggregated*, they would show about the same w_{agg}/p_{agg}, but a smaller slack. One solution would be to aggregate jobs and solve for aggregate priorities, but this gets a little compli-cated. A simpler procedure is to use an idea similar to the biased random sampling idea. In general, our slacks are biased too large due to both discreteness and aggregation problems. The larger the possible bias we consider, the smaller the probability that correcting for a bias that large is necessary. Suppose we choose an exponentially decreasing set of weights adding to 1.0 as before, and average the time urgencies accordingly. Then we can obtain a rule of the following form.

Heuristic 3 for Weighted Tardiness. (U corrected for discreteness and aggregation problems)

$$\pi_j = (w_j/p_j)\ [\exp\{-(S_j)^+/kp_{av}\}]$$

The graph of π_j for Heuristic 3 is shown in Figure 6.7.

This heuristic (along with COVERT) is state of the art for accurate practical priority heuristics for the weighted tardiness problem. We shall report computation-al testing in some detail in the next chapter. Note that k is a free parameter that should be experimentally fit for each problem class. Fortunately, good solutions are generally attainable for k between 1.0 and 3.0 (for simple one-machine problems.) (Remember these priorities are recalculated as time advances.)

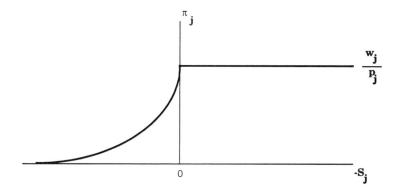

Figure 6.7. Case 2C: weighted tardiness, fully corrected (R&M).

Insight can also be gained by considering the logarithm of the priority, which itself preserves the priority ordering:

$$\ln(\pi_j) = \ln(w_j) - \ln(p_j) - (k/p_{av})(S_j)^+$$

Thus there is a form of the priority rule that is linear in the slack.

6.5A Numerical Exercises

6.5A.1   Create three 6-job problems for T_{wt} objective: one for which the average due date is about 80% of total processing time, one for which it is about 50%, and one for which it is about 20%.

(a) Explain why average due date over total processing time is interesting (be specific).

(b) Solve each problem by the software manual interchanger or another simple helper, so that you will understand the solution thoroughly. Try EDD, WSPT, and the three heuristics from this section.

6.5A.2   For one of the three problems, try Heuristic 3 for several different k values. How sensitive is the solution?

6.5A.3   For one of the three problems, try Heuristic 2 with different ways of estimating p_{av}: (a) the actual processing time of the job with which it is being switched, (b) unweighted average of all other candidates, (c) average of top three candidates in terms of due date, and (d) average of top two candidates in terms of w/p.

6.5B Software/Computer Exercises

6.5B.1   Use the software package to study various heuristics for the T_{wt} problem on a set of 20 problems of about 15 jobs each generated statistically.

6.5B.2 Find the "optimal" solutions to the same problems using (truncated) branch-and-bound.

6.5B.3 Try your problems using tabu search, beam search, or some type of neighborhood search.

6.5C Thinkers

6.5C.1 Develop a heuristic for the unit-preemptive T_{wt} problem that schedules jobs in order of decreasing w_j/p_j, working backward from the due date, assigning to vacant unit capacity on the resource where it occurs. If all units of the job have not been assigned by the time its arrival has been reached, the remainder is assigned forward from the due date. (The job is tardy.) Evaluate this heuristic versus Heuristic 3 on some problems that have several jobs due at the same point in time.

6.5C.2 Try to create and test a priority rule that corrects itself for the discreteness/aggregation problems in a more complex dynamic fashion.

6.5C.3 Jobs have zero penalty until their due date. Then the penalty is w_j from the due date until a second drop dead date dd_j, after which it rises to W_j. Fashion and test a heuristic for this problem.

6.5C.4 Prove Proposition 1.

___7
SINGLE MACHINE: SEQUENCING

7.1 INTRODUCTION

In Chapter 4 we began to develop sequencing results for regular static one-machine problems for which exact results not requiring advanced methods are available: makespan, weighted flow, weighted lateness, maximum lateness, maximum tardiness, number of tardy jobs, and weighted number of tardy jobs (preemptive case). In addition, we derived partial results (dominance properties, etc.) for weighted number of tardy jobs (N_{wt}), weighted tardiness (T_{wt}), and the mixed weighted flow/weighted tardiness problem (FT_{wt}).

Recognizing that we could not proceed with developing exact and/or accurate heuristic solutions to these and other sequencing problems without additional mathematical techniques, we devoted the next two chapters to developing the fundamentals of techniques that are useful not only for sequencing but for all scheduling problems. In Chapter 5 we discussed such classical techniques as transitive pairwise interchange, convex dispatch priorities, neighborhood search, dynamic programming, and branch-and-bound. In Chapter 6 we investigated, in addition, such newer techniques as beam search, approximate dynamic programming, and extended neighborhood search, including tabu search, simulated annealing, genetic algorithms, Lagrangian relaxation, and the single-resource part of bottleneck dynamics.

While any or all of these methods are applicable to the sequencing problems we shall discuss, we selectively limit ourselves to those that have received the most attention for a given problem and/or those that seem to have the most potential for that particular problem. In Section 7.2 we discuss methods for regular static problems for which we already have some dominance results from Chapter 4. These include weighted number of tardy jobs (N_{wt}), total weighted tardiness (T_{wt}), and weighted flow/tardy jobs (FT_{wt}). In Section 7.3 we discuss methods for nonregular static problems. Here the primary example that has been studied is the weighted early/tardy problem (ET_{wt}), which has the complication that priorities are basically negative before the due date (the activity does not want to be scheduled) and positive after the due date. However, we also discuss briefly the general case in which costs are simply convex in lateness. This leads to a picture of monotonically increasing priorities in lateness, so that priorities will also be negative up to some point and positive thereafter.

In the remainder of the chapter, we move past the case in which all activities are available at the start (static) and begin to deal with the more realistic dynamic case (each activity arrives at a different time). In Section 7.4 we develop exact results for some easier cases of dynamic arrivals, namely, the nonpreemptive makespan case, as well as the preemptive cases for weighted flow and maximum lateness, and partial preemptive results for more difficult problems such as weighted number of tardy jobs. In Section 7.5 we develop heuristics and/or combinatorial search methods for the nonpreemptive, no-inserted-idle-time version of these problems, together with methods for other problems such as weighted tardiness, weighted flow/tardiness, and weighted early/tardiness. In Section 7.6 we discuss some methods for the nonpreemptive case with inserted idleness. Finally, in Section 7.7 we discuss some related results for the static and dynamic one-resource problem where some of the variables are probabilistic.

7.2 REGULAR STATIC PROBLEMS: GENERAL

7.2.1 Weighted Number of Tardy Jobs (N_{wt})

We are aware of no major prior analytic or computational studies on the *weighted* number of tardy jobs problem. However, recall that Proposition 17 of Chapter 4 provides us with a great deal of structure for helping to solve the one-machine weighted number of tardy jobs problem, whether by branch-and-bound, economic heuristics, or beam search. As a quick summary, Proposition 17 says: (a) order all jobs EDD; (b) find first tardy job k; (c) find "best" job to remove; (d) if job k is still tardy, repeat (c); (e) else return to (b). The "best" job to remove is not obvious but the following is true:

The "best" job to remove for N_{wt} is

(a) job k itself, or

(b) nondominated j meaning there is no i with $w_i \leq w_j$ and $p_i \geq p_j$.

This leads us immediately to a practical exact algorithm using branch-and-bound.

Procedure 1: Branch-and-Bound for N_{wt}

Step 1: First, implement Proposition 17.

Step 2: When branching on a given subproblem:

(a) identify the first tardy job k;

(b) identify all possible "best" jobs to remove;

(c) each represents a potential subproblem; and

(d) the lower bound is simply the sum of tardy weights to date.

Note that this problem is much easier to solve than, say, the weighted tardy problem, because only a relatively small number of choices are likely to be "nondominated." (Nondominated jobs are on the trade-off frontier between weight and tardiness; they are roughly either those of very low weight, or very high processing time, or in between, as shown in Figure 7.1.

For this reason, it is to be expected that rather large problems should be solvable exactly (e.g., perhaps 100 jobs, with 30 jobs dominant, and 15 actually tardy). Proposition 15 of Chapter 4 makes it clear that for the unit-preemptive case it is always exactly optimal to remove the joblet with the lowest value of w_j/p_j. Intuitively, this makes sense in general: we have a certain amount of processing time that it is necessary to remove, say, p_r, and we would like to remove the smallest total weight possible to obtain this amount of removed processing time. The only danger in choosing the lowest w_j/p_j is that we might end up removing somewhat more than p_r, and thus more weight than necessary. There are a couple of obvious ways to make a correction for this effect, leading to several possible heuristics.

Procedure 2: Economic Heuristic I for N_{wt}. Always choose to remove the job with lowest w_j/p_j.

This heuristic is very easy to understand, intuitive, and likely to be quite accurate. It would seem the apparent standard of comparison.

But we can improve it, at the cost of some simplicity.

Procedure 3: Economic Heuristic II for N_{wt}. Always choose to remove the job with lowest w_j/p_j, unless that removal produces no further tardy jobs in the problem.

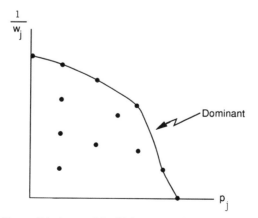

Figure 7.1. Low weight, high process time frontier.

In that case, remove instead the smallest weight job in the dominant set which produces no further tardy jobs in the problem. Terminate.

Finally, the most complicated economic heuristic recognizes that low w_j/p_j jobs and low p_j jobs have different claims to be chosen and attempts to compromise between them.

Procedure 4: Economic Heuristic III for N_{wt}. Choose a parameter α. Always choose to remove the job with lowest value of $(1 - \alpha)(w_j/p_j) + \alpha(1/p_j)$. Make a terminal correction for the last job as in Procedure 3.

There are a number of issues with respect to choosing α. Should one simply run a number of problems and choose the apparent best α? Should α change as one gets to the end of the problem?

Our students have, in past years, done some limited testing, which indicates Procedure 2, the simple w_j/p_j heuristic, is quite robust and accurate. More testing on these heuristics would be useful.

We turn finally to sketching a beam search procedure for N_{wt}.

Procedure 5: Beam Search for N_{wt}. This procedure is similar to branch-and-bound, except that each time a subproblem is branched on, a trial solution is generated for each subproblem corresponding to all dominant set members. The trial solution is generated using Heuristic I. At any level, the top b (beam width) solutions are saved for expansion, the rest discarded.

Other possibilities here include:

(a) No trial solution, just expand the top b lower bounds at each level.

(b) Use lower bounds to obtain depth first trial solutions; otherwise as in (a).

7.2.2   Weighted Tardiness (T_{wt})

We turn next to more completely specifying procedures for the static weighted tardiness problem.

(A) Procedure 6: Dynamic Programming for T_{wt}. Rachamadugu and Morton (1982) developed a modified hybrid dynamic programming approach for solving the weighted tardiness problem. While it is too complicated to present here, it has several interesting features. First, upper and lower bounds on the optimal solution are available at each stage. Second, there is a stopping rule for identifying the first job in an optimal procedure. (This allows starting the procedure again on the remaining jobs, and so on.) It also uses extensive dominance procedures to curtail the number of subsets that must be considered.

To understand the power of the exponential growth law in the increasing cost of solving such problems as the number of jobs is increased, their DP procedure was able to solve all 10-job problems, over 99% of the 20-job problems, about one-half of the 30-job problems, and no 40-job problems, before running out of computation time and/or storage space on the computer. (Improvements in computation speed and storage since that time are considerable, but would not likely add more than 5 or 10 jobs to this performance.)

Procedure 7: Branch-and-Bound for T_{wt}. Potts and Van Wassenhove (1985) have fairly recently produced an improved branch-and-bound procedure (utilizing Lagrangian relaxation for the lower bounds) for the weighted tardiness problem. Again, it is too complicated to present here, but, basically, neighborhood search is performed on partial solutions in the tree. This produces paths in the tree, which can be shown to dominate many others, and so pruning is much facilitated.

Their procedure can solve perhaps a 50-job problem. The ability to deal with this problem size is solid enough to say that branch-and-bound is reasonably adequate to handle the static problem. Unfortunately, the dynamic problem, rather than the static problem, is most directly generalizable to full job shop problems, and it seems to be considerably harder to solve. We discuss this issue later.

We turn now to Montagne (1969) for the weighted tardiness problem.

Procedure 8: Montagne's Heuristic for T_{wt}

$$\pi_j = (w_j/p_j)[1.0 - (d_j/\, \Sigma_i\, p_i)\,]$$

Note that this heuristic has the right overall form that we discussed in our bottleneck dynamics section: basic full priority WSPT, multiplied by a slack factor. The slack factor is close to 1.0 for very early due date jobs, and very close to 0.0 for very late due date jobs.

A glaring problem with the slack factor is that it is not dynamic. After an item is overdue, it is still not given the full WSPT priority. This heuristic is similar to the weighted critical ratio heuristic (see Procedure 10 below), except that the latter is dynamic and more accurate.

Two other heuristics that are often employed are EDD and WSPT. We know, for example, that EDD minimizes the maximum job tardiness and hence should do well for lightly loaded shops. WSPT works well if most jobs have to be tardy and hence should do well for heavily loaded shops.

Rachamadugu and Morton (1982) developed the following heuristic (called R&M) for the T_{wt} problem from bottleneck dynamics procedures, as discussed in Section 6.5. We reproduce it here for convenience.

Procedure 9: R&M Heuristic for T_{wt}

$$\pi_j = (w_j/p_j)[\exp\{\, -(S_j)^+/kp_{av}\, \}\,]$$

A typical value of k for the static one-machine problem is $k = 2.0$. Intuitively, the formula may be explained as follows. If the slack is negative, the job is sure to be tardy and has full priority. If it is two average processing time lengths until tardy, the slack factor is exp[-0.5(2)] or exp[-1] = 0.4. At four average time lengths the slack factor is 0.16, and so on. Thus close to the tardiness point only, the priority builds up. (A note on solving the unit-preemptive problem: suppose the original problem had $p_{av} = 5.0$. Now when jobs are divided into joblets, k should not remain 2. k should be 10. Another way to put it is that p_{av} remains at 5 even after forming the joblets.)

Rachamadugu and Morton also conducted an extensive experimental study to compare the performances of EDD, WSPT, Montagne, and R&M with their hybrid dynamic program as the benchmark for comparison. They did a factorial experimental design, involving four levels of shop congestion, two levels of variation in processing time, two levels of correlation between processing times and due dates, two levels of variability of due dates, three problem size levels (10, 20, 30 jobs), and 20 random replications of each condition, giving a total of 1920 problems. Each problem was solved by the four heuristics and their hybrid DP procedure. Problems that could not be solved by DP were benchmarked with lower bounds. (A lower bounding procedure finds solutions that may violate some constraints but have a lower cost than the optimal. Thus if a heuristic differs, say, 5%, from a lower bound, it is known also to be within 5% of optimal.)

Because optimal tardiness can be zero for some problems, Rachamadugu and Morton realized that percent deviation of a heuristic from optimum might give (0.0/0.0). So they measured the error in a heuristic by the "excess fraction of a weighted average job length that a weighted average job is tardy in the heuristic." Table 7.1 summarizes briefly the overall results of the study.

Some overall conclusions from the original detailed tables may be drawn: (a) weighted tardiness increases strongly in the shop congestion factor (tardiness factor); (b) weighted tardiness decreases slightly in the spread of the due dates; (c) the most difficult problems to solve to optimality are at the intermediate tardiness factors; (d) EDD deteriorates markedly at high tardiness factors; (e) WSPT competes fairly well with EDD, even at low tardiness factors, and dominates it at higher tardiness factors; (f) WSPT, Montagne, and R&M do worst at intermediate tardiness factors; (g) Montagne does best with low numbers of jobs; and (h) R&M is the outstanding performer and behaves robustly across all regimes.

There are two other heuristics not studied by Rachamadugu and Morton that deserve mention here. They were tested in multi-resource job shop experiments to be described later in the book. These experiments were carried out by authors such as

TABLE 7.1 Comparison of Heuristics for the Weighted Tardiness Problem

Jobs	Optimum Value	EDD Error	WSPT Error	Montagne Error	R&M Error
10	0.788	0.800	0.165	0.073	0.031
20	1.276	1.008	0.336	0.226	0.036
30	1.854	1.839	0.550	0.268	0.031

Rachamadugu and Morton (1982): 1920 problems tested; 2 of 640 20-job problems omitted, 369 of 640 30-job problems omitted, since optimal solution could not be obtained. Remaining problems were tested against lower bounds; see study.

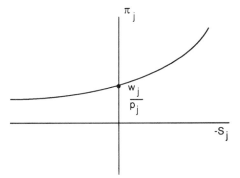

Figure 7.2. Weighted critical ratio heuristic.

Vepsalainen, Lawrence, Pentico, and Morton. Both heuristics were originally developed for the unweighted tardiness problem T_{av} and adapted by the authors mentioned for the weighted tardiness problem T_{wt}.

The critical ratio rule is used widely in production scheduling. It uses the ratio of required lead time to current slack to determine the priority of a job or activity.

Procedure 10: Weighted Critical Ratio Rule for T_{wt}

$$\pi_j = (w_j/p_j)[1/(1 + (S_j/p_j)]$$

This rule has the "proper form" of WSPT times a slack factor and is dynamic. It has three defects: (a) slack is allowed to go negative, so that the priority continues to rise rapidly after the job is overdue; (b) the priority goes down too slowly; there is still a noticeable priority at slacks of 5 to 10 times processing time; and (c) slack is normalized by the processing time of the job itself, rather than of competing jobs. (Rachamadugu and Morton found that the exponential decay in slack behaved better than the 1.0/slack type. See Figure 7.2.)

Procedure 11: Weighted COVERT Rule for T_{wt}

$$\pi_j = (w_j/p_j)[\ 1 - (S_j)^+/kp_j]^+$$

The weighted COVERT rule is a weighted improvement of the unweighted COVERT rule for total tardiness objectives first suggested by Carroll (1965) in a classic job shop scheduling environment. Furthermore, we have specialized the COVERT rule to the one-resource case, to facilitate comparison with other heuristics. Later, Vepsalainen and Morton (1987) tested weighted COVERT and demonstrated it worked well for weighted tardiness flow shop problems. (See Chapters 14 and 16.) Note that this rule also has the proper form of WSPT times a slack factor. In fact, the rule is very similar to Heuristic 2 of Section 6.5, with an additional parameter k; hence it should perform similar to the R&M heuristic for one-resource

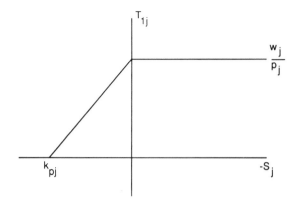

Figure 7.3. Weighted COVERT heuristic.

problems. We defer further discussion of weighted COVERT to more complicated problems where it is more differentiated from the R&M heuristic. (See Figure 7.3.)

7.2.3 Weighted Flow/Tardy Problem (FT_{wt})

To illustrate how our bottleneck dynamics methods can be useful for more complex regular objectives, consider the weighted flow/tardy objective discussed in Section 4.5.2. Remember that each job has processing time p_j as before. However, the job's flow time is penalized w_j per unit time, while tardiness is penalized in addition w_{T_j} per unit time. Thus the total objective is given by $FT_{wt} = F_{wt} + T_{wt}$. We propose a heuristic whose priorities are simply the sum of those for F_{wt} and T_{wt}. (This works because both heuristics have an economic base.)

Procedure 12: Bottleneck Dynamics Heuristic for FT_{wt}

$$\pi_j = (w_j/p_j) + (w_{T_j}/p_j) \exp[\ -(S_j)^+/kp_{av}]$$

This formula satisfies all the properties we proved about FT_{wt} in Section 4.5.2. Priorities for very slack jobs are given by WSPT1 or (w_j/p_j). Priorities for guaranteed tardy jobs are given by WSPT2 or (($w_j + w_{T_j})/p_j$) for WSPT2 (see Figure 7.4). This heuristic has not been studied thoroughly experimentally.

7.2A Numerical Exercises

7.2A.1 Create a few 20-job N_{wt} problems, which require that four to six jobs be tardy. (This can be checked on the software by running the problems through Hodgson's algorithm with N_{av} as the criterion.)

(a) Solve them by truncated branch-and-bound.

(b) Solve them by beam search.

(c) Solve them by Procedure 2 (WSPT–Hodgson's in the software).

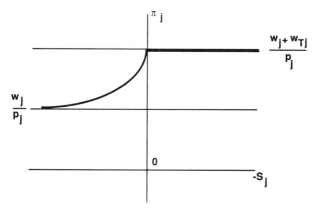

Figure 7.4. Bottleneck dynamics heuristic for F_{wt}.

7.2A.2 For the same job sets, solve by heuristic Procedures 3 and 4 and compare.

7.2A.3 Create (or get from the software) a 10-job T_{wt} problem, for which Hodgson's reveals several jobs must be tardy.
(a) Solve by beam search.
(b) Solve by R&M heuristic.
(c) Solve by Montagne's heuristic (hand).
(d) Solve by the critical ratio heuristic (hand).

7.2A.4 Create a 5-job FT_{wt} problem, which requires that one or two jobs be tardy. Solve it by WSPT1, WSPT2, and Procedure 12.

7.2B Software/Computer Exercises

7.2B.1 Do 7.2A.1 by computer.

7.2B.2 Extend 7.2B.1 to include the three heuristics by computer.

7.2B.3 Do the same for 7.2A.3.

7.2B.4 Do the same for 7.2A.4.

7.2C Thinkers

7.2C.1 Develop a mixed best first search branch-and-bound/beam search algorithm for T_{wt} or N_{wt}, which develops subproblems via a best trial solution but eliminates branches via lower bounds. Can your procedure say whether or not an optimal solution is known to have been obtained?

7.2C.2 What can you say about the relation between the optimal k in the R&M procedure and the optimal k for the COVERT procedure on any given set of problems?

(A) 7.3 NONREGULAR STATIC PROBLEMS

To date we have studied only regular static problems in any detail: in fact, primarily those that have additive increasing functions of the completion time of each job. But many practical situations are just-in-time (JIT) in nature: we don't want the job to be tardy, but neither should it be early. Tardy jobs cause customer dissatisfaction at the missed deadline, but early jobs will not be paid for until the due date anyway and accumulate the higher storage, obsolescence, and deterioration costs of finished goods inventory. (It may not be obvious why an order could become obsolete but, in practice, the longer a finished order sits, the higher the probability the customer will desire some modification.)

It is possible in this case that if all jobs are quite early and the shop is not too heavily loaded, then the resource should simply sit idle so that all jobs can get closer to their desired due date. This causes some difficulty. One way to deal with this is to schedule with no idleness allowed, which works reasonably well for a heavily loaded shop. All results in this section are assumed to have no inserted idleness except where expressly stated. A second approach is to estimate a priority function and to not schedule any job until at least one priority is positive. This approach is discussed briefly in this section also. The third approach is to estimate a release time for each job, after which it may be scheduled. This approach is dealt with in Chapter 20. A fourth approach would be to schedule with no inserted idleness, and then add a heuristic to add idleness after the fact.

The nonregular objective function that has received by far the most study to date is the weighted early/tardy objective: $ET_{wt} = \Sigma_j (w_{E_j}E_j + w_{T_j}T_j)$, where E_j and T_j are the earliness and tardiness of job j, and w_{E_j} and w_{T_j} are the earliness and tardiness weights for job j. For each job j, the marginal costs of completion at slack S_j are shown in Figure 7.5.

Heuristics for this problem were studied in some detail by Ow and Morton (1989). We first present some of their results, and then review other work in the literature.

Definition. WSPT2 sequencing is by $(w_{T_1}/p_1) \geq K (w_{T2}/p_2) \geq \cdots$

Definition. WLPT1 (longest process time) sequencing is by $(w_{E_1}/p_1) \leq (w_{E_2}/p_2) \leq \cdots$.

Proposition 1. If WSPT2 results in a schedule that does not have any early jobs, then this sequence is optimal among no idleness schedules.

Proof. Left as an exercise. ∎

Proposition 2. If WLPT1 results in a schedule that does not have any tardy jobs, then this sequence is optimal among no idleness schedules.

Proof. Left as an exercise. ∎

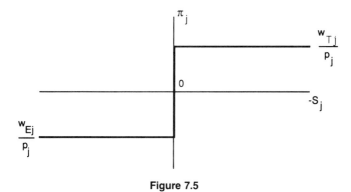

Figure 7.5

Ow and Morton (1989) use Propositions 1 and 2, together with another result we do not give here, to derive a heuristic that correctly predicts the effect of interchanging any two neighboring jobs. (We add the freedom of a parameter k to improve the performance.)

Procedure 13: Linear Heuristic for ET_{wt} Problem (LINET). Choose the job to be processed with highest value of

$$\pi_j = [w_{T_j} - \min((1, S_j^+/kp_{av})(w_{T_j} + w_{E_j}))]/p_j$$

The LINET priority function is shown in Figure 7.6. For very early jobs the priority is $-w_{E_j}/p_j$; for tardy jobs the priority is $+w_{T_j}/p_j$; for small tardinesses it rises linearly from the negative priority to the positive priority.

It is also easy to modify Procedure 13 for the inserted idleness case.

Procedure 14: LINET for Inserted Idleness. Utilize Procedure 13, but whenever all priorities are negative schedule no job.

(Analogous to the R&M procedure for the T_{wt} problem, we may also derive a double exponential procedure for the ET_{wt} problem, which the authors called EX-PET. Now congestion may affect the problem forward into the tardy portion, or back into the early portion. It is perhaps not worthwhile to develop the exact functional form of this priority function; it involves estimating several parameters. It is shown on the same diagram as LINET.)

Ow and Morton also investigated neighborhood search and beam search. The neighborhood search used the EXPET solution as the initial seed and adjacent pairwise interchange. The beam search method produced trial solutions using the EXPET heuristic. Two beam widths, 3 and 5, were investigated. A variation called filtered beam search was employed, in which only subproblems passing a crude preliminary filter were actually explored.

A computational study somewhat similar to that for R&M was performed. Pa-

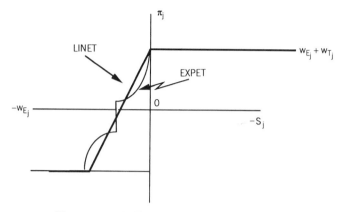

Figure 7.6. LINET and EXPET heuristics for ET_{wt}.

rameters varied included the tardiness factor (rough percentage of jobs tardy), due date range factor (variability of due date), correlation between processing times and due dates, earliness and/or tardiness cost per unit of processing time, and number of jobs. Twenty problems were generated for each combination of parameters, yielding 1440 test problems. A preliminary pilot study set parameter values. Performance was measured as the percentage excess cost of the heuristic from the optimal (or lower bound) cost (see Table 7.2). The optimal cost was found using branch-and-bound; lower bounds were found from the unit-preemptive relaxation by solving a linear assignment problem. Optimal solutions were found for all 8-job problems, but only some 15-job problems, and no 25-job problems. (Thus ET_{wt} seems to be much more difficult than T_{wt}.)

Table 7.3 shows the average computation time in seconds of the search methods

TABLE 7.2 Comparison of Heuristics for the Weighted Early/Tardy Problem[a]

Jobs	LINET	EXPET	NBHD	BEAM-3	BEAM-5
8	29.65	18.94	5.84	0.37	0.00
15 (l.b.)	29.88	25.95	18.29	10.46	8.62
25 (l.b.)	37.11	25.73	15.26	10.24	7.42

[a] Values given are average percent error from optimum or lower bound.

TABLE 7.3 Comparison of Computation Times per Problem

Jobs	NBHD	BEAM-3	BEAM-5	OPT/LB
8	0.01	0.15	0.29	147.49 (opt)
15	0.02	0.67	1.71	80.89 (l.b.)
25	0.05	2.76	7.36	266.57 (l.b.)

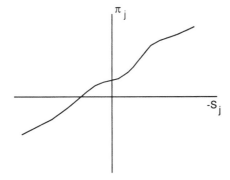

Figure 7.7. Marginal cost for convex objective.

when executed on a VAX 7880. From the data, it seems that using neighborhood search to improve a dispatch heuristic is worthwhile, although beam search at the two settings is a quite inexpensive way of obtaining consistently outstanding results.

The results here are disappointing in one way: this is the first time in our investigations to date that carefully crafted dispatch procedures have not produced outstanding results. Two possible solutions to this problem seem to be: (a) continue research into more refined dispatch procedures and (b) accept a hybrid mix of dispatch and beam search, which will in practice cause some degradation in the ability to solve reactive-scheduling corrections in real time.

We turn briefly to a look at more general additive objectives that are convex in each job's completion time. This means that the marginal cost as a function of slack is increasing over time as shown in Figure 7.7.

Note that, in the general nonregular case, we have a region to the left that has negative marginal costs in increasing completion time, and a region to the right that has positive marginal costs. This can be the basis for constructing a heuristic, as before. LINET may be considered the priority function arising by averaging the marginal cost curve between t and $t - kp_{av}$ for each value of t to produce the priority function.

(A) 7.3A Numerical Exercises

7.3A.1 Consider the following ET_{wt} problem:

Job j	1	2	3	4	5
p_j	4	3	7	2	2
d_j	5	6	8	8	17
w_{E_j}	1	1	1	1	1
w_{T_j}	4	4	4	4	4

Solve by LINET and EXPET. See how much adjacent pairwise interchange improves your result. (You may wish to develop computational aids.)

7.3A.2 Solve 7.3A.1 by exact and/or approximate DP. You may use any computational aids you desire.

7.3A.3 Change the objective function for each job to be $w_j(S_j)^2$. (Cost is proportional to the square of slack.) Try the analogue of LINET on the problem of 7.3A.1.

7.3B Software/Computer Exercises

7.3B.1 Write a branch-and-bound code (or modify an existing one) to solve your 7.3A.1 problem exactly. What would you use for lower bounds?

7.3B.2 Write a larger problem set for 7.3B.1 and solve it both by interesting heuristics and by branch-and-bound.

7.3B.3 Write a beam search code (beam width = 2) for the same problems.

7.3B.4 Write some sort of extended neighborhood search routine for these same problems.

7.3C Thinkers

7.3C.1 Why are the methods of this chapter less useful for nonconvex objective functions?

7.3C.2 What problems does the greater inaccuracy of heuristics for the nonconvex systems cause for building practical heuristic scheduling systems?

7.3C.3 Prove Propositions 1 and 2.

7.4 DYNAMIC PROBLEMS: EXACT CASES

7.4.1 Overview

The static version of a single-resource problem refers to the artificial situation in which all jobs are simultaneously available for processing. However, most single-resource problems seen in the context of being embedded in a larger multi-resource shop have activities arriving at different times to the local resource. (In many ways the static problem may be considered a simplification studied for the purpose of understanding solution methodologies and heuristic principles.)

In the case of differing arrival times, the set of tasks to be scheduled changes over time, giving rise to a *dynamic* single-resource problem. An immediate consequence of dynamic arrivals is that *inserted idle time* may now be necessary, and *Cost-at-End* (*CAE*) *preemption* may be better than no preemption where allowed. (And, of course, *Spread-Cost (SC) preemption* may be better than nonpreemption as in the static case.)

Remember from Section 4.1 that dynamic problems are most easily studied over a period of time for which the resource ends with an empty queue. In the following we assume this to be the case.

To illustrate the role of inserted idle time, CAE preemption, and SC preemption for dynamic problems, consider the two-job example shown below in which T_{wt} is the criterion, with weights $= 1.0$.

Job j	r_j	p_j	d_j
1	0	5	7
2	1	2	2

There is only one sequence that avoids all inserted idle time and preemption: the sequence 1–2, with total tardiness of 5 (Figure 7.8). When inserted idle time is permitted, the sequence 2–1 (Figure 7.9) gives rise to a total tardiness of 2. Furthermore, if a job can be preempted, either CAE or SC, then a total tardiness of 1 can be achieved (Figure 7.10). (In general, SC preemption will be less costly than CAE, since part of the job finishes earlier and receives a partial reward or a reduced penalty.)

The dynamic case is much harder than the static case, and fewer exact results are available. Exact dynamic results are available for makespan, as well as several preemptive cases.

7.4.2 Some Exact/Lower Bound Nonpreemptive Dynamic Results

Define the queue as items waiting to be processed at a resource and "system queue" as the queue plus the item being processed. It is useful to define the minimum possible backlog of processing (minimum system queue processing time content) as a function of time in a dynamic problem.

Definition. Let $SQ(t)$ be the *minimum possible remaining processing time in queue* on an interval ending with empty system queue.

We are especially interested in the first time past the start t_s for which SQ is zero and the first subsequent arrival time.

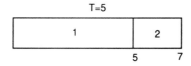

Figure 7.8. Solution: no inserted idleness or preemption.

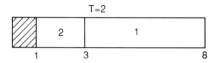

Figure 7.9. Solution: inserted idleness allowed.

Figure 7.10. Solution: both idleness and preemption allowed.

Definition. For a given schedule:

(a) Let t_b, the *minimum first busy period end*, be the minimum t for which $SQ(t) = 0$.

(b) Let t_c, the *minimum first busy cycle*, be the time of the first subsequent arrival of a job after t_b, or the end t_e if there is no subsequent arrival.

That is, a busy period is a period during which the resource is being used; a busy cycle is a busy period plus the following empty period until the next time it is being used.

To illustrate, suppose that in the above example, we add a third job arriving at time 9 with processing time 5. Then the minimum first busy period is from 0 to 7, while the minimum first busy cycle is 0 to 9. The minimum possible second busy period is from 9 to 14; the minimum second busy cycle is from 9 to 14.

Definition. For any objective function, the *static relaxation of a busy period problem* is the relaxation obtained by setting all job arrival times to the start t_s and solving by methods of earlier chapters.

Proposition 3. For regular objective functions not involving the arrival times, the optimal solution to the static relaxation of a single busy period provides a lower bound to the optimal solution to the original problem. In particular, if it should happen to be feasible, then it is optimal.

Proof. Left as an exercise. ∎

(Note that Proposition 3 does not hold for F_{wt}, for example, since moving the arrival times back increases the flow. The second part of the proposition remains true.)

Proposition 4. Dynamic makespan can be optimized by simply processing any available job at all times; the makespan achieved is simply the sum of the lengths of the minimum possible busy cycles (last empty period is not included).

Proof. Use Proposition 3; left as an exercise. ■

Above, the first busy cycle has length 9, the second 5; so the makespan is 14. In general, it is worthwhile to break the problem into minimum busy cycles, in case solutions close to optimal for the static version should be feasible. In particular, the first busy cycle is of special interest, since in a rolling horizon procedure we are most concerned about the correctness of the first few decisions.

7.4.3 Exact Dynamic Preemptive Results

In this section we develop some elementary results for CAE preemption and SC preemption. (The two versions of SC, depending on whether due dates are also spread, give the same policy. Hence we tend to use whichever assumption is most convenient in a proof.) Remember that preemptive solutions are not only useful in actual situations where preemption is allowed. In addition, they provide good lower bounds in branch-and-bound and similar procedures for the nonpreemptive case. Finally, they provide insights for building heuristics for nonpreemptive situations, as we shall see.

Baker (1974) talks about preempt repeat mode. This means that if a job is interrupted, the partial processing is lost, and the job must start over. We defer discussion of this to a later section for two reasons. First, this is simply one version of a number of difficult and realistic assumptions that are of the form: when a job is preempted, a setup time or cost must be paid and part of the work done over when the job is started again. Thus such assumptions should be considered together. Second, these problems are almost always difficult and do not yield to exact methods such as those being considered here.

Weighted Flowtime and Weighted Lateness. For the preemptive cases we consider, and for objective functions that are globally stable and transitive, the types of proofs we used for the static case still work, although they are somewhat more complicated. Consider first the problems of minimizing F_{wt} or L_{wt}.

Definition. The PDWSPT (preemptive dynamic weighted shortest processing time) priority rule is given as follows:

(a) When a job finishes, start the currently available job with highest w_j/p_j (in SC discipline).

(b) When a job i arrives with job j in process, start job i provided $w_i/p_i > w_j/p_j$; otherwise continue processing job j (in SC discipline).

(c) For CAE discipline, replace p_j in (a) and (b) by the remaining process time p_{rj}.

Proposition 5. The PDWSPT rule is optimal for the dynamic weighted flow problem F_{wt} and the dynamic weighted lateness problem L_{wt}.

Proof. The full proof is left as an exercise, but we sketch the subpropositions to be proved here. We limit ourselves to the unit preemptive case. ∎

Subproposition 5.1. Consider two jobs i and j both available at the beginning. These jobs may be preempted by later jobs, but there exists an optimal solution in which one is completed before the other starts.

Subproposition 5.2. Of those two jobs, i will be scheduled first if $w_i/p_i > w_j/p_j$.

Subproposition 5.3. There is an optimal schedule that is WSPT until a new job arrives.

Subproposition 5.4. When a new job arrives, the priority of the partly processed job is w_j/p_{rj} for CAE and is w_j/p_j for SC.

Subproposition 5.5. Now reapply subproposition 5.3 starting at the arrival of that job.

Thus we see here that the optimal rule is to always keep the machine assigned to the available job with maximum weight per unit of *remaining* processing time for CAE and per unit of *total* processing time for SC. It is not always easy to decide in practice whether the discipline should be CAE or SC. If the longer a job has been processing, the more we resist taking it off, then we are probably in a CAE situation, such as producing a single large order. If the longer a job has been processing makes no difference to us, then we are probably in an SC situation, with the job perhaps being a number of individual units all going to final inventory.

It is also quite easy to find optimal preemptive rules for the dynamic programming of minimizing maximum lateness L_{max}, or tardiness T_{max}. Remember that the EDD ordering was optimal in the static case.

Definition. The PDEDD (preemptive dynamic earliest due date) discipline has two parts:

(a) Always schedule the job (or joblet) with the earliest due date.
(b) If a new job arrives with an earlier due date than the one being processed, it preempts the current job.

Proposition 6. PDEDD provides an optimal policy for either L_{max} or T_{max} objectives for either CAE or SC preemption.

Proof. Left as an exercise. ∎

Note that in the dynamic versions of both EDD and WSPT, the implementation of an optimal scheduling rule requires no look-ahead features (i.e., it remains dispatch) even though jobs become ready intermittently. No forecast is required of the nature of jobs that have not yet arrived. There are many other advantages of dispatch methods that we have described elsewhere. (Given a set of prices and local arrival times, bottleneck dynamics depends on dispatch characteristics to get rapid and easily updated solutions.)

Weighted Number of Tardy Jobs. We turn now to obtaining some exact, but partial, results for some more difficult dynamic preemptive problems. We turn first to the weighted number of tardy jobs, N_{wt}. These types of problems appear to be of exponential difficulty even in the preemptive case.

Proposition 7: Algorithmic Structure for Preemptive N_{wt}

Step 1: Order joblets in PDEDD order.

Step 2: Find the first tardy joblet k, or else terminate.

Step 3: (a) For SC, remove the "best" joblet between start and k, including k; (b) for CAE, remove the "best" total job.

Step 4: Order remaining joblets in DEDD order.

Step 5: Return to step 2.

Proof: Left as an exercise. ∎

Note that under CAE there is no point in removing part of a job. There is no credit unless the whole job is done. Again note for CAE that the job removed may be in several pieces and may extend past job k. Note that step 4 requires little recomputation. "Best" is not specified and may be difficult to determine. The algorithm simply says: "if the right jobs are removed, an optimal solution will result." "Best" is quite a complicated question, even for heuristics. Note, for example, that even if all weights are 1.0, the longest job may exist mostly after k; removing it may not help k much. At the same time, removing the job with the most content up to k may not be optimal either. Note also that removing even a joblet may leave internal idle time. However, it seems quite possible to use Proposition 7 to fashion branch-and-bound or beam search routines. We leave this as an exercise. Turning to the dynamic weighted tardiness problem T_{wt}, we can prove exact partial results analogous to Propositions 18 to 24 in Section 4.5. We state the analogous propositions here briefly; proofs are left to the exercises.

Proposition 8. If DEDD produces preemptive $T_{wt} = 0$, then this is optimal.

Proposition 9. If DEDD produces one minimum-weight tardy job for CAE or one minimum-weight tardy joblet for SC for preemptive T_{wt}, then this is optimal.

Proposition 10. For dynamic preemptive SC T_{wt}, if all jobs have the same due date, then T_{wt} is minimized by PDWSPT.

Proposition 11. For dynamic preemptive SC, where PDWSPT makes all jobs tardy, then T_{wt} is minimized by PDWSPT.

Proposition 12. For dynamic preemptive SC or CAE, if no possible sequence can make any job (joblet for SC) nontardy, then T_{wt} is minimized by PDWSPT.

Proposition 13. For dynamic preemptive SC or CAE T_{wt}, if a job is not tardy when scheduled other than first, it need not be scheduled first.

Proposition 14. For dynamic preemptive SC or CAE T_{wt}, if the job with the overall highest priority by PDWSPT is tardy even if done first, then it should be scheduled first and not be interrupted by arriving jobs.

Similarly, we can generalize Propositions 25 to 28 in Section 4.5 for the FT_{wt} (flow/tardiness) problem.

Proposition 15. For dynamic preemptive SC or CAE FT_{wt}, if PDWSPT1 produces no tardiness, then PDWSPT1 is optimal; on the other hand, if PDWSPT2 produces no earliness, then PDWSPT2 is optimal.

Proposition 16. For dynamic preemptive SC or CAE FT_{wt}, if no possible sequence can make any job (joblet for SC) tardy, then PDWSPT1 is optimal.

Proposition 17. For dynamic preemptive SC or CAE FT_{wt}, if no possible sequence can make any job (joblet for SC) nontardy, then PDWSPT2 is optimal.

Proposition 18. For dynamic preemptive SC or CAE FT_{wt}, if the job with the highest priority by PDWSPT2 is tardy even if done entirely first, then it should be done first without interruption.

Note that Proposition 15 does not really depend on the early weights being positive. Thus we also have directly the generalization of Propositions 1 and 2 for Section 7.3 for the early/tardy problem.

Proposition 19. For dynamic preemptive SC or CAE ET_{wt}, if PDWSPT1 (early weights) produces no tardiness, then PDWSPT1 is optimal; on the other hand, if PDWSPT2 produces no earliness, then PDWSPT2 is optimal.

7.4A Numerical Exercises

7.4A.1 Consider the following dynamic problem with makespan objective:

Job j:	1	2	3	4	5	6	7	8
p_j	3	7	4	5	4	7	8	3
r_j	17	6	0	37	26	20	33	0

(a) Make a busy periods/idle periods analysis to predict total idleness and thus the minimum makespan.

(b) Use the software to run a number of random dispatch solutions, and compare the answers you get.

(c) Use the software to run a number of other dispatch heuristics, and compare the answers you get.

7.4A.2 Make up two or three dynamic problems with five or six jobs each and processing times of small integers (1 to 4). Let the problems have varying degrees of tardiness ($1 - d_{av}/np_{av}$). (You may also select these problems from those provided by the software.) Consider the T_{max} criterion. Find the best nonpreemptive solution you can by inspection, arguing for optimality if obvious. Also solve these problems preemptively by DWSPT (both for CAE and for SC).

7.4A.3 Repeat 7.4A.2 for the F_{wt} criterion.

7.4A.4 For Proposition 8 on preemptive N_{wt}, come up with a heuristic definition of "best". Repeat 7.4A.2 using your heuristic.

7.4B Software/Computer Exercises

7.4B.1 Construct a program that is an aid to finding solutions for dynamic non-preemptive F_{wt} or T_{max}. The program should first find the preemptive solution and then allow the user to recombine job pieces one by one to eliminate pieces. For each such change the computer should refuse illegal changes and report the associated increase in the objective function.

7.4B.2 Write a program to investigate various heuristics for preemptive N_{wt}. It should have a number of subroutines involving a heuristic for "best" to remove. It should be easy to add new subroutines.

7.4B.3 Using your program for 7.4B.2, do experimental research into what constitutes a good heuristic.

7.4C Thinkers

7.4C.1 Prove Propositions 3 and 4.

7.4C.2 Prove Proposition 5, giving a short proof of each of the five subpropositions.

7.4C.3 Prove Proposition 6.

7.4C.4 Prove Proposition 7.

7.4C.5 Can you derive any properties of the "best" job to remove in Proposition 7, perhaps for some special case(s)?

7.4C.6 Discuss using Proposition 7 to fashion branch-and-bound and/or beam search routines.

7.4C.7 Briefly discuss proofs of Propositions 8 to 11.

7.4C.8 Briefly discuss proofs of Propositions 12 to 14.

7.4C.9 Briefly discuss proofs of Propositions 15 to 19.

7.5 DYNAMIC PROBLEM: NO INSERTED IDLENESS

7.5.1 Overview

The dynamic one-resource sequencing problem with no inserted idleness is basically the same as the corresponding static sequencing problem in several important intuitive ways. It is more difficult in some other ways, and easier in still others.

(a) It is the same because choices for the next job to be sequenced are always between jobs that have already arrived. Hence the static results for pairwise interchange of two such jobs work here exactly, whether EDD, WSPT, or R&M. Thus *myopic priorities* (best current results, considering only the two jobs) are unchanged.

(b) It is more difficult as soon as we look beyond the two jobs being interchanged to jobs soon to arrive. It requires more computation to check feasibility for an interchange or partial solution and/or to calculate the effect on the objective function.

(c) For the reasons in (b), it is more difficult because exact results are not available (except for makespan) for dynamic nonpreemptive models and are only available for a smaller subset of dynamic preemptive models.

(d) It is much easier in that search methods usually have only a few choices at each stage. (The choices are among those jobs that have arrived up to now but haven't yet been processed.)

(i) The neighborhood is much smaller in neighborhood search.

(ii) There are fewer choices to schedule next in branch-and-bound and/or beam search.

(iii) There are fewer priorities to calculate in priority dispatch.

(iv) Initial subsets are limited by job arrival in DP.

"Easier" dominates for heuristics based on interchange such as neighborhood search, dispatch heuristics, and bottleneck dynamics. For exact and very high-quality approximate methods, "easier" tends to dominate if arrivals are spread out to keep the choice queue small, "Harder" tends to dominate if the queue can build up to be large fairly often.

Many shops have the discipline that resources are never held empty for a length of time to facilitate a soon-to-arrive "hot job." This leads to the "no inserted idleness" case of this section. A different interest in this case is that good, simple priority dispatch heuristics are available, as we have suggested above. We discuss this case in this section. Note that feasible solutions are the feasible subsets of all permutations of the jobs. The valuation of a given solution will be exactly the same as for the static case.

Many other shops have the discipline that a long low-priority job should not be started if it will block a soon-to-arrive "hot job." We discuss this inserted idleness case in Section 7.6.

7.5.2   No Inserted Idleness: Classic Methods

Neighborhood Search. It is important to use the extra constraints produced by dynamic arrivals and no inserted idleness to limit the choices in the neighborhood that are feasible before going through expensive value-of-interchange computations. For example, suppose at time t we have scheduled job i followed directly by job j. If j does not arrive by time t and there are other jobs that could be scheduled, we need not consider j followed by job i since this would lead to inserted idleness.

The neighborhood search method will still converge to a local optimum in the search space, but the more complicated constraints often will have many more local optima, so that the quality of a single attempt may well be lower. This indicates that methods such as tabu search, simulated annealing, and beam search, which are not stopped as easily by local optima, may be of greater interest.

Branch-and-Bound. Branch-and-bound is best worked forward because then, at a point in time, there will be a branch only for every job sitting in the queue as possible next choice. Because branches are fewer, it is basically feasible to look at much larger problems. However, it is important to have a procedure to get an efficient and useful lower bound. Note that preemptive solutions to the problem provide a lower bound. We have analytic preemptive solutions available for F_{wt}, L_{wt}, F_{max}, and T_{max}. One should therefore expect to be able to solve reasonably large-sized dynamic versions of these problems.

As a matter of fact, Baker and Su (1974), using slower computing equipment, perused this reasoning for the T_{max} problem and tested 90 problems with 10, 20, or 30 jobs. Best first search was used, truncated after the creation of 1000 nodes. More than 93% of the problems were solved to optimality. Moreover, the total number of nodes created for these solved problems was quite small on the average, as shown in Table 7.4.

Exploiting special structure in a problem of this sort can lead to even more efficient solutions. Carlier (1982) developed a special purpose procedure for the dynamic T_{max} problem, which is able to solve up to, perhaps, 100 jobs and hence represents an extremely fast and efficient method for this problem. Adams et al. (1988) developed a method for solving the makespan problem for large job shops by repeatedly solving dynamic T_{max} subproblems as part of the algorithm and thus required such a specialized algorithm. (See a fuller discussion in Section 10.4.)

TABLE 7.4   Nodes Investigated to Find the Optimum Solution

	Number of Jobs, n		
Problems	10	20	30
High tardiness problems	14.8	40.4	84.9
Medium tardiness problems	11.1	35.4	91.9
Low tardiness problems	8.9	39.5	72.8

T_{max} problem (Baker and Su, 1974).

On the other hand, harder problems such as dynamic weighted tardiness T_{wt} do not possess such an easily calculated preemptive lower bound. While it is always possible to calculate the preemptive lower bound by solving the associated linear assignment problem, as was done by Ow and Morton (1989), this problem is usually very large and not easy computationally.

(A) Dynamic Programming. Forward dynamic programming can be simplified as follows. Since job j is not available for scheduling until r_j, a minimum of a certain number of jobs will have to be processed before it. If the first position it is allowed is, say, j^*, then subsets with fewer than j^* elements cannot contain j. This will restrict the number of subsets to be considered. Similarly in doing the recursive calculations, certain elements cannot be placed last.

7.5.3   No Inserted Idleness: Dispatch/Bottleneck Dynamics

The basic dispatch priorities, including the newer ones given by bottleneck dynamics, are very simple:

Dynamic Dispatch Procedure

Step 1: The choice-set at time t is unprocessed jobs available at t.
Step 2: Priorities of choice-set jobs are calculated as in the static case.
Step 3: The job with the highest priority in the choice-set is processed.

In particular, this means we still use WSPT for L_{wt} and F_{wt}, EDD for T_{max} and L_{max}, and R&M (bottleneck dynamics) for T_{wt}. The dispatch procedure is a *myopic* procedure that makes the best current choice, without worrying about the future.

As mentioned before, dynamic one-resource problems provide useful inputs in solving large job shop problems. Several studies have shown that such myopic procedures work well in larger problems (Lawrence and Morton, 1989; Morton and Pentico, 1993). The robustness of myopic dynamic procedures is very important; it will be investigated in the exercises.

It is possible to see where myopic procedures can go wrong. Suppose at $t = 0$

that j has priority 8 and process time 10. So we start it, ignoring another job i with priority 6 and process time 5. At $t = 4$ a hot job arrives with priority 15 and process time 6. It would probably have been better to do first the priority 6 job, which will not delay the hot job nearly as much when it arrives. However, corrections get quite complicated and can dilute the advantages of dispatch procedures; hence such corrections are problematical.

Complex Heuristics Based Partly on Dispatch. Remember that to minimize the number of tardy jobs, we used *Hodgson's rule:*

Step 1: Order the list in EDD order to minimize maximum tardiness.
Step 2: Find the first tardy job.
Step 3: Remove the job up to that point with the longest process time.
Step 4: Repeat steps 2 and 3 until there are no more tardy jobs.

In the dynamic case, the extension, while not exact, is quite obvious.

Dynamic Hodgson Heuristic

Step 1: Order the list by dynamic dispatch EDD, in order to approximately minimize maximum tardiness.
Step 2: Find the first tardy job.
Step 3: For each job j up to that point, find the reduction in partial makespan if removed; term this p_{jk} if k is the tardy job.
Step 4: Remove the job $1 \leq j \leq k$ with largest p_{jk}.
Step 5: Repeat steps 2, 3, and 4 as necessary.

The intuitive reason for this heuristic is that in Hodgson's rule removal of the longest job helped all jobs from k on the most. Here, the greatest reduction in partial makespan will do just that.

The extension to weighted number of tardy jobs is also obvious:

Weighted Dynamic Hodgson Heuristic

Step 1: Order the list by dynamic dispatch EDD, in order to approximately minimize maximum tardiness.
Step 2: Find the first tardy job.
Step 3: For each job j up to that point, find the reduction in partial makespan if removed; term this p_{jk} if k is the tardy job.
Step 4: Remove job with smallest w_j/p_{jk}; repeat until k is removed or not tardy.
Step 5: Repeat steps 2, 3, and 4 as necessary.

Rolling Horizon N_{wt}. The dynamic Hodgson's heuristic, as just given, is somewhat unusual in that we cannot simply calculate the priority of each available job at time t and choose the highest priority. Rather, we must forecast the arrival times and due dates of future jobs into the near future, up to the point that a forecast tardy job will arrive. Then we apply the dynamic Hodgson's heuristic to find the job actually to be scheduled currently. The job the heuristic schedules first is then scheduled. When it is time to actually schedule the next job (rolling horizon), the whole procedure must be repeated, to allow forecasts to be updated.

This requires that we forecast arrival dates for a number of future jobs, which may or may not be a reasonable task. A heuristic requiring less forecast information could be devised in a number of ways. One such rule might be the following:

Simplified Rolling Horizon N_{wt} Rule

Step 1: Order currently available jobs in EDD order.

Step 2: If the earliest due date job i has negative slack, delete it permanently; return to step 1.

Step 3: If no critical-slack high-value job is known in the "near" future, schedule job i; leave the routine.

Step 4: Otherwise temporarily veto jobs one by one in order of increasing w_j/p_j until the critical job is no longer critical.

Step 5: If the earliest due date job is not vetoed, schedule it; leave the routine.

Step 6: If it is vetoed, remove it from the available set (permanently) and return to step 1 (removing vetoes, etc.).

7.5.4 No-Inserted-Idleness: Other Procedures

Beam search using dispatch heuristics to obtain trial solutions is likely to be especially effective for the dynamic no-inserted-idleness case. It enjoys a minimum number of branches, as does branch-and-bound, but also has available easy, accurate trial solutions obtained by dispatch heuristics. We are aware of no major studies in this area. Tabu search, simulated annealing, and genetic algorithms should also be very helpful to neighborhood search, since local optima will be more common. These methods are specially adapted to this situation.

7.5A Numerical Exercises

7.5A.1 Generate 50 dynamic 15-job problems statistically in the software package, with F_{wt} criterion.

(a) Study one of them in detail in the software's manual interchange routine, first using the appropriate dispatch heuristic, and then trying to improve the solution.

(b) Solve the same problem using beam search with width 20.

(c) Run the 50 problems through the heuristic routine for various heuristics including WSPT.

(**d**) Discuss the comparison with the benchmark WSPT followed by general pairwise interchange.

7.5A.2 Repeat 7.5A.1 with T_{wt} criterion.

7.5A.3 Repeat 7.5A.1 with T_{max} criterion.

7.5B Software/Computer Exercises

7.5B.1 Write a branch-and-bound code with preemptive lower bounds to solve the problems in 7.5A.1.

7.5B.2 Write a beam search routine using dispatch generated trial solutions for the same problems.

7.5C Thinkers

7.5C.1 Develop a heuristic for N_{wt} (all weights $= 1$) for the dynamic nonpreemptive no-inserted-idleness case. Can you argue that it should be optimal, or show by counterexample that it is not?

7.5C.2 Construct an example to demonstrate that EDD sequencing will not guarantee minimum T_{max} in the dynamic nonpreemptive no-inserted-idleness case.

7.5C.3 Consider the following 3-job example of a dynamic single-machine with preempt-repeat processing:

Job j	1	2	3
r_j	0	2	3
p_j	4	1	2

For any regular measure of performance it is sufficient in seeking an optimum to consider only the six permutation schedules. Show that regardless of the regular measure involved, there is in fact a dominant set in this problem containing fewer than six schedules.

7.6 DYNAMIC PROBLEM: INSERTED IDLENESS

7.6.1 Overview

Although, as we have stated, many shops have a discipline that resources will never be held idle while jobs are waiting to be processed, other shops choose to specify that a long, low-priority job should not necessarily be started soon before the arrival of a "hot job." The central issues are: (a) How much is the difference in priority between the two jobs? and (b) How long will the resource be idle? For regular objectives, it is easy to see that one need never wait longer than the shortest job currently waiting, since that job could then be processed entirely within the idle time. Thus the feasible solutions again are a subset of all permutations of the jobs, with feasibility not determined by already having arrived by time t, but by projected arrival before $t + p_{min}$. (Of course, if the queue is empty, there is no choice but to wait for the next arrival.) Now the evaluation of a solution will not be the same as for the static case. Also, the effect of an interchange is more complicated, since one or both of the jobs being considered for interchanging may be soon-to-arrive instead of already-arrived. We look at this issue next.

7.6.2 Pairwise Interchange: Inserted Idleness

Seeing how pairwise interchange is modified for the inserted-idleness case can be very useful in understanding how to modify priority calculations for other extensions of dispatch heuristics and bottleneck dynamics heuristics.

To simplify our calculations, let us assume that at time $t = 0$ we are trying to calculate the effect of interchanging two jobs i and j. Both jobs are assumed to arrive very shortly, at times r_i and r_j, respectively. (Again to save notation, if one has already arrived just consider r_i to be 0. In general, we would need to replace r_i by $(r_i - t)^+$ in the formulas.) We assume that if either is scheduled first at arrival, the other arrives before it is finished. Now if i is scheduled first, we have

$$C_i = r_i + p_i; \quad C_j = r_i + p_i + p_j$$

whereas if j is scheduled first, we have

$$C_i = r_j + p_j + p_i; \quad C_j = r_j + p_j$$

What is the effect of scheduling j first rather than i? Three things:

(a) i is later by $p_j + (r_j - r_i)$.
(b) j is later by $-p_i + (r_j - r_i)$.
(c) All subsequent jobs are later by $(r_j - r_i)$.

Now, by the discussion of resource pricing in Section 6.5.2, the cost per unit time of processing j later is $w_j U_j - IRp_j$ and similarly for i. Also, the price of losing the resource for all subsequent jobs per unit time is just R.

Summing (a), (b), and (c), dividing by $IRp_i p_j$, and noting that our standard job priority is given by $\pi_j = (w_j U_j / IRp_j) - 1.0$, we obtain

$$\text{Net Cost} = [-\pi_j + \pi_i] + [r_j - r_i] [\, 1/Ip_i p_j + \pi_j/p_i + \pi_i/p_j \,]$$

From this, we can separate out a revised priority function for any job j:

$$\pi_{j\text{ idle}} = \pi_j - r_j[\ 1/Ip_i p_j + \pi_j/p_i + \pi_i/p_j\ ]$$

Unfortunately, this priority depends on the other job i in the switch. If we assume $\pi_j \gg \pi_i$ and approximate the other processing time by p_{av}, we have a somewhat simpler approximation depending only on j (we correct the formula for arbitrary starting t).

Bottleneck Dynamics Inserted Idleness Heuristic

$$\pi_{j\text{ idle}} = \pi_j(1 - (r_j - t)^+/p_{av}) - ((r_j - t)^+/p_j)(1/Ip_{av})$$

This formula, although relatively untested, should work reasonably well where simplicity and reasonable performance are desired. Note that the traditional priority is decreased by a fraction depending on the lost processing r_j as a fraction of an average job, and further decreased by a second term also depending on lost processing time. The formula does not depend on the resource price R, but does depend on I, the rate of conversion between "buying resource time" and "renting resource time."

An even simpler approximation, which we term "x-dispatch" seems to work quite well and to be very intuitive, is

X-Dispatch Bottleneck Dynamics Heuristics (XBD)

$$\pi_{j\text{ idle}} = \pi_j(1 - B(r_j - t)^+/p_{av})$$

Here B is a constant, which seems to fit well experimentally as

$$B = 1.3 + \rho \quad (\rho \text{ is the average utilization of the machine})$$

7.6.3 Inserted Idleness: Bottleneck Dynamics

It is instructive to briefly look at the bottleneck dynamics inserted-idleness heuristic we have just developed.

Consider makespan. Here there is no time pressure on the jobs; that is, $\pi_j = 0$. Thus the idleness heuristic says that the priority is negative for any job that hasn't arrived yet, and 0 for any available job. This then yields the optimal policy: always start processing any available job.

Consider an additive objective such as F_{wt}. Here $\pi_j = w_j/p_j$. Thus the formula says that the basic priority is degraded by the fraction of the processing time waited for the job. The same fraction times the resource price (spread out over an average job) for the lost time is also subtracted from the priority.

Consider a maxi objective such as L_{max}. Here we have a problem, because this is not an economic objective, and it is being mixed with the cost of the machine, which is. If we recognize, however, as we have explained, that this objective is usually concerned with coordinating a group of items downstream, the latest one of which determines the lateness of the shipment, we see that it is reasonable to modify the objective to minimize WL_{max}, where W is the importance of the overall shipment.

We know the solution to the static problem is to prioritize by due date, so by analog with our work with the R&M heuristic for weighted tardiness, it is reasonable to try as a priority for job j:

$$\pi_j = W \exp\{ -(d_j - d_{min})/kp_{av}\}$$

That is, the priority of the earliest due date job is the full W, but jobs with later due dates have exponentially smaller priorities depending on how much later they are. This priority can now be inserted in the idleness heuristic and tested. (This has not been done to date.)

7.6.4 Inserted Idleness: Modification of Other Methods

Now, instead of checking in neighborhood search whether the interchanged job arrives on time to be available, we ask whether it is within one minimum processing time (of other schedulable jobs) of being available. (If not, there is no use wasting that much time; schedule some other short job first.) This typically does not expand the neighborhood very much. The remarks on branch-and-bound are essentially unchanged, as are those for dynamic programming.

Bottleneck dynamics methods are also very similar. We now use the modified priorities discussed in the last subsection, and consider jobs that will arrive within one minimum processing time of now. Similar remarks apply to beam search.

7.6.5 Computational Study: Inserted Idleness, Weighted Tardiness

As part of a much larger study of new heuristics for the job shop Morton and Ramnath (1992) investigated a wide range of heuristics for the 100-job dynamic weighted tardiness problem. The Phase I starter heuristics included RANDOM, FCFS, LEAST SLACK, EDD, WSPT, R&M, and X-DISPATCH R&M. The latter, discussed in Section 7.6.2, is the only nondispatch method and requires little extra computation time.

Jobs have a normal distribution of processing time with mean of 1.0 and standard deviation of 0.5. Jobs arrive with a Poisson distribution with parameterized utilizations of 0.2, 0.5, and 0.8, with $n = 100$. Jobs have due dates set at arrival time plus $(k + error)$ times their processing times. k is set to approximate parameterized tardiness factors of 0.4, 0.6, and 0.8. The error is parameterized at a standard deviation of 1 or 2. Any given problem is first run through one of the basic Phase I heuristics to establish (or determine) the seed. Then simple neighborhood search is performed by k-move parameterized at $k = 1$ (adjacent pairwise) or $k = 5$. Finally, the neighborhood search result is refined by tabu search for 100 moves, then 100 more, then 100 more. The tabu list was parametrically kept at 50, 100, or an infinite number of moves.

Using the best of the seven heuristics as the benchmark, and RANDOM −
BENCHMARK as the normalization factor in the denominator, without separating
the two neighborhood sizes, the results were:

Heuristic	Phase I	NBHD	TS-100	TS-200	TS-300
RANDOM	79.	12.	11.	11.	11.
FCFS	74.	8.0	7.8	7.8	7.7
SLACK	71.	7.6	7.4	7.3	7.3
EDD	69.	7.7	7.6	7.6	7.6
WSPT	8.7	1.6	1.5	1.5	1.4
R&M	4.8	1.5	1.4	1.3	1.3
XR&M	3.5	0.80	0.77	0.77	0.76
BEST OF 7	2.9	0.07	0.01	0.006	0.000

Several facts are readily apparent.

(a) The quality of the Phase I heuristic strongly affects the quality of the final
solution after neighborhood/tabu search.
(b) X-dispatch R&M is clearly the best Phase I heuristic.
(c) k-move neighborhood search is very effective, even for $k = 1$ or $k = 5$.
($k = 5$ is about 10% the cost of general pairwise interchange.)
(d) Tabu search, for this problem set, saves only 5–10% in addition.

Although tabu search turned out not to be very effective for this problem set, it
was evident in looking at individual results that the size of the k-move factor was
important. A second study used FCFS as the "bad benchmark," with R&M and
XR&M as the apparent best heuristics. About 100 problems were run for 100-job
problems, and 100 for 200-job problems. Each problem was then improved in turn
by $k = 1$ search, $k = 5$ search, $k = 10$ search, and $k = 20$ search. The benchmark
was the "best of 3."

Heuristic	Phase I	$k = 1$	$k = 5$	$k = 10$	$k = 20$
		100 Jobs			
FCFS	100.	17.	5.5	4.9	4.8
R&M	8.4	2.0	0.81	0.74	0.74
XR&M	5.2	1.3	0.22	0.19	0.18
		200 Jobs			
FCFS	100.	15.	6.4	6.0	5.7
R&M	11.	3.2	1.0	0.93	0.86
XR&M	6.1	1.5	0.26	0.22	0.19

Again some facts are obvious:

(a) The 200-job problems are very similar to 100-job problems, although somewhat "harder."
(b) The best Phase I heuristic gives the best final result.
(c) XR&M produces about 60% of the error of R&M in Phase I, but only about 25% as much error after neighborhood search.
(d) Adjacent pairwise interchange reduces errors by a factor of 3 to 4.
(e) The 5-move interchange produces another factor of 2.5 to 6.
(f) Larger k's are much less effective in producing further savings.

7.6A Numerical Exercises

7.6A.1 Consider the following F_{wt} problem with dynamic arrivals and inserted idleness allowed:

Job j:	1	2	3	4	5	6	7	8
p_j	4	3	7	3	7	4	8	5
r_j	0	1	6	17	20	26	33	35
w_j/p_j	1	2	5	2	6	1	2	6

(a) Solve this problem by the inserted idleness heuristic.
(b) Find the best solution you can using the manual interchange routine of the software.

7.6A.2 Improve the solutions in 7.6A.1 (using the software) by using them as seeds for:

(a) Adjacent pairwise search
(b) General pairwise search.

7.6B Software/Computer Exercises

7.6B.1 Modify your code in 7.5B.1 to handle inserted idleness as an option, and solve some problems.

7.6B.2 Modify your code in 7.5B.2 in a similar fashion.

7.6C Thinkers

7.6C.1 Modify your heuristic for N_{wt} (all weights = 1.0) for the inserted idleness case.

7.6C.2 Construct an example to show that our modified priority function for, say, F_{wt} may not work well if there are two hot jobs about to arrive.

7.6C.3 Reconsider 7.5C.3 for the inserted idleness case.

(A) 7.7 RELATED PROBABILISTIC RESULTS

7.7.1 Overview

In some cases optimal sequencing depends relatively little on the accuracy of future information; in other cases it is strongly dependent. For example, in the ordinary preemptive case, transitive optimal sequencing rules can be adopted directly just considering currently available jobs, with *no information about future arrivals*. On the other hand, if the discipline is preemptive repeat (the preempted job must repeat all processing), with low information we may tend to prefer short jobs to minimize the unknown chance of interruption.

This suggests that, in general, different techniques will be required to analyze models in which the future behavior of jobs is uncertain. Our purpose here is to survey some probabilistic results that are, in fact, quite similar to the deterministic results. This gives us hope that deterministic results will prove useful in general in evaluating probabilistic extensions.

We shall simply state these random results without proof, since the necessary proof methods are somewhat outside the scope of this book.

7.7.2 Probabilistic Static Results

Much of the material in this section is from Baker (1974). Let us first turn to the analogue of WSPT scheduling for the F_{wt} problem. Suppose each processing time p_j is random, but with known expectation $E[p_j]$. Many early probabilistic results can be found in Banerjee (1965), Crabill and Maxwell (1969), Fife (1965), and Roth-kopf (1966).

Proposition 20. For the static problem with random processing times and objective min $E[F_{wt}]$, the optimal policy is WSEPT:

$$w_1/E[p_1] \geq w_2/E[p_2] \leq w_3/E[p_3] \geq \cdots \geq w_n/E[p_n]$$

Intuition. While we don't give a formal proof here, intuitively it follows from the linearity of the objective function and the fact that, from the law of large numbers, in the long run the average will be "essentially" achieved.

Now suppose due dates are also random and we wish to minimize the maximum expected lateness $\max_j\{E[L_j]\}$.

Proposition 21. For the static problem with both processing time and due dates random and objective min $\max_j\{E[L_j]\}$, the optimal policy is EEDD; that is,

$$E[d_1] \leq E[d_2] \leq E[d_3] \leq \cdots \leq E[d_n]$$

If job times are random, but due dates are fixed, it can be shown that EDD minimizes the maximum probability of being late.

Proposition 22. For the static problem with random processing times and objective to min $\max_j\{\Pr[L_j \geq 0]\}$, the optimal sequence is given by $d_1 \leq d_2 \leq d_3 \leq \cdots \leq d_n$.

All these results are direct generalizations of the deterministic results and this fact gives us some comfort on the importance of doing a thorough deterministic analysis prior to moving to stochastic models.

7.7.3 Probabilistic Dynamic Results

If job arrival times are dynamic and not completely known in advance, the situation is much more complicated. Perhaps the simplest appropriate stochastic model would be to assume that interarrival times are stochastic and drawn from a distribution that leads us to generalized single-channel queuing theory. (In practice, there may be a great deal of forecast information modifying this simple arrival process.)

Typical queuing theory concerns itself primarily with first-come first-serve sequencing rules, which are not of great interest here. The topic of optimizing sequencing rules for queues has been investigated in some studies of *priority* or *normative queues*, but few results are available. In Poisson queues with no preemption allowed, it is known that WSPT minimizes both weighted flowtime and weighted queue length (WIP). In general queues operating under preemption, the optimal rule for CAE is WSRPT (weighted shortest remaining process time) and for SC it is WSPT. We are not aware of similar extensions for the weighted tardiness problem. Details on the results listed above can be found, respectively, in Bagga and Chakravati (1968), Brown and Lomnicki (1966), Campbell et al. (1970), Gupta (1972), McMahon (1969), and Mitten (1959).

7.7C Thinkers

7.7C.1   Consider the static problem with stochastic processing times and expected weighted tardiness as the objectives. Propose a heuristic for this problem and try to justify it as well as you can.

7.7C.2   Test your heuristic.

____8
SINGLE MACHINE: TIMING

8.1 INTRODUCTION

Remember that we have defined a single-resource problem not by whether there is one machine or many machines, but by the existence of a single queue to the resource, with decisions of interest being restricted to the system queue:

(a) When to remove the current job from the processor.
(b) Which job to start next.
(c) When to start it.

The second question is the sequencing question, with which we have mostly been preoccupied until now. For simple one-machine static problems a job is released from the processor when finished or, in the preemptive case, when a waiting job attains higher priority. Jobs are always started exactly when the current job is taken off. Thus these three questions reduce to the question of the *optimal sequence*.

When we moved on to considering dynamic arrivals, the situation became slightly more complicated. Removal logic is not changed, but a job should start only after both the resource becomes available and the job arrives. This is still not too complicated; the problem still reduces to the question of the optimal sequence, which is constrained in the no-inserted-idleness case to jobs that have already arrived, and to those arriving within one job processing time (there is no point waiting for more than the minimum processing time of a waiting job) for the inserted-idleness case. Thus, although we have been considering *timing* issues, they (fortunately) have been reducible to *sequencing* issues.

In this chapter we continue developing these types of issues. In Section 8.2 we consider the situation in which the time to process a job depends on which job was processed last. We model this as a setup between jobs k and j requiring a *setup time* T_{kj} in addition to the usual processing time p_j. Thus j cannot start at the usual time t, but must wait until $t + T_{kj}$. We can see that, once again, this will be reducible to a (more complicated) sequencing problem. A special case of the sequence-dependent setup times is when jobs may be divided into classes. Within a class setup times are small (*minor setup*); between classes setup times are large (*major setup*). This problem is especially interesting when arrival times are dynamic; we consider this problem in Section 8.3. (In some situations it may be possible to begin setting up the machine for job i before i arrives. This case can easily be accommodated but is not treated here.)

A different kind of timing issue arises when certain activities cannot start until other activities have finished. This is common when the activities are part of one or more projects. We use the terms *projects* or *dependent activities* or *precedence constraints* to denote this case. The one-resource version of this problem is discussed in Section 8.4.

The preemptive problems we have considered to date are relatively straightforward, in that the job may be restarted without penalty. More commonly, some part of the processing time may be lost and the resource may have to be set up again to start the processing a second time. We discuss such *complex preemptions* in Section 8.5.

Finally, in Section 8.6 we discuss related probabilistic results. In particular we look at the general precedence constraint problem with some relatively simple objectives. It is easy to show that with probabilistic processing times, EDD still minimizes $E(L_{max})$ and $E(T_{max})$. For the criterion of $E(F_{max})$ or $E(L_{wt})$, the fundamental idea of finding sets A where aggregated weight divided by aggregated expected processing times is a maximum is still useful. However, useful theorems and heuristics are somewhat more complicated. These are briefly sketched.

8.2 SEQUENCE-DEPENDENT SETUPS

8.2.1 Overview

Sequence-dependent setups are commonly found where a single facility produces several different kinds of items or where a multi-purpose machine carries out an assortment of tasks. The use of a single system to produce different chemical compounds may require that some amount of cleansing be carried out between process runs on different compounds to ensure that tolerably low impurity levels are maintained. Furthermore, it should not be hard to envision situations in which the extent of the cleansing depends on both the chemical most recently processed and the chemical about to be processed. Similar setup properties can be found, for instance, in the production of different colors of paint, strengths of detergent, and blends of fuel. The same observations apply to certain assembly lines where retooling, inspection, or rearrangement of work stations represents the setup activity.

TABLE 8.1 Setup Cost Problem

Set Up to: Job: Process Time:	Racing (1) 150	Premium (2) 800	Regular (3) 450	Lead-free (4) 1250
Set Up from:				
Racing	—	30	50	90
Premium	40	—	20	80
Regular	30	30	—	60
Lead-free	20	15	10	—

For example, suppose that a process line manufactures four types of gasoline: racing fuel, premium, regular, and lead-free. The processing times p_j and setup times T_{ij} might represent that shown in Table 8.1.

For a makespan objective, solving this problem amounts to minimizing the amount of nonproductive time, that is, minimizing the sum of the setup times. For example, if one assumes regular was being produced before the start of the problem, then there are six distinct sequences for producing racing, premium, and lead-free following the given initial regular production. The six distinct sequences possible for this small problem are

$$(3)\text{--}1\text{--}2\text{--}4 \quad 30 + 30 + 80 = 140$$

$$(3)\text{--}1\text{--}4\text{--}2 \quad 30 + 90 + 15 = 135$$

$$(3)\text{--}2\text{--}1\text{--}4 \quad 30 + 40 + 90 = 160$$

$$(3)\text{--}2\text{--}4\text{--}1 \quad 30 + 80 + 20 = 130$$

$$(3)\text{--}4\text{--}1\text{--}2 \quad 60 + 20 + 30 = 110$$

$$(3)\text{--}4\text{--}2\text{--}1 \quad 60 + 15 + 40 = 115$$

Thus the minimum makespan is provided by schedule (3)–4–1–2. It might be argued that the ending conditions are a little off. Ending with job 4 is a little better than shown, since it has low setup costs to whatever the next job will be. Corrections for this are not important when scheduling large numbers of jobs. For small numbers of jobs we consider a correction scheme in the exercises.

Classical texts such as Baker (1974) consider only the makespan objective and assume that products are produced in a cycle over and over. The principal virtue of these assumptions is that the problem is then formally equivalent to a *traveling salesperson* problem, which has been studied a great deal. We are interested here in more general objective functions with dynamic ending conditions. However, we shall look at the traveling salesperson problem in Section 8.2.4 due to its historical importance and the wide availability of software for it.

Suppose job k is currently on the resource and will finish at time t. Job i is to be processed next. It arrives at time r_i, and the setup preparation starts at max(t, r_i), and finishes T_{ki} later. At that point job i will begin and will be completed at max(t, r_i) $+ T_{ki} + p_i$. Once all jobs in the problem have been processed and the resource becomes empty, the problem terminates. (There will be n jobs processed, and n preceding setups included in our formal problem.) Our heuristics can often be simplified by the assumption of symmetry: $T_{ki} = T_{ik}$; the assumption will be indicated if made. Many of our other techniques do not need this (reasonable) simplification.

Our strategy will be to develop bottleneck dynamics heuristics for a number of objectives first. In particular, for the usual makespan objective, this will reduce to a new myopic heuristic, which will be compared with traditional myopic heuristics. We give some conditions under which myopic heuristics will be "good" heuristics. Next, we briefly discuss other techniques such as beam search, branch-and-bound, dynamic programming, and neighborhood search. Beam search especially is facilitated by using our bottleneck dynamics heuristic to find quick reasonable trial solutions. Finally, we give a brief discussion of the most well known approach to the classic traveling salesperson problem as a solution to the cyclic makespan version of the problem.

8.2.2 Setups: Bottleneck Dynamics

We initially restrict our attention to the static case. (Development of the dynamic case will be left as an exercise.) Consider the following pairwise interchange situation, involving jobs k, i, and j. Job k is currently processing and will finish at time $t = 0$. What is the cost differential between processing job i next and then job j, or job j next and then job i?

If job i is processed first, completion of i is

$$T_{ki} + p_i$$

and completion of j is

$$T_{ki} + p_i + T_{ij} + p_j$$

If job j is processed first, completion of i is

$$T_{kj} + p_j + T_{ji} + p_i$$

and completion of j is

$$T_{kj} + p_j$$

Recall that the net marginal cost economic advantage to job j of completing a unit of time sooner is $MC_j = w_j U_j - IRp_j$; similarly for i, while the cost of a wasted unit of resource is R. Then, if we let Δ_{kji} be the net savings of sequence kji over kij, we see that

$$\Delta_{kji} = (T_{ki} - T_{kj})R + (T_{ji} - T_{ij})R + [(T_{ki} - T_{kj}) + (T_{ij} + p_i)]MC_j$$
$$- [(T_{kj} - T_{ki}) + (T_{ji} + p_j)]MC_i$$

As before, we wish to simplify this and try to make it separable in i and j. We assume the following:

(a) Nearly symmetric case: $T_{ji} - T_{ij}$ approximately 0.
(b) Resource cost much higher than activity delay costs: $R >> MC_j$; $R >> MC_i$.
(c) Process times much larger than setup times.

With these simplifications, we have approximately that

$$\Delta_{kji} = \{T_{ki}R - p_j MC_i\} - \{T_{kj}R - p_i MC_j\}$$

Dividing everything by $IRp_i p_j$:

$$\Delta_{kji}/IRp_i p_j = \{T_{ki}/Ip_i p_j - \pi_i\} - \{T_{kj}/Ip_i p_j - \pi_j\}$$

Now we are almost separated into priorities, with one final approximation of p_{av} for the processing time of the other job:

$$\pi_{kji} = \pi_{kj} - \pi_{ki}$$

where we have the following heuristic:

Setup Cost Heuristic Using Bottleneck Dynamics (General)

$$\pi_{kj} = \pi_j - (T_{kj}/p_{av})(1/Ip_j)$$

This heuristic states that the priority of a job being considered when k is finishing is the same as without setups, except that a penalty must be paid for a longer setup in proportion to the normalized size of the setup. (Note that the cost of the setup gets spread over all p_j units of job j.)

We turn now to a brief discussion of how this heuristic can be applied to the *makespan* objective problem. In a makespan problem the resource has some price R, but the jobs have zero delay costs. Thus our heuristic simplifies to the following:

Setup Cost Heuristic Using Bottleneck Dynamics (Makespan).

$$\pi_{kj} = -T_{kj}/p_j$$

That is, process the job next whose setup time is the smallest percentage of its processing time.

The bottleneck dynamics procedure for the general case works best when different jobs have different priorities π_j in the absence of setup times. The correction term would then be an approximate minor correction for setup times. Thus there is no particular guarantee that the idea works well when all π_j are identically zero as in makespan.

Nevertheless, the heuristic is very simple and can be compared with other common myopic heuristics for the setup problem.

Myopic 1: Smallest Setup

$$\pi_{kj} = -T_{kj}$$

Just choose the smallest setup. Gavett (1965) speculated that this simple heuristic might generate total setup times within 10% of optimum for up to, say, 20 jobs.

Myopic 2: Smallest Normalized Setup

$$\pi_{kj} = -T_{kj}/T_{(av)j}$$

Find the average setup from any job to j, and express this setup as a fraction of the average setup from any job to j.

Myopic 3: Smallest Two-Job Setup

$$\pi_{kj} = \min_{j'}\{ (T_{kj} + T_{jj'}) \}$$

That is, find the cheapest total setup for putting on two jobs, and put on the job resulting, using a rolling horizon procedure. This can also be done with three, or even with k jobs, but the computational cost becomes of the order of n^k.

There are many other heuristics for the traveling salesperson problem, such as Karg and Thompson (1964). Since their version of the problem is not our primary focus, we do not consider this further here.

Any of the myopic heuristics work very well if jobs are clustered into minor and major setups. There may be 40 types of jobs divided into five clusters. Within each cluster setup times are almost zero, between clusters setup times are mostly a function of which cluster one starts in and which cluster one finishes in.

Proposition 1. Consider the makespan criterion; let n jobs be grouped into m clusters. If within a cluster setup times are 0, and between any clusters A and B the setup time is T_{AB}, independent of jobs chosen, then any of the myopic proce-

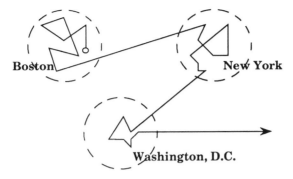

Figure 8.1. Major and minor setups in traveling salesperson problem.

dures above behave effectively as myopic procedures for the aggregated m cluster problem.

Proof. Left as an exercise.   ■

(Since myopic procedures generally work much better for a smaller number of jobs, myopic procedures should be considered in this situation.)

The idea of Proposition 1 is illustrated in Figure 8.1 where major setups correspond to distances between cities, and minor setups to distances between customers in cities.

If we now turn to interpreting the bottleneck dynamics heuristic for the objective of weighted flow time, the priority becomes the following:

Bottleneck Dynamics Setup Heuristic for F_{wt}

$$\pi_{kj} = [w_j - T_{kj}/Ip_{av}]/p_j$$

The nature of this priority function clearly depends very much on the relative sizes of the job urgency w_j and of setup times T_{kj} as a fraction of process times. If w_j dominates, then the priority will look very much like simple WSPT. If T_{kj} dominates, then the heuristic will primarily try to minimize setup times.

We could continue to investigate the bottleneck dynamics heuristic for other objectives but will leave this for the exercises.

8.2.3 Setups: Beam Search, Branch-and-Bound, Dynamic Programming, and Neighborhood Search

Beam Search. To be most effective, beam search requires a fast and reasonably accurate heuristic to obtain trial solutions for any subproblem being considered. Beam search does not require lower bounds. Thus, given the availability of the bottleneck dynamics setup cost heuristic for essentially any objective function desired, beam search can be expected to be reasonably fast and very accurate for any of these problems. As a brief review:

Step 1: Start with a given job finishing processing at time t.
Step 2: Branch on all possible next jobs to schedule.
Step 3: Obtain trial solutions with the bottleneck dynamics heuristic.
Step 4: Prune all but b of these branches.
Step 5: Expand and repeat.

Branch-and-Bound. On the other hand, while branch-and-bound can make use of trial solutions available from the bottleneck dynamics heuristic, it is most dependent on a strong lower bounding procedure. Setup costs make it difficult to obtain sharp lower bounds. A reasonable and general lower bounding procedure is the following:

Lower Bounding Procedure

Step 1: Replace each T_{kj} by $T_{\min j} = \min_k(T_{kj})$.
Step 2: Replace each p_j by $p_j + T_{\min j}$.
Step 3: Solve the preemptive version of the resulting problem.

We solve the branch-and-bound forward in time, applying the lower bounding procedure only to the part of the solution that is not yet fixed.

(A) Dynamic Programming. Our forward dynamic programming method described in Section 5.4 still works very nicely. The complexity, software running time, and storage space are about doubled or tripled, basically because besides keeping track of the optimal cost $G(J)$ for all possible initial sets of jobs J, we also must keep track of $C(J)$, the optimal completion time for the set J, and $j^*(J)$, the optimal last job in J (to allow finding the setup cost when the next element is added). We briefly give the modified equations. (The reader may wish to review Section 5.4 before studying the modification.)
    Let:

$C(J)$ = optimal completion time for J

$G(J)$ = optimal cost for J

$j^*(J)$ = optimal last job for J

$C(J, j)$ = optimal completion time for J conditional on j last

j^{**} = shorthand for $j^*(J - \{j\})$; that is, the optimal second-to-last job if j is last

$f_j(C_j)$ = objective function for job j given its completion time

The recursive equations are

$$C(J, j) = C(J - \{j\}) + T_{j**j} + p_j$$

$$G(J) = \min_{j \in J}[G(J - \{j\}) + f_j(C(J, j))]$$

$$= [G(J - \{j*\}) + f_{j*}(C(J, j*))] \quad \text{(defines } j*(J)\text{)}$$

$$C(J) = C(J, j*(J))$$

Thus dynamic programming for static setups is about as attractive or unattractive as for the simpler static case. As discussed previously, for the dynamic case the number of sets to consider can be much reduced by considering which jobs can have arrived to be scheduled.

Neighborhood Search. Neighborhood search, tabu search, simulated annealing, and genetic algorithms are very practical for this problem. If there is a large matrix of setup times, however, the cost of computing all pairwise interchanges from a given solution may get rather large. A compromise procedure would be to use the bottleneck dynamics approximations as a gross filter to make a preliminary estimate of unlikely interchanges. Then do exact pairwise interchange computations only on the reduced set of "reasonable" choices. (Filtering with bottleneck dynamics is also likely to be necessary in job shops and project management with non-makespan criterion.).

(A) 8.2.4   The Traveling Salesperson Problem

The special setup cost case with makespan objective, and with the last job processed being the same as the last job processed before time t (cyclic case), is formally equivalent to the traveling salesperson problem. In that classic problem a salesperson starting at a certain city must visit all cities once and return to the original city, at minimum total distance. For our problem the distances are equivalent to the setup times.

Of many classic approaches to the traveling salesperson problem, we review a branch-and-bound approach due to Little et al. (1963) because it indicates the flexibility of the branch-and-bound approach. (Note it does *not* use a forward approach, such as we have emphasized in this book.)

The method branches on whether a particular setup is forced to be part of the solution or forced to be not part of the solution. For example, as shown in Figure 8.2, a partition of the original problem might require that setup T_{13} be part of the problem on one branch, but be excluded as part of the problem on the other branch.

In a matrix (or submatrix) of T_{ij}, a feasible solution clearly requires n cells to be chosen, one in each row and one in each column. Finding such an assignment at minimum cost would be easy; we simply use the linear assignment algorithm. Unfortunately, most such assignments would break into smaller cycles instead of one large cycle; the salesperson would make several subtours with no provision for

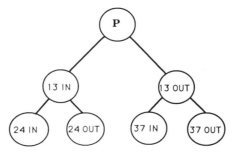

Figure 8.2. Traveling salesperson branching.

getting from one subtour to another. This does, however, provide a lower bound on the desired solution.

Lower bounds for a given T_{ij} matrix may also be calculated by a method called *reduction*. Since any feasible solution contains exactly one element in each row, it is possible to subtract a constant from any row without altering the relative desirability of any feasible solution. In other words, this subtraction reduces the lengths of all tours by the same constant and, in particular, would not affect which of the feasible tours is optimal. In the reduction process, the minimum row element is the constant chosen for subtraction from each row. Then, similarly, the minimum element can be subtracted from each column. The matrix that emerges has at least one zero element in every row and in every column, and the sum of the subtraction constants is a lower bound on the optimal solution. In this way, the reduction process identifies a distance that must be part of any feasible tour and therefore is a lower bound on the optimum tour. To illustrate these steps specifically, consider the traveling salesperson problem associated with Table 8.2.

By subtracting the minimum element in each row, and then subtracting the minimum element in each column of the row-reduced matrix, the original Table 8.2 matrix is reduced to the one shown in Table 8.3. The sum of the elements subtracted is 20, which is a lower bound on the optimal solution. At this point there is at least one zero in every row and column.

The algorithm next partitions the problem by forcing an apparently attractive element into the solution on one branch and out on the other branch. (If it is attractive enough, we hope the other branch will have unattractive lower bounds and

TABLE 8.2 Traveling Salesperson Problem: Original Data

—	4	8	6	8
5	—	7	11	13
11	6	—	8	4
5	7	2	—	2
10	9	7	5	—

TABLE 8.3   Traveling Salesperson: Reduced

—	0	4	2	4
0	—	2	6	8
7	2	—	4	0
3	5	0	—	0
5	4	2	0	—

be quickly pruned away.) To make it the most attractive, it makes sense to choose that zero element in the reduced matrix which, when prohibited, would give the other branch the highest lower bound. Therefore we label each zero element with the sum of the minimum element remaining in the row and the minimum element remaining in its column, as shown in Table 8.4. Element (2,1) is chosen. The original problem is partitioned into two subproblems, one containing and one not containing (2,1).

The reduction procedure can now be applied to each subproblem. In the first subproblem, we modify Table 8.4 further as follows. Since (2, 1) is in the solution, we circle that. Since (1, 2) must be prohibited, as well as elements (2, j) for j not 1, and elements (i, 1) for i not 2, we cross out these elements. (Or think of them as changed to very large.) After reduction this lower bound increases to 24 (by subtracting 2 from the second column, and from the second row).

In the second subproblem, a similar process causes us to circle (1, 2) as in the solution, prohibiting (2, 1) as well as other elements in row 1 and column 2. Then further reduction causes subtracting 3 from column 1 and 2 from row 2, giving a bound of 25 as anticipated. This procedure can then be imbedded in any strategy for expanding subproblems such as depth first search, best first search, and so on. The optimal solution turns out to have a cost of 25. An optimal tour is 4–1–2–3–5–4. Depth first involves expanding six subproblems. The first trial solution obtained is after three subproblems, with a value of 26.

Other useful references for this section are Flood (1955) and Lawler et al. (1985).

8.2A Numerical Exercises

8.2A.1   Solve the following setup cost problem for the makespan problem, using bottleneck dynamics and the other myopic heuristics suggested. Assume

TABLE 8.4   Traveling Salesperson: Branching Choice

—	0^4	4	2	4
0^5	—	2	6	8
7	2	—	4	0^2
3	5	0^2	—	0^0
5	4	2	0^4	—

job 6 was processed before problem start, that all six must be processed in any order, and that no cyclic processing is presumed.

Job j	1	2	3	4	5	6
p_j	6	2	8	4	6	2
d_j	14	6	8	12	11	10
w_j	12	4	16	4	6	0.5
T_{1j}	0	1	2	6	10	8
T_{2j}	1	0	1	8	4	6
T_{3j}	2	1	0	6	6	10
T_{4j}	6	8	6	0	0	2
T_{5j}	10	4	6	0	0	0
T_{6j}	8	6	10	2	0	0

8.2A.2 Repeat 8.2A.1 for the weighted flow objective.

8.2A.3 Repeat 8.2A.1 for the weighted tardiness objective.

8.2A.4 Improve 8.2A.1, 8.2A.2, and 8.2A.3 with some sort of approximate adjacent pairwise interchange. (Use some method for screening unlikely interchanges.)

8.2A.5 To check how stable your results are to the exact problem, invent two more data sets for 8.2A.1, solve them for makespan, and run pairwise interchange.

8.2A.6 In the optimal solution to the traveling salesperson problem for the matrix shown below (which includes Table 8.2 as a subproblem), does the optimal tour for the Table 8.2 problem appear intact as a part of the larger optimum?

—	4	8	6	8	2
5	—	7	11	13	4
11	6	—	8	4	3
5	7	2	—	2	5
10	9	7	5	—	2
8	4	6	3	5	—

(If you are lazy and can figure out how to solve the problem by inspection rather than branch-and-bound, feel free to do so.)

8.2B Software/Computer Exercises

8.2B.1 Code the setup cost problem with dynamic arrivals and the bottleneck dynamics heuristic, with the flexibility of choosing one of several objectives. (R and I can be input.)

8.2B.2   Run a computational study with your code to check the sensitivity of the solution to varying importances of setup time versus activity delay cost.

8.2B.3   Modify any earlier code to allow solution of general setup time problems by branch-and-bound and/or beam search.

8.2B.4   Write a code for the branch-and-bound approach to the traveling salesperson problem given in the text.

8.2C Thinkers

8.2C.1   Prove Proposition 1.

8.2C.2   Do you think there is an analogous result to Proposition 1 if within-cluster setups are "small" rather than zero?

8.2C.3   Why is it easier to get lower bounds for makespan setup costs than for the general case?

8.2C.4   If you modify the general case setup cost problem to be cyclic, can the "forced in/forced out" method of branch-and-bound for the traveling salesperson problem be modified? (If you cannot do this completely, discuss your progress/difficulties.)

8.2C.5   The problem of correcting ending conditions for the dynamic setup problem to penalize ending at a "bad" job is often solved as follows. (A "bad" job in this context means that costly setups will be incurred in the next part of the problem past the end of the current horizon.) If job k is last, add to the costs the average setup cost of going to the unknown next job. That is, add a dummy $(n + 1)$st job with 0 processing times, and setup time $T_{k\text{dum}} = (1/n)\Sigma_j T_{kj}$. Critique this correction and/or experiment with it in small problems.

8.3   LOT SIZING

8.3.1   Overview

A *lot* is what we get in a setup cost problem when we sequence a number of jobs (which represent the same product or similar products) in a group that has zero or low setup costs between members of the lot (*batch*). We can see from Proposition 1 that for the static case with makespan objective we should always make our batches as large as possible. That is, every member of one class should be run, then every member of a second class, and so on. The classes could simply be aggregated, and a simple sequencing problem would result.

However, neither assumption is particularly realistic. The average flowtime in a batch is proportional to the size of the batch; since there is one setup per batch, the setup cost per job is proportional to one over the size of the batch. Thus problems in which there is any emphasis on timeliness will tend to produce a balance between lot size and setup savings. [Compare with the classic economic order quantity (EOQ) model, which will turn out to be a special case of our results here.]

Similarly, in a dynamic arrival case, trying to make a lot more than a certain size may require extensive waiting for the arrival of further members of that class. The resulting resource idleness can easily more than offset the setup cost savings.

We consider the dynamic arrival case here, although the static results are easily given as a special case. We limit ourselves to bottleneck methods. Lagrangian relaxation methods for this problem are discussed by Karmarkar (1983) and Dobson and Karmarkar (1986), among others

(A) 8.3.2 Lot-Sizing: Bottleneck Dynamics

Setup costs employed in lot-sizing models are usually setup times multiplied by some accounting price for the machine. Instead, we multiply the setup time by the machine's current dynamic price, so that, in particular, lot sizes will go up and down with congestion at the machine.

Suppose there is a major setup time T_{IJ} between current lot I and next lot J of different type. Assume the minor setup time between two jobs (or lots) of the same type may be approximated as constant within that type. Thus we may add this minor setup time into the process times and, in the sequel, treat these interval setup times as zero. For the major setups we calculate an imputed setup cost $R_{IJ} = RT_{IJ}$, where R is the current imputed price of the resource.

Given that lot type I is currently finishing processing, we wish to know (a) which type J to process as the next lot, and (b) what the lot size Q should be.

We do this in three steps:

Step 1: For each possible J not equal to I, choose the best $Q(J)$.

Step 2: Aggregate these lots into "aggregate" jobs, and select the one to go first.

Step 3: Repeat the entire procedure with updated information when this lot finishes.

In step 1, for each lot type we calculate the opportunity costs of a lot of Q individual jobs (or transfer lots) and then divide by the time over which the lot arrived to get an average cost per unit time. We do this successively, increasing Q until no more members for the lot are currently available (Q_{max}) and then stop, choosing the quantity with minimum average cost AC*. That is, we are employing the Silver–Meal dynamic lot-sizing heuristic (Silver and Meal, 1973). In step 2, we aggregate (momentarily) each lot into a single job and use standard sequencing rules to choose which lot to do next. These two steps are described in greater detail below. We suppress the lotting subscripts, since this causes no real confusion and adds clarity.

Step 1: Lot Sizing. At the moment of finishing the last lot, the remaining queue at the resource is simply ordered by their bottleneck dynamics benefit/cost sequencing priorities for the associated non-setup-cost problem. Now we wish to imagine the effectiveness of choosing a given job type M to be the next lot. Thus we imagine:

(a) Paying the lot setup cost.

(b) Moving items of the given type one by one to the front to the forming lot.

(c) Paying the opportunity costs for moving them in front of their natural priority.

(d) Adding items that have not yet arrived, not going past the first item requiring inserted idle time (this determines Q).

(e) Holding the items in the lot after processing until all are finished.

(An alternate model without this last assumption and cost is easily carried through, but we do not do this here.)

We now need a number of definitions:

TC   is the total cost in the model per unit of time (to be derived)

TQ   is the time from when the first item of the lot arrived until the first item of the type not included in the lot arrived

i indexes (by ranking) jobs within the lot from 1 to Q

p_i processing time for job i

X_i is the net cost per unit time for job i to be delayed or expedited

E_{ij} is the direct cost of exchanging i and j in the line (i.e., extra delay cost of putting j later, minus delay cost saved by putting i earlier, see below)

$n(k)$ is the position of unit k not yet moved to the lot, counting back from the forming lot (i and all units ahead of i not yet part of the lot)

We note from our earlier work that

$$X_i = w_i U_i - IRp_i; \quad E_{ij} = p_i w_j U_j - p_j w_i U_i$$

Using these definitions, we note the following (remember we are suppressing lot subscripts; e.g., T is short for T_{IJ}, the setup cost for switching from lot I currently on the machine to lot J, the currently forming lot):

(a) The setup cost is RT.

(b) Given a job k, the cost of moving it $[n(k) - 1]$ positions to the front is $\sum_{j=1,n(k)-1} E_{kj}$.

(c) The cost of holding unit k until the entire lot is finished is $U_k \sum_{j=k+1,Q} w_j p_j$.

Thus the cost per unit time as a function of Q is

$$TC(Q) = \frac{1}{TQ(Q)} \left[RT + \sum_{k=1,Q} \left(\sum_{j=1,n(k)-1} E_{kj} + U_k \sum_{j=k+1,Q} w_j p_j \right) \right]$$

Finally, repetitively solve this last equation for Q from 1 to $Q_{\max}$, choosing the lot size Q^* that minimizes this average cost per time unit for this type of unit. (It is easily seen that the calculations for $Q + 1$ can be done in such a fashion as to utilize those for Q.)

Step 2: Lot Sequencing. Aggregate process times, direct costs, and tardiness costs for each aggregate job (lot) to be sequenced are calculated as follows:

$$W_{agg}U_{agg} = \sum_{i=1,Q} w_i U_i$$

$$p_{agg} = T + \sum_{i=1,Q} p_i$$

and thus the overall priority is given by:

$$\pi_{agg} = [(W_{agg}U_{agg})/IRp_{agg}] - 1$$

That is, the benefit of expediting for the aggregate job is just the sum of these benefits for the individual jobs. The process time for the aggregate job is the lot setup time plus the sum of the individual processing times. Schedule the first lot. After this lot is finished, redo the whole lotting procedure with updated information on a rolling horizon basis.

We now work through an approximation to this analysis involving jobs arriving rather uniformly in time, to show how these formulas relate to the ordinary square root formula (economic order quantity). Assume demand is large and uniformly spaced, so that the total time for the arrival of the lot is Q/γ, where γ is the demand rate. Suppose all E_{ij} are similar and may be estimated by their average value E_{av}, and that U_{av} and p_{av} similarly replace U_i and p_i. Further suppose total demand outside the lot is m times that for the lot, and also arrives at a uniform rate. Then $n(i) - 1$, the number of non-lot jobs ahead of i, may be estimated as follows. Suppose the Q lot members are uniformly distributed throughout mQ non-lot members. Thus, for large numbers, this means that $n(i) - 1$ is about mi. This gives us the approximate formula

$$TC(Q) = (\gamma/Q)[RT + (mE_{av} + p_{av}U_{av})(Q^2/2)]$$

which can be solved for the optimal lot size

$$Q^* = \sqrt{2RT/(mE_{av} + p_{av}U_{av})}$$

Note that mE_{av} and $p_{av}U_{av}$ each can serve the role of dynamic holding costs—the cost of expediting the lots and the cost of holding the lot until complete, respectively. (Note that E_{av} and U_{av} are functions of the resource price R, however.)

We will be talking more in later chapters about the process for estimating the resource price R. It is worth noting that R should not change a great deal between the model with setup cost and that without setup cost. (Busy period lengths should not change dramatically.) Whatever the method used, it may be easier to estimate R for the non-setup cost model and then use it as an approximation for the setup cost model.

Other references for this section include Afentakis (1985), Jacobs and Bragg (1988), and Kono and Nakamura (1990).

(A) 8.3A Numerical Exercises

8.3A.1 Set up a problem with 10–15 jobs and only one job class, F_{wt} objective, all jobs with differing w_i and p_i but the same priority w/p. Try both a small and a large setup cost. Discuss how reasonable your answers seem.

8.3B Software/Computer Exercises

8.3B.1 Set up a beam search program to solve the problems in 8.3A.1.

8.3A.1 Try both small and large beam widths. How accurate were the approximations of this section for your problem?

8.3C Thinkers

8.3C.1 Can you give examples where the approximations used in this section do not work too well? Can you say why? Can you give generalizations as to when these approximations work well?

8.4 DEPENDENT JOBS, PRECEDENCE CONSTRAINTS

8.4.1 Overview

We turn next to constraints common in completion of complex projects, which require that certain activities must be complete before certain others can start. For example, in building a house, the basement must be excavated before the foundation can be poured. Perhaps the sidewalk could be poured either before or after, and so on. The presence of such constraints reduces the number of feasible sequences for the jobs, just as arrival times do. Similarly, these constraints may make the problem easier or harder: easier in that fewer problems must be considered; harder in that each is more complicated.

Precedence Relations. The existence of technological restrictions on job sequences can be interpreted mathematically as the existence of a partial ordering among the jobs based on the *precedence relation*. The notation $i < j$ denotes the fact that job i precedes job j; that is, job j may not begin until job i is complete. When $i < j$, job i is said to be a *predecessor* of job j, and job j is said to be a *successor* of job i. Job i is called a *direct predecessor* of job j, if no job comes logically between them, that is, there is no k for which $i < k < j$.

As an example, consider the computer programs submitted for processing by a payroll department. Program A reads daily employee time cards, sorts the information, and updates the monthly records that are maintained on tape. Program B reads the tape and prints out paychecks. On the last day of the month, both programs are submitted, but B cannot be run until A is complete. Therefore job A precedes job B.

To illustrate the effect of adding precedence constraints to a sequencing problem, consider minimizing weighted mean flowtime, with all weights equal to 1.0, in a single-machine problem with three jobs. Suppose the jobs are labeled a, b, and c, and suppose that $p_a < p_b < p_c$. Without precedence constraints, the optimal sequence is clearly a–b–c. Now suppose there exists one precedence constraint: $c < a$. Although job b "ought" to follow job a and precede job c, it is not immediately clear in such a situation whether sequence c–a–b or sequence b–c–a is more desirable. (We can, however, rule out the possibility that c–b–a is best with a simple interchange argument.) The specific properties of this problem can be examined more closely, but the main point is that the existence of precedence constraints can complicate even the simplest scheduling problems.

Initial Sets. In Section 8.4.2 we develop the concept of an *initial set* as a set of jobs that also contains all the predecessors of those jobs. A *simple initial set* consists of a given target job and all its predecessors. In particular, if a job has a very high weight (in a project we call this a milestone), it may still not be appropriate to give a very high priority to doing this job early if there are many long predecessors to be done first. We are intuitively interested in the aggregate weight compared with the aggregate processing time, where we aggregate over the job and all predecessors. We develop exact and heuristic results for the L_{max}, T_{max}, F_{wt}, L_{wt} problems. Throughout, unless explicitly mentioned, we are assuming the static case where all jobs arrive in advance.

8.4.2 Classical Exact Results and Extensions

Maximum Lateness and Tardiness. The initial set concept immediately allows us to generalize the EDD result for T_{max} and L_{max} to the precedence constrained problem. We use the notation $I(j)$ to mean the initial set of job j.

Proposition 1. In the static single-resource problem with objective T_{max} or L_{max} and with arbitrary precedence constraints, let j^* be the job with earliest due date.

(a) Setting the due dates of predecessors to j^* equal to d_{j^*} will not affect either objective.

(b) There will be an optimal solution where $I(j^*)$ is scheduled first.

Proof. (a) As long as j^* has the earliest due date, it will have lateness larger than any of its predecessors. Setting their due dates equal to d_{j^*} preserves this fact.

(b) By arguments in Chapter 4, all items k that are not predecessors to j can be moved later than $I(j^*)$, since it is earliest. ■

Corollary. An algorithm for this problem is simply to do pairwise interchange until all interchangeable pairs are in due date order.

Proof. Left as an exercise. ■

Definition. The *derived due date* dd_j of a job is the minimum of the due dates of activities j and all successors.

Intuitively, if a successor job must be done by time 5, and this job must be done first, then this job must really be due by time 5 also. This leads to Proposition 2.

Proposition 2. An optimal sequence for $L_{\max}$ or $T_{\max}$ satisfies

$$dd_1 \le dd_2 \le dd_3 \le dd_4 \cdots$$

Proof. Left as an exercise. Can be proved directly from Proposition 1, or directly from pairwise interchange arguments. ■

Weighted Flow/Weighted Lateness. We turn now to developing results for the weighted flow problem/weighted lateness problem. A subset of jobs, which is known to be scheduled as a block in the unknown larger schedule, can be aggregated into a single *aggregate job,* with appropriate aggregate processing time, weight, normalized processing time, and precedences, as shown in Figure 8.3. (We will not prove this fairly obvious fact, but we used it in the lot-sizing development.)

$$p_A = \Sigma p_j; \quad w_A = \Sigma w_j; \quad \pi_A = w_A/p_A$$

We can derive a more general aggregated version of WSPT for the F_{wt} problem when the precedence structure forms a *tree.* To be specific, a set of jobs in which

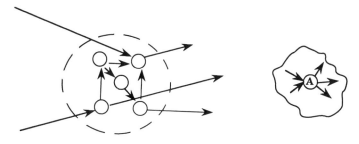

Figure 8.3. Aggregate job.

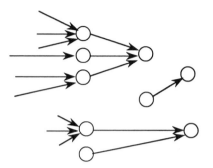

Figure 8.4. Assembly tree precedence structure.

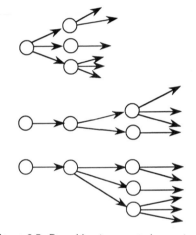

Figure 8.5. Branching tree precedence structure.

each job has at most one direct successor is called an *assembly tree* (Figure 8.4). Assembly tree structures are common in shops that assemble products/projects, for example.

A set of jobs in which each job has at most one direct predecessor is referred to as a *branching tree* (Figure 8.5). Shops (such as oil refineries or chemical plants) that make several products out of one raw material, and then several products from each of those (also called disassembly shops), are good examples of branching trees.

Definition. $\pi_j = w_j/p_j$

Let j^* be the job for which π_j is maximized.

Let j^{**} be the job for which π_j is minimized.

Let n^* be the number of direct predecessors of j^*.

Let n^{**} be the number of direct successors of j^{**}
Let k^* be the direct predecessor of j^* with lowest π_j.
Let k^{**} be the direct successor of j^{**} with highest π_j.

Proposition 2. For a static F_{wt} problem with arbitrary precedence structure, an optimal solution can be chosen for which:

(a) j^* is scheduled next to a direct predecessor.
(b) j^{**} is scheduled next to a direct successor.

Proof. Use pairwise interchange to move j^* as close to the beginning as possible without violating precedence constraints, since it has the highest priority of all. And so on.   ∎

Proposition 3. Given the correct direct predecessor, j^* can be aggregated with it, summing weights and process times, preserving all predecessors and successors. Similarly for j^{**}.

Proof. Left as an exercise.   ∎

Now if a rule could be found for choosing the correct predecessor and/or successor to aggregate with, an exact algorithm could be constructed. Failing this, the following heuristic suggests itself. k^* is a good choice to aggregate with j^*, since it would be intuitive to schedule the smallest π_k as late as possible. Similarly, k^{**} is a "good" choice to aggregate with j^{**}.

Morton and Dharan (1978) gave the following procedure (which they dubbed an *algoristic*) for the static general precedence structure T_{wt} problem. (The name is a combination of "exact algorithm" for trees, where it is optimal, and "heuristic" for more general problems.)

Tree-Optimal Algoristic

Step 1: Update j^*, j^{**}, n^*, n^{**}, k^*, k^{**}; if there remains a single job go to step 9.

Step 2: If j^* has no predecessors, schedule it first; remove it from the current set of jobs; go to step 1.

Step 3: If j^{**} has no successors, schedule it last; remove it from the current set of jobs and go to step 1.

Step 4: If j^* has a single direct predecessor, go to step 5; otherwise go to step 6.

Step 5: Combine k^* and j^*; go to step 1.

Step 6: If j^{**} has a single direct successor, go to step 7; otherwise go to step 8.

Step 7: Combine j^{**} and k^{**}; go to step 1.

Step 8: If $n^* \le n^{**}$ go to step 5; otherwise go to step 7.

Step 9: Reconstruct the sequence by disaggregating jobs.

Proposition 4. The tree-optimal algoristic provides an optimal solution for branching trees or assembly trees, and whenever the algorithm is executed without doing step 8.

Proof. Left as an exercise. ∎

Note that by changing step 8, the algoristic may either be made into branch-and-bound (e.g., try all different predecessors) or beam search. (See the examples.) There is a quite different way of attacking the same problem using initial sets.

Proposition 5. (Sidney, 1975). For assembly trees, the simple initial set with maximum $\pi_I = w_I/p_I$ may be scheduled first.

Proof. See the reference. ∎

To actually execute Proposition 5, we would execute the Sidney-type algoristic given below after Proposition 7, where an example is also given.

Proposition 6. (Horn, 1972). For branching trees, the simple final set (reverse of initial set) with minimum $\pi_F = w_F/p_F$ may be scheduled last.

Proof. Mirror image of Proposition 5. ∎

Finally, Sidney also gave a similar result for general precedence structures.

Proposition 7. (Sidney, 1975). For general precedence structures, the (not-simple) initial set with maximum $\pi_I = w_I/p_I$ may be scheduled first.

Proof. See the original paper. ∎

There are two problems with this more general procedure:

(a) There are an exponential number of (not-simple) initial sets to consider (exponential in the number of jobs).

(b) The maximal set may turn out to be the original set itself, so that one cannot progressively look at smaller and smaller sets.

For these reasons, the full Sidney algorithm is rarely implemented. Morton and Dharan (1978) suggested and tested what they called the Sidney-type algoristic.

(A) Sidney-Type Algoristic

Step 1: Set the job set at all unscheduled jobs.

Step 2: If the set is empty, terminate.

Step 3: Find the simple initial set with maximal π_I.

Step 4: If the resulting set has more than one element, go to step 5; otherwise schedule the job and go to step 1.

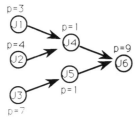

Figure 8.6. Sidney assembly tree example (all weights = 1, processing times given).

Step 5: Set the current job set to the current initial set less its last element; go to step 3.

As we shall see in Proposition 8, this algoristic is optimal for assembly trees. We first give an example of how the algoristic works.

The first real work is to implement step 3 of the algoristic on the original problem (Figure 8.6):

$$\pi\{J_1, J_2, J_3, J_4, J_5, J_6\} = \text{(sum weights)/(sum times)} = 6/25 = 0.24$$

$$\pi\{J_1, J_2, J_4\} \qquad\qquad\qquad = 3/8 = 0.38$$

$$\pi\{J_1\} \qquad\qquad\qquad\qquad = 1/3 = 0.33$$

$$\pi\{J_2\} \qquad\qquad\qquad\qquad = 1/4 = 0.25$$

$$\pi\{J_3, J_5\} \qquad\qquad\qquad\quad = 2/8 = 0.25$$

$$\pi\{J_3\} \qquad\qquad\qquad\qquad = 1/7 = 0.14$$

Since 0.38 is the highest WSPT value, we choose the initial set $\{J_1, J_2, J_4\}$. We proceed to steps 4 and 5, and remove J_4 (the job with no successors, which hence *must* be scheduled last, and return to step 3, with the new set $\{J_1, J_2\}$.

Now there are two initial sets, $\{J_1\}$ and $\{J_2\}$ of which $\{J_1\}$ has the highest WSPT value, and hence J_1 should be scheduled.

Next return to step 1 on the reduced problem with J_1 deleted and repeat. The reader might verify that the complete optimal solution is $J_1, J_2, J_4, J_3, J_5, J_6$.

The Sidney-type algoristic as given is easy to explain compactly but does not take full advantage of possible computational efficiencies. This is left as an exercise.

(A) Proposition 8. The Sidney-type algoristic is optimal for assembly tree structures.

Proof. Follows directly from Proposition 7.   ■

While the tree-optimal heuristic is known to be optimal for a larger class of problems than the Sidney-type heuristic, the latter algoristic is simpler to program and appears to be almost as accurate as the tree-optimal approach. (The tree-optimal approach seems easier to prove and to execute by hand.)

Morton and Dharan (1978) did a large computational study, comparing a first come first serve rule (FCFS), a myopic rule (most valuable job currently available) (MYOPIC), the Sidney-type algoristic (SIDNEY), and the tree-optimal algoristic (TREE) with from 10 to 50 jobs. Their results were that if the cost of FCFS = 100, then the cost of MYOPIC = 80, the cost of SIDNEY = 76, and the cost of TREE = 76. (All results are an average of the problems studied.) Where branch-and-bound was able to solve a given problem, it averaged only a very small fraction of 1% better.

Thus we may conclude that myopic scored very well, that the two algoristics picked up another 4%, and that branch-and-bound did not add much accuracy even when computationally practical.

8.4A Numerical Exercises

8.4A.1 Create a data set of about five 10-job problems with precedence constraints, weights, process times, and due dates. Graph them each on a large sheet of paper so that you can work directly on it. Make copies. Find the sequence that optimizes L_{max} for each problem.

8.4A.2 For the same data set and the F_{wt} objective, solve each problem by the myopic heuristic.

8.4A.3 Solve each problem by the tree-optimal algoristic.

8.4A.4 Solve each problem by the Sidney-type algoristic.

8.4A.5 Design other problems specifically to bring out differences in the capabilities between the myopic, tree-optimal algoristic, and Sidney-type algoristic procedures.

8.4B Software/Computer Exercises

8.4B.1 Code the tree-optimal algoristic with more than one possible subheuristic for which predecessor or successor to choose. Test your different subheuristics.

8.4B.2 Turn your tree-optimal algoristic code into a good beam search code (branching on choice of predecessor/successor).

8.4B.3 Turn your beam search code into a good branch-and-bound code.

8.4C Thinkers

8.4C.1 Why do you think the myopic procedure does so well on average? Why does it get worse as the number of jobs increases? Invent problems designed to make the myopic heuristic do very badly.

8.4C.2 Try to develop a heuristic for precedence constraints and the weighted tardiness objective. (If it helps, you may assume that most of the weights are 0, that is, only a few of the jobs actually are done for their own sake.) What problems do you run into?

8.4C.3 Try to develop a heuristic for precedence constraints, dynamic arrivals, and makespan objectives. What problems do you run into?

8.4C.4 Prove Proposition 4.

8.4C.5 Consider various computational efficiencies for the Sidney-type algoristic.

    (a) Keep track efficiently of whether or not the WSPT for a given set of jobs has already been computed.

    (b) Rather than returning to step 1 when an individual job is scheduled, backtrack to the next largest initial set.

8.5 COMPLEX PREEMPTION

8.5.1 Overview

All the preemptive models that we have studied to date centered on the assumption that the preemption involved no waste of the resource; that is, 30% of the job could be run now and the other 70% later. (The preemptive models we studied differed only in the question of whether individual pieces were allocated portions of the cost or benefit of the activity and in how to assign due dates.) This assumption is tremendously convenient mathematically; it means both that the preemptive case is easier to solve than the original problem (e.g., assignment algorithm) and that the preemptive case provides a lower bound for costs on the original problem.

Real situations where activities are taken off a resource before completion are often much more complicated. The preempted job is likely to need a second setup when it is put back on the resource. Furthermore, if the activity is interrupted at an inconvenient place, part of the first processing may need to be repeated. The extreme case of this is that the activity must be started over and treated as a new job. It is also possible that some or all of the raw material may have to be scrapped. (We will not consider this final issue here, since we have not yet included material costs in our models.)

In Section 8.5.2 we shall consider several simple examples of complex preemption and illustrate how bottleneck dynamics can deal with them. We do not attempt to cover all cases. Bottleneck dynamics is quite flexible. The user may need to design new uses in practice as needed. In Section 8.5.3 we discuss briefly how complex preemption may be treated by the other methods we have discussed.

8.5.2 Complex Preemption via Bottleneck Dynamics

Consider the following simple situation. A single machine has an initial set of jobs to be processed. Job j has setup time T_j, processing time p_j (p_j includes T_j) and weight w_j. It is desired to minimize weighted flow time F_{wt}. Thus the initial solution is to order by WSPT, or

$$w_1/p_1 > w_2/p_2 > \cdots > w_n/p_n$$

Now let us suppose a "hot job" arrives during the middle of processing job 1, that is, at time T, where $T_1 < T < p_1$. This job m has a much higher priority w_m/p_m than any other job. If it is allowed to preempt at time T, the setup time T_1 must be repeated when job 1 goes back on (CAE preemption).

Case I. If it does not preempt:

1 finishes at	p_1
m finishes at	$p_1 + p_m$
2 finishes at	$p_1 + p_m + p_2$
etc.	

Case II. If it preempts:

1 finishes at	$T_1 + p_m$	later than Case I
m finishes at	$-(p_1 - T)$	later than Case I
all others finish at	T_1	later than Case I

The net advantage of moving job 1 a unit forward (or back) is $w_1 - IRp_1$; for job m, it is $w_m - IRp_m$, while the cost of moving all others back one unit totals R. After some simplification we find the net value of the preemption to be calculated as follows:

Preempt with Second Setup Time

$$\pi_{pre} = \pi_m(1 - T/p_1) - \pi_1(1 + T_1/p_1) - T_1/(Ip_1p_m)$$

The value of the interchange is fundamentally the difference $\pi_m - \pi_1$ with some modifications:

(a) π_m is decreased depending on the time into the first processing as a fraction of the job being processed.

(b) π_1 is increased depending on the extra loss to job 1 as a fraction of its processing time.

(c) A priority correction must be made for the extra resource time T_1 used. The correction is proportional to that lost time and inversely proportional to the product of the two processing times.

The preempt repeat case sounds somewhat different, but the modification of priorities is similar. Now, after elapsed time T, the job must repeat the entire time T when it reprocesses. Thus in effect we have the previous case with a variable resetup time. We may substitute T for T_1 in the original formula. We may then solve for the maximum fraction of the time p_1 that may be wasted and still make preemption worthwhile. Call that time T^* and the maximum fraction $f^* = T^*/p_1$. Set $\pi_{\text{pre}} = 0$ in the original formula, changing the formula to

$$0 = \pi_m(1 - f^*) - \pi_1(1 + f^*) - f^*/(Ip_m)$$

Break-even Time for Preempt Repeat

$$f^* = (T^*/p_1) = [1 - \pi_1/\pi_m] / [1 + (\pi_1/\pi_m)(p_1/p_m) + 1/I\pi_m p_m]$$

Bottleneck dynamics can be used to analyze many other kinds of complex preemption. Some of these will be examined in the exercises.

8.5.3 Complex Preemption via Other Methods

It is relatively straightforward to analyze complex preemption via (extended) neighborhood search, branch-and-bound, and/or beam search, although the computational practicality of these methods has not yet been tested.

(Extended) Neighborhood Search. The first difficulty in developing neighborhood search for this problem is to clarify what a job sequence means, and therefore what pairwise interchange means.

The main thing that needs to be clarified is the notion of how preemption corresponds to a given job sequence. First, the ordinary job sequence can be used to indicate in what order jobs are started, irrespective of whether or not they are later preempted. Thus $[6, 1, 9, 3, \cdots]$ indicates that jobs are started in order $6, 1, 9, 3, \cdots$ irrespective of subsequent preemption. Second, split each position in this sequence into an ordered pair. $(1, 0)$ means job 1 is not preempted; $(0, 1)$ means it is preempted by the job immediately to the right. Thus $[(6, 0)\ (1, 0), (9, 0), (3, 0), \cdots]$ means there is no preemption and this is equivalent to the original meaning of $[6, 1, 9, 3, \cdots]$. Also, $[(6, 0), (0, 1), (9, 0), (3, 0), \cdots]$ means job 9 interrupts job 1 as early as feasible, but no other jobs preempt. Finally, $[(0, 6), (0, 1), (0, 9), (3, 0), \cdots]$ means that job 1 preempts 6, and then 9 preempts 1. As an example of a neighborhood, general pairwise interchange consists of two possible operations:

(a) Interchange any two pairs, without changing the internal order of the pairs.

(b) Interchange no pairs, but interchange the two members of a single pair.

This allows us to do neighborhood search, tabu search, simulated annealing, or genetic algorithms. Note, however, two caveats:

(a) Calculating the cost effect of a pairwise interchange may be quite complicated due to the extra constraints.

(b) There may be a great many undesirable local minima.

Branch-and-Bound/Beam Search. Similarly, branch-and-bound is straightforward, except that obtaining lower bounds is more difficult. As in the setup section, lower bounds can be obtained by reducing all setups to the minimum setup for a job, dropping all extra resource usage from extra setups and repeats of processing, and then forming the preemptive relaxation of the resulting problem. It is not clear how good the resulting bounds would be.

Beam search seems more promising, since again all that is required are decent "guide" heuristics to obtain trial solutions. These are available from the general bottleneck dynamics approach.

8.5A Numerical Exercises

8.5A.1 Consider a static F_{wt} problem, with independent setup included in the processing time, but a new setup required to restart a preempted job:

Job j	1	2	3	4	5	6
d_j	14	6	7	12	12	27
p_j	6	8	2	4	6	4
w_j/p_j	2	1.8	6	1.6	1.4	10
r_j	0	0	5	0	0	23
T_j	1	1	3	1	4	1

(a) Solve this problem via bottleneck dynamics.

(b) Find the optimal solution to this problem by inspection.

(c) How sensitive is your answer to the exact time of the arrival of the two hot jobs?

8.5A.2 Repeat 8.5A.1 for the T_{wt} criterion.

8.5B Software/Computer Exercises

8.5B.1 Modify one of your neighborhood search routines to provide advanced heuristics for 8.5A.1.

8.5B.2 Modify one of your beam search routines to provide advanced heuristics for 8.5A.1.

8.5C Thinkers

8.5C.1 Modify the bottleneck dynamics analysis of this section to handle sequence-dependent setups.

8.5C.2 Modify the bottleneck dynamics analysis of this section to handle dynamic arrivals/precedence constraints.

(A) 8.6   RELATED PROBABILISTIC RESULTS

Probabilistic results are in general much more difficult. We present some relatively straightforward extensions for the case of general precedence constraints with random processing times.

Precedence Constraints with Probabilistic Processing Times: Due Date Rules. It is very easy to generalize the EDD rule for minimizing L_{max} and T_{max} to stochastic processing times.

Proposition 8. L_{max} and T_{max} are minimized by EDD for stochastic processing times as well.

Proof. Consider any particular realization of processing times. L_{max} and T_{max} will be minimized by EDD, irrespective of the processing times. [Of course, $E(L_{max})$ and $E(T_{max})$ will be minimized as well.] ∎

PART III
MULTI-MACHINE PROBLEMS

In Part II we have studied one-machine problems in some detail. Our purpose in Part III is to the following:

(a) To develop basic techniques for dealing with multi-machine problems.
(b) To consider multi-machine problems such as parallel machines, which can be treated as a single resource (solved by a permutation schedule).
(c) To consider simple routing problems.

One straightforward way to create tools to simplify multi-machine scheduling is to take a complex multi-resource shop and simply prespecify sequencing rules/ decisions for all resources but one. We say that resource has been *embedded* in the larger problem. Embedded problems are clearly much easier to solve than the full problem; in fact, techniques we have already learned will often suffice. But more importantly, embedded problems can be treated as subproblems to be solved in heuristic approaches to the full problem.

Chapter 9 introduces and develops some procedures for solving embedded problems. Chapter 10 develops some new heuristic approaches, so called "bottleneck approaches." Both OPT-like and shifting bottleneck approaches use the embedding idea we have mentioned. Finally, we develop bottleneck dynamics much more fully than previously, with emphasis on accurate estimation of prices and lead times.

Chapter 11 develops models for equal parallel machines, proportional parallel machines, and batch machines, which are mathematically similar. In all cases there is a single queue, and there are rather obvious ways for determining the next machine to schedule, so that the problem reduces to determination of priorities for the queue as before. Chapter 12 develops simple routing problems including the generalized parallel machine problem.

____9
EMBEDDED PROBLEMS

9.1 INTRODUCTION

Here we consider simple one-resource problems. Consider a multi-resource complex shop. Focus on one critical resource. Pretend the other resources are in large supply, so that queues will not develop for them. Thus there is only one queue and hence, by our definition, only one effective resource. The resulting problem is called the *simple embedded one-resource problem*. It is usually much simpler to deal with (exactly or approximately) than the full problem. The principle used in OPT-like procedures suggests that if the most critical resource is embedded, then the solution obtained to this simple embedded problem may be a good approximation to the full solution. (If this resource is by far the most important, pretending the others are "free" may not be too far off.)

An extension to this idea is called the *complex embedded one-resource problem*. We model some of the other resources to be in limited supply but specify exact sequencing rules for them in advance. The resulting problem again has only one decision queue and hence is easier to analyze. The shifting bottleneck method (SBM) is a sophistication of OPT-like procedures, which solves simple and complex embedded problems in an iterative "neighborhood search" way. (See Chapter 10.) Thus we will be especially interested in developing ways to solve simple/complex embedded problems very quickly and easily, since such problems need to be solved repeatedly as subproblems in the SBM. The simplest case of interest is the case where each job is a set of activities to be performed in series, the so-called classic job shop. However, we also may consider cases of scheduling projects or multiple projects, relaxing all but one resource in the same way.

A problem in studying one resource embedded in a complex shop at this point is that complex shops are multi-resource problems and thus will not be fully discussed until Part IV. On the other hand, if we put off the discussion of the multi-processor single-resource case until then, we would miss the development of much intuitive insight, which will help us gain initial insights on multi-resource models.

We make a compromise on this conflict by developing one example carefully and just sketching more complex cases, indicating later sections of the book where that case will be studied in greater detail.

In Section 9.2 we study embedding for serial jobs, that is, jobs where each activity must be done in order and each on a specified different machine. This is also called the "classic job shop." We discuss solving the resulting equivalent one-machine problems, first for makespan (which is the easiest) and then for other objectives. We discuss both simple embedding and complex embedding.

In Section 9.3 we sketch a number of model extensions:

(a) Reentrant shops, where a job may visit a machine more than once (such as an annealing furnace).

(b) Single and multiple project shops, where the activities of the job may have complex precedence constraints, with reentrant behavior.

(c) Complex resources and multiple resources per activity.

9.2 SERIAL JOBS (CLASSIC JOB SHOP)

9.2.1 The Classic Job Shop Model

A job shop has m ($k = 1, \ldots, m$) machines, each of which can do several types of operations but can process only one activity at a time. At time zero there are n ($j = 1, \ldots, n$) jobs available for processing. Each job consists of q ($i = 1, \ldots, q$) operations in series. An input table, such as Table 9.1, must be given of p_{ji}, the processing time of operation i of job j. A second input table must be given of k_{ji}, the machine k on which operation i of job j is to be run. (Only one activity of a job may be assigned to a machine; that is, this is the nonreentrant case.)

The pair ji represents operation i of job j and thus represents a particular activity. If we wish to specify a priority ordering for activities on machine k, we would specify $ji(k, 1)$ as the first activity to go on k, $ji(k, 2)$ as the second, and so on. In the

TABLE 9.1 Processing Times p_{ji}

	Operation		
Job	1	2	3
1	4	3	2
2	1	4	4
3	3	2	3
4	3	3	1

TABLE 9.2   Machine Routing k_{ji}

	Operation		
Job	1	2	3
1	1	2	3
2	2	1	3
3	3	2	1
4	2	3	1

machine routings in Table 9.2, for example, we see that the activities to be scheduled on machine 3 include 13, 23, 31, 42. A particular fixed sequencing rule might be

$$ji(3, 1) = 13; \quad ji(3, 2) = 23; \quad ji(3, 3) = 31; \quad ji(3, 4) = 42$$

Note that the sequence rule can select only from activities assigned to k. The basic constraints for assigning an activity to start are as follows:

(a) It cannot start until the job's previous operation is finished.
(b) It cannot start until the machine is free.
(c) It cannot start until any decision or sequencing rule gives the green light.

In equations we might write

$$C_{ji} = S_{ji} + p_{ji} \tag{1}$$

This reads: the completion time of operation ji is its start time plus its processing time.

$$S_{ji} \geq C_{j(i-1)} \quad \text{(or 0 for } i = 1\text{)} \tag{2}$$

This reads: it cannot start until the previous operation for that job has completed.
    Suppose that $ji(k, q^*) = ji$. That is, activity ji is scheduled to be on machine k and in position q^*.

$$S_{ji} \geq C_{ji(k, q^* - 1)} \quad \text{(or 0 if first priority)} \tag{3}$$

This reads ji cannot start until the completion of the jobs with priorities before this job. (Either a fixed number of jobs to go first, or all available with higher priority, etc.)
    If every machine has given priorities, we may solve (1), (2), and (3) for completion times for all operations and jobs via simulation. Start with an event list and a clock.

Step 1: $t = 0$.

Step 2: Start any feasible operations.

Step 3: Calculate their finish times and enter as future events in the sorted event list.

Step 4: Increment time to the next event in the list.

Step 5: Return to step 2.

It is possible to set up sequencing rules for which our simulation would jam (this is usually called *gridlock*). For example, consider a problem with two machines and two jobs, each with two operations. Job 1 has two operations 11 and 12; job 2 has 21 and 22. Operation 11 is done on machine 1, operation 12 on machine 2. Similarly, operation 21 is done on machine 2, operation 22 on machine 1. Now the corker: our sequencing rule for each machine is that second operations of jobs must be done before first operations! Thus both machines will wait forever for second operations to process, which can never happen because the predecessor operations will never get processed.

There are a number of ways out of this impasse. We list a few of the most important.

Proposition 1. Gridlock cannot occur under any of the following situations:

(a) Machines are dedicated to first, second, kth operations.

(b) Any first operation has priority over any second operation, and so on.

(c) Priorities are dispatch (choose the highest priority job currently available).

Proof. Obvious. ∎

We call the case of dedicated machines a *dedicated job shop*. Note that a flow shop is a special case where there is just one machine dedicated to the kth operation. A somewhat different way to deal with this problem that is useful in search is to find a starting solution that is not blocked, and then search in such a way as to look only at nonblocking changes.

We are almost ready to consider solving some embedding problems for the classic job shop. First, we need to talk about the amount of time that must expire in a shop before a given operation can start, and the amount of time that must expire after the operation finishes and until that job finishes.

Definition. The *head time* of operation ji is the minimum time before ji can start if no previous operations are required to wait; that is,

$$HT_{ji} = \sum_{i'=1,i-1} p_{ji'}$$

We define the tail time of operation ji similarly.

Definition. The *tail time* of operation ji is the minimum time before job j can finish after ji finishes if no subsequent operations are required to wait; that is,

$$TT_{ji} = \sum_{i'=i+1,q} p_{ji'}$$

Definition. The *make time* of job j is the total time required to complete j if there were no waiting; that is,

$$MT_j = \sum_{i'=1,q} p_{ji'}$$

Definition. The *span time* of the job set is the maximum time of all jobs to complete if there were no waiting; that is,

$$ST = \max_j \{MT_j\}$$

If the shop is actually simulated for a given set S of sequencing rules for machines, the head time and tail time will be larger, and will depend on S.

Definition. The *head lead time* $HLT_{ji}(S)$ of operation ji is the actual time before operation ji can start given policy S.

Definition. The *tail lead time* (also just called the lead time) $TLT_{ji}(S)$ of operation ji is the actual time before job j finishes after ji finishes, given policy S.

We turn to defining and solving embedded problems in classic job shops.

9.2.2 Simple Embedded One-Resource Problems

Remember that in simple embedded problems we create a one-resource problem by assuming that all resources except the one of interest are in great enough supply so that no queues will form at the other activities. Thus activities before or after this machine experience no waiting, and the arrival time of operation ji at the current machine k is just the head time; that is,

$$r_{ji} = HT_{ji}$$

Now, if the final due date of job j is d_j, and operation j requires precisely the tail time after finishing this operation until the job finishes, then the effective due date for operation ji to finish is

$$d_{ji} = d_j - TTji$$

Finally, some jobs may not process on this machine at all. Since such jobs do not have to wait on any other machine, they will simply have a fixed completion time equal to the make time; that is,

$$C_j = MT_j$$

Now we have simply a standard one-resource problem with dynamic arrivals to solve. That is, optimize the objective function, for a group of jobs ji, with arrival times r_{ji}, due dates d_{ji}, and completion time of some of the jobs fixed. This sort of a problem can be solved by the methods of Chapter 7.

One minor problem is that for at least two objective functions, makespan and weighted flowtime, there are no due dates. An alternative formulation of the minimum makespan problem can easily be constructed with due dates. Set all the due dates equal to any arbitrary time, say, the span time. (Since the span time is the makespan with no waiting at *any* machine, it is a lower bound for the makespan in any event). Change the objective function to "minimize the maximum lateness."

Proposition 2. Consider the one resource dynamic makespan problem. Modify the problem to add due dates at time $= 0$ for all jobs, and modify the objective to "minimize maximum lateness." The two problems have equivalent optimal sequences and equal optimal objective functions.

Proof. Left as an exercise. ■

Setting due dates for the weighted flow case is even easier. Set all due dates at some common time ($t = 0$?) before any jobs arrive. Thus all jobs will be tardy, and measuring weighted tardiness will also measure weighted flow/weighted lateness, just by using the flow weights in place of the usual tardy weights.

The makespan objective is by far the one most studied for either simple or complex embedded one-resource problems. We see that it reduces to finding fast solutions for the T_{max} problem. Remember for this problem that a strong lower bound is easy to construct: order available jobs by EDD, preempting as necessary if an earlier due date job arrives. Similarly, a good heuristic is simply to order available jobs by EDD. It should be easy to construct branch-and-bound to handle large problems, and several heuristic methods would be expected to give good results.

Designing a special purpose algorithm to further exploit such special structure can lead to even more efficient solutions. Carlier (1982) developed a special procedure for the dynamic T_{max} problem, which is able to solve up to perhaps 100 jobs and hence represents an extremely fast and efficient method for this problem. We shall see in Chapter 10 that this allows an effective procedure for solving makespan problems in large shops, using the shifting bottleneck procedure.

The dynamic weighted flow problem would also likely be quite easy: good heuristics and quick and efficient lower bounds should be just as easy to produce. The weighted tardiness problem is probably somewhat more difficult, but it would likely be possible to solve 20- to 25-job problems.

9.2.3 Complex Embedded One-Resource Problems

In complex embedded problems we complicate the issue by allowing there to be one or more other resources, which are in limited supply and which have waiting lines. But to keep it a one-resource problem for decision purposes, a sequencing rule is prespecified at each of the other waiting lines, so that there remains only one queuing discipline decision in the problem.

Now our arrival times r_{ji} will not be determined simply by the head time HT_{ji} but somewhat more complexly by the head lead time HLT_{ji}, which is the sum of operation times and waiting times of the activities preceding ji. Similarly, our due dates d_{ji} are not determined by the tail time TT_{ji} but somewhat more complexly by the tail lead time (lead time) TLT_{ji}, which is the sum of operation times and waiting times of the activities after ji.

For most objective functions we are presented with a chicken-and-egg problem. We cannot optimize our one resource problem without knowing arrivals r_{ji} and local due dates d_{ji}; but we can most easily determine the latter by simulating the full system with *all* policies, which requires that we already know the optimal queuing rule we are trying to determine!

One possible way out is to use an iterative method.

Iterative Method for the Complex Embedded Problem

Step 1: Solve the simple embedded problem, using head times and tail times to estimate r_{ji} and d_{ji}.

Step 2: Simulate the full system using the current estimate of the optimal policy for the sequencing of the active resource.

Step 3: Calculate the objective function arising from this current estimate of the optimal policy.

Step 4: Get new r_{ji} and d_{ji} arising from the simulation.

Step 5: If unchanged since last iteration, terminate with the locally optimum solution.

Step 6: If a different termination criterion is met, terminate with the best solution to date.

Step 7: Calculate a new optimal policy based on current r_{ji} and d_{ji}.

Step 8: Return to step 2.

This procedure is really quite general and is used many times throughout the book. However, it remains a heuristic rather than an algorithm and needs to be tested.

Proposition 3. If the iterative method above terminates with repeating r_{ji} and d_{ji} for two successive iterations, then a local optimal solution has been obtained.

Proof. Left as an exercise. ∎

For the *makespan* criterion, Adams et al. (1988) give a faster and more straightforward method for solving the r_{ji} and d_{ji} problem. Our method is, of course, more general. An interesting question for the simpler makespan criterion is to find under what conditions the above iterative method will terminate finitely with an optimal or near-optimal solution.

9.2A Numerical Exercises

9.2A.1   The *classic flow shop* is a special example of the classic job shop for which there are exactly as many machines as operations, and every job goes through the same set of machines in the same order. Consider a static version of such a problem, with five jobs, three machines, and a makespan criterion, given by the following table:

Job j	1	2	3	4	5
p_{j1}	6	4	3	9	5
p_{j2}	8	1	9	5	6
p_{j3}	2	1	5	8	6

(a) Decide which is the dominant machine. Explain.

(b) Form the simple embedded machine problem for it, and solve it the best you can. (Use the software package if you like.)

(c) Sequence the other machines in the same order, and solve the resulting problem to obtain a legitimate makespan for the problem.

(d) Do you think your makespan is optimal for the problem? Why or why not?

9.2A.2   Formulate a small classic (interesting) job shop problem with makespan criterion, with three machines, six jobs, and two operations per job. Find the best solution you can by inspection.

9.2A.3   For the same example, choose the machine that seems to be the bottleneck. Specify the T_{max} that needs to be determined to solve the simple embedded problem for this resource.

9.2A.4   Find the optimal solution to this T_{max} problem. (You need not prove that it is optimal, unless you use the software package.)

9.2B Software/Computer Exercises

9.2B.1   Write a routine to solve the dynamic T_{max} problem.

9.2B.2   Use 9.2B.1 to program the iterative method for the complex embedded problem for makespan. What is your experience about convergence?

9.2C Thinkers

9.2C.1   Give an interesting new condition under which gridlock cannot occur.

9.2C.2   Prove Proposition 2.

9.2C.3   Prove Proposition 3.

(A) 9.3   OTHER MODELS

9.3.1   Introduction

In Section 9.2 we developed the subject of embedding for the classic job shop. This was perhaps a little confusing, considering that the flow shop and job shop materials aren't really developed until Part IV. We justified this in terms of motivating some of the bottleneck methods to come in Chapter 10, in particular, OPT-like methods and shifting bottleneck methods. In many cases, however, the reader may obtain a stronger understanding of this material after first studying Part IV, and then coming back and fully mastering Section 9.2.

There is even less justification for carefully developing embedding for other models at this point, since intuition sufficient for Chapter 10 has perhaps already been developed. However, the alternative of developing this embedding material in scattered ways throughout the text is not very attractive either. We have finally chosen the alternative of marking this section as (A), showing it can be omitted at first reading if desired, although it may give some intuition even if not understood in every detail. In any event, the reader may find it profitable to reread the associated material here after mastering the corresponding model later in the book.

In Section 9.3.2 we discuss embedding for reentrant shops. Later material that will help in understanding this section includes Section 14.5 (real-world case with reentrant machines), Chapter 16 (heuristics for job shops), and Sections 18.2 and 18.3 (exact and heuristic methods considering embedding for reentrant shops as a special case of single resource project job shops).

In Section 9.3.3 we discuss embedding for single and multiple project job shops. Chapters 15 and 16, which discuss job shops, give a good foundation. Again, Sections 18.2 and 18.3 give exact and heuristic embedding for project shops by treating an embedding problem as a single-resource constrained project problem.

In Section 9.3.4 we discuss complex multi-machine resources and multi-resource activities. Material relating to this includes Sections 14.4, 14.5, 18.1, and 18.2 and Chapter 19.

Finally, Section 9.3.5 briefly looks at some rather different embedding issues. Parallel resources, including parallel machines, and routing situations are discussed, with related material available in Chapters 11 and 12. (Here, for one reason or another, the parallel resources are not aggregated.) We also briefly talk about embedding dominant jobs rather than dominant machines.

9.3.2   Embedding in Reentrant Shops

Consider a classic job shop again, as in Section 9.2, except that now jobs may return to some machine k more than one time; that is, more than one activity of a job may be processed on the same machine. For the sake of this section as well as the next few following it, if we are embedding on some machine k, we suppress k for simplicity.

For example, instead of talking about operation jk on machine k, we simply talk of j, which is understood to be on the embedded machine for operations $j1, j2,\dots,$

jn_j. The other machines may all be relaxed, giving the simple case, or some may not be relaxed but may have a fixed sequence rule. In any case, we suppress this overall policy S; head times, tail times, and delay times (which we introduce) will all be understood to depend on S, however.

What is the total completion time of job j?

Job j will arrive at the embedded machine at time HLT_j, which is simply the arrival time in the shop plus the time in the shop until all previous operations are finished, which may include some waiting. This time is easily calculated, since the rest of the shop has fixed rules.

Then $j1$ will wait for the processing of higher priority jobs, which depends on the policy we are trying to determine.

Then $j1$ will process, taking time p_{j1}.

Then there will be a delay $\text{TLT}_{j1,j2}$, which represents time taken outside the machine to process the operations between $j1$ and $j2$, which depends on the sequence on the embedding machine, which is not fixed, and that off the machine, which is.

At this point $j2$ arrives, and the process repeats.

Note how much more complicated this is than the classic job shop problem and how complicated it looks to solve. However, the iterative method of Section 9.2.3 can be used to give approximate solutions with only minor changes.

Changes to Iterative Procedure

Step 1: Find the initial estimates of arrival times $r_{j1}, r_{j2}, \ldots$ for all jobs j simply as the sum of all previous processing time. (Assume 0 waiting even on the embedded machine.)

Step 2: Similarly, find initial estimates of due dates $d_{j1}, d_{j2}, \ldots$ for all jobs j simply as d_j less all processing time after $j1, j2, \ldots$ (Assume 0 waiting even on the embedded machine.)

Step 3: Solve the one-machine problem with those arrival and due dates and processing times.

Step 4: Simulate the full system with given policy S off the machine and current solution on the machine.

Step 5: Calculate the objective function value arising from this current full policy.

Step 6: Save actual arrival times, lead times, and due dates implied from the simulation, and use these in the next iteration.

Step 7: If the embedded policy is repeated, terminate with a local optimum.

Step 8: If other termination criteria are met, terminate.

Step 9: Return to step 3.

It is important to note in step 6 above that the lead times to save from the previous iteration for any activity on the embedded machine are exactly the processing times plus the waiting time for every successor activity to that activity irrespective of whether or not that successor runs on the embedded machine or not.

9.3.3   Embedding in Single and Multiple Project Shops

Embedding in a single project shop presents many of the complexities of the reentrant shop. Whereas in the reentrant shop we may have as many as n delays to consider between the finishing of an activity and the start of another due to external processing, for the project shop this may rise to more like n^2 delay relationships for a single project, and that many for each project in the problem.

For the makespan objective, all the projects may be considered a single project to be finished. Also, any regular single project problem can be solved as a makespan problem, since the objective is certainly to finish the project as soon as possible.

Since makespan problems are often much easier than other problems, there is hope that the one-machine procedure, including all the delays, is exactly solvable. This would be the approach of the original shifting bottleneck procedures, to be discussed in Chapter 10.

On the other hand, suppose the shop is doing five projects, with weighted tardiness criterion. It is easy to show by example that the projects cannot be aggregated and that solving a makespan criterion as an approximation may behave arbitrarily badly.

Under these circumstances, we may realistically solve an embedded problem by the same iterative procedure given earlier in this chapter. It remains only to specify a starting values and an iterative procedure for estimating arrival times r_{jh}, lead times LT_{jh}, and local due dates d_{jh}, where j is the project with due date d_j and h represents one of the activities to be performed on the embedded machine k.

Project Estimation Procedure (Initial)

Step 1: Relax *all* resource constraints, and solve the unconstrained project problem for each project.

Step 2: r_{jh} are the corresponding early start times.

Step 3: d_{jh} are the corresponding late finish times, corrected to be later by the amount the project would be early, or to be earlier by the amount late.

Step 4: LT_{jh} is project completion less late finish.

Project Estimation Procedure (Iteration by Iteration)

Step 1a: From the previous iteration, augment each activity processing time by the waiting time for resources in the previous activity.

Step 1b: Solve the augmented process time problem as a simple unconstrained project.

Steps 2–4: As before.

There are also a number of important variants, such as smoothing values between iterations. See Chapter 21.

9.3.4 Complex Resources, Multiple Resources per Activity

Complex Resources. As we shall discuss in Chapters 11 and 19, there are many cases where a group of machines only requires a single input queue, and thus heuristics can often be developed treating it as a dynamic one-machine problem.

An example is a grouping of five equal parallel machines. Another example is a group of five parallel machines that differ only by a simple speed factor. Another example is a flexible manufacturing center (FMC) with a single loading and unloading station, seven or eight machines, and a conveyor belt.

If there is a shop made up of such complex resources, it makes sense to talk of the bottleneck complex resource and to consider embedding it in various ways. For example, suppose there are just three sets of equal parallel machines in the shop and one is definitely the bottleneck. An embedded problem around the bottleneck can be solved, as long as we have a procedure for approximately solving a set of parallel machines with dynamic input and local due dates. In fact we shall see that there are reasonable heuristics for solving this one-resource problem, which are little more difficult than solving a one-machine problem, so that doing either a full OPT-like analysis or a full shifting bottleneck analysis is quite feasible.

Multiple Resources per Activity. It is very common in a shop for an activity to require several resources. For example, an NC drilling machine requires an operator. If the operator is dedicated to that machine, the operator need not be considered a separate resource. The machine and operator have a common queue. However, very often the operator is required to keep track of perhaps four machines. If all four machines would like to process an activity at once, the four activities are the queue for the operator, quite distinct from the queues for the machines. In a similar fashion, the machine usually requires a tool to do the particular drilling. If there is an adequate set of tools always available that can be changed very rapidly, then the tools can be ignored as a separate resource. But if tools must be inserted from a tool crib or an automatic tool exchanger, then three resources will now be required for the activity.

In large-scale project management, activities are usually quite aggregate and will almost always require several resources. Pouring the basement of a school may require a cement mixer, cement, three trucks, a crew of eight, and a supervisor. Our resource-constrained project models of Chapter 18 handle such aggregate situations rather well.

Nonaggregate multi-resource problems can be quite difficult, and a very common solution is to do some embedding. For example, if labor and tools are treated as if in large supply, the resulting ordinary job shop problem may become tractable. Of course, it is important to check via sensitivity whether the shop is the true bottleneck.

(A) 9.3A Numerical Exercises

9.3A.1 Consider a simple reentrant shop with three machines and four jobs. Jobs either go through machine A and then machine C twice or through machine B and then machine C twice. The objective is makespan. Make up some data for four static jobs, and find the best answer you can. (If this is too easy, make up harder data.)

9.3A.2 In 9.3A.1 solve a simple embedding of machine C. Try to verify whether your answer is optimal. How does it compare with your answer in 9.3A.1?

9.3B Software Exercise

9.3B.1 Write a software program to generate dynamic data for a problem of the form of 9.3A.1, which also does a good heuristic solution of the simple embedding of machine C.

9.3C Thinkers

9.3C.1 Sometimes it may be useful to think of a particular *job* as a bottleneck rather than a machine. Consider a bottleneck job to be embedded in the shop if all other jobs are removed and that job is processed efficiently by all resources. When would this be useful? Can you give an example. (Compare giving the job a high priority.)

———10

MATHEMATICAL SOLUTION METHODS: BOTTLENECK APPROACHES

10.1 INTRODUCTION

In Chapters 5 and 6 we developed a number of mathematical methods, both exact and heuristic, for dealing with scheduling problems. These included exact methods, such as branch-and-bound and dynamic programming, and such heuristic methods as priority dispatch, neighborhood search, tabu search, simulated annealing, genetic algorithms, beam search, and approximate dynamic programming. For ease of exposition, they were first developed in the easier context of single-machine models, although we shall see that they are adaptable to simple multi-machine problems (Part III) and multi-resource problems (Parts IV, V, and VI).

There are a number of additional techniques, which we term *bottleneck approaches*, that could not be discussed then, since clearly it is not possible to have a bottleneck machine when there is only one. These methods include:

(a) The forecast myopic dispatch approach.
(b) OPT-like methods.
(c) The shifting bottleneck algorithm.
(d) Multi-resource bottleneck dynamics.

We restrict our attention in this chapter to illustrating these methods for the classic job shop, which has already been introduced. We will develop these methods for other models as they are introduced in later chapters.

10.2 THE FORECAST MYOPIC DISPATCH APPROACH

10.2.1 Overview

The forecast myopic dispatch approach simulates the job shop forward in time, using an event list. Whenever a machine becomes free, all operations available to be scheduled at that time are evaluated by a priority heuristic *for that machine* (hence the name "myopic"). For example, suppose the objective function for the shop is F_{wt}, weighted flow time. Then the WSPT ratios w_j/p_{ji} are calculated, and the highest priority operation is run on the machine next. We also enter in the event list the later time $t + p_{ji}$ at which point the machine will again be free for a new operation.

In many cases, of course, the machine's heuristic requires knowledge of a due date. It is necessary to estimate a local due date for the job at the local machine. We know the final due date d_j for job j. If we can somehow estimate the tail lead time TLT_{ji}, we could estimate the local due date by $d_{ji} = d_j - \text{TLT}_{ji}$. In Section 10.2.2 we discuss methods for estimating the tail lead time (usually just called the lead time). In Section 10.2.3 we discuss intuitively strengths and weaknesses of the method.

Terming the myopic approach a "bottleneck method" may seem somewhat questionable. It is partly a question of convenience; we could not introduce the method in Chapters 5 and 6 because it is not a single-machine method. But it can be partly considered that the method treats the current machine as a (flow) bottleneck. Expediting flow at the current machine will be helpful downstream, and information for making better use of the current machine is certainly more reliable and available than downstream.

10.2.2 Lead Time Estimation

There are a number of ways of estimating the lead time from operation ji until the finish of job j, which differ in ease of calculation and accuracy. They divide into roughly three classes: (a) historical fixed parameter, (b) historical varied parameter, and (c) iterative estimation.

Historical Fixed Parameter. The simplest possible useful procedure is to estimate lead time as a fixed multiple of the remaining process time TT_{ji}.

Simple Historical Lead Time Estimation

$$\text{TLT}_{ji} = k\text{TT}_{ji}$$

k can be estimated very easily historically by finding the average time through the shop divided by the average process time for a large number of jobs. k is often taken as being between 3.0 and 6.0.

Historical Varied Parameter. Unfortunately, in practice k is highly variable, depending on the current shop loading. An improvement would be to estimate k historically for a lightly loaded, moderately loaded, and heavily loaded shop. Then the loading condition of the shop would be an input parameter.

Just as one could try to correct for overall shop loading, one could also try to correct for priority of the job, loading on the machine, and average loading on following machines. To do this, one could run regressions on past shop performance.

Historical Estimation

$$(k \text{ factor}) = A_0 + A_1(\text{shop load}) + A_2(\text{job priority}) + A_3(\text{machine load})$$

This is a research question at this point.

Iterative Estimation. So far, we have talked about estimating k historically and simply using that. However, the computer gives us a lot of power. We could also choose to run the whole dispatch procedure for a number of values of k (3.0, 3.5, 4.0, . . .). Or we could try some variations in the regression formula.

However, if we are willing to run the simulation more than once, lead time iteration is likely the method of choice.

Lead Time Iteration for Myopic Dispatch

Step 1: Choose any initial estimate of lead times (e.g., $\text{TLT}_{ji}(0) = 3\text{TT}_{ji}$).
Step 2: Perform a myopic dispatch simulation.
Step 3: Record the objective function obtained, $F(n)$ for iteration n.
Step 4: Record actual (ex post) lead times, $\text{ATLT}_{ji}(n)$ for iteration n.
Step 5: Make smoothed new estimates:

$$\text{TLT}_{ji}(n + 1) = (1 - \alpha)\text{ATLT}_{ji}(n) + \alpha\text{TLT}_{ji}(n)$$

Step 6: If the termination condition is satisfied, go to step 8.
Step 7: Go to step 2.
Step 8: Report the objective value and scheduling policy for the iteration giving the best value of the objective function.

The lead time iteration method given here has been used extensively by the authors and others in testing both myopic methods and bottleneck dynamics, with consistently good results. These results are reported fairly completely in Chapters 14, 16, and 18.

We explain the method somewhat more intuitively in words: First, we choose some initial lead time estimates, such as three times remaining processing time, allowing us to run the myopic dispatch simulation. This gives us a value of the objective function, telling how well we are doing, and a set of actual lead times resulting from our myopic policy. Second, we don't wish our lead time estimates to change too fast and so we smooth the new estimate and the old estimate together. (A typical smoothing constant might be 0.5). The objective function usually improves quite fast during the first few iterations, and then it begins to flip flop around due to discreteness. A termination rule might be: "stop when there has been no improvement in the objective function for five iterations. (This might typically involve 10 total iterations.) Save the best results to date as the output of the procedure." A simpler rule, which is actually also quite good, is: "take the best run of the first five iterations."

It is of interest that such iterative procedures are more commonly used for convex programming problems than in providing an approximation to integer programming. The intuition here is that for scheduling problems with many smaller pieces, the underlying problem is much like a convex problem (but with some medium size nonconvexity thrown in, which acts a lot like noise). Thus exponentially smoothing the lead time estimates is intended to reduce the effect of the noise. Similarly, exact convergence cannot be expected due to discreteness. We simply wait until the best solution to date does not improve. It is then an empirical question whether such a procedure obtains good results or not. (Again, see studies reported in Chapters 14, 16, and 18.)

10.2.3 Strengths and Weaknesses

The forecast myopic dispatch procedure shares many strengths with its cousin, the bottleneck dynamics procedure.

(a) *It is a dispatch procedure.* The cost of a dispatch procedure is very low, essentially that of running a simulation. Feasibility is automatically maintained in the simulation process. Reactive scheduling to changes during implementation of a dispatch plan is extremely inexpensive. Implementing full lead time iteration adds a factor of 5–10 to the weekly scheduling run, but no extra effort to the reactive scheduling. The latter can usually continue to use the established priorities except for changes that can be entered manually. A final strength of dispatch procedures is that they are understood well by human schedulers.

(b) Benefits of expediting an activity are accurately estimated. The use of lead time iteration to update estimated completion time, and hence update the marginal penalty of delaying a current activity, has widely been documented to be an accurate procedure in studies of both myopic procedures and bottleneck dynamic procedures. Research continues into procedures for estimating lead times with somewhat lower accuracy but much lower cost.

(c) Myopically minimizing local resource costs rather than remaining shop costs seems to be very robust. Myopic procedures set priorities with a benefit/cost ratio using costs at the local machine. For example, for F_{wt} the priority of ji at the local machine would be w_j/p_{ji}. An OPT-like procedure might set the priority at w_j/p_{ji*}, where p_{ji*} is the processing time used on the bottleneck machine downstream. A bottleneck dynamics procedure might estimate total resource usage on all remaining machines, weighted by some prices to be estimated. The prime weakness of the myopic procedure is that it is easy to design shop situations where thinking locally is inappropriate. In particular, if there is a strong bottleneck downstream, then jobs that have low usage of the bottleneck rather than the current machine should be given higher priority.

(d) Dispatch procedures are easily extended to allow reasonable amounts of inserted idleness. We term these x-dispatch procedures. Basically, one extends the concept of activities to be chosen from among just those available at the time of choice to include those in process on their previous machine. These must be given priorities biased somewhat downward to account for wasted processor time if they are chosen. (See Section 7.6.)

A shortcoming of the myopic dispatch approach is that it gives no attention to aiding a downstream congested area. However, initial experiments conducted by Morton and Pentico (1993) indicate that the myopic procedure is surprisingly robust, even when there are strong bottlenecks downstream. There seem to be two reasons for this:

(a) The myopic procedure expedites flow at the current machine. This increased flow will help downstream machines somewhat.

(b) This help is independent of any fancy but imprecise forecast of downstream needs and hence is stable and guaranteed.

10.2A Numerical Exercises

10.2A.1 Set up two problems with three machines, three operations, and six jobs. Let each job go to the same machines in the same order (flow shop). Solve the problems for the F_{wt} objective by the myopic dispatch method.

10.2A.2 Using your problems from 10.2A.1, assume the lead time is approximated as some fixed multiple of remaining process time. Solve the problems for C_{max} and T_{wt} by the myopic dispatch method.

10.2A.3 Using a spreadsheet or some such help, ascertain the effect of varying the multiple assumed in 10.2A.2.

10.2B Software/Computer Exercises

10.2B.1 Write a simple simulator for a three-machine flow shop, with a very simple lead time estimation procedure. Test a larger set of problems.

10.2B.2 Add lead time iteration to your simulator. Fine-tune the iteration smoothing constant. How much does lead time iteration improve your results from 10.2A.2?

10.2C Thinkers

10.2C.1 One of the advantages of the myopic method over bottleneck dynamics is that since only the current machine is considered, its price cancels out of priority formulas and need not be considered. Can you think of models where the myopic approach would still need to estimate the resource price of the current machine?

10.3 OPT-LIKE APPROACHES

10.3.1 Overview

In the late 1970s Eliyahu Goldratt, an Israeli physicist, was approached by a friend for help in scheduling his large-scale chicken ranch operation. Goldratt felt there must be some relatively simple principle in the operation that he could exploit to come up with a better scheduling system than those currently available. He quickly came up with a simple idea that helped his friend, but he realized it had broader potential. Within a few months the basic OPT package was written.

While the details of OPT remain proprietary, it is known to involve four basic stages:

(a) Determine the bottleneck resource in the shop.
(b) Schedule to use the bottleneck resource most effectively.
(c) Schedule the remainder of the shop up to the bottleneck.
(d) Schedule the remainder of the shop after the bottleneck.

Although OPT was an extremely successful and widely used system, it suffered a number of legal problems and is no longer being actively sold in the United States. However, a number of newer systems mentioned in Chapter 2, including OPIS, MICRO-BOSS, the shifting bottleneck algorithm, and bottleneck dynamics, incorporate and expand on many of OPT's principles.

We develop here an OPT-like procedure that is a dispatch procedure and closely related to the forecast myopic dispatch procedure and to bottleneck dynamics. First, we describe our dispatch OPT-like procedures; second, we discuss strengths and weaknesses of OPT-like procedures.

10.3.2 The OPT-like Method

We work through each of the four stages of our OPT-like method in turn.

(a) *Determine the Bottleneck Resource.* Three methods for determining the bottleneck suggest themselves immediately. A quick, but crude, method is simply to add up the processing time to be done on each machine and choose that machine with the greatest processing time. That is, *choose the machine with the greatest utilization.* (Alternatives would be past utilization or forecast utilization.)

A much better but more time-consuming way is to solve the *simple embedded one-resource problem* for each possible choice of the bottleneck machine. The simple embedded problem that gives the highest cost defines the bottleneck resource. (To see that this is correct, this is the machine—the only one limited in quantity—that would most limit our throughput.)

It is probably not necessary to solve such a problem for every possible bottleneck resource; probably only three or four will seem like reasonable candidates. We could take the resources with the four highest utilizations, for example.

A third method involves being able to solve the problem by a faster, somewhat lower quality, heuristic. Solve the problem in full by the heuristic. For each of the few most likely bottlenecks in turn, double the amount of that resource (two machines in parallel) and solve the problem again with the heuristic to see how much cost is saved by the relaxation. The machine whose expansion saves the most is the bottleneck.

(b) *Schedule the Bottleneck Resource.* Here the method of choice would be to solve the simple embedded problem for the bottleneck resource by the methods of Section 9.3. [We will already have this from part (a) with one of the methods there.] It is not necessary to solve this problem optimally. An accurate one-resource heuristic should be adequate.

(c) *Schedule up to the Bottleneck.* Consider the problem up to but not including the bottleneck as a separate subproblem. Set the due dates for this problem as the actual starting times to the bottleneck from the simple embedded solution. Solve this restricted problem by myopic dispatch, using lead time iteration if desired.

(d) *Schedule the Remainder.* Using the actual completion times up to the bottleneck from part (c) as the arrival times into the bottleneck, schedule the bottleneck through the end. Use myopic dispatch and lead time iteration if desired.

(Note that the procedure we have described does not solve a general classic job shop, since it depends on part of the shop being before the bottleneck and part after. However, our only purpose here is to illustrate in general how OPT-like procedures might work.)

When there are no due dates and lead times to estimate, an OPT-like procedure can be considerably simpler. One can make some use of secondary bottlenecks:

OPT-like Procedure for Weighted Flowtime

Step 1: Schedule by a dispatch procedure; that is, choose the highest priority available activity to be scheduled next.

Step 2: The priority is given by w_j/p_{ji*}, where p_{ji*} is the processing time for the downstream successor operation $ji*$ on the bottleneck machine.

Even if we are past the main bottleneck, this procedure suggests that we look ahead to the most important smaller bottleneck remaining.

10.3.3 Strengths and Weaknesses of the OPT-like Method

In comparing the OPT-like and myopic methods, the following hypotheses emerge:

   (a) The OPT-like method is only good if there is a single strong stable bottleneck with an adequate queue.

   (b) The myopic method is robust: it is more difficult to find cases where it behaves very badly than for the OPT-like method.

   (c) The myopic method is poorest when a strong downstream bottleneck has priorities very different from local priorities.

Studies by Morton and Pentico (1993) tend to confirm these and other hypotheses. We will discuss these issues and computational support when we discuss flow shops in a later chapter.

Other helpful references for this section include Jacobs (1984), Lundrigran (1986), Meleton (1986), Glassey and Petrakian (1989), Pence et al. (1990), Goldratt (1990), and Fawcett and Pearson (1991).

10.3A Numerical Exercises

10.3A.1 Consider again the two problems you generated in 10.2A.1. Increase all the processing times sufficiently on machine 4 so that it becomes the bottleneck; the objective is still F_{wt}. Solve these problems by the myopic dispatch method; solve by the OPT-like method for F_{wt}.

10.3A.2 Add due dates to the problems in 10.3A.1 and assume lead time is approximated as some fixed multiple of remaining process time. Solve the problems for C_{max} by the myopic dispatch method and by the OPT-like method.

10.3B Software/Computer Exercises

10.3B.1 Modify the simple simulator of 10.2B.1 to be able to compare myopic dispatch and OPT-like methods on a wider set of problems.

10.3C Thinkers

10.3C.1 Can you construct some small examples to show why OPT-like methods might work better for one strong bottleneck than for two or more nearly equal bottlenecks?

10.3C.2 Can you construct some small examples to show why OPT-like methods might work better for a bottleneck with twice the utilization of the other machines than for, say, 10% more?

10.4 SHIFTING BOTTLENECK ALGORITHM

10.4.1 Overview

A classic idea in nonlinear programming is to hold all but one variable fixed and then optimize over that variable. Now hold all but a different one fixed, and so on. For a discrete problem where there are only a finite number of choices for any variable, this procedure is guaranteed to increase the objective function monotonically, and thus to converge to at least a local optimum in a finite number of iterations. Furthermore, if we can somehow do the one-variable optimizations in order of decreasing importance, there is better hope that the local optimum so found will be the global optimum or close to it. This is the essence of the shifting bottleneck procedure.

In our classic job shop problem, the sequence in which jobs are to be processed on a given machine corresponds to fixing the value of a variable. We hold the sequence (order in which to process operations) on all machines but one fixed, and optimize the objective function by optimizing the sequence on the remaining machine. (It is just a little more complicated than this, because one of our choices for "fixed sequence" is to increase the capacity of the machine enough so that sequence is immaterial; all jobs can be processed without waiting.) A fundamental bottleneck idea is that often only one machine is very important to the objective function. Thus, if the problem is solved with all other machines fully relaxed, a near optimum will be achieved in the objective function, and a near optimal sequence for that machine for the original nonrelaxed problem. Actually, all the bottleneck approaches use this idea.

10.4.2 Basic Shifting Bottleneck Algorithm

The basic shifting bottleneck algorithm may be stated as follows:

Shifting Bottleneck Algorithm

Step 1: Solve the simple embedded problem for each machine in turn.

Step 2: The machine in the problem with the highest cost is designated the *primary bottleneck*.

Step 3: If the primary bottleneck problem was solved exactly rather than heuristically, this cost represents a *lower bound* on the full original problem we are trying to solve.

Step 4: Solve the complex embedded problem holding the primary bottleneck fixed, and all but one of the remaining machines relaxed. Do this for each remaining machine in turn.

Step 5: The machine in the problem with the highest resulting cost is designated the *secondary bottleneck*.

Step 6: Holding the secondary fixed, reoptimize the primary, then hold the primary fixed and reoptimize secondary, and so on until there is no further improvement or, say, three times. (In the following let us call this *balancing the bottlenecks..*)

Step 7: In a similar fashion, find the *third-level bottleneck.*

Step 8: Balance over the three bottlenecks.

Step 9: Now find the fourth-level bottleneck, the fifth-level bottleneck, and so on, finally ranking all the machines.

Step 10: If the final heuristic solution obtained has the same cost as the lower bound obtained from the first bottleneck solution, then it is guaranteed optimal.

Adams et al. (1988) term this procedure *shifting bottleneck I* (*SBI*). They worked only with the makespan criterion. They found that the procedure often obtains such good answers that they can be guaranteed to be optimal. They also develop a partial enumeration extension (*SBII*), which is even more computationally intensive. This extended procedure is a bit like beam search added to the original procedure. Each time the bottleneck is searched for, the top several candidates are saved, and each defines a branch on the tree. The beam width decreases in the depth of the tree; also, an evaluation penalty function prunes by penalizing both depth and deviation of the given machine from the bottleneck. The authors do not give many details of this procedure.

Note that the simple embedded problems are readily solved as dynamic one-machine sequencing problems. Also, the complex embedded problems with makespan objective are readily solved. Complex embedded problems for other objectives will probably require some kind of iterative approximation. Actually, since the shifting bottleneck algorithm is a heuristic in any event, it is not always necessary to solve the embedded problems exactly. In particular, exactness of solution is most useful to get the initial lower bound, and in the very late stages of the algorithm if a refining of the solution is desired.

10.4.3 Extension: A Stronger Lower Bounding Procedure

Although the complete relaxation of all resources but one is very simple and easy to work with, it is a very weak relaxation in the sense that solving a problem with all machines but one completely relaxed may give a poor lower bound—too far below the actual optimum (much more so, for example, for tardiness than for makespan). Suppose we work with a unit preemption instead. We could hope to get a much tighter relaxation and a much better lower bound on the objective function.

We could solve for the preemptive version of the entire problem directly by LP. However, it would be much faster (and probably very accurate) to look at the preemptive problem as a whole new problem and solve *it* by the shifting bottleneck procedure. (We term an embedded problem using preemptive relaxations a *quasi-simple* embedded problem.) Note that, like the complex embedded problem, the quasi-simple embedded problem must typically be solved approximately by iterating the starting and stopping points. However, the preemptive sequences for the other machines are at least easily available (e.g., for makespan the preemptive policy is just: schedule the available operation with the earliest local due date).

Stronger Lower Bound, Shifting Bottleneck Procedure

Step 1: Approximate the solution to the preemptive problem by the shifting bottleneck procedure.

Step 2: Solve the quasi-simple embedded problem for each machine in turn.

Step 3: Pick the machine for the problem with the highest cost, and designate it as the *primary bottleneck*.

Step 4: The corresponding cost will be based on a stronger relaxation. (We will have calculated an approximation to this tighter relaxation.)

Step 5: Finish the procedure as before. This raises two interesting questions: (a) Under what conditions will the optimal preemptive solution be obtained in step 1? (b) Can the procedure in step 1 be improved to guarantee the optimal preemptive solution?

We will discuss other extensions in later chapters.

10.4B Software/Computer Exercises

10.4B.1 Develop a shifting bottleneck routine for a three-machine flow shop.

10.4B.2 Check whether or not your program seems to find the optimal solution to preemptive flow shops.

10.4C Thinker

10.4C.1 What are some of the advantages and disadvantages of the shifting bottleneck procedure versus various types of dispatch procedures, such as forecast myopic?

10.5 BOTTLENECK DYNAMICS

10.5.1 Introduction

In Section 6.5 we developed bottleneck dynamics for a single resource. Such heuristic pricing methods for a static single-resource (one-machine) problem are relatively easy because:

(a) The unknown resource price enters all priorities equally for static sequencing and hence need not be determined for this case.

(b) The lead time from the completion of the activity until the completion of the job (TLT_{ji}) need not be determined, it is zero.

(c) Estimation of the economics of a decision is generally much easier for a one-resource problem.

We have already covered (b), the estimation of lead times, in Section 10.2. In this section, we generalize the heuristic pricing definitions given in Section 6.5. In Section 10.5.2 we will give a broad overview of various methods of estimating resource prices in both single and multiple resource problems. In Sections 10.5.3 and 10.5.4 we will discuss extending bottleneck dynamics decision methodologies to the multiple-resource case.

Heuristic Pricing Definitions

(a) Let $R_k(t)$ be the (implicit) price per unit time of using resource k at time t (the dual price of postponing all activities in the busy period).

(b) Let I be the interest rate derived from the firm's cost of capital.

(c) Therefore $IR_k(t)$ is the premium *per unit time shifted* of using a unit time's worth of k either earlier or later. (That is, I has units of "per unit of time" and so does R_k.)

(d) Let jk be the activity of interest, so named because it is processed on machine k and is required to be done before a downstream job j can be finished.

(e) Suppose the completion of jk has no influence on the objective function except via the completion of j.

(f) Define the nontime-based importance of job j by the weight w_j, where $w_j = (A_j)(V_j)$, A_j is value added in shop, and V_j is customer importance.

(g) Let $L_{jk}(t)$ be the forecast lead time from the time activity jk is finished until job j is finished, given jk is started at time t.

(h) Let the forecast slack (depending only on the forecast lead time) similarly be given by

$$S_{jk}(t) = d_j - L_{jk}(t) - p_{jk} - t$$

where d_j is the due date for job j and p_{jk} is the processing time of activity jk.

(i) Let the job–time urgency $f_j(S)$ be the marginal customer cost of making the completion later, conditional on the slack S. (That is, f_j is simply one of the additive components of the objective function.)

(j) Let the activity–time urgency $U_{jk}(t)$ be the job time urgency for the slack $S_{jk}(t)$, that is, $U_{jk}(t) = f_j(S_{jk}(t))$, the marginal cost of delaying jk in terms of the ultimate delay of job j.

(k) Then the delay price or delay cost of the activity jk is given by $w_j U_{jk}(t)$.

10.5.2 Estimating Resource Prices

The question of how bottleneck dynamics estimates resource prices in a scheduling problem has two levels:

(a) At the *strategic* level all bottleneck methods are based on a busy period analysis: if a resource is shut down for a day, the cost is that all operations

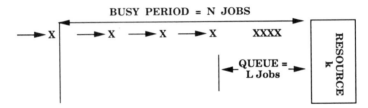

Figure 10.1. Busy period analysis of resource *k* pricing.

waiting or arriving in the current busy period are delayed by a day. We will develop this busy period analysis first.

(b) At the *tactical* level a number of variations are possible:
   (i) Estimated prices can be *static* or *dynamic*.
   (ii) Price estimation methods can be *simple*, *moderate*, or *complex*.
   (iii) Estimated prices can be by *fitted formula*, *iterative*, or *exhaustive search*.

We will discuss the strategic busy period analysis of prices first and later discuss the more tactical issues.

Busy Period Analysis.   Consider a resource *k* within a larger scheduling problem. Suppose the overall problem has been scheduled so that at a time *t* we not only know the current queue in front of the resource and the priorities within that queue, but also have good forecasts of the arrival time and priorities of arriving jobs. We can imagine simulating the resource: jobs are processed; new ones arrive; more are processed. At some point after a total of *N* jobs have been processed (including one partially processed at the start), the resource will first become idle. The time up to that point is called the *busy period*. We show this situation in Figure 10.1.

Number these jobs from 1 to *N* in the order in which they will actually be processed. We assume that job *j* has weight w_j and forecast urgency U_j, which has been estimated currently by some lead time estimation method we have talked about. To allow for a partially completed job 1, we assume w_1 has already been decreased in proportion to the part of it already processed. (That is, we are pricing as if we were using SC preemption.)

Now if the resource were shut down for an hour, what would be the differential cost; that is, what is the approximate resource price? The fundamental insight is that the optimal scheduling would not change much, so that the busy period and priorities would be relatively unchanged. The principal effect would be that all operations in the busy period would also be delayed for an hour. Thus we have the following equation:

Fundamental Resource Price for Resource *k*

$$R_k = \sum_{i=1,N} w_j U_j(t)$$

where N is the number of operations in the busy period, w_j is the weight of job j, and U_j is its urgency.

Note from the formula that it is clear that the price in general will vary over time and will be roughly proportional to the current busyness of the resource. It is also clear that good estimation of prices requires working with a solution that is reasonably good. (Sensitivity of prices to poor estimates of scheduling solution and/or sensitivity of bottleneck dynamics solution quality to pricing quality will be discussed in a later chapter.)

Recall from our earlier discussions that, for an objective of minimizing weighted flowtime or lateness, the U_j are just constant; in fact, $U_j = 1.0$. Recall also from Section 6.5 that, for due date based objectives such as weighted tardiness, estimating U_j depends on estimating the slack of j by: (a) a fixed multiple of remaining processing, (b) regression fit on shop load factors, (c) lead time iteration, or (d) human judgment.

One situation in a more complex shop, which would complicate this analysis, is if resource k has a finite queue, and arriving jobs first block a previous resource, then fill its queue, then fill an even earlier resource, and so on. These cases can still be handled by a busy period analysis. The resource doing the blocking will get credited with all jobs until it is next free, even though many are sitting at other resources. The resource being blocked should be shown as having a price of zero, since delaying it one hour will have no effect, even though it has jobs in its queue.

Price Estimation Methods. We next turn to describing individual methods available to implement the busy period idea.

Dynamic and Iterative Pricing Method

Step 1: Run the full sequencing problem by some heuristic (e.g., myopic).

Step 2: Record the detailed trace of that run.

Step 3: Analyze actual busy periods and w_j and U_j for each resource at a number of points of time. (Each time an activity begins processing would probably be adequate.)

Step 4: Use these to compose ex post price profiles for each resource.

Step 5: Run a new simulation, making decisions with bottleneck dynamics and the newly estimated prices.

Step 6: If the convergence criterion is satisfied, stop.

Step 7: Return to step 2.

Several points deserve note. First, the lead times can be iterated simultaneously with the prices. Second, as for lead time iteration, the convergence criterion is likely to be of the form: "stop if current best solution hasn't been improved for X iterations." The iterative pricing method is actually the original method employed for bottleneck dynamics and has been tested extensively in job shops using the weighted tardiness criterion. (See Chapter 16.)

One interesting thing found in these initial empirical studies is that extremely accurate prices do not seem to be necessary to get good results. Since iterative methods are quite complex and there may be many problems where direct investigation of busy periods is difficult, a number of simpler methods were also explored.

Static and Moderately Simple Methods

Method A: Price the resource proportional to its actual utilization.

Method B: Simulate using a good heuristic. Then simulate increasing the resource amount and estimate the sensitivity. The price is then taken as proportional to the improvement in the objective per unit relaxation of the resource.

Method C: Estimate the long run average busy period by queuing theory, making a number of simplifications:

$$R_k = (wU)_{av}(\rho_k/(1 - \rho_k)^2)$$

[In this last formula, $(wU)_{av}$ is the average delay price of activities at the resource, and ρ_k is the long-term utilization rate of resource k. The last part of the formula expresses the (exact) unconditional average length of the busy period for an $M/M/1$ queue as a function of the utilization rate.]

Method A is very simplistic. It just says to price resources by their utilization and linearly, independent of the weights of the jobs going through the resource. Method B is quite a good method theoretically since it experimentally measures the price directly. But it is quite expensive considering that it does not yield dynamic prices. Finally, method C uses the formula for the average busy period length of a simple $M/M/1$ queue, and scales it up or down by the average urgency of the jobs it faces.

Static and Exhaustive Search Methods

For smaller problems with, say, two or three resources, it is feasible to simply try "all combinations" of possible prices. For each set of prices simulate the system and record the value of the objective function, trying all sets of prices via a coarse search. (For three machines, 19 runs with all combinations of values of prices at 0.01, 0.5, or 1.0 proved quite adequate.)

This method gives excellent results as far as static pricing models go and can serve as a benchmark for evaluating simpler procedures. (These studies will be discussed in detail later when we discuss flow shops.)

Dynamic and Moderately Simple Methods

Method A: Price proportional to the actual current length of the queue.

Method B: This method is analogous to the earlier queuing theory result, except that it is based on the actual jobs currently in the queue, rather than the average line. This dynamic approximation is

$$R_k(t) = \sum_{i=1,L} w_{ki}(t)U_{ki}(t) + (wU)_{av}L_k(t)(\rho_k/(1 - \rho_k))$$

Here $(wU)_{av}$ is the long-term average marginal delay cost of a job in queue at this machine; ρ_k is the long-term utilization of the machine; L_k is the current line length (it could be reduced by the fractional job remaining on the machine if desired); and w_{ki} and U_{ki} are the weight and urgency factors for each job currently in the line.

Basically, the summation term of the formula adds up the actual delay costs currently in the line. The right side estimates the delay costs for the rest of the busy period, based again on exact results for an M/M/1 queue. The expected number of jobs to arrive in the rest of a busy period for an M/M/1 queue conditional on current line length L is exactly $(L)(\rho/(1 - \rho))$. Since the urgency for incoming jobs is unknown, we multiply by an average urgency. This rule has had some testing and seems to behave well. Again, we defer discussion of the results until the flow shop chapter.

10.5.3 Bottleneck Dynamics Sequencing Rules

In Section 6.5 we developed a bottleneck dynamics formula for evaluating the net effect of expediting an activity i on a resource k whose only purpose is to help complete job j downstream. (We used the subscript jk to indicate this activity i as a convenient shorthand, since it references both resource k and job j.) We assume job j has a weight w_j and a current urgency (depending on the due date and our current estimate of the lead time) of U_j. Thus the total value of expediting activity i by a time amount Δ is $\Delta w_j U_j$. Expediting one activity at resource k, however, causes all other activities to suffer; this is summarized in the price R_k. If the time Δ were lost, the cost would be ΔR_k. However, the actual situation is that p_{jk} job pieces with processing time $= 1.0$ each get pushed forward in time an amount Δ, for which they would have to pay interest. Thus the net savings (loss) for processing the job earlier would be

$$\Delta w_j U_{jk} - \Delta IR_k p_{jk}$$

(Actually for a serial job j, U_{jk} will typically not depend on the machine. We leave the formula general so that it applies in project shops as well.)

Now let us generalize this formula to the case where there are many resources and activities required to complete the job. We will limit ourselves here to the case where each activity requires one resource and there are never two different activities needing the same resource on one job. This is basically only for notational convenience. The idea generalizes easily.

The basic approximation (principle) we use is that if want to expedite the job, we need not only expedite this activity by Δ, but all activities by Δ.

Bottleneck Dynamics Sequencing Principle. The resource cost of expediting a job currently waiting at a machine by a time Δ can be estimated by weighting all remaining resource usage of the job by the appropriate prices and multiplying by $I\Delta$.

This principle is not exact, but it works extremely well in practice and simplifies the whole bottleneck dynamics theory considerably. The basic intuition is as follows. There is no real point in (expediting) making this activity first on this machine unless it is (expedited) made first on downstream machines also. We only expedite to make the final job earlier; not expediting downstream would not in fact make it any earlier. This principle holds even for a more complex job with precedence constraints.

Now consider job j, which currently has activities $i, i + 1, \ldots, M$ as yet unstarted. Activity i is currently waiting at resource $k(i)$, and we are considering expediting it by a time amount Δ. Let m index the remaining operations, m going from i to M. Operation m will be at resource $k(m)$ and have resource usage there of $R_{k(m)}P_{jk(m)}$. Then the principle just discussed says that the formula for the value of expediting activity i becomes

$$\Delta \left[w_j U_{jk(i)} - I \sum_{m=i,M} R_{k(m)}P_{jk(m)} \right]$$

Now finally, as before, we normalize and ignore factors the same for all activities to obtain the following:

Bottleneck Dynamics Sequencing Priority

$$\pi_{jk(i)} = \frac{w_j U_{jk(i)}}{\displaystyle\sum_{m=j,M} R_{k(m)}P_{jk(m)}}$$

where w_j is the job weight, $U_{jk(i)}$ is the urgency of the particular current activity i in incurring cost to j, $p_{jk(m)}$ are the processing times of remaining activities for j on machines $k(m)$, and $R_{k(m)}$ are the resource prices to be charged for these times. We leave it for the exercises to generalize this formula to the case where several resources are needed per activity, or more than one activity use the same resource.

Note that it is not so important for individual prices R_k to be accurate as for there to be a reasonable estimate of *aggregate* resource usage in cost terms. Note also that some of our other simple dispatch methods may be seen as special cases:

(a) Myopic is a special case where

$$\text{Current machine price} = 1.0$$

$$\text{Future machine prices} = 0.0$$

(b) OPT-like is also a special case where we estimate the bottleneck in advance; then

$$\text{Bottleneck price} = 1.0$$
$$\text{Other prices} \quad = 0.0$$

(A) 10.5.4 Full Net Present Value Sequencing Analysis

We have waved our hands a bit about charging some interest rate I for using resources earlier or later. It may be useful to show a slightly more careful (although still approximate) net present value analysis in which I has a direct interpretation as the cost of capital to the firm in a theoretical finance sense, plus an interest-like correction for the percentage rate change in resource prices. In the process, we show how to handle mixed objective functions that involve direct cash outlays at the time an operation is processed. These would include direct materials added, labor (if not a sunk cost), and machine rental (if not a sunk cost). They, of course, include R_k, the imputed cost of postponing other work. They definitely do not include depreciation or other sunk amortization of the machine/resource.

We discount costs using a continuous interest rate I', the firm's cost of capital. We consider only the costs accruing to job j for activities at each of the k resources. At each resource there is the imputed processing cost $R_k p_{jk}$ and the direct processing cost c_{jk}. There is also the lateness based cost $g_j(C_j)$ when job j completes at C_j. (This could be negative, for example, payment for the order.)

$$\text{Cost}_j = \sum_{k=k(i),k(M)} \{e^{-I'(t_k+\Delta)}[R_k(t_k + \Delta)p_{jk} + c_{jk}] + e^{-I'(C_j+\Delta)}g_j(C_j + \Delta)\}$$

Now suppose that the first of these serial activities i is expedited by some amount of time. The basic assumption is that, in order to be effective, all activities and completion time will be (on average) expedited by the same amount. So take the derivative with respect to Δ:

$$\frac{d}{d\Delta}(\text{Cost}_j) = \sum_{k=k(i),k(M)} e^{-I'(t_k+\Delta)}\{(-I')[R_k(t_k + \Delta)p_{jk} + c_{jk}] + [R_k'(t_k + \Delta)p_{jk}]\}$$

$$+ e^{-I'(C_j+\Delta)}\{(-I')g_j(C_j + \Delta) + g_j'(C_j + \Delta)\}$$

Take time to be measured in weeks. If I' might be, say, about 0.005–0.010 per week and the horizon is 4 or 5 weeks, the exponential terms may be approximated by 1.0 to within 2 or 3%. We identify the derivative of the cost at finishing with $w_j U_j$:

$$w_j U_j = -I'g_j + g'_j(C_j)$$

For example, suppose the only effect at delivery is to receive a payment P_j. Then $g_j = -P_j$, $g'_j = 0$, giving $w_j = -I'P_j$, and $U_j = 1.0$. This would represent a weighted lateness objective with the only penalty per unit time being the lost interest on payment.

Next, we define an augmented interest rate $I_k(t)$, which is the sum of the cost of capital and the percentage rate of change of the resource cost: $I_k(t) = I' + R'_k(t)/R_k(t)$, which allows us to simplify the expression above to

$$\frac{d}{d\Delta}(\text{Cost}_j) = - \sum_{k=k(i),k(M)} [I_k R_k p_{jk} + I' c_{jk}] + w_j U_j$$

This formula is now simple enough to be useful. However, keeping track of dynamically varying I_k may still be more effort than justified by the impreciseness and uncertainty in the analysis. Thus, for everyday use, we recommend approximating both I_k and I' by some common "fudged" interest rate I containing some average correction for the rate of change of the resource cost. In practice, it may be better to estimate the best I by extensive simulation rather than by tying it directly to estimates of R'_k, since estimating derivatives in practice can be quite unstable. Making this approximation we have that

$$\frac{d}{d\Delta}(\text{Cost}_j) = -I \sum_{k=k(i),k(M)} [R_k p_{jk} + c_{jk}] + w_j U_j$$

Or expressed in terms of a sequencing cost/benefit ratio

$$\pi_{ji} = \frac{w_j U_{ji}}{\displaystyle\sum_{k=k(i),k(M)} [R_k p_{jk} + c_{jk}]}$$

This heuristic is just our usual sequencing heuristic generalized to allow some direct costs at each resource. We may still read it as follows:

Sequencing Heuristic Principle. The sequencing priority for an operation that is part of a larger job is the benefit to the larger job of expediting the operation, divided by the sum of all explicit and implicit resource costs remaining to complete the job.

(A) 10.5.5   Bottleneck Dynamics: Other Decision Rules

Here we briefly discuss a number of other decision situations and sketch how bottleneck dynamics deals with them. More detail will be found in the sections actually discussing these types of rules.

Job Release. There are a number of motives for not releasing raw material to the production floor until just before it is needed:

(a) Work in process (WIP) on the floor incurs inventory charges at a higher rate than raw material.

(b) There is limited space on the floor for WIP.

(c) Obsolescence, damage, and confusion cause high WIP to be expensive.

We may roughly model any of these by charging an inventory rate w_{I_j} per unit time from the moment the job is released on the floor until it is completed. (If the customer does not accept the job before it is due, this might need to be modified slightly.) w_{I_j} is the *excess* cost of carrying inventory on the floor rather than as raw materials, and hence w_j, which always at least contains the *full* cost of inventory, will always be larger. Now the decision to release a job to the floor Δ time earlier should mean roughly that the job will be scheduled at every machine Δ earlier and finished Δ earlier. Thus the marginal benefit to the job at the end remains $\Delta w_j U_{jk(1)}$; but now there is a marginal cost to the job at the beginning of $-\Delta w_{I_j}$. Thus the value of releasing is very similar to the earlier calculated value of expediting, and if we normalize as before we obtain the following:

Bottleneck Dynamics Release Priority (Rate of Return on Release)

$$\pi_{j\,\mathrm{rel}} = \frac{(-w_{I_j} + w_j U_{jk(1)})}{I \displaystyle\sum_{k=k(1),k(M)} R_k p_{jk}} - 1.0$$

This priority represents a rate of return for the release decision. Jobs with a negative priority should not be released yet; those with 0 or positive priority should be released. (As a secondary control on the crowding in the shop, jobs can be released in order of decreasing priority to a desired crowding.) Several things are worth noting. First, for projects/dependent jobs, activities that are critical in finishing the job will be released earlier than those that are less critical. (See Chapters 17 and 18.)

Second, the release value equation may be solved to directly give the time to release. This is especially useful if the prices $R_k(t)$ are changing much more slowly than the urgency factor. In this case the release time may be found from

$$U_{jk(1)}(t^*) = (1/w_j) \left[w_{I_j} + I \sum_{k=k(1),k(M)} R_j p_{jk} \right]$$

For example, for the T_{wt} objective and the R&M heuristic, the urgency at time t^* is given by the formula

$$U_{jk(1)}(t^*) = e^{-(d_j - L_{jk} - p_{jk} - t^*)^+ / 2p_{av}}$$

Thus we can equate the right-hand sides, take logarithms of both sides, and solve for t^* directly. This is left as an exercise.

Note finally that charging for inventory from release gives a nonregular objective function. Heuristics not designed for job release release jobs to the floor long before needed and hence are often extremely poor.

Job Routing. Suppose there are several different ways to perform a job. We are then concerned with more than just the interest on the implicit and explicit resource cost by doing it earlier or later. We are also concerned with different main costs for two different routes and also completion times that may be very different. Suppose we label a route by r, and label the kth machine to be visited on route r by k_r (note k is no longer the machine number, but the machine sequence number in a route) and try to write down the total costs and total interest costs.

Now in the net present value analysis discussed above, we noted that in expanding $\exp\{-I_t\} = 1.0 - I_t + O((I_t)^2)$ there is a main term 1.0 not depending on I_t, a linear term in I_t, and higher order terms that are small. In calculating the effect of interchanging two jobs in sequencing, the main term is unaffected and can be ignored, and the change in t is the same for both jobs, so we only need consider derivatives on the first order term.

In routing, however, the size of the main cost will typically depend on the route we take, and t will usually differ in a major way between the two routes rather than differentially. Thus we would estimate

$$\text{Cost}_r = \left[\sum_{k=1,K_r} (1 - It_k)(R_k p_k + c_k) \right] + wU_r(t_r - t)$$

A term in the summation tells us the implicit resource cost and the explicit resource cost for each resource, each roughly discounted back to the present. As for the last term, wU_r gives the marginal cost of job lateness, while $(t_r - t)$ basically gives us how much later the route makes us.

(We suppress the job subscript j throughout the formula, not so much because adding a j subscript everywhere would be too complicated, but for a deeper reason. The routes r and the corresponding machines k are all functions of j! The resulting notational mess would make routing seem much more complicated than it is.)

The optimal route r^* is the one that minimizes this cost.

An important special case is when there is only one operation, which may be performed on one of several machines. Ignoring interest issues as of second order we can approximate

$$\text{Cost}_{jr} = R_{jr} p_{jr} + c_{jr} + w_j U_{jt_r}$$

Deteriorating Jobs. In many situations jobs take longer to process the later they are processed. For example, if hot steel slabs wait in front of the roller, more heat has to be supplied to get them up to the proper temperature before they can be rolled. Thus we may consider an ordinary static one-machine problem where the processing time for job j is $p_j(t)$. Kunnathur and Gupta (1990) and Alidaee (1990) have obtained partial results for this problem for the makespan objective. Here we develop accurate bottleneck dynamics heuristics for more general objectives.

The expression for the cost rate of delaying a job now just has an extra term involving the extra job length, which is paid for at the machine price $R(t)$:

$$\text{Cost rate} = w_j U_j(t) - IR(t)p_j(t) + R(t)p'_j(t)$$

As usual, we normalize by the resource usage (second and third terms) and cancel the common price term affecting all priorities equally:

$$\pi_{jt} = w_j U_j(t)/[-Ip_j(t) + p'j(t)]$$

This result is exact for local pairwise interchange of differential length preemptive joblets. If we note that the usual priority for constant-length jobs would be

$$\pi^*_j(t) = w_j U_j(t)/p_j(t)$$

we can show the corrected priority in terms of the original priority:

$$\pi_j(t) = \pi_j^*(t)[-I + (\ln p_j(t))']$$

In most cases of interest $(\ln p_j(t))' \gg I$, so that to a very good approximation (canceling I across comparable priorities) is

$$\pi_j(t) = \pi^*_j(t)(\ln p_j(t))'$$

Thus all we need are the nondeterioration penalties and the current percentage rates of change of processing times for all jobs. The priority for current scheduling is then normal priority times percentage rate of deterioration.

As an example, for makespan normal priorities are all equal. Thus for makespan, the bottleneck dynamics rule is to schedule the fastest deteriorating job; for weighted flow, multiply the WSPT rule by the percentage rate deterioration, and so on. This (myopic) rule can be expected to be accurate when neither the percentage rate of deterioration nor percentage rate of change of urgency changes "too fast."

We develop these models in more detail in later chapters.

10.5.6 Advanced Pricing Methods

The busy period methodology for determining resource prices that we have studied in this chapter work well for short run and very short run situations (Levels 4 and 5), where the jobs to be processed are given, along with due dates and other job characteristics. However, at a higher level such as MRP (Level 3) or even higher levels, other options are available to reduce the busy period and thus reduce the resource price.

Suppose, for example, that we have a one-machine dynamic makespan problem with arrival r_j and process time p_j, with current machine busy period price of $R(t)$. Suppose now we are allowed to accept or reject jobs for processing at our discretion. Suppose that the customer for j is offering V_j (value added) for processing the job. If we process it, our net implicit profit would be

$$V_j - Rp_j$$

If this is positive we clearly want to take the job. However, if it is negative it is not clear we should reject the job, since as we reject the worst jobs, the price R, dependent on the busy period, will decrease also. Thus we would look for an equilibrium set of jobs and resource price, where the sets showing a profit at that price are kept, and their busy period determines the price. One way to do this is start at a low imagined price and gradually throw away jobs as it is increased. When the increasing price equals the decreasing busy period price, an equilibrium is obtained.

In the absence of such equilibration, suppose we had a very large number of jobs wanting to be done, leading to an inflated resource price. Most jobs would wait a long time, and most jobs would show a loss for us. As in typical economic theory, a good balancing price can solve this problem.

In a similar vein we might negotiate higher V_j in order not to reject some jobs. Or we might do a mixture of the two techniques.

When prices and negotiations are flexible, the price of the resource times the processing time of the marginal job will equilibrate to zero profit. This gives some credence to the popular accounting practice of valuing machine time at the added value of lost production.

These advanced resource pricing ideas will not typically be needed at Levels 4 and 5, where pricing decisions are assumed to have already been made. They will be discussed again in Chapters 19, 20, and 21.

10.5A Numerical Exercises

10.5A.1 Set up an example problem with an existing queue and jobs to arrive at given times. The objective is weighted flow. Each job j in the queue or to arrive has parameters p_j, w_j, and r_j.

    (a) Calculate the busy period. (If necessary modify your data so that you have a full busy period.)

    (b) Calculate the current resource price R_k.

    (c) Calculate the resource price $R_k(t)$ over the course of the busy period and plot. (Partial completion of a job is treated as SC preemption.)

10.5A.2 In 10.5A.1 try pricing by three other methods:

    (a) Constant price proportional to long-term utilization ρ.

    (b) Constant price estimated by queuing theory using ρ.

    (c) Dynamic price estimated by queuing theory.

Explain why, although the method of 10.5A.1 is more theoretically correct, one of these methods might be used in practice.

10.5A.3 Set up a small flow shop example with weighted flow objective. Use some very simple method to estimate static prices for each machine. Solve the problem by bottleneck dynamics, working forward through the problem by dispatch simulation, and making sequencing decisions by the bottleneck dynamics sequencing priority rule.

10.5A.4 Repeat 10.5A.3 for the same problem using the dynamic queuing methods for estimating prices. Compare your results.

10.5B Software/Computer Exercises

10.5B.1 Via either spreadsheet aids or writing a computer program, repeat 10.5A.1 and 10.5A.2 for a larger set of problems. Try to draw some experimental conclusions.

10.5B.2 Via either spreadsheet aids or writing a computer program, repeat 10.5A.3 and 10.5A.4 for a larger set of problems. Try to draw some experimental conclusions.

10.5B.3 Program lead time iteration/price iteration for a three-machine flow shop and weighted flow criterion. Try to fine-tune your parameters on a small pilot study, and study your results on a set of generated problems. You will want to have some simpler heuristics to compare against as benchmarks.

10.5C Thinkers

10.5C.1 Since actual prices behave as a sawtooth, rising to the maximum at the beginning of a busy period and going down to zero at the end, and since our forecasts for start and end of busy periods at some point in the future shop are very imperfect, especially as to timing, it can be argued that forecast prices should be smoothed, as a crude way of handling the stochastic issue. Develop some simple ideas for doing this.

10.5C.2 Given that heuristic pricing is somewhat inaccurate, it might be proposed to solve the problem more than once with some kind of randomization superimposed. Carry this idea as far as you can, and discuss strengths and weaknesses.

10.5C.3 For the best release time formula specified for the T_{wt} objective and R&M heuristic, expand out the formula, take logarithms of both sides, and solve for the release time t^* directly.

10.5C.4 Develop a release priority formula for the weighted flow/weighted tardiness mixed objective problem.

10.5C.5 A kaleidoscope can either be manufactured, polished, and boxed in one operation by a futuristic robot with a high heuristic price, or these steps may be done by three machines in sequence, which are relatively slow and low to moderately priced. Work out a formula to find which is better, choose some numbers to illustrate the computations.

─── 11
PARALLEL MACHINES/
BATCH MACHINES

11.1 INTRODUCTION

11.1.1 Overview, Notation, and Terminology

Chapters 9–12 deal with models having more than one machine and/or more than one job processing simultaneously. Hence they do not seem solvable by sequencing methods alone but seem to require routing decisions as well. However, some of these models can be simplified by taking advantage of the fact that (as suggested in Figure 11.1) there is really just one input queue whose priority sequence need only be determined to solve the problem. Hence we term them multi-process one-resource problems. (There may be several machines, but the decision process acts as if they are aggregated into one, with one queue.)

It is not at first obvious that the parallel machine problem fits this characterization. We will prove that parallel machine problems can be considered as one-dimensional sequencing problems as one of our first tasks.

Consider m parallel machines indexed by $k = 1, .., m$. There is a set of n jobs $j = 1, \ldots, n$, which are each a single operation and can typically be processed on any machine with processing time p_{jk}. (To forbid job j on machine k, simply make p_{jk} very large.) Job j arrives in the system at time r_j. As in the one-machine case, we may have any objective function with appropriate due dates d_j, weights w_j, and so on. Machine k is first available for processing at time a_k. (For example, it initially may be occupied finishing a previous job.) If all r_j are 0, we speak of *static job arrivals*. If all a_k are 0, we speak of *static machine availability*. If both are static we speak of the *static case*. If both are dynamic we speak of the *dynamic case*. In a mixed case, we specify which is static and which is dynamic. If all machines are equal, that is, all $p_{jk} = p_j$, we talk of the *equal parallel machine case*. If all

Figure 11.1. Parallel machines.

machines differ from each other only by constant speed factors, that is, $p_{jk} = p_j/s_k$, where s_k is the *speed factor* for machine k, we talk of the *proportional parallel machine case*. Finally, if the p_{jk} are arbitrary, we talk of the *general parallel machine case*.

We shall usually mean by the *preemptive* case that a job can be split arbitrarily but may not be processed on more than one machine simultaneously. If we wish to allow simultaneous processing of different parts of the same job on different machines as well, we speak of *free-preemption*. For the equal machine and the proportional machine cases, utilizing the free-preemption assumption makes it easy to aggregate machines and derive results almost identical to the one-machine case. The general case is more complicated, but preemptive results may still be derived by linear programming.

For mathematical convenience, early analysis of the parallel machine problem tended to concentrate on the equal machine static case with a makespan criterion (with some analysis also of unweighted flow problems). This leads to a preoccupation with accidents of *fitting*: packing jobs into machines like groceries into bags. That is, pack in larger things first, and save small items to fill in the chinks to get the bags as close to evenly filled as possible. (This is often called a multiple bin-packing problem.)

This preoccupation is an artifact of the static fixed horizon formulation. In point of fact, more work will be arriving all the time. Shops for which utilization is the proper objective rarely run out of work even after the problem horizon is over. We will deal with these issues both by presenting exact and heuristic results for the dynamic case, and by developing a new measure of utilization called *economic makespan*. Economic makespan tries to avoid internal machine wastage, but not the artificial "wastage" at the end of the problem.

11.1.2 Chapter Summary

In Section 11.2 we first prove that the general parallel machine problem can be characterized as a one-dimensional (one-machine) sequencing problem, that is, as a multi-processor one-resource problem. We then develop a number of exact results for the preemptive equal machine case including makespan with independent jobs, makespan with assembly tree structure, weighted flow, and maximum lateness. We also develop an exact result for the nonpreemptive equal machine case for minimizing unweighted lateness.

The results of Section 11.2 are most helpful in developing heuristics in Section 11.3. We use the fact that the problem may be reduced to sequencing a list and suggest priority rules given by the preemptive results to construct a rather general procedure for developing heuristics for equal or proportional problems. We give procedures to solve such problems heuristically by bottleneck dynamics priority rules. We also sketch how to solve them by branch-and-bound, beam search, extended neighborhood search, or the shifting bottleneck algorithm. The general nonproportional case is discussed again in Chapter 12.

We also show that the concept of makespan as a surrogate for utilization is somewhat limited for parallel machines. The basic point is that credit should be given for early completion of other than the latest completing machine. Machines finishing earlier in a highly loaded shop should not be assumed to be idle subsequently; new work not in the problem statement will usually have arrived by then and such machines can begin to process that work. This leads to a new concept of utilization called *economic makespan*, which is discussed briefly.

Section 11.4 develops the idea of *batch machines*. Here again up to m jobs are processed at one time. The difference is that they are being handled by a single machine and thus must start and stop at the same time. There may be a common setup cost. With dynamic arrivals, sometimes a decision must be made as to whether to make a run partly loaded, or to lose processor time by waiting. Also, sometimes jobs may fall in Q different classes, and jobs of two different classes cannot be run at the same time. A few batch machine cases can be handled exactly. Bottleneck dynamics is naturally suited to provide economic heuristics in more difficult batch machine cases.

11.2 PARALLEL MACHINES: EXACT RESULTS

11.2.1 Reducibility to a One-Resource Problem

Early exact and heuristic results for parallel machines bear very little resemblance to similar results for one-machine problems. This is primarily because parallel machine problems seem to involve both routing and sequencing rather than simply sequencing. We demonstrate, however, that, in fact, parallel machine problems can be reduced to the problem of sequencing the list of jobs just as for one-machine problems. Although such operations as simple pairwise interchange are more difficult for the parallel machine case, we are provided an overall framework for applying all the methodologies given in Chapters 5 and 6 directly, as we shall see later in Section 11.3.1.

While most results for nonequal and nonproportional machines are deferred until Chapter 12, the following result is *general*.

Proposition 1. Consider a general parallel machine problem with dynamic arrivals, dynamic availability, and regular objective function. There is an optimal solution of the form: sequence the n jobs in an optimal order (method for obtaining optimum order is not known) in a list, and schedule each in turn on the machine that can finish it first. For the dynamic arrival case, this remains true whether the discipline is dispatch or inserted idleness is allowed. A job need not be considered for first in the sequence if scheduling it to the machine finishing it first leaves a gap large enough to allow another job to be feasibly scheduled.

Proof. We show first at the start that a job 1 can be chosen and scheduled on the machine that can finish it first and be part of an optimal solution.

Consider some optimal solution. (How to obtain this is not given!) Restrict attention only to those jobs scheduled first on one of the machines. (Some machines may be empty.) Now, if any job is on the machine finishing it earliest, we are done: it could have been put first in the list. Otherwise, imagine a pointer pointing from each machine to the machine that would have finished that job earliest. If an empty machine is pointed to, that job could be moved there, improving the objective function (regularity)—a contradiction. Otherwise, a subgroup of machines will form a cycle with the pointers. These jobs can simply be moved around the cycle, completing them all earlier, and allowing later jobs on each machine to be completed earlier as well. Thus a contradiction.

However, scheduling job k is no different than scheduling job 1. After $k - 1$ jobs are already scheduled we have a smaller dynamic problem for which we know a first job can be placed on the machine finishing it first. ∎

Note that since the existence of a list ordering in Proposition 1 is derived indirectly, we are given no help at all in how to derive the list ordering in advance. Intuitively, we would like to order the list using priority rules similar to those for the one-machine problem. We will show that this is the correct idea for equal and proportional machines for the static case and discuss how to do this properly. Then we show how to extend the ideas to cope with dynamic arrivals. Deriving a list ordering for the general case is much more difficult; we defer discussion of this until Chapter 12.

Although this result, that parallel machines may be considered a single resource, is rather simple to prove, it is powerful in practice. We next prove a number of free-preemptive and preemptive results suggesting a rather general priority dispatch heuristic procedure for equal and proportional parallel machines.

11.2.2 Free-Preemptive (SC) Results

The main result of this subsection is that solving the free-preemptive version of an m-machine problem is equivalent to solving a one-machine problem that is m times as fast in the case of equal machines, or that is $m' = \Sigma_{k=1,m} s_k$ times as fast for m proportional machines. (m' might be called the *effective number of machines.*) This is not important directly in practice, since free-preemptive situations are not terribly common. It is, however, very important for three other reasons:

(a) As a lower bound for interesting problems, it provides a valuable input for branch-and-bound procedures and relaxation procedures.

(b) We have previously in this book found preemptive solutions useful in building accurate heuristics.

(c) The idea of approximately aggregating resources to simplify large problems is very attractive.

We first analyze the unit-preemptive case, because it provides a good introduction to the ideas. However, the fact that the number of pieces a job is cut into may not be exactly divisible by m obscures the simplicity of the result we are trying to get. Thus we move on to the free-preemptive case, which may be considered a limiting case with each job divided into a very large number of pieces.

The Free (SC) Unit-Preemptive Case. We discuss primarily the equal machine case; however, we briefly explain how the results would extend to proportional machines. Recall that in a spread cost unit-preemptive problem, job j with weight w_j, process time p_j, and due date d_j is divided into p_j jobs, each with weight w_j/p_j, process time 1.0, and due date d_j. For parallel machines different units of the same job may be done on different machines, but not simultaneously. However, in a free-preemptive case units may also be done simultaneously on different machines. The preemptive solution is a relaxation of the original problem; the free-preemptive solution is a further relaxation on that. There are real situations where free preemption is permissible. More important, exact solutions are often easily obtained, which gives us insight into heuristics for the nonpreemptive problem. Proposition 2 gives us several exact results.

Proposition 2. Consider the equal parallel machine problem with free (SC) unit preemption, dynamic arrivals, and dynamic availability. The optimal solution is given by prioritizing a list and scheduling on the machine that is first available. Priority rules are given by:

(a) Makespan—order irrelevant.

(b) Weighted flow—WSPT.

(c) Maximum lateness—EDD.

(*Note*: Since order is irrelevant for the free-preemptive case, we may as well specify LPT, which works rather well for the nonpreemptive case.)

Proof. Because all job lengths are equal, interchange of two jobs in the schedule affects no other jobs. Rather than interchange increasing one job's completion time by one and decreasing the other job's completion time by one, however, as in the one-machine problem, the amount added and subtracted may be zero, one, two, and so on. In particular, if the first job is sequenced in position i, and the second job is sequenced in position $i + k$, then interchange will add k to the completion time of the first and subtract k from the completion time of the second. However, the condition for the objective function change of an interchange being positive is unchanged, and thus interchange proofs from Chapter 4 still go through. ■

We leave as an exercise to decide how to handle the weighted number of tardy jobs problem for free preemption, in analogy to Proposition 2.

To give a sketch of the proof idea for the proportional machine case, consider the speeds s_k. Pretend that the speeds of all machines are given by integers. Divide the jobs into units such that the machine with speed 3 can process all three in a time unit; thus we treat this faster machine as three machines in parallel, and so on. Finally, the resulting problem can be solved by the equal machine methodology.

The Free-Preemptive Case. By the simple device of cutting up jobs into a very large number of very small pieces, Proposition 2 is easily shown to be true for the (non-unit) free-preemptive case. The highest priority available job is simply spread over all machines to "fill" them to the same time point. However, instead of showing this, we proceed directly to a more powerful result. (We do need to limit ourselves to static availability.)

Proposition 3. Consider the equal or proportional parallel machine problem with free (SC) preemption, static or dynamic arrivals, and static availability. The optimal solution is given by solving an equivalent one-machine problem with m times the speed in the equal case and m' times the speed in the proportional case, where $m' = \Sigma_{k=1,m} s_k$.

Proof. We look initially at the equal machine problem. We need first to show that there is a feasible solution to the m-machine problem of equal cost to an optimal solution to the one-machine problem, and, finally, that there is a feasible solution to the one-machine problem of equal cost to an optimal solution to the m-machine problem.

First, take any optimal solution to the one-machine problem and duplicate it m times, for the m machines that are only $1/m$ times as fast. This is feasible and has the same cost.

For the other half, take any optimal solution to the m-machine problem and consider it to be many rows of infinitesimal unit jobs. Consider the first row at the earliest time. We claim we can interchange so that the first row consists entirely of the same job. For if there were a unit of j in the first row, it would be higher in value than any other job, or it could be pairwise interchanged out. Hence we can pairwise interchange enough of j to fill the first row. By repeating this argument, all rows can be assumed to be of an identical job. Thus we have m identical solutions, which remain optimal, and can construct an equivalent single-machine solution that has jobs m times as fast to compensate.

Note that we have glossed over necessary limiting arguments, which would prove we can ignore a few leftover pieces.

For the proportional case, assume the s_k are integers. Treat a machine of speed s_k as s_k machines of speed 1.0. Now apply arguments for the equal machine case. Finally, we would again need some limiting arguments if speeds were not exactly integer multiples of a common speed. ∎

Note now that, for static machine availability, Proposition 2 results are a special case. We leave as an exercise the case of dynamic machine availability.

11.2.3 Preemptive Makespan Results

As we shall argue in the next section, makespan is usually not the best surrogate for resource utilization except in the one-machine case. We shall discuss there a better method arising from bottleneck dynamics: economic makespan. Nevertheless, it is important that we study makespan for several reasons:

(a) Makespan as a measure of utilization is the only one in wide use (economic makespan is quite new).
(b) For multi-resource problems more exact results and procedures are known for makespan than for all other objectives put together.
(c) Understanding makespan results intuitively is useful in developing economic makespan results.
(d) There are situations where makespan is more appropriate than other utilization measures.

We turn first to developing a lower bound on preemptive equal or proportional machine makespan, which is rather obvious and quite general.

Proposition 4. For rather general assumptions, a lower bound on preemptive equal or proportional machine makespan is given by

$$C^*_{\text{max lb}} = \max\{ (\Sigma_{j=1,n} p_j)/(\Sigma_{k=1,m} s_k) , \max_j[p_j]/\max_k[s_k]\}$$

Proof. The first main expression is just the total processing time on a standard machine divided by the effective number of machines. This is clearly the makespan for the free-preemption static problem and clearly cannot be improved on; it assumes no machine idle.

The second term says the makespan must be at least as much as the processing time of the longest job on the fastest machine. This is basically true because we do not allow two machines to process a job at the same time. ■

This lower bound remains true for dynamic arrivals, precedence constraints, sequence-dependent setups, and so on, although stronger bounds can then be given. We leave some of this for the exercises.

We turn next to an early result, due to McNaughton (1959), which shows the lower bound is actually the same as the optimal solution for the simplest case. A simple but important result was given by McNaughton (1959) for preemptive makespan.

Proposition 5. (McNaughton's Algorithm). For the preemptive static equal parallel machine problem, the minimum makespan C^*_{max} is given by

$$C^*_{max} = \max\{ (1/m)\Sigma_{j=1,n}\,p_j,\ \max_j[p_j] \}$$

The procedure achieving this makespan is as follows:

Step 1: Order jobs arbitrarily, and number the machines.
Step 2: Assign the next job to the first available time on the lowest number machine with total assignment less than C^*_{max}.
Step 3: When a machine becomes filled past C^*_{max}, preempt and assign the remainder to start the next machine.

Proof. The fact that the formula is a lower bound follows from Proposition 3. The idea of the algorithm is to lay out a plan out to time C^*_{max} and find a way to schedule where no job goes on two machines at once. Clearly, for any job not split there is no problem. But we have split any job that is last on one machine and first on the next. There can be no time overlap since C^*_{max} is known to be at least as long as the job being split. ■

It is tempting to try to apply this construction method to the proportional machine case; it is instructive to see what goes wrong. This is left to the exercises.

It is perhaps worthwhile to work through an example. Consider the following job set for the case of $m = 3$ machines. There are eight jobs with processing times 1, 2, 3, 4, 5, 6, 7, 8. We determine from the formula that $C^*_{max} = 12$. We arbitrarily decide to assign the jobs in order of increasing size, as shown in Figure 11.2.

It is important to note that this problem does not usually have a unique solution, and so the construction method produces only one of potentially many optimal schedules. In particular, it may not minimize the number of preemptions. To the extent that a small setup might actually be involved with each preemption, the construction method would have to be modified to produce truly optimal schedules.

There are other things wrong with this procedure as well. We have come to expect free-preemptive and/or preemptive solutions to provide a basis for designing

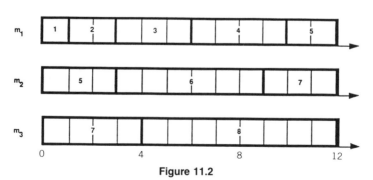

Figure 11.2

dispatch heuristics/bottleneck dynamics heuristics for more complicated problems. The solution here is not dispatch; we assign again and again to the same machine rather than to the first available one.

We turn therefore to developing an optimal dispatch procedure that does generalize better and provides the basis for heuristics. We limit ourselves to the unit-preemptive case for independent jobs. Thus each job of integer length p_j may be considered to be a chain of p_j one-unit jobs, attached sequentially by precedence constraints. (This may seem a bit clumsy; we wish to clarify that Proposition 6 is a special case of Proposition 7 for assembly trees rather than just chains.)

Proposition 6. Consider the unit-preemptive static equal parallel machine problem. Makespan can be minimized by the following dispatch procedure:

Step 1: Assignments for a given time unit are called a "round."

Step 2: Start a new round, making all jobs and machines unassigned; if all jobs are finished, exit.

Step 3: Assign any job and any machine at most once per round.

Step 4: If no unassigned jobs or machines remain in the round, go to step 2.

Step 5: Assign a unit from longest (remaining) unassigned job to any unassigned machine.

Step 6: Decrement the length of the job, and the unassigned job and machine lists; go to step 4.

Proof. This is a special case of Hu's algorithm (Proposition 7 below), which we do not prove here. However, the intuition is as follows. There are basically two motives for doing a particular job unit next. (a) The associated job is long and should not be left until near the end. (b) Doing the job unit makes new job units available to work on.

In this case the two motives work together. Working on the longest remaining job also assures a new unit will be available when this one is done. Thus we would expect optimality. ∎

Note that this dispatch algorithm can be stated as: work with the job that currently has the least slack for meeting a common due date, which is the dispatch heuristic we suggested for makespan in the one-machine case.

After Hu's algorithm we show cases where the two motives work against each other and the problem of assuring optimality becomes more complex.

Conjecture. The same algorithm remains optimal for dynamic arrivals and availability.

Reasoning. The same two motives operate, and there still seems to be no conflict. Now a longer job may arrive and we shift attention to it, but there seems to be no loss. ∎

If we return to the problem illustrated in Figure 11.2, we may verify that the following solution obtained from Proposition 6 is also optimal. (Rounds are indicated by parenthesis, machines 1, 2, and 3 in order; numbers represent jobs worked on.)

$$(876)(876)(875)(876)(854)(876)(543)(876)(543)(287)(654)(321)$$

It is also not hard to see ways to rearrange the solution to produce less preemption:

$$(876)(876)(876)(876)(876)(872)(875)(854)(456)(453)(453)(321)$$

Note that by simple relabeling of rounds we have achieved an optimal solution with rather little preemption.

(A) Proposition 7. (Hu, 1961). Consider the unit-preemptive static equal parallel machine problem, with each job having at most one direct successor (one or more assembly trees). Makespan can be minimized by the following dispatch procedure.

Step 1: Assign each unit joblet a "level" equal to the number of joblet successors of the job, including itself (sometimes called the "depth" of the joblet).

Step 2: Assignments for a given time unit are called a "round"; only joblets with all predecessors finished at the start of the round may be assigned in round; each machine may be assigned only once in the round.

Step 3: Start a new round, making all machines unassigned; if all jobs are finished, exit.

Step 4: If no available unassigned joblets or machines remain in a round, go to step 3.

Step 5: Assign any highest level available joblet to any unassigned machine.

Step 6: Remove the assigned joblet from unfinished predecessors, decrement unassigned machines, and go to step 4.

Proof. While we do not formally prove this result here, the intuition is the same as for Proposition 6. Deep level joblets correspond to long jobs and should be done early. At the same time we wish to create multiple new jobs to do, but this will not happen anyway due to the assembly tree structure, so that conflicts do not arise. ■

We give an example of the algorithm for a six-job and three-machine problem, with the eight jobs comprised of 17 unit joblets. The precedence structure and solution procedure are given in Figure 11.3. The 17 joblets are labeled 1 to 17. The 8 jobs are given by $A = \{17, 12\}$, $B = \{13\}$, $C = \{15, 10\}$, $D = \{16, 11\}$, $E = \{8, 5, 2\}$, $F = \{14, 9, 6, 3\}$, $G = \{7, 4\}$, $H = \{1\}$. The precedences are $A < E$, $B < E$, $C < G$, $D < G$, $E < H$, $F < H$, $G < H$. Figure 11.3 shows the precedence diagram, with equal level joblets shown in vertical columns, and the round-by-round solution, with rounds marked in dotted lines and roman numerals. It simultaneously shows the

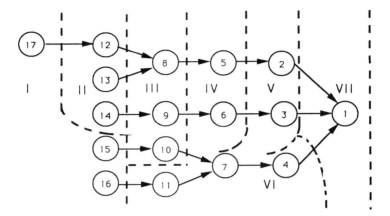

Figure 11.3. Hu's algorithm example: three machines.

solution. However, more insight into rearranging the solution for less preemption can be obtained by a unit-preemptive Gantt chart.

The unit-preemptive Gantt chart for this solution is

```
A  A  E  E  E
C  B  F  F  F
D  F  C  D  G  G  H
```

If we look for preemptive interchanges to reduce preemption, we can immediately produce

```
A  A  E  E  E
C  F  F  F  F
D  B  C  D  G  G  H
```

Now B and D can be interchanged, and the second C inserted before the first F to produce a nonpreemptive optimal schedule:

```
A  A  E  E  E
C  C  F  F  F  F  H
B  D  D  G  G
```

It is an interesting question whether an improved heuristic could be found, which would not employ more structured pairwise improvement. This is left as an exercise.

It is also useful to give an example of a more general precedence structure where the algorithm can be seen not to be optimal (Figure 11.4).

The point in this example is that we have a conflict. Scheduling all at the greatest

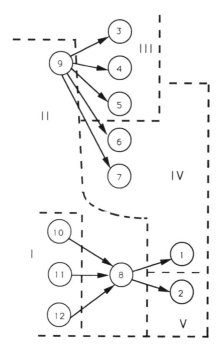

Figure 11.4. Counterexample for general precedence structure: three machines.

depth doesn't create enough available new jobs and the machine starves. Scheduling at the lesser depth with many branches helps to keep the machine from starving. A good guess is that these problems are difficult. We suggest heuristics in the next section.

Other work on equal parallel machines with precedence constraints includes Muntz and Coffman (1969, 1970).

11.2.4 Unweighted Flowtime Results

Other than for makespan, there are surprisingly few exact results for parallel machine problems. Important exceptions are the following dispatch results for unweighted flowtime.

Proposition 8. Consider the static arrival, dynamic availability, equal parallel machines problem with unweighted flowtime, F_{av}, as the objective function. The "m longest job" procedure is optimal:

Step 1: Assign the m longest jobs to the m machines in any order.

Step 2: Repeat until jobs are exhausted (may not come out even).

Step 3: Sequence jobs on each machine in SPT order.

Proof. Each job scheduled i positions from the end of the line in front of a machine adds exactly i times its processing time to the total flowtime (once for itself, once for each successor). Thus the optimal solution will fill all machines the same number of times, except for a leftover. (For example, if one line has eight in the line and another six, then moving the front job from the longer line to the front of the other reduces its count from eight times to seven times; nothing else changed.) Then if the number of jobs n is divided by the number of machines m, and rounded up, let N be the result. Add enough dummy jobs of processing time 0, so that there are exactly N jobs per machine. This will have no effect on the problem. Let $p(i,k)$ be the processing time of the job assigned to the ith position from the back on machine k. First row jobs get counted N times in the total flow, second row $N - 1$, and so on. Thus the total flow to be minimized is

$$N \sum_{k=1,m} p(1, k) + (N - 1) \sum_{k=1,m} p(2, k) + \cdots + (1) \sum_{k=1,m} p(N, k)$$

It is obvious that this expression can be minimized by scheduling the m longest jobs in any order the first round, the next longest m the second round in any order, and so on. (There are many alternate optima.)   ∎

There is no advantage gained from preemption for this problem.

For example, consider six jobs with processing times 1, 2, 3, 4, 5, 6 and two machines. The algorithm would assign 5 and 6 last on different machines, 3 and 4 next last, 1 and 2 first. One possible assignment is shown in Figure 11.5. In general the algorithm can construct many alternative choices with the same F_{av}.

With a little insight we can see how the m-jobs-at-a-time procedure may be interpreted as our more usual dispatch procedure, as described by Proposition 9.

Proposition 9. Consider the static arrival, dynamic availability, equal parallel machine problem with unweighted flowtime, F_{av}, as the objective function. An optimal solution is given by the following:

Step 1: Order the jobs in SPT order.

Step 2: Assign the jobs in turn to the first available machine.

Proof. Apply Proposition 8, using the discretion of interchanging jobs at a given level.   ∎

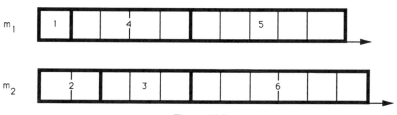

Figure 11.5

Except for ties, this dispatch algorithm will produce a unique schedule, and, of course, it will be one of the schedules that might be produced by the m-jobs-at-a-time approach. The procedure of Proposition 9 has several advantages, however. First, it is a dispatch procedure, so that scheduling decisions may be implemented in the order they are made. Second, it corresponds with our intuition gained from the free-preemptive case. Third, it is really a bottleneck dynamics procedure and hence can be generalized to good heuristics for dynamic arrivals and so on. It also leads to natural heuristics for the weighted tardiness case, as we shall see in the next section.

Note that Proposition 9 gives the same exact procedure that would be suggested as a heuristic by Proposition 3. This is the only case for which the heuristic is exact; however, the procedure turns out to be robust in general.

11.2A Numerical Exercises

(Use the provided software, where helpful.)

11.2A.1 Consider the following static three-machine problem with 15 jobs, and F_{wt} as the criterion:

Job j	d_j	p_j	w_j/p_j
1	10	17	0.5
2	8	6	1.1
3	17	4	1.5
4	6	3	3.0
5	11	3	2.7
6	9	2	2.6
7	14	2	1.5
8	13	1	3.2
9	6	1	3.0
10	18	1	2.5
11	4	1	2.2
12	5	1	2.0
13	15	1	1.4
14	8	1	1.0
15	4	1	0.6

Try to approximately solve this problem in two ways:

(a) Assign randomly to machines; schedule each optimally.

(b) Optimize the queue as for one machine, and then assign each to keep the load as level as possible.

(c) Which works best? Can you argue why?

11.2A.2 Repeat 11.2A.1 for the same problem and the T_{max} problem.

11.2A.3 Repeat 11.2A.1 for the same problem and the T_{wt} objective (You may wish to use a less computationally demanding heuristic than R&M.)

11.2A.4 Form the unit-preemptive version of the problem in 11.2A.1 and use Proposition 2 to optimize free unit preemption for makespan, weighted flow, and maximum lateness.

11.2A.5 **(a)** Solve the same problem for preemptive makespan, using McNaughton's algorithm (Proposition 5).

   (b) Solve the same problem by Proposition 6 (Hu's algorithm).

11.2A.6 Create an interesting assembly tree structure to be scheduled on three machines with makespan criterion, and schedule it using Proposition 7 (Hu's algorithm).

11.2B Software/Computer Exercises

11.2B.1 Computerize Hu's algorithm and try it on a number of problems both with and without assembly tree structure. Discuss your results.

11.2B.2 Try to modify your program from 11.2B.1 to produce only small numbers of preemptions. Try this both by (a) intuitive choices during the algorithm when there are multiple choices and (b) search procedures added to the end of the algorithm.

11.2C Thinkers

11.2C.1 Try to extend Proposition 2 to cover the weighted number of tardy jobs case.

11.2C.2 Discuss extension of Proposition 3 or a related idea to the case of dynamic machine availability.

11.2C.3 What kinds of stronger bounds can you give than Proposition 4 for dynamic arrivals, precedence constraints, and/or sequence-dependent setups?

11.2C.4 What goes wrong in attempting to extend the proof method of Proposition 5 to the proportional machine case?

11.2C.5 Try to prove or disprove the conjecture following Proposition 6. If this proves difficult, try enough examples to try to decide whether or not the conjecture is likely to be true.

11.3   PARALLEL MACHINES: GENERAL PROCEDURES AND HEURISTICS FOR EQUAL/PROPORTIONAL CASES

11.3.1   Overview

The fact (Proposition 1) that scheduling for parallel machines can be reduced to sequencing a list of jobs has implications for all sorts of mathematical approaches to the problem. In Section 11.3.2 we give basic dispatch heuristics that come from our free-preemption results in Section 11.2 In Section 11.3.3 we consider general purpose methods using the basic sequencing list result and dispatch and/or search heuristics now available. These general purpose methods include branch-and-bound, beam search, dynamic programming, and extended neighborhood search methods. In Section 11.3.4 we give a general purpose description of how bottleneck dynamics handles parallel machines, including aggregation and pricing, again building on our results in Section 11.2. We also discuss the economic makespan criterion for parallel machines. Finally, in Section 11.3.5 we consider embedding for parallel machines.

Other early work on the parallel machine problem includes Root (1965), and Rothkopf (1966).

11.3.2 Basic Dispatch Heuristics for Equal and Proportional Parallel Machines

We are ready to put several of the results of the last section together in order to derive a number of effective heuristics.

Definition. Let m' be defined as the "effective number of machines"; that is, $m' = m$ in the equal case and $m' = \Sigma_{k=1,m} s_k$ in the proportional case.

To summarize our situation:

(a) Proposition 1 says there is an optimal sequence for the jobs such that each can be scheduled in turn on the first available machine for the equal case, or the machine that can first finish the job in the general case.

(b) Earlier chapters suggest that rules optimal for the preemptive case make good heuristics for the nonpreemptive case.

(c) Proposition 3 suggests that free-preemption problems may be aggregated into a one-machine problem with the machine having m' times the speed.

(d) Propositions 2 and 6 suggest a longest processing time (LPT) ordering for a makespan heuristic, WSPT for weighted flow, and EDD for maximum lateness.

This all leads us to a general heuristic design for the nonpreemptive problem and independent jobs:

Equal and Proportional Parallel Machine Dispatch Heuristics: Static Arrivals, Static Priorities

Step 1: Prioritize the jobs.

Step 2: Schedule the highest priority job on the machine capable of finishing it first.

Step 2a: For a simpler heuristic, schedule on the first available machine for proportional machines also.

Step 2a works well unless machines have markedly different speed factors.

Addition of dynamic arrivals complicates our life considerably, since placing the highest priority *available* job on the first finishing machine in dispatch mode may leave an actually available machine unfilled. We present a simple and a more complicated heuristic for dealing with this problem.

Equal and Proportional Parallel Machine Dispatch Heuristics: Dynamic Arrivals, Dynamic Priorities

A. Simple Dispatch Heuristic

Step 1: Wait until some machine M becomes free.

Step 2: Calculate the priorities of currently available jobs if placed on M.

Step 3: Assign the highest priority job of these to M.

Since in many cases the first available machine will in fact be the first finishing machine, this algorithm should be robust.

B. Complex Dispatch Heuristic

Step 1: Wait until some machine M becomes free.

Step 2: Calculate the priority of each currently available job if placed on the machine that would finish it first.

Step 3: Tentatively assign the highest priority job to the machine that could finish it first (whether that machine is actually currently available or not), then the second highest, and so on until either a job *J* has been assigned to M, or until there is none left.

Step 4 If some *J* has been tentatively assigned to M, make this assignment permanent, and delete other assignments; go to step 1.

Step 5: Otherwise assign the job with lowest priority to M and delete the other tentative assignments; go to step 1.

Note that Step B.5 forces the procedure to be a dispatch procedure.

Dispatch methods have a number of practical advantages. In a rolling horizon problem with future events rather uncertain, all we really wish to do at a point in time is to actually start a fairly appropriate job on the machine currently available. We basically need to go through the tentative assignments to ensure that highest priority jobs are going to tend to get the fastest machines.

This is complicated enough that in many practical situations we are likely to simply assign the highest priority job to the first available machine, that is, the first heuristic above.

Note that we are also not considering the case of waiting for hot jobs, which would involve costing out inserted idleness by bottleneck dynamics.

Priority Rules for Equal and Proportional Parallel Machines

A. *Makespan.* largest p_j/m' first (same as LPT).

B. *Weighted Flow.* Largest $w_j/(m'p_j)$ (same as WSPT).

C. *Maximum Lateness.* EDD.

D. *Weighted Tardiness.* Modified R&M.

E. *Number of Tardy Jobs.* Modified Hodgson's.

For the makespan, weighted flow, and maximum lateness heuristics, the fact that the machine's speed has been multiplied by m' affects all priorities equally and thus cancels out.

However, in the exponential R&M heuristic, processing time enters in three places in the formula:

$$\pi_j = (w_j/p_j)\exp\{-(d_j - p_j - t)^+/p_{av}\}$$

We want the machine time multiplied by m' for the tardiness priority in the front, and the p_{av} in the back. However, the estimation of the slack should be the actual time used by machine k to do the processing. Thus we have the following heuristic:

Weighted Tardiness Heuristic for Equal/Proportional Machines (Modified R&M)

$$\pi_j = (w_j/(p_j/m'))\exp\{-(d_j - p_j/s_k - t)^+/(p_{av}/m')\}$$

(p_j gets divided by m' in the first and third instance, but by s_k in the second; thus it does not cancel out in comparing jobs.)

For static Hodgson's for parallel machines, list the jobs by the EDD heuristic as before. When the first tardy job is found, look for the longest job to remove as before.

As a final argument in favor of modifying one-machine dispatch heuristics for equal/proportional machines when priorities are constant or changing fairly slowly over time, we note that since such heuristics are optimal for the free-preemptive case, if they are used for the nonpreemptive case and relaxed to the free-preemptive case, the most favorable thing that will happen to a job's $f_j(C_j)$ is that the job will finish possibly almost instantly, saving at most its processing time on the slowest machine, p_j. Thus the most this job could be aided would be $f_j(C_j) - f_j(C_j - p_j)$. If f_j is convex and has a quadratic approximation this would be approximately $p_j f'_j(C_j)$ + $0.5p_j f''_j(C_j)$. While the average job in the relaxation will finish at about the same time, the worst case would be if all finished this much earlier.

Since the unit-preemptive case is a lower bound on the optimal, this implies crudely that the dispatch policy cannot exceed the optimal by more than

$$\sum_{j=1,n} [p_j f_j'(C_j) + 0.5 p_j f_j''(C_j)]$$

This argument is much too crude to be called a theorem, but a similar worst case bound could probably be constructed. Note that the result says roughly that a dispatch procedure aggregating machines appropriately cannot produce a worse result than shifting all the completion times by a processing time. Thus for a large number of jobs compared to the number of machines the error bound will be fairly good.

11.3.3 Search Methods for Equal and Proportional Parallel Machines

The results of this subsection hold in broad outline for the general nonproportional parallel machine case, as well as for proportional machines; hence by strict mathematical principles we could postpone this discussion until Chapter 12 and thus avoid a second short repetitious discussion there. However, for some pedagogical reasons we discuss search methods now:

(a) The proportional case is easier to understand intuitively.
(b) The equal/proportional case can be studied in the software.
(c) The equal/proportional case is most used practically.

The fundamental result that makes search methods quite tractable for any kind of parallel machine assumptions is again Proposition 1. Remember that it states that, under very general conditions, any parallel machine problem is solvable by finding the best permutation sequence (among those feasible) by routing each job in turn to the machine that can finish it first.

Thus all the search methods we studied in Chapters 5, 6, and 10 for the one-machine problem may be applied directly to any parallel machine problem. The search part of the method is unchanged. Evaluating a given solution may be more expensive. Unless special structure is available, it is likely to involve simulating the given permutation sequence (job order) and then calculating the resulting objective function. (This may be much cheaper for the makespan objective, since the makespan for a given policy may be evaluated by solving a much faster maximum path problem.) Similarly, evaluating LP lower bounds for forward branch-and-bound and myopic probes as estimates of evaluation in beam search are typically longer. Extended neighborhood search methods and the shifting bottleneck method require primarily the cheapest methods for doing a preliminary screening of the neighborhood and for evaluating candidate solutions. Dispatch/bottleneck dynamics methods remain very fast since they simply simulate the problem one to five times in any event. It is still easy to solve quite large problems by any of these methods, as will become apparent in using the supplied software, which uses this permutation sequence method.

A side point is that if the problem involves only one to three machines, there will be some cases for which the permutation sequence method will be inferior to searching over both the machine and the sequence on a machine directly.

11.3.4 Bottleneck Dynamics: Aggregation, Pricing, and Economic Makespan

Let us investigate our idea of a group of parallel machines as being more or less equivalent to a single aggregated faster machine as suggested by Proposition 3. It states that this idea is exact for parallel or proportional machines for the free-preemptive case with the speed of the faster machine being the equivalent number of machines $m' = \Sigma_{k=1,m}s_k$. Since we have been quite successful by basing our heuristics on the preemptive case, we may try to see how far this idea will take us.

We decide to estimate the price of an aggregate machine in the same way as for a simple machine. Suppose there are N_{agg} jobs in the busy period of the aggregate machine. (We know we may treat the parallel machines as having a single queue. The busy period is the time until the aggregate queue is empty.) Then the price of the aggregate machine is simply

$$R_{agg} = \sum_{i=1,N_{agg}} w_j U_j(t)$$

Note that the price does not depend on how fast the aggregate machine is processing these jobs. This is at first puzzling, since intuitively we expect the aggregate machine to have a much higher price. The difficulty is easily explained. In a busy period of the same time length, the aggregate machine can process m' times as many jobs, and thus will have a higher price than a slower machine with the same busy period in time length. In practice, a fast machine will have a much larger number of jobs in its busy period; it processes them rapidly, is more desirable, and has a higher price.

Determining the price of an individual machine in terms of the aggregate price for parallel/proportional machines is quite easy. Let us do it for equal machines first. Suppose machine k is shut down for a time period Δ. Then this is the same as if the aggregate machine (m individual machines) were shut down for a time period Δ/m. (This idea is exact for free preemption and an approximation for the non-preemptive case with a reasonably large number of jobs.) Thus the price of the individual machine is just $R_k = (1/m)R_{agg}$. For the proportional cost case, the individual machine's capacity as a fraction of the total is $s_k/\Sigma_{k'=1,m}s_{k'}$ or more simply s_k/m'.

Proposition 10.

Pricing Principles for Equal/Proportional Parallel Machines

(a) Price the aggregate machine by $R_{agg} = \Sigma_{j=1,N_{agg}}w_j U_j(t)$.
(b) Price an individual machine by $R_k = (s_k/\Sigma_{k'=1,m}s_{k'})R_{agg}$.
(c) The resource cost of a job does not depend on which machine it is run on.

Proof. To see point (c), the resource cost of job j on machine k is given by

$$(RC)_{jk} = R_k p_{jk} = (s_k/m')R_{agg}(p_j/s_k) = R_{agg}p_j/m'$$

which is not dependent on k. These results do not hold for nonproportional machines, although they may hold approximately for machines that are "close to" proportional. We return to this issue in the next chapter. ∎

Although all machines have the same resource cost usage, it is not true that it is irrelevant on which machine a job is run. Clearly the job, for timeliness considerations, will prefer to be run on the machine that finishes it first, as suggested by Proposition 1. Thus bottleneck dynamics agrees with the heuristics of the previous subsection: for the dynamic problem, sequence as for a single faster machine, and then schedule on the machine that can finish first.

We will use bottleneck dynamics extensively in our development of the nonproportional machine case in Chapter 12, but it may be useful to give some taste of the ideas here.

Example: Weighted Flowtime with Setups. Consider first the problem of equal parallel machines with weighted flowtime as the objective, but with the added complication that there are sequence-dependent setup times that don't vary by machine but differ by which job was last. Suppose that job i will finish on machine k at time t_{ik}. If job j is scheduled next on machine k, it will have setup time T_{ij} and processing time p_j, finishing at $C_{jk} = t_{ik} + T_{ij} + p_j$. The total (myopic) cost of scheduling job j on machine k at this time will then be

$$(\text{Myopic Cost})_{jk} = (R_{agg}/m)(T_{ij} + p_j) + w_j(t_{ik} + T_{ij} + p_j)$$

That is, the myopic estimate of the cost is the resource cost of placing j on k plus the weighted flow penalty of placing j on k. The first is the price of machine k, which is the aggregate machine price divided by the number of machines, times the sum of the setup time and the processing time. The weighted flow penalty is the weight w_j times the completion time, which is the sum of the start time, the setup time, and the processing time.

This myopic cost idea might be used in various ways in building a more complex heuristic for the problem. One approach would be to sequence the jobs by some other heuristic, and then assign job j, which was first in that sequence, to a machine in such a way as to minimize the myopic cost over k.

A different approach would be to consider every possible job to be first, and assign it to the machine that could finish it first. The opportunity cost for job j would be the myopic cost regret if job j is not assigned first and to this machine but to the best alternative. The job j with the maximum regret for not being currently assigned would be assigned first. (This is basically Vogel's heuristic for the transportation model.) Both of these ideas will be explored further in Chapter 12.

Example: Makespan Versus Economic Makespan. We use our bottleneck dynamics pricing ideas to evaluate the accuracy of using the makespan criterion for parallel machines. The makespan criterion is mostly useful for a highly utilized shop. In this case each machine price must be high and wasting resource time costly. Yet the makespan criterion only considers the utilization of the latest finishing machine and gives no credit for amounts of time by which other machines finish earlier. In fact, in a highly loaded shop there will probably be new jobs arriving to use this spare capacity, and therefore it should be valued.

An example may clarify things. Suppose two equal machines have a price R_{agg} = 10. Both machines are evaluated starting at time $t = 0$. Suppose the solution minimizing makespan finishes machine 1 at $t = 90$ and machine 2 at $t = 90$, for a makespan of 90, and a total resource usage of $(90 + 90)(10/2) = 900$.

However, suppose an alternative solution finishes all the jobs for machine 1 at $t = 65$ and machine 2 at $t = 95$, for a makespan of 95, but a total resource usage of $(65 + 95)/(10/2) = 800$, much better than the optimal makespan solution.

Note that the solution that minimizes makespan puts a high premium on equalizing the finishing times for the two machines, while resource usage minimization puts little premium on such exact balancing.

We call any objective that essentially tries to minimize resource usage an *economic makespan* objective (Kumar and Morton, 1991). We consider two types of resource usage (a) that directly due to processing and (b) internal wastage (inserted idleness). We do not consider idleness at the end of the horizon to be wastage, unless the model specifically states that new jobs will not be available.

For simple problems where resource usage for processing a job is independent of the machine and there is no internal wastage, economic makespan is independent of the sequencing decisions.

Proposition 11. For equal/proportional parallel machines with static job arrivals, economic makespan is minimized independent of how the jobs are processed.

Proof. Obvious ∎

11.3.5 Embedding for Parallel Machines

Recall our embedding idea, which we developed in Chapter 9, and the application of embedding to OPT-like and shifting bottleneck procedures, which we discussed in Chapter 10. Basically, we focus on one critical resource. We consider cases in which all the other resources are either in large enough supply so that queues are unnecessary, or for which we specify exact decision rules for arriving jobs. These are called the simple/complex embedded one-resource problem, respectively. In either case the resulting problem is a (rather complex) one-resource problem, which we may hope to solve exactly or approximately by the one-resource methods we have developed thus far.

Solving all possible simple embedded problems and choosing the most costly to define the bottleneck machine can form a basis for OPT-like procedures. The shifting bottleneck method is a sophistication of OPT-like procedures that solves simple and complex embedded problems in an iterative way, seeking an improvement each time.

The essence of these procedures is that the simple/complex embedded one-resource problem be relatively easy to solve. Until this chapter, we mostly studied one-machine versions of one-resource problems, which can be reduced to optimizing the sequence of jobs. Hence these problems are amenable to most of our techniques, although ideas for iterating starting and local due date times for the embedded problem have had to be developed.

It is encouraging in this chapter, via Proposition 1, to discover that aggregations of parallel machines may also be treated as a single resource and optimized by optimizing a single sequence as before. While these problems are somewhat more computationally intensive than one-machine problems, it seems clear that makespan problems may still be solved optimally, and other embedded problems for other objective functions can be solved accurately with good heuristics. Thus OPT-like and shifting bottleneck procedures seem viable at reasonable computational cost.

11.3A Numerical Exercises

11.3A.1 Consider the 15-job problem given in 11.2A.1. Using the parallel machine portion of the software, try a number of heuristics on each of the following objective functions, and summarize the results: weighted lateness, weighted tardiness, number of tardy jobs, and maximum lateness.

11.3A.2 Try 11.3A.1 also for makespan and economic makespan. (A common heuristic for makespan is LPT.)

11.3A.3 Make some interesting changes in the data set of 11.2A.1 and re-solve 11.3A.1 and 11.3A.2.

11.3A.4 Redo these problems for some proportional machine case.

11.3B Software/Computer Exercises

11.3B.1 Redo 11.3A.1 and 11.3A.2 for larger problems and multiple problem sets via computer, using proportional speeds. Add some type of benchmark for your heuristics (lower bound, optimum solution, or more expensive heuristic).

11.3C Thinkers

11.3C.1 After you talk to management about economic makespan to measure utilization, they agree that giving idle capacity at the end zero value is too conservative, but they feel that giving it full value is too liberal. Create a *compromise economic makespan* measure to satisfy them.

11.3C.2 Develop weighted and/or unweighted versions of minimizing the number of tardy jobs for equal parallel machines.

11.3C.3 Do the same for proportional parallel machines. What difficulties do you run into if any?

11.3C.4 Suppose that, due to inefficiencies, three equal parallel machines will effectively run jobs a little less than three times as fast as a one-machine problem. Incorporate this into the bottleneck dynamics formulation.

11.4 BATCH MACHINES

11.4.1 Overview

Consider a machine that processes up to m jobs simultaneously, with common beginning and end times, such as a steel melting furnace, a preheat furnace, an annealing furnace, a paint room, or a transporter vehicle. (We use the same symbol m as for parallel machines, to emphasize how similar the models are.) This is a very common situation in practice, but not one that has been treated very well mathematically, probably due to its difficulty.

The problem bears much resemblance to an equal parallel machine problem, except that jobs have a common processing time and are constrained to start and stop together. In the simplest one-class case, any two jobs may be processed together. In the multi-class case, only jobs of the same "type" may be processed together. We consider the two cases separately.

11.4.2 One-Class Case

Suppose that n jobs arrive dynamically with processing time p, arrival time r_j, due dates d_j, weights w_j, and slack factors $U_j(t)$; the machine has m slots and currently has price $R(t)$. Initially, assume all jobs occupy exactly $1/m$ of the machine.

The following general heuristic form suggests itself:

Simple Batch Heuristic Approach. Whenever the machine becomes free, order the available jobs in order of priority:

$$\pi_j = (w_j U_j)/(I(R/m)p)$$

Step 1: If m or more jobs are available, load the top m (and move ahead by time p to the next scheduling decision).

Step 2: If less than m are available, but inserted idleness is not allowed, load those available and proceed.

Step 3: If less than m are available, and inserted idleness is allowed, perform the economic balancing below. If positive, wait for the job, and return to step 1; otherwise move ahead to the next scheduling decision.

(It may be iteratively possible to wait for two or more jobs.)

Batch Economic Balancing (One Class)

(a) Waiting a time Δ_1 for the next arrival costs the resource $R\Delta_1$.

(b) It fills the resource better, saving $(R/m)p$.

(c) The job expedited saves $\Delta_2 w_j U_j$, where $\Delta_2 = p - \Delta_1$ is the amount of time it is expedited.

Looking at just the first two terms, we see a useful rule of thumb: for the one-class case, always wait if the waiting time is less than the process time divided by the number of slots in the machine; wait longer for a higher priority job.

This is only a quick myopic approximation. Note, in particular, that if the first job does not have a high enough priority to justify waiting, the first two may *together* justify waiting, and so on. It is interesting to note that for an annealing furnace at one real company a standard rule was "don't wait over 2 hours." In this case p was 12 hours, and m was 10. So our rule would be "wait at least 1.2 hours, more depending on urgency." This is fairly decent agreement.

If different jobs occupy different amounts of space v_j, and can be accommodated as long as $\Sigma v_j \leq V$, where V is the total batch space, a myopic procedure (greedy) is to assign items to the batch in order of decreasing priority per unit volume π_j/v_j.

This method could be improved in several ways, although it is often a good heuristic. First, some clever packing method (this is basically a bin-packing problem) might fill the batch fuller and improve utilization. This sophistication may or may not be worth the considerable trouble to pursue in a dispatch-oriented system. Second, considered as a knapsack problem, some method might be able to pack more total delay cost into the batch, irrespective of utilization. This issue also will not be pursued.

In certain other situations the jobs may be of irregular size and shape, so that fitting becomes all important. Or different types of pieces may be preferred to be in different parts of the furnace, and so on. These issues are beyond our scope.

11.4.3 Multi-Class Case

Now suppose the jobs divide into classes, such that jobs within a class can be processed together, those outside cannot. As a quick notation, say j belongs to class J, and so on. Different classes may comprise different job processing times, sizes, and so on. Up to m_J jobs of class J may be processed simultaneously, with processing time p_J.

Consider first the possibility that type J jobs should be processed next. (Ignore inserted idleness for the moment.) Order them in order of decreasing priority $w_j U_j/p_J$ as before, and consider the top m_J as a batch. We form an aggregate priority for this batch:

$$\pi_{J1} = (\Sigma_{j \in J1} w_j U_j)/p_J$$

(where $J1$ is the proposed batch). Now repeat this procedure for each class. The batch class with the highest priority is then scheduled. The entire procedure is repeated when the machine next becomes free.

Suppose now that the top batch for class J has been chosen for scheduling. It is not full, and it is desired to consider the economic benefit of waiting for a member of the class to arrive.

The myopic economics are really quite similar to the non-multi-class case, except that:

(a) The waiting time Δ_1 until a suitable arrival may be much larger, since the job to be expedited must be of the same class.

(b) The savings in time Δ_2 for the expedited job may also be much larger, since the machine may process a number of batches before returning to one of this type. Thus Δ_2 may be more like Kp_{av} where K is the number of classes and p_{av} is the average process time over the different classes.

The multi-class case does not seem a priori more or less likely to involve inserted idleness. The case is much harder to evaluate accurately, in that decent estimation of Δ_2 may involve some kind of forecasting or iteration of the problem. Also, the myopic heuristic itself seems less likely to be accurate given the longer horizons involved.

In some cases it will be worthwhile to compute an inserted idleness-adjusted priority for every possible class to schedule next. The apparently top-priority batch, which involves excessive waiting, may, after adjustment, score lower than an apparently lower priority batch, which can be filled immediately.

A useful reference for this section is Ikura and Gimple (1986).

11.4B Software/Computer Exercises

114B.1   Write a routine to schedule a batch machine with dynamic arrivals and a single class. Set up a test set of different objective functions and problem data.

11.4B.2   Extend the program in 11.4B.1 to include multiple classes and different job volumes.

11.4C Thinkers

11.4C.1   Suppose there is one class of jobs, but they have different two-dimensional shapes and must be fitted like cookies on a cookie plate, while keeping track of differing priorities. What sorts of principles can you think of to make heuristics for this problem? (This situation comes up in curing layered airplane parts, for example.)

11.4C.2   Suppose there were two batch machines in parallel. What heuristics could you work out for this situation?

_____12
SHOP ROUTING

12.1 INTRODUCTION

In Sections 12.2 and 12.3 we treat the case of general parallel machines. This is actually a single-resource problem in the sense that there is a single queue. We demonstrated in Proposition 1 of Chapter 11 that for regular objectives the problem may be optimized by optimizing the job sequence in that queue and assigning each job to the machine that can finish it first. However, from a different point of view, if processing times by machine differ for each job and different machines have a comparative advantage for different jobs, then there is an interesting routing question. (This issue does not arise for parallel equal or proportional machines since, as we have shown, the resource usage of a job is independent of which machine it is scheduled on.)

Turning next to more general routing issues in Section 12.4 we consider activity routing. Activity routing just means that the same activity has a choice of several resources (or machines) to be assigned. The case of general parallel machines (which can be modeled as a single queue in accordance with Proposition 1) is one example. However, in general, the different machines that could be chosen would have different queues and would not be aggregatable into a single machine for other jobs. For example, if the activity is getting to school, a person might choose to walk, take a cab, or take a bike. The different resources involved are not aggregatable into a single resource.

In Section 12.5 we discuss some advanced routing issues. First, we consider process routing. Here there may be several different overall procedures with associated resources for performing the same job. For example, we might manufacture a job by a new high-tech FMS cell, the usual high-speed transfer line, or by starting up the obsolete line in the corner. Making the right decision is more complex than if only one activity has resource choice. In this section we also discuss cost/resource trade-offs, which may be considered a type of routing issue. Suppose that an activity can be done in several ways; each uses different amounts of resources involving implicit costs and other raw materials involving explicit costs. How should the correct method to do the activity be chosen?

12.2 GENERAL PARALLEL MACHINES: UTILIZATION

12.2.1 Overview

Recall the model for general parallel machines. There are m machines $k = 1, \ldots, m$, and n jobs $j = 1, \ldots, n$. Machine k first becomes available at a_k; job j arrives at r_j. For any regular objective, we know there is an optimal solution that sequences the n jobs in some unknown fashion and then assigns each in turn to the machine that can finish it first. Simple utilization is a reasonable objective in a highly loaded shop, but not otherwise. Classically, makespan has been the dominant utilization objective considered; however, we argued in Chapter 11 that economic makespan (bottleneck dynamics resource usage) is a superior utilization measure. We also showed, for the objective of economic makespan on equal and/or proportional machines with static job arrivals, that the sequence in which jobs are assigned is irrelevant. This is an intuitively satisfying result, since the machines have no comparative advantage. (This is not true for the makespan objective.)

In Section 12.2.2 we give some exact results for minimizing makespan on general parallel machines, including linear programming solutions of the free-preemption version of the problem that are necessary for good branch-and-bound (B&B) solutions. In Section 12.2.3 we give corresponding heuristic results. In Section 12.2.4 we develop bottleneck dynamics (BD) results for economic makespan, including machine aggregation and pricing issues.

12.2.2 Exact Makespan Methods

It will be useful to solve the free-preemptive version of the general parallel machine makespan problem, both to provide strong lower bounds in doing a forward branch-and-bound procedure and later to provide resource prices for the bottleneck dynamics economic makespan objective. We assume dynamic availability and arrivals.

Proposition 1. The optimal solution to the free-preemption version of the general dynamic parallel machine problem with makespan objective is the solution to the linear program

$$\min\ C_{\max}$$
$$\text{s.t.}$$
$$a_k + \sum_{j=1,n} x_{jk} p_{jk} \le C_{\max} \quad \text{for each machine } k$$

$$\sum_{k=1,m} x_{jk} \;=\; 1 \qquad \text{for each job } j$$

$$r_j + \sum_{j'=j,n} x_{j'k} p_{j'k} \;\leq\; C_{\max} \qquad \text{for each job } j \text{ and machine } k$$

$$x_{jk} \;\geq\; 0 \qquad \text{for each assignment } jk$$

where jobs j are numbered in order of arrival, $C_{\max}$ is the makespan, a_k is the first availability of machine k, x_{jk} is the fraction of job j assigned to machine k, p_{jk} is the processing time of job j on machine k, r_j is the ready time of job j, and j' is an index representing jobs arriving no earlier than j.

Removal of the equations with r_j produces the equations for the static arrival case.

Proof. Each machine equation simply states that the completion time of that machine must be less than or equal to the makespan. Each job equation simply says that all of that job must be assigned somewhere. Each job/machine equation states that all jobs arriving after j and assigned to machine k cannot be more than the makespan. The assignment inequalities simply state that assignments can be fractional but not negative. ∎

Proposition 2. If in Proposition 1 the integrality constraints

$$x_{jk} \in \{0,1\}$$

are substituted for the nonnegativity constraints, then the resulting integer program solves the nonpreemptive version of the problem.

Proof. Obvious. ∎

We present a forward branch-and-bound approach, which allows us to obtain planning horizon results and also to modify the procedure for heuristics such as beam search. For general objective functions we know that an optimal solution may be found by optimally ordering the list of jobs, and then assigning each to the earliest available machine, involving at most $n!$ sequences. However, whenever an optimal ordering rule for each machine is available and the number of machines is much smaller than the number of jobs, an alternative approach is to pick an arbitrary job and assign it to one of m machines; doing this n times gives m^n choices. Then optimally order each machine.

Forward Branch-and-Bound Procedure for Makespan on Parallel Machines

Step 1: Order the main list in arrival order.

Step 2: Simulate; when a machine becomes free, branch on the set of available jobs.

Step 3: Develop lower bounds for subproblems by Proposition 1.

Step 4: Use best-first search or depth-first search (or other preferred method), as desired.

In tests, this sort of procedure has proved capable of solving problems with 10 to 15 jobs and two to eight machines in reasonable computation times.

12.2.3   Heuristic Approaches to Makespan Problems

We defer discussion of bottleneck dynamics heuristics until the next section, since they lead rather naturally to a consideration of the economic makespan criterion as an improvement over makespan.

Classic Approaches to Heuristic Makespan.   Ibarra and Kim (1977) develop five makespan heuristics for general parallel processors with dynamic machine availability but static arrivals. We first give the five heuristics, and then briefly discuss advantages and disadvantages.

Heuristic A

Step 1: Order the job list randomly.
Step 2: Schedule the next job on this list to the machine that can finish it first.

Heuristic B

Step 1: Order the list by decreasing $p_{j\ min} = \min_k\{p_{jk}\}$.
Step 2: Complete as in Heuristic A.

Heuristic C

Step 1: Order the list by decreasing $p_{j\ max} = \max_k\{p_{jk}\}$.
Step 2: Complete as in Heuristic A.

Heuristic D

Step 1: Find j^* and k^* satisfying $\min_{j,k}\{C_k + p_{jk}\}$, where C_k is the current partial completion time of k.
Step 2: Schedule job j^* next on machine k^*.

Heuristic E

Step 1: Find j^* and k^* satisfying $\max_j[\min_k\{C_k + p_{jk}\}]$, where C_k is the current partial completion time of k.
Step 2: Schedule job j^* next on machine k^*.

De and Morton (1980) developed a new heuristic by analyzing the strengths and weaknesses of Heuristics A to E. They saw three partial objectives that one would like to meet in order to solve the makespan problem on unequal processors:

Makespan Principles

(a) Equalize the loads on different processors as much as possible.

(b) Avoid "lumpiness" near the end (i.e., avoid processing an excessively long job late in the schedule).

(c) Process each job on the processor that has a "comparative advantage" for that job.

Note that principle (c) applies for general machine problems only. In the proportional machine case a processor does not provide any comparative advantages to a particular job over others. If it is fast for one job, then it is fast for all jobs. (Note also that in discussing economic makespan for proportional machines, we argued that avoiding lumpiness near the end was really unnecessary. However, we postpone further discussion until the next subsection on economic makespan.)

We next give a brief summary of the analysis the authors gave of these five heuristics in terms of the three makespan principles. All five heuristics equalize the load adequately by focusing on the machine that can finish a job first. (We have seen this allows an optimal schedule.) All five heuristics have at least a minor tendency to help the machine with the comparative advantage, again by focusing on the machine that can finish a job first. (Note, however, that if the machine was available early, the main effect is load equalization; if it was not, the main effect is comparative advantage.)

By ordering randomly, Heuristic A makes no attempt at avoiding end "lumpiness," which for the simple makespan objective is a serious disadvantage. In fact, Heuristic A did not perform well in the pilot study and was eventually discarded.

Heuristics B and C attempt to handle the lumpiness by scheduling short jobs last. The problem is that with general p_{jk} there is no longer a unique meaning to "shortest job" or "longest job." The processing time to be used should be the time for the processor to which the job is finally assigned. However, this is unknown until the job is scheduled.

Heuristic B estimates this by the minimum time for the job on any machine. The idea is that jobs will be placed on machines that process them quickly, making this a good estimate of actual processing time. There are two things wrong with this. First, the last jobs to be scheduled "fill the chinks" and hence the idea that their length is represented by the time on the fastest machine for them seems weak. Second, when the minimum processing time for a job is extremely long there is less likelihood that that machine will have much, if any, comparative advantage for a job. Also, Heuristic B still makes no strong attempt to deal with comparative advantage directly.

We turn to Heuristic C. It differs only in choosing the machine with the maximum processing time for the job. Note that the processors underloaded near the end are, indeed, likely to be the slowest processors overall (jobs are usually placed on machines processing them quickly), and thus having $\max_k\{p_{jk}\}$ small is likely to be important in filling the chinks at the end. A disadvantage is that the maximum time is a good estimate for later jobs scheduled but not for the early ones. Otherwise, Heuristic C has the strengths and weaknesses of Heuristic B.

Heuristic D gives up the longest processing time idea entirely to concentrate on comparative advantage scheduling. Other things being equal, it finds and schedules the shortest processing time of all, which is very likely to represent a comparative advantage as well. Thus this heuristic should be expected to perform poorly in the proportional case when there is no comparative advantage to be found, and quite well for problems that are highly nonuniform with large comparative advantages. It handles lumpiness perversely, if anything.

Finally, Heuristic E is an attempt to combine the advantages of Heuristics B and D: scheduling the longest jobs first, but on the fastest machine for them. It turns out to be an unfortunate compromise. The minimizing processor may well be determined by the most underloaded processor. The job that maximizes completion time on this processor will then tend to be the most inefficient one for that processor.

It occurred to De and Morton that a compromise way of estimating the "longest" job, using the *average* processing time over the machines, might be useful.

Heuristic F

Step 1: Order the list by decreasing $p_{j\ av} = (1/m)\Sigma_{k=1,m} p_{jk}$.
Step 2: Same as Heuristic A.

In preliminary studies, they found that none of Heuristics B through F were dominant. Since they are all of low cost, this led to a composite heuristic.

Heuristic G

Step 1: Run Heuristics B to F on the problem of interest, and simply choose the solution yielding the best makespan.

Although Heuristic G performs fairly well, it is inelegant and difficult to sharpen naturally. They therefore designed their own heuristic, using the lessons learned.

Heuristic H

Step 1: Estimate the makespan (from running another heuristic) as M.
Step 2: Choose a fill parameter f, such as 0.7.
Step 3: Order the list in decreasing $p_{j\ av} = (1/m)\Sigma_{k=1,m} p_{jk}$.
Step 4: When machines are relatively unfilled, assign job j from the list to that machine k^* that minimizes p_{jk}, without regard to which machine is the most filled.
Step 5: If an assignment would fill k^* to more than fM, declare the machine "closed" and assign the job instead to the open machine fastest for it.
Step 6: Once all machines are declared closed, finish using Heuristic F, that is, basically LPT.
Step 7: The procedure is run with several fill parameter values, and the best result is chosen.

This heuristic is built on a compromise among the three objectives. Although the concentration on avoiding end lumpiness is always maintained, while the machines are "open" the main emphasis is on achieving comparative advantage. Toward the end, when the machines are "closed," the emphasis shifts toward equalizing the load.

This is an excellent heuristic, as test results described below indicate. The main remaining defect is that the idea of "comparative advantage" is not perfectly spelled out. The heuristic clearly tends to fill the fastest processor first. (This does not necessarily seem to be a particularly good idea in general. An exception is the proportional machine case, where no machine has comparative advantages and placing the longest job on the fastest machine reduces lumpiness effectively.) Heuristic pricing ideas in the next section allow us to make the comparative advantage idea more precise by making it identical with "resource usage." This could be used to give an even better heuristic for makespan, by employing economic resource usage as the criterion in Heuristic H. However, we wish there to focus on economic makespan in any event. We therefore leave this last makespan improvement for the exercises.

We now very briefly review the computational study presented in detail by De and Morton (1980). An initial test study (Table 12.1) compared all the heuristics for 80 proportional machine problems and 80 general machine problems with n of 5, 10, 20, 30 and m of 2, 3, 4, 5, standard deviation of job length of 20%, and five replications.

The pilot study makes it rather clear that Heuristic D is by far the best performer among Heuristics B to F, that taking Heuristic G as the best of Heuristics B to F is a major improvement, and finally that Heuristic H is somewhat superior to the combination Heuristic G.

For the main study they therefore restricted attention to Heuristics G and H (Table 12.2).

The study makes it clear that the new heuristic is superior to choosing the best of the older heuristics. The new heuristic performs especially well for the proportional case, for reasons discussed above.

Morton and De (1982) also present another similar study for the objective of minimizing maximum lateness on general parallel machines.

Newer Approaches to Heuristic Makespan. De and Morton chose depth-first branch-and-bound, truncated after expanding up to 1000 nodes, as their high com-

TABLE 12.1 Makespan Pilot Study: 80 Problems, General Case

Heuristic	B	C	D	E	F	G	H
Percentage above benchmark[a]	9.12	9.00	4.65	18.75	10.14	2.81	2.19

Source: De and Morton (1980).

[a]The benchmark was truncated branch-and-bound.

TABLE 12.2 Makespan Main Study: 2400 Problems

Variability:	Proportional			General		
	0.1	0.2	0.4	0.1	0.2	0.4
Heuristic G	1.14	1.82	1.60	0.65	.59	2.64
Heuristic H	0.11	0.30	0.52	0.49	1.06	1.58

Source: De and Morton (1980).

Note: Percentage error above truncated B&B benchmark.

putational benchmark. (Makespan has the special property that the ordering of the jobs on each machine is unimportant. Thus one can branch on which of the m machines is assigned a given job, facilitating solution of problems with small numbers of machines.) Today we have available some new choices, which we sketch briefly.

Using the fact (not known then) that scheduling for general machines and any regular objective function can be reduced to determining a permutation sequence of the jobs, all our approximate search methods are directly applicable.

To do neighborhood search, for example, we would want to start with a good trial solution. Pairwise interchange involves about one-half of a problem simulation per interchange. Tabu search should cost little extra, since the cost of a pairwise interchange is high, so that the cost of checking repetition of previous solutions should be relatively minor. Similar remarks apply to simulated annealing.

To do beam search, we again need a good initial permutation schedule as solution, together with a heuristic. If there are 20 jobs, for example, with a beam width of 5, we would first try 20 choices of jobs to run first, using, say, Heuristic F to get a trial solution. Pick the five best, and expand into $19 \times 5 = 95$ subproblems, run a trial solution for each, and so on. The number of subproblems run would be about the same as for a one-machine problem; however, the cost of running a trial solution would be perhaps m times as high.

The approximate dynamic programming approach is not likely to work so well for the makespan problem. The main problem is that there are a great many quite different looking solutions that come up with much the same result. Thus, just because a heuristic schedules a job 13th, one may not have much confidence in the idea that the job is guaranteed to be scheduled between 9th and 17th, for example.

12.2.4 General Parallel Machines: Economic Makespan

Recall from Chapter 11 that we define the economic makespan problem as that of minimizing the value of resource usage on the machines, defined by appropriate resource prices.

Economic Makespan Criterion. Minimize the value of resource usage, penalizing for internal slack, but with no penalty for final slack:

$$\min \sum_{k=1,m} R_k(C_k - a_k)$$

where R_k is the machine price, a_k its availability time, and C_k its completion time.

Note that $C_{k(j)}$ means literally "the final completion time after all jobs are assigned for the machine on which j is assigned." Where necessary to avoid confusion, we use the notation $C_{k[j]}$ as "the partial completion time of the machine on which j is assigned only for the jobs assigned before j."

Equivalent Economic Makespan Criterion. Let $k(j)$ be the machine assigned to j; j will start at $\max\{C_{k[j]}, r_j\}$ (the larger of machine availability and the arrival time), giving a criterion of

$$\min \sum_{j=1,n} R_{k(j)}[p_{jk(j)} + I_j]$$

where the idleness caused by j is $I_j = [r_j - C_{k[j]}]^+$.

In particular, since economic makespan applies typically only to shops that are heavily loaded, and for which jobs have no significant time priority, the need for inserted idleness is likely to be small, and we may approximately simply minimize the sum of direct resource usage. This will form the basis of our heuristics.

Static Arrivals. We must find an efficient way to estimate prices R_k to use this method. The linear program of Proposition 1 solves the free-preemption version of the ordinary makespan problem. Consider the problem restricted to all jobs available from the start. The dual prices for this linear program, in fact, give the marginal decrease in makespan that would be caused by decreasing a_k, that is by increasing the availability of machine k. We are thus led to use these dual variables as estimates for R_k. Kumar and Morton (1991) prove the following results, which we simply state and develop intuitively here.

Proposition 3. Consider the problem of minimizing economic makespan for a static general parallel machine problem. Define R_k as the machine duals obtained from the first set of constraints in Proposition 1. Then the static economic resource problem can be stated as

$$\min \sum_j \sum_k R_k x_{jk} p_{jk} \quad \text{(economic resource usage)}$$

s.t.
$$\sum_k x_{jk} = 1 \quad \text{for each } j$$
$$x_{jk} \geq 0 \quad \text{for each } j, \text{ each } k$$

Proposition 4. For a given problem set, the static economic makespan problem (Proposition 3) and the relaxation of the static makespan problem (Proposition 1) have the same objective value. In addition, the solution for the economic makespan problem can be built up from the solution to the relaxed makespan problem as follows:

(a) Retain all integer assignments from the makespan relaxation solution.

(b) Collect fractional jobs to the machine assigned to any one of them.

This integer solution is optimal to the integer version of Proposition 3.

The proofs of Proposition 3 and 4 use duality theory. Intuitively, in the linear programming solution, jobs will be assigned to the machine with the least resource cost. If a job is split, the resource cost on the two machines must be identical, and thus it can be moved either way, since we do not insist on balanced completion times.

Although arbitrary fixing of the split jobs does not affect optimality, we might wish a solution leaving the machines relatively balanced, with near equal finishing times. (Then no machine is likely to run out before further work arrives.) We now look at this issue a little further.

Proposition 5. The total number of jobs that will be split in the optimal solution to this preemptive makespan problem is no more than $m - 1$. Also, for the split jobs, the total number of fractions cannot exceed $2(m - 1)$.

Proof. From elementary LP the total number of nonzero variables will not exceed $m + n - 1$. The nonzero variables in excess of n denote split jobs. Then the maximum number of split jobs is $m - 1$. It is fairly easily seen that the maximum number of split jobs occurs when each job is split in two. ∎

Thus for, say, three machines and 50 jobs, balancing is not a serious issue. In general, it seems possible to round off so that no machine is loaded with more than two jobs over the minimum load.

Dynamic Arrivals. Let us turn next to the dynamic job arrival case. Let the jobs be indexed in order of the arrival times.

BD Heuristic 1 (Dynamic Economic Makespan)

Step 1: Estimate R_k from the static job arrival preemptive solution.

Step 2: Assign jobs in order of arrival to the machine $k(j)$ that minimizes $R_k p_{jk}$.

This heuristic can depart from optimality by at most the amount of inserted idleness involved. This should be small in a truly heavily loaded shop. However, this heuristic might leave some machines much more heavily loaded than others. While that (it has been argued) does not matter, a large amount could cause scheduling gaps before the arrival of jobs past the end of the horizon.

This issue will become important in more complex shops where prices are being set in a more complex and less exact fashion. Heuristic 1 is likely to work better with pure LP prices than in a more complex environment.

This leads to the next heuristic.

BD Heuristic 2 (Dynamic Economic Makespan)

Step 1: Determine R_k and a lower bound M on the makespan from the static job arrival preemptive solution.

Step 2: Determine a fill parameter, such as 1.1.

Step 3: Assign jobs to machines as in Heuristic 1.

Step 4: If a machine loading exceeds $1.1M$, do not assign further jobs to it.

Step 5: Continue with the remaining subset of machines.

Heuristic 2 appears to work fairly well, but it is somewhat clumsy. A more sophisticated heuristic (which, however, does not allow inserted idleness) can be based on Proposition 1 of Chapter 11. That is, the heuristic uses the fact that there exists some list sequence for which an optimal policy is simply to assign the next job on the list to the machine that will finish it first. Note that this automatically ensures balance, since seriously underloaded machines are automatically scheduled.

We do not try to order the list in advance here. Rather, at any point we consider every possible available job as a candidate to be immediately scheduled on the machine that finishes it first. We also determine the "regret" for that job, the excess in cost if that job is forced to its next best alternative. The job that has the largest regret if not scheduled is the one scheduled.

(A) BD Heuristic 3 (Dynamic Economic Makespan)

Step 1: For each job yet to be assigned, determine $k(j)$, the machine that would finish it first, considering only those machines not available until after r_j (to avoid inserted idleness).

Step 2: In spite of the no-inserted-idleness restriction, we assume some j is to be assigned to its $k(j)$. We decide which one by the principle of minimizing regret. For each j and every machine k, calculate the cost of an alternative assignment by $R_k p_{jk}$.

Step 3: Define the normalized regret at not making the assignment of j to $k(j)$ by

$$N_j = \min_{k \neq k(j)} [(R_k p_{jk} - R_{k(j)} p_{jk(j)}) / R_{k(j)} p_{jk(j)}]$$

Step 4: Define j^* as the job with the highest normalized regret; that is, $N_{j^*} = \max_j [N_j]$.

Step 5: Assign j^* next to machine $k(j^*)$.

Step 6: Repeat.

Note that the procedure automatically produces a well-balanced line. If there are jobs available that minimize resource usage on the machine needing filling, one will automatically be assigned. In the unusual event that all regrets are negative, the least negative one will be assigned.

The idea of minimizing regret is a sort of "second-order myopic" procedure. It has been used widely and successfully as a heuristic for transportation and assignment problems (Vogel's method). The intuition is as follows. It is not so much the cost of assigning this job to this machine that is important. The assignment may be cheap on all machines or expensive on all machines, so that most of the cost is really sunk. A more important question is: If this job is not assigned right now on this machine, how much extra will it cost to assign it to the second cheapest place?

Here it is important to normalize and calculate the regret per unit of the resource under consideration. This is similar to the preemptive case calculations we have made for bottleneck dynamics in dividing the savings by the resources used. This heuristic seems likely to be robust, but it has not yet been tested.

Kumar and Morton (1991) tested a heuristic similar to BD Heuristic 1 against branch-and-bound in a rolling horizon environment. The idea of testing in a rolling horizon environment is as follows. For each time increment, called the review period (perhaps 1 week), job arrivals are forecasted over a longer forecast horizon (perhaps 3 weeks). The 3-week problem is solved either by the exact makespan procedure or by the approximate but better objective economic makespan procedure. The first week of the solution is implemented. One week later a new 3-week problem is forecasted and solved and the first week is again implemented and so on.

The experiment is then run for a long period, say, 100 weeks, and the makespan of the two procedures for the long horizon is compared. This simulates the way a shop might actually use a finite horizon deterministic model.

The Kumar–Morton results were quite encouraging for the new heuristic. The heuristic almost always outperforms the "optimal" procedure, while requiring only about 1% of the computation time. The heuristic appears particularly suited to situations in which comparative advantages can be exploited, in which case it appeared up to 6% better in their small study. They also exhibited a simple example to show that the advantage of the heuristic over the "optimal" procedure could be up to 25%. More study is warranted.

Finally, BD Heuristic 3 appears suitable for economic makespan heuristics incorporating such extensions as sequence-dependent setups and precedence constraints. In particular, in a myopic approach to setups, the cost of assigning job j to machine k would be

$$K_{jk} = R_k[\ T_{i[k]jk} + p_{jk} + (r_j - C_{k[j]})^+\ ]$$

(T_{ijk} means the setup cost on k if job i is already scheduled and followed by j. Thus if $i[k]$ is the last job already scheduled on k, the setup would be $T_{i[k]jk}$. Similarly, $C_{k[j]}$ is the partial completion time on k before j, so that $(r_j - C_{k[j]})^+$ gives the enforced idle time before the setup can start.)

Using a minimax regret approach here would probably provide useful (myopic) results.

Another useful reference for this section is Kellerer and Woeginger (1992).

12.2A Numerical Exercises

12.2A.1 Consider a makespan problem with static arrivals and five jobs on three unequal parallel machines. The machine/processing time matrix is given as follows:

	Jobs				
Machine	1	2	3	4	5
1	2	4	6	8	10
2	1	3	2	5	4
3	0.5	2	2	4	5

Test Heuristics A to H on this problem set. Use fill parameters of 0.5, 0.7, and 0.9 for Heuristic H.

12.2A.2 Invent four or five makespan problems on parallel machines with the help of the software. These should have static arrivals, 10 jobs, and two or three unequal machines.

12.2A.3 Test Ibarra and Kim heuristics A to E on your problem set.

12.2A.4 Test Heuristics F and G on the same problem set.

12.2A.5 Test Heuristic H on the same set with two or three fill parameters. How do your results compare with the reported studies?

12.2B Software/Computer Exercises

12.2B.1 Use a linear programming package to estimate machine prices for the problems in 12.2A.1 and 12.2A.2.

12.2B.2 How sensitive are the prices to minor changes in the problem data input?

12.2B.3 Write a simple beam search program for the two-machine makespan problem.

12.2C Thinkers

12.2C.1 Develop an improved (or at least more complicated) version of makespan Heuristic H. Test it. How much improvement did you make?

12.2C.2 Develop a version of Heuristic H in which the assignment minimizes economic resource usage. Test it. How good is it?

12.2C.3 Makespan minimizes the maximum machine completion time. Suppose instead a weight W_1 is assigned to the maximum time, W_2 to the second maximum time, and so on, weights adding to 1.0. Can you think of modifying any heuristic methods you have learned to deal with this problem?

12.3 GENERAL PARALLEL MACHINES: OTHER OBJECTIVES

12.3.1 Overview

For objectives other than makespan and economic makespan, there seems to be no easy analogue to the linear programming solution to the free-preemptive version of the general parallel machine problem with static job arrivals. This is basically because the sequencing decisions on each machine are not important for the makespan objective.

Forward branch-and-bound is more difficult for other objectives because the linear programming relaxation is not available to provide good lower bounds when the sequence on each machine is important. Lower bounds can typically still be constructed, although they are not usually as strong. We will discuss this issue briefly in Section 12.3.2.

The lack of the good LP relaxation also means that relative prices are not directly available as an input into heuristic pricing for the machines. We argue that relative prices from the makespan relaxation will often provide a fairly good approximation for this and suggest other possible approximations. In addition, we discuss how to obtain the total aggregate price for the group of general parallel machines from the relative prices and arrival data. We discuss this issue also in Section 12.3.2.

In Section 12.3.3 we use these ideas to construct bottleneck dynamics procedures for other objectives, based on the three heuristics we developed for makespan/economic makespan. Finally, in Section 12.3.4 we discuss an important case of the general parallel machine problem which is intermediate in complexity between the proportional and general case and rather easy to handle. Roughly, this case assumes that the overall parallel machine cluster may be divided into clusters that are themselves solvable. This case is important in practice, as we shall see in the nuclear fuel tube plant case in Section 14.5.

12.3.2 Lower Bounds and Resource Prices

Although LP relaxations do not seem to be available in general, other lower bounds on the objective function are available. For example, consider the objective of weighted flowtime. Take the total weighted flowtime accumulated to date for the partial solution as part of the lower bound. Now consider a problem where every machine is speeded up enough to be as fast for every job as the fastest machine for that job in the original problem. This is now a problem involving equal machines, for which a trivial preemptive solution is available. Finally, add these two pieces together (flowtime to date + lower bound from faster machines) to obtain a lower bound on the total flowtime for the problem.

Fairly good approximations for prices in the general case seem somewhat more difficult. We use the insight that relative prices for the proportional parallel machine case depend only on the relative speed of the machines and not on the objective function. Also, the dual prices from our LP relaxation for the makespan problem can easily be seen to have an interpretation as the "average relative speed of the machines for the jobs that can be done on more than one machine."

This leads us to suggest: "irrespective of the objective function, solve the static LP, and use the resource dual prices as approximations to the relative machine prices for the current problem." (As indicated earlier, the total price of the machine group is estimated by treating the group as an aggregate single machine.)

A simpler approximation would be to estimate average processing speeds of the machines from average processing speeds over a subset of "typical jobs" (deciding which jobs are atypical might be a bit sticky, however).

The total aggregate price of a machine is not needed for scheduling it but may be useful as part of a larger system. Once prices have been determined, the machines and job set may be scheduled over time. Then the overall time–price profile of the aggregate machine may be determined by a standard busy period analysis.

12.3.3   Bottleneck Dynamics Heuristics for General Parallel Machines

We turn now to the regular objective analogue of BD Heuristic 1 for the economic makespan problem.

BD Heuristic 1 (Dynamic Regular Objective)

Step 1: Find average process times $p_{j\ av} = (1/m)\Sigma_{k=1,m}p_{jk}$.

Step 2: Find the optimal job sequence for the aggregate machine with processing times $p_{j\ av}/m$. Order jobs j in this sequence.

Step 3: Assign the next job j to the machine k that minimizes

$$R_k p_{jk} + (C_{jk} - t)(w_j U_j - IR_k p_{jk})$$

where C_{jk} is j's completion time if it is scheduled on k (after waiting for jobs previously scheduled), w_j is the weight, U_j is the urgency factor, and I is the interest rate.

To understand the heuristic, note that the cost of placing j on machine k is the usual resource usage $R_k p_{jk}$, but there is now, in addition, a time cost for delaying the job. If there are enough jobs already on k to delay completion to C_{jk}, then for each time unit delayed we should assess the standard bottleneck penalty delay factor $(w_j U_j - IR_k p_{jk})$.

Once again, since the prices may not be expected to be perfect, this heuristic may leave some machines more heavily loaded than others. We can make the same rough correction as before.

BD Heuristic 2 (Dynamic Regular Objective)

Step 1: Find approximate R_k and makespan M for the preemptive problem.
Step 2: Select a fill parameter $(1 + b)$.
Step 3: Assign jobs to machines as in BD Heuristic 1.
Step 4: If a machine loading exceeds $(1 + b)M$, do not assign further jobs to it.
Step 5: Continue with the subset of currently open machines.
Step 6: If all machines are closed, open them and finish with LPT.

We turn finally to BD Heuristic 3, which is somewhat more complicated but assures natural, balanced loading. One again, we restrict ourselves to the no-inserted-idleness case.

(A) BD Heuristic 3 (Regular Objective)

Step 1: For each job yet to be assigned, determine $k(j)$, the machine that would finish the job first among those machines available after the job is.
Step 2: Some j is assumed to be assigned to its $k(j)$. We decide which by the principle of minimizing regret. For each j and every machine k, calculate the cost of an alternative assignment by

$$K_{jk} = R_k p_{jk} + (C_{jk} - t)(w_j U_j - IR_k p_{jk})$$

where t is the current time.
Step 3: The normalized regret of not choosing $k(j)$ is then

$$N_{jk} = [(K_{jk} - K_{jk(j)})/R_{k(j)} p_{jk(j)}] , \quad k \neq k(j)$$

Step 4: But we also have the choice of scheduling on machine $k(j)$ one average processing time $p_{k(j)\,av}$ later at a normalized regret of

$$N_{jk(j)} = [p_{k(j)\,av}(w_j U_j - IR_{k(j)} p_{jk(j)})/ R_{k(j)} p_{jk(j)}]$$

Step 5: Define the normalized regret for not currently assigning j to $k(j)$ by $N_j = \min_k(N_{jk}, N_{jk(j)})$.
Step 6: Define j^* as the job with maximum normalized regret for being passed over, that is, $N_{j^*} = \max_j[N_j]$.
Step 7: Assign job j^* to machine $k(j^*)$.
Step 8: Repeat.

12.3.4   Clustered General Parallel Machines

There is an important special case of the general parallel machine problem that is intermediate in complexity between the proportional case and the truly general case, and that can be handled by the methods appropriate to the proportional case (see Chapter 11). Although it is relatively easy to understand and might even seem obvious, it is worthwhile to develop here for several reasons. For one thing, it shows that it is often important not to use a difficult general model where special structure can be recognized. For another thing, it is an important case in practice, as will become obvious in the nuclear fuel tube case in Section 14.5.

Suppose the m machines, $k = 1, \ldots, m$, can be grouped into M clusters $K = 1, \ldots, M$, cluster K having m_K machines. Suppose that any arriving job j can be processed only by one cluster. That is, there are K job families, each of which can only be processed by one cluster.

Furthermore, suppose that each cluster acts as proportional parallel machines for the appropriate family of jobs. To summarize, as seen in Figure 12.1, there is one incoming stream of jobs, which sorts without a decision being made into K queues for K parallel machine problems. Then the outputs unite to form a single output stream.

Each subproblem is independent and easy to analyze. For subproblem K we add speeds of machines to get the effective number of machines m_k', and replace K by one aggregate machine with processing times $1/m_k'$ of the original. For heuristics that require pricing, the aggregate problem can be analyzed easily by a standard busy period analysis, where nonarrived jobs are also assigned to the Kth busy period. Any heuristic applicable to a one-machine problem can then be employed.

A mildly nonproportional subproblem may often be well approximated by a proportional one by averaging relative processing times on the different machines. Although the idea here is fairly straightforward, it can simplify complicated problems greatly. See the nuclear fuel tube plant example in Section 14.5

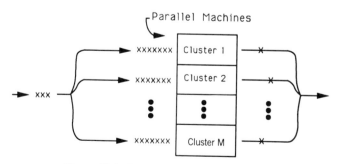

Figure 12.1. Clustering in parallel machines.

12.3A Numerical Exercises

12.3A.1 Using each set of problems given in (a) and (b), and changing the objective function to economic makespan, solve each by Bottleneck Dynamics Heuristic 1 (using the LP prices): (a) the problem given in 12.2A.1 and (b) the problems you generated in 12.2A.2.

12.3A.2 Solve each by BD Heuristic 2; compare with BD Heuristic 1.

12.3A.3 Solve each by BD Heuristic 3; compare with BD Heuristics 1 and 2.

12.3A.4 Change the objective to weighted flowtime, and again solve the problem by the three appropriate BD heuristics.

12.3A.5 Do the same for the weighted tardiness criterion.

12.3B Software/Computer Exercises

12.3B.1 Program a makespan heuristic for general machines not using heuristic prices. Test it extensively; discuss your conclusions.

12.3B.2 Program an economic makespan heuristic; test it.

12.3B.3 Program a general purpose bottleneck dynamics procedure for two or three objectives for general machines; test it.

12.3C Thinkers

12.3C.1 Sketch some bottleneck dynamics ideas for adding setup costs to the general parallel machine problem with makespan and/or other objective functions.

12.3C.2 Sketch some bottleneck dynamics ideas for adding precedence constraints to these problems.

12.4 ACTIVITY ROUTING

12.4.1 Overview

We say that an activity or operation has *multiple routing* if there is more than one resource that can be chosen to perform the activity. For the case of general parallel machines, each activity j can be performed on any machine k (of m machines) with processing time p_{jk}. In this case the machines are in proximity to each other, with a single queue (i.e., the transportation times to and from the machines may be considered independent of the choice). In general, however, each activity might be done on a different set of machines, and these sets might overlap in a rather haphazard fashion. There would not be a single physical queue from which activities would be chosen and routed on one of the physically adjacent machines. The routing choice might be made when the workpiece leaves its previous machine; thus the routing choice could noticeably affect the transportation time. Because the concept of the decision queue becomes more diffuse and routing choices overlap, it is no longer possible to aggregate the machines into a single aggregate machine for pricing algorithms and similar purposes and procedures.

To give an example, if one's activity is getting to school, the resource choices might be to walk, take a cab, or fix and take one's bike. Someone else's activity might be to get groceries with choices to take a cab or drive a car. The different resources involved cannot be aggregated into a single resource.

In Section 12.4.2 we show that the activity routing problem has the identical heuristics for sequencing and routing (properly interpreted) as for the general parallel machine problem, provided we know the correct heuristic prices for the activities and the resources and that transportation times are taken into account. We consider only the case where each activity requires exactly one resource. (The multiple-resources-per-activity problem is considered in the next chapter.) Next, we consider an iterative procedure for obtaining resource and activity prices and routings for this problem. We also briefly consider more complicated cases (see Chapter 19), where each "machine" may be a more complex resource such as a line or FMS cell.

12.4.2 Activity Routing with Known Prices and Lead Times

Consider a situation where a shop is made up of a number of large work centers (or departments). Each job visits a given work center at most once. Jobs may enter/exit the work center at many different physical locations; there are more than one physical storage area; and most activities cannot be performed on most resources. Nevertheless, we intend to treat the entire work center as a group of parallel machines by suitable extension of that methodology.

(a) There are m resources and n activities currently in the work center.
(b) Most activities cannot be performed on most resources.
(c) There are transportation times (assumed fixed) for transporting j to resource k of τ_{jk1} from its entry point into the work center and for transporting it from resource k to the exit point to its next work center of τ_{jk2}.

(The subscripts 1 and 2 imply there is a single choice for transportation time into k and out of k for activity j. The more difficult case of multiple routing choices into/out of k for activity j will not be considered here, although it is only complicated, not difficult.)

We also assume in this section that we have a lot of good forecast information:

(a) Approximately what the resource prices are at any point in time (how critical a given resource is).
(b) Approximately when a given activity's predecessors will be finished, giving its available start time.

(c) How long it will take to finish the successors of the activity, that is, its effective lead time.

In practice, while we need these to solve the problem, we can find them out only by solving the problem! This is not a new difficulty. We discussed lead time iteration and pricing methods for dealing with this problem in Sections 10.2 and 10.5, respectively. We will consider these issues as applied to the routing problem in Section 12.4.3 below.

We intend to "shoehorn" this problem into a format that can be solved by the BD Heuristic 3 (Dynamic Regular Objective) of Section 12.3.4 by several devices.

Routing Heuristic

Step 1: Treat the entire shop as one large parallel machine problem where every activity j can potentially be done on any resource k.

Step 2: Initially allow only jobs that are simple chains of activities (job shop).

Step 3: Initially do not allow setup times.

Step 4: If an activity j cannot be processed by a resource k, simply set p_{jk} at a very large number (the "big M" referred to in OR).

Step 5: Consider assigning j first. Let $k^*(j)$ be the machine that can start it first, and let $a(k^*)$ be that time. (The machine that can start j first is usually the same as the more correct one that can finish it first, and makes the heuristic tractable.)

Step 6: Activity j can start at max $(a(k^*)$, $r_j + \tau_{jk*1}$).

Step 7: The urgency (U_j), which depends on the estimated lead time, now depends on the choice k^*, due to postoperation transportation time τ_{jk*2}. That is, we must consider U_{jk}, which depends both on the lead time through the rest of the shop and the transportation if this machine is chosen:

$$U_{jk} = U_{jk}(LT_{jk}) = U_{jk}(\tau_{jk*2} + LT_{rest})$$

(That is, the delay cost for the activity depends on the total lead time on leaving k, which is the sum of the exit transportation time and the lead time through the rest of the shop.)

Step 8: Now if prices, lead times, and therefore activity delay costs are known, the BD Heuristic 3 of Section 12.3.3 can be used to solve the routing and sequencing problem.

(Note in step 5 that it is important we simply focus on jobs that can be processed on the machine that becomes available next. Otherwise, considering all combinations of machines or activities for an entire shop at each time interval would quickly become unwieldy.)

12.4.3 Estimating Prices and Lead Times

The problem of estimating prices and lead times occurs throughout bottleneck dynamics and is, in many ways, one of the most interesting parts, since there is always a trade-off between (a) spending large effort in the estimation, giving more safety but less practicality for the system, and (b) making intuitive, "quick-and-dirty" estimates that are easy to understand and practical but provide less safety.

Let us apply the methods of Sections 10.2 and 10.5 to provide an iterative approach to solving our problem without initially knowing prices and lead times.

Step 1: Choose an initial starting heuristic. This can be extremely simple: whenever a machine becomes available, schedule the job first available (FCFS).

Step 2: Run the heuristic and sensitivity runs. Make a full run of the heuristic via simulation; record the value of the objective function and all actual lead times and arrival times. Set a machine to unavailable and repeat. The change in the objective function divided by the change in resource amount gives a rough estimate of the price of that machine. Do for each resource in turn. (Note that identical resources do not necessarily all have the same price. Different transportation times can make a machine more or less desirable.)

Step 3: Run the routing heuristic of Section 12.4.2. We now have estimated prices and lead times that we may use. Save the lead times, arrival times, and objective function value that will be available when the routing heuristic is finished.

Step 4: Make sensitivity runs. Repeat the running of the routing heuristic with the same prices, shutting down one resource at a time as in step 2, giving us improved estimates of prices.

Step 5: Repeat until termination. Return to step 3. Repeat steps 3 to 5, saving the solution with the best objective function value to date. If no improved solution has been obtained for perhaps five iterations, terminate and report the solution with the best objective function.

This is only one possible way we might have estimated lead times and prices. We do not discuss the issue fully here. It has been developed fairly carefully in Sections 10.2 and 10.5. Furthermore, we shall have ample opportunity to wrestle with the same issue when we deal with flow shops, job shops, and project management in later chapters.

Useful references for this section include Arbib and Perugin (1989), Agnetis et al. (1990), Arbib et al. (1991), Schweitzer et al. (1991), Sengupta and Davis (1992), Wang and Wilhelm (1992), and Younis and Mahmoud (1992).

12.4A Numerical Exercises

12.4A.1 Consider a shop with two equal machines. Jobs come from one of three entry points to the shop. They are transported just before actual processing to whichever machine is chosen; then the job is processed and transported to a unique exit point from the shop. Completion time for a job is a sum of shop arrival, entry transport, waiting at the machine, processing, and exit transport time. The jobs have equal weights and unit processing time and the objective is F_{wt}. Thus the only data that are needed are a 2×4 matrix of transportation times and a list of arrival times at each entry point. The data are as follows:

Transport Times and Job Arrival Times

	ENT1	ENT2	ENT3	EXIT
Machine 1	3	1	2	1
Machine 2	1	3	2	2
9-Job arrival times	(0, 0, 1)	(0, 1, 1)	(0, 0, 1)	

Make an approximate solution by hand, as follows:

(a) Solve the problem first by relaxing the machine capacity constraints. (No job waits in any queue.)

(b) Identify the bottleneck machine and the critical time period on it.

(c) Experiment repeatedly with shedding load to the other machine until you are satisfied with the solution.

(d) Can you argue whether or not your solution is optimal?

12.4A.2 Solve 12.4A.1 approximately by brute force bottleneck dynamics, by guessing (repeatedly) static prices for the two machines. Could you match or beat your solution in 12.4A.1?

12.4A.3 For the same conditions of 12.4A.1, see how difficult or easy you can make this problem by making the transportation times strange, or the arrival times difficult, or whatever.

12.4A.4 Make the approximation that both machines should have the same price, and work through the iterative procedure of this chapter to get solutions to some of the same problems you looked at previously.

12.4A.5 Can you show for some extreme transportation times that your procedure in 12.4A.4 is inaccurate?

12.4B Software/Computer Exercises

12.4B.1 Program the procedure of 12.4A.4, except allow one price to be some multiple of the other. Rather than doing iterative pricing, simply solve the problem for a number of possible ratios of prices and pick the best solution. Does this give good solutions?

12.4B.2 Test your procedure of 12.4B.1 under a variety of transportation times and arrival patterns. Does it seem to work well under pressure?

12.4B.3 Make your program more general: more machines, structure of the shop, and so on.

12.4C Thinkers

12.4C.1 Suppose in our routing model there was more than one way for activity k to arrive and/or leave; that is, there may be other routing problems going on in the shop simultaneously. Come up with any ideas you can for dealing with this.

12.4C.2 Simplify the model for the economic makespan case. Are there any practical advantages for dealing with this special case?

12.4C.3 Suppose routing choices are not between machines as much as between groups of parallel machines. How might you model this situation, without destroying the information inherent in each grouping?

12.5   ADVANCED ROUTING ISSUES

12.5.1   Overview

In this section we discuss two more advanced routing issues. First, in Section 12.5.2 we discuss process routing. This is basically a generalization of simple activity routing. Now there may be alternate groups of resources used in sequence as choices to perform a particular sequence of activities. For example, we may perform the central four activities of a job either by a new high speed transfer line or by using a group of obsolete tools pressed into service to meet an emergency. When should one be done, when the other?

We first look at economic makespan as a criterion, and then at more complicated cases. The problem becomes more complicated if one of the process routes has subroutes. We consider this briefly.

Next, in Section 12.5.3, we talk about cost/resource trade-offs. A basement for a house may be dug by 10 workers in 4 days for $6000, or by 3 workers and a large steam shovel in 1 day at, say, $9000. Which is preferable, depending on deadlines, shortage of labor for other jobs, and so on? We recognize that this problem can be formulated as a routing problem; that is, consider alternate plans as alternate routes in the problem. (Of course, the example given involves more than one resource per activity; see Chapter 19.) Such problems can also be formulated alternatively as "crashing" the activity. This alternative will be developed in Chapters 17 and 18.

12.5.2   Process Routing

Let us illustrate a more concrete example of process routing, shown in Figure 12.2. A certain job has six activities performed in sequence. There is only one possible resource to do the first activity, and one resource to do the last. However, the four intermediate activities may be handled by one of three process routes marked $r = 1$, $r = 2$, and $r = 3$. Route 1 uses machines 1 and 2, each doing two of the activities. Route 2 uses machines 3, 4, 5, and 6, each doing one of the activities. Finally, route 3 has a single multi-purpose machine 7 that does all four of the activities.

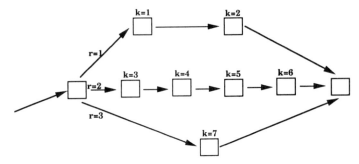

Figure 12.2. Process routing example (three routes and seven machines).

We actually already developed a formula for determining the best process route in Section 10.5, where we illustrated a number of possible bottleneck dynamics decision rules. That formula is given next (again we suppress the job j for simplicity).

Process Routing Formula

$$\text{Cost}_r = \sum_{k=1,K_r} (1 - It_{kr})(R_{kr}p_{kr} + c_{kr}) + wU_r(t_r - t)$$

Here $r = 1, 2, 3$ are the routes being considered; $k = 1, .., K_r$ are the resources (machines) on route r; R_{kr} is the price of machine k on route r; p_{kr} is the processing time of the job on machine k on route r; c_{kr} is the direct cost of the job using machine k on route r; t_{kr} is the expected time that the job would finish machine k on route r; w is the importance of the job; U_r is the urgency for finishing the job later using route r; and $t_r = t_{kr}$ (k last machine of route) is the time at which r would be finished.

Intuitively, the formula says the following: "the cost of each route is the cost of each resource on the route plus the cost of completing the job earlier or later using that route. The cost of each resource is the imputed price of the machine times the processing time on that machine plus direct costs of using that machine, all times an interest discount that is approximated as linear in time. The extra cost of late finish due to a route is the weight of the job times the marginal cost of lateness for the job times the finish time if that route is used."

If all the times of finishing at each machine given a route r were known and all the prices were known, this cost would easily be evaluated for each possible route and the cheapest process route chosen. For the simpler activity routing problem we discussed in the last section, we could handle the problem by repeatedly updating prices, lead times, and decisions in an iterative process. Unfortunately, for the more general process routing case, simple iteration does not suffice. If we are trying to find estimates for t_{jkr} for every possible route r, simply looking at the results of the last iteration will not work, since only one process route was actually chosen, and therefore last iteration values will not be available for the unused routes.

We will first discuss an important special case for which this problem does not arise. Then we will give a reasonable approach for the general case. Now our problem is more difficult for estimating future lead times involved with t_{kr} than for estimating prices R_{kr} and U_r. There are multiple chances for estimating the price of a machine since many jobs use this machine in a relatively short period of time. Hence some value for this price close to this time will be available from the last iteration. Similarly, the rate at which U_r changes with time is known. Therefore t_{kr} estimation remains central.

There is an important special case for which the exact completion times are not important. Consider a highly loaded shop, with a criterion of minimizing economic makespan. Then there is no penalty for jobs being later, so that $U_r = 0$. Suppose too that the interest charges due to changing the timing of using resources is much less critical than the amounts being used. Then we would have a simpler formula.

Process Routing for Economic Makespan

$$\text{Cost}_r = \sum_{k=1,K_r} (R_{kr}p_{kr} + c_{kr})$$

Intuitively, this formula says that, in a highly loaded shop, one should choose the process route with the lowest total resource usage, including both implicit (congestion) usage and explicit usage.

Suppose, on the other hand, one of the following conditions hold:

(a) The shop is not highly loaded.
(b) Interest due to resource usage timing is important.
(c) Delay costs for jobs are significant.

Then the full formula must be used, so that some method to estimate the t_{kr} is necessary.

Now we know the process times p_{kr}, which will be incurred on the route. If we knew the waiting times to come were some known multiple, say, β_{kr}, then we could estimate

$$t_{kr} = t_{\text{now}} + \sum_{k'=1,k} (1 + \beta_{k'r})p_{k'r}$$

Reasonable estimates of this multiple at a machine may be estimated by adding up all waiting time at the machine for the last iteration and dividing by all processing time. (Note that we are ignoring the fact that β_{kr} will vary with machine loading and the job priority. These kinds of sophistications could be added.)

12.5.3 Cost/Resource Trade-offs

Now consider a problem that is superficially a little different but really has a similar structure. Suppose a machine with price R can be set up to run an activity in one of three ways r with process time p_r and direct cost c_r depending on the way. Then, utilizing our process routing formula (ignoring interest), we obtain

$$\text{Cost}_r = Rp_r + c_r + Up_r = c_r + (R + U)p_r$$

That is, we choose the method with the lowest cost defined as the direct cost plus total implicit cost per unit of processing time multiplied by the processing time.

It is also possible that we might have an infinite number of choices for setting the machine. For example, suppose the machine tool can be set at any speed, but if processing time p is chosen, then tool wear costs of $w(p)$ result, which represent the direct costs. In this case, we can find the optimal processing time by simple calculus:

$$\text{Cost} = w(p) + (R+U)p$$

$$\frac{d}{dp} \text{Cost} = w'(p) + (R + U)$$

Finally, $w'(p^*) = -(R + U)$.

That is, we equate the marginal tool wear for shortening process time to the sum of the price of congestion on the machine and the price of additional lateness for the job. The fact that all three factors—marginal tool wear, the price of the machine, and the price of job delay—have equal importance for this simple model is interesting.

If we look at economic makespan (i.e., set $U = 0$), it is not difficult to look at resource/cost trade-offs at several successive machines for a job simultaneously, since the problems will not interact with each other. We leave this for the exercises. On the other hand, if completion times are important, then the subproblems are not independent, and our main problem is to not have combinatorially many routes to consider.

Useful references for this section are Blazewicz et al. (1992) and Jafari (1992).

12.5A Numerical Exercises

12.5A.1 Consider the process routing shown in Figure 12.2, that is, three routes and seven machines. Consider the simplified formula for process routing for economic makespan. Assume all machines have price 100 and that direct costs are zero. Choose a small set of jobs with different processing time data and simulate how they would move through the system. Can you tell if the prices assumed are good ones or not?

12.5A.2 Repeat 12.5A.1 with different sets of prices (one possibility might be prices of 0.04, 0.20, 1.0, 5.0, 25.0) and see how it affects congestion in the system.

12.5A.3 Set up two or three cost/resource trade-off problems with finite numbers of choices, and solve them, looking for patterns.

12.5A.4 Do the same with the continuous version of the problem.

12.5A.5 Suppose there are two activities in series and that each one can have its speed varied, but the same speed must be chosen for both machines. Set up this problem and solve by calculus.

12.5B Software/Computer Exercises

12.5B.1 Set up and program an iterative solution procedure for the process routing problem for economic makespan. Debug with a very small program.

12.5B.2 Set up an experimental design for doing some scheduling experiments using your program. Carry them out.

12.5C Thinkers

12.5C.1 In the simple tool wear model, given in the text, suppose that the average quality of the part produced is also a function of the machine speed. Create a model optimizing machine speed. (You may want to assume quality can be quantified in something like sales price.)

12.5C.2 What can you say about cost/resource trade-off models if jobs have more complicated precedence constraints?

12.5C.3 For longer horizon problems where linear approximations to interest are not accurate, what would be the appropriate form for the process routing model?

12.5C.4 Consider a shop in which jobs go through two resources in series. Jobs are sequenced the same on both machines. Resource 1 has C_1 possible settings; resource 2 has C_2 possible settings. Each setting produces a different machine speed and a different cost per hour of maintenance and wear. The objective is to minimize economic makespan. For given resource prices (sum of delays to other jobs), develop bottleneck dynamics formulas to decide what the settings should be for a single job coming through with known availability times on both machines.

PART IV
FLOW SHOPS AND
JOB SHOPS

In this part, we consider more complex multi-machine problems by considering flow shops and job shops. Chapters 13 and 14 develop flow shops, including modeling, model extensions, and the case of a nuclear fuel tube shop. Chapters 15 and 16 develop job shops, including modeling, testing, and a second look at the nuclear fuel tube shop.

____13

SCHEDULING FLOW SHOPS: MAKESPAN

13.1 INTRODUCTION

13.1.1 Overview

In a simple flow shop each job is processed by a series of machines in exactly the same order. There are also some simple variations, which might be called "skip shops," "reentrant flow shops," "compound flow shops," and "finite queue" flow shops. (See Figure 13.1.)

In skip shops (e.g., flexible cells) some jobs may skip some machines. (This may actually be treated as giving the job zero process time and top priority at those machines.)

In reentrant flow shops (e.g., annealing furnaces in a metal shop) some machines may be visited more than once. This is difficult to model analytically but relatively easy to deal with in many heuristic approaches such as the shifting bottleneck method or bottleneck dynamics, for example.

In a compound flow shop each machine in the series may be replaced by a machine group. Most commonly, the groups would be parallel machines or a batch line followed by parallel machines. Compound flow shops are extremely common: for example, paper mills, steel lines, pilgering shops, bottling lines, and food processing lines. They are not always recognized due to their physical complexity. They will be more common in the future as automation causes more subsystems to take on the characteristic of a single group. Again, heuristic systems such as bottleneck dynamics can represent the groups of machines as an aggregate machine and solve the resulting simpler flow shop problem.

In a finite queue flow shop there is limited storage in front of machines other than the first. An important special case is when no storage is allowed except at the first

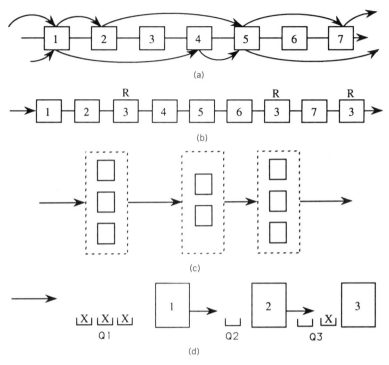

Figure 13.1. (a) Skip flow shop, (b) reentrant flow shop, (c) compound flow shop, and (d) finite queue flow shop.

machine. For example, such a requirement is frequently encountered in the metal processing industries, particularly where metal is rolled while it is hot. Delays between operations result in cooling that makes the rolling operation prohibitively difficult.

Most of the classic literature on flow shops is for simple flow shops for the makespan criterion. Even here, most of the exact results are available for the two-machine case (Johnson's rule). Section 13.2 develops exact and heuristic methods for two- and three-machine problems. Section 13.3 develops exact and heuristic methods for the general m-machine problem.

In Chapter 14 we complete our discussion of flow shops by discussing objectives such as economic makespan, weighted flow, and weighted tardiness. We also discuss how to develop heuristics for model extensions and give a concrete example of a complex flow shop: a nuclear fuel tube shop.

13.1.2 Possibility of Inserted Idle Time

Recall that for the regular static one-machine problem it was proved that for regular objectives there were optimal schedules where the machine would never be kept idle when work was waiting. In the flow shop case, however, it may be necessary to

**TABLE 13.1 Data for Two-Job,
Four-Machine Example**

Machine Number	Operation Times for Job 1	Operation Times for Job 2
1	1	4
2	4	1
3	4	1
4	1	4

sometimes provide for inserted idleness to achieve optimality. (This is not really surprising; we know that with dynamic arrivals on the one-machine case we may wish to wait for a propitious job. In our case here (Figure 13.1) machines after the first have, in a sense, dynamic arrivals. Consider the two-job, four-machine problem described in Table 13.1. Suppose that F_{wt} is the measure of performance with all weights equal to 1. The two schedules shown in Figure 13.2a and 13.2b have no inserted idle time, and for both $F_{wt} = 24$. The schedule in Figure 13.2c is an optimal schedule, with $F_{wt} = 23$. Note that in this third schedule, machine 3 is kept idle at time $t = 5$ (when operation 3 of job 1 could have been started) in order to await the availability of the "high priority" operation 3 of job 2, which should be finished quickly to facilitate starting operation 4 of job 2.)

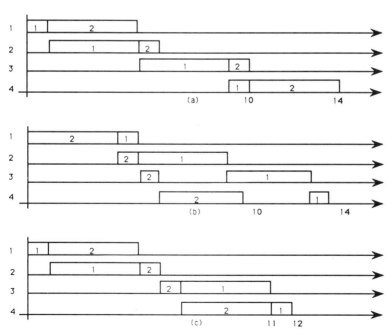

Figure 13.2. Three schedules for the example: (a) and (b) have no inserted idle time.

13.1.3 Permutation Schedules

Recall for the one-machine problem that there was a one-to-one correspondence between possible sequences and permutations of n numbers. That is, there were $n!$ choices to be considered in the worst case. If there are now m machines in the flow shop and $n!$ choices at each, the total possible choices are $(n!)^m$. Clearly, it would be helpful to be able to ignore most of these. We already know some ways to do this: dominance properties, exact results, implicit enumeration, and heuristics. Here we discuss a couple of important dominance properties and leave the other issues until later in the chapter. The results in this subsection and the next are taken from Baker (1974).

Proposition 1. For any regular measure of performance, with dynamic availability but static arrivals, there is at least one optimal schedule on an m-machine problem for which machines 1 and 2 have the same job sequence.

Proof. Otherwise, somewhere there must be a job j and $j + 1$ adjacent on machine 1 and reversed on machine 2 as shown in Figure 13.3. But then it is clear that j and $j + 1$ can be reversed on machine 1 without affecting the rest of the diagram. Note that the only activity made later is (j, 1). (Remember $(j,1)$ means job j on first machine). But a regular objective function only rewards completion times on the final machine! ∎

Because, for the makespan objective, individual jobs do not have time priorities, a stronger result can be given for this case.

Proposition 2. For the makespan objective, with dynamic availability and static arrivals, the reversed problem with last operation and last machine first, and time reversed, has the same optimal makespan as the original problem.

Proof. Left as an exercise. ∎

Note that the partial makespan in the interchange in Proposition 1 is not affected on either machine. We shall use this below.

Proposition 3. With respect to the makespan measure on m machines, it is sufficient to consider only schedules in which the same job sequence occurs on machines $m - 1$ and m.

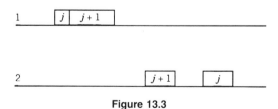

Figure 13.3

Proof. The proof uses Propositions 1 and 2.   ■

A *permutation schedule* is simply a schedule with the same job order on all machines—a schedule that is completely characterized by a single permutation. Another way to interpret these propositions is that permutation schedules are sufficient for the following:

(a) Regular measures, static, $m = 2$.
(b) Makespan, static, $m = 3$.

We might also be led to guess that permutation schedules may be interesting to test as heuristics for the general case, especially for makespan. The exercises give counterexamples to show that permutation schedules are not exactly optimal for any other cases.

13.1A Numerical Exercises

13.1A.1 Consider a case where five jobs are to be run in the same permutation sequence on each of the machines in a three-machine static flow shop. Calculate the makespan for the following input data for the permutation sequence 1–2–3–4–5:

	Job				
	1	2	3	4	5
M1	6	7	3	9	7
M2	9	10	6	3	7
M3	3	7	9	6	12

13.1A.2 Create a spreadsheet to represent and calculate the makespan for 10 jobs input in a given permutation sequence for three machines. Set up four or five examples. Find optimal solutions by trial and error. Is there typically more than one optimal solution?

13.1A.3 Try to create a hand example where inserted idleness is necessary and more important than in the book's example.

13.1B Software/Computer Exercises

13.1B.1 Program an m-machine flow shop that, given n jobs of data and a given input sequence, will evaluate the makespan. Use your program to try to find some near optimal answers by trial and error. Are there any principles apparent?

13.1B.2 Add a makespan heuristic of your own devising to your program. How well does it perform?

13.1C Thinkers

13.1C.1 Construct a three-machine flow shop problem for which F_{wt} (weights = 1) is minimized only by a schedule that is not a permutation schedule.

13.1C.2 Construct a four-machine flow shop problem for which M is minimized only by a schedule that is not a permutation schedule and that contains no inserted idle time.

13.1C.3 Prove Proposition 2.

13.1C.4 Give a more careful proof of Proposition 3.

13.1C.5 Sequential crew systems consist of two or more crews following one another in a fixed sequence to complete a particular task on a unit being processed. (Such systems are commonly found in the construction of aircraft and ships, with crews that can work continuously; the first crew prepares new units upon entry into the system, and the remaining crews process these units as they become available. Crews perform their assigned work in a fixed sequence with respect to each other, and units are processed by each crew in the same order as their order of entry. Show how a flow shop model can be used to describe the behavior of such a system. What is the role of permutation schedules in the analysis of this kind of system?

13.2 TWO- AND THREE-MACHINE MAKESPAN PROBLEMS

13.2.1 Overview

The makespan problem for flow shops and job shops has been the most studied by far in the literature. This is partly because makespan (and now economic makespan) is a simple and useful criterion for heavily loaded shops when long-term utilization should be maximized. However, it is also because makespan is the only objective function simple enough to have available some analytic results for multi-machine problems, and simple enough to make some branch-and-bound methods practical for medium-sized problems. Even here there are few results except for two- and three-machine problems.

In Section 13.2.2 we study the static arrival two-machine flow shop with makespan objective, also called "Johnson's problem." This leads to an easy exact procedure called "Johnson's rule." In Section 13.2.3 Johnson's rule is extended to a more complex two-machine problem, as well as to some special three-machine problems. In Section 13.2.4 the ideas in Johnson's rule are used to provide heuristics for general two-machine problems and a foundation for later heuristics for the m-machine problem.

13.2.2 Two-Machine Makespan Problem

Consider a two-machine flow shop with makespan criterion. There are n jobs, static arrivals ($r_j = 0$). However, it is not necessary that the two machines be available at zero; there may be positive a_1 and/or a_2. We have already shown there is an optimal rule that sequences each job in the same order on both machines. We will state and show an example of Johnson's rule first. [The proof is a little long and nonintuitive, and may be omitted for the first reading (Johnson, 1954).]

Proposition 4. For the two-machine flow shop makespan problem with dynamic availability but static arrivals, Johnson's rule is optimal:

Step 1: Schedule the group of jobs U that are shorter on the first machine than the second; $U = \{ j \mid p_{j1} < p_{j2} \}$ as the first priority group.
Step 2: Schedule the group of jobs V that are shorter on the second machine; $V = \{ j \mid p_{j2} \leq p_{j1} \}$ as the second priority group.
Step 3: Schedule within U by SPT on the first machine.
Step 4: Schedule within V by LPT on the second machine.

Proof. Since this proof is a bit long, we postpone it until we first give an example. ∎

To illustrate the algorithm, consider the eight-job problem shown in Table 13.3A. It shows how an optimal sequence is constructed.
The solution is constructed as follows:

Step 1: The set U of jobs shorter on the first machine is $U = \{2, 3, 6\}$; these will be the first block of jobs to be run.

TABLE 13.3A   Example of Johnson's Rule

					Problem			
Job j:	1	2	3	4	5	6	7	8
p_{j1}:	5	2	1	7	6	3	7	5
p_{j2}:	2	6	2	5	6	7	2	1
				Optimal Solution				
Job j:	3	2	6		5	4	7	1	8
p_j1:	1	2	3		6	7	7	5	5
p_j2:	2	6	7		6	5	2	2	1
Finish on M1	1	3	6		12	19	26	31	36
Finish on M2	3	9	16		22	27	29	33	37

MAKESPAN = FINISH ON M2 = 37

Step 2: The set of jobs shorter on the second machine is $V = \{1, 2, 4, 5, 7, 8\}$.

Step 3: Within U order by increasing time on the first machine: 3–2–6.

Step 4: Within V order by decreasing time on the second machine: 5–4–7–1–8.

As shown by the first line of the optimal solution, the complete optimal sequence is $\{3, 2, 6, 5, 4, 7, 1, 8\}$. Next, copy down the processing times in the optimal order on the first and second machines, as shown by the second and third lines. Then calculate the finish times on machine 1 in the next line. Since this is the static case and all jobs are always available, this is just the cumulation of machine 1 processing times, seen on line 4. Finally, calculate the finish times on machine 2 in line 5. Here each job starts at the maximum of the finish times of the same job on machine 1 and the previous job on machine 2. For example, the start time of job 1 on machine 2 is $\max(31, 29) = 31$. (This leaves a gap of 2 on the second machine.) The makespan is simply 37, the finish time of the last job on machine 2.

Now we are ready for a proof of Johnson's rule.

Proof of Johnson's rule. Since we know, by Proposition 1, that the jobs may be sequenced in the same order on the two machines, we can apply pairwise interchange ideas. There are two cases. In Case I the shortest operation of the four for the two jobs is on the first machine; in Case II, on the second machine. We work through the three subcases of Case I in some detail and then argue that Case II follows directly from the symmetry discussed in Proposition 2.

Remember that jobs on the first machine will be packed without gaps (we are using the static assumption). Thus interchange will not affect the completion time on the first machine. On the other hand, if the two jobs complete on the second machine at M, they may complete after interchange at M'. Define $\Delta = M' - M$. Then clearly $\Delta < 0$ means the pairwise interchange made the overall makespan (weakly) better, while $\Delta \geq 0$ means the overall makespan may be worse.

Let t_a be the time the first of the two jobs may start on machine 1; and similarly $t_a + t_b$ be the time the second machine is available (Figure 13.4). Then

$$M = t_a + p_{j1} + \max[\, p_{j+1,1}, (t_b - p_{j1})^+ + p_{j2}] + p_{j+1,2}$$
$$M' = t_a + p_{j+1,1} + \max[\, p_{j1}, (t_b - p_{j+1,1})^+ + p_{j+1,2}] + p_{j2}$$
$$\Delta = M' - M = (p_{j2} - p_{j+1,2}) + (A)^+ - (B)^+$$

where we define

$$A = (t_b - p_{j+1,1})^+ + p_{j+1,2} - p_{j1}$$
$$B = (t_b - p_{j1})^+ + p_{j2} - p_{j+1,1}$$

Case I: $p_{j+1,1} \leq \min(p_{j1}, p_{j2}, p_{j+1,2})$. That is, the shortest of the four operations occurs on the first machine, and in the wrong order for Johnson. We wish to show that Δ is less than or equal to zero for every possible subcase.

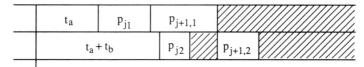

Figure 13.4

Subcases

(a) $A \geq 0, B \geq 0$ So that $\Delta = \max(t_b, p_{j+1,1}) - \max(t_b, p_{j1})$; since $p_{j+1,1} \leq p_{j1}$, clearly $\Delta \leq 0$, as desired

(b) $A \leq 0, B \geq 0$ So that $\Delta = (p_{j+1,1} - p_{j+1,2}) - (t_b - p_{j1})^+$; since $p_{j+1,1} \leq p_{j+1,2}$, clearly $\Delta \leq 0$

(c) $B < 0$ But since $p_{j+1,1} \leq p_{j2}$, this case is impossible

Thus for Case I, we have proved that pairwise interchange to enforce Johnson's rule cannot hurt the makespan.

Case II: $p_{j2} \leq \min (p_{j+1,1}, p_{j+1,2}, p_{j1})$. Now we apply Proposition 2. In the reversed problem this interchange would be Case I, which we have just proved. This proves Case II.

Finally, the conditions of the cases make clear that priorities do not depend on the current position in the sequence; that is, we have transitivity. Thus repeated pairwise interchange will produce the optimal sequence, which will also obey Johnson's rule. ∎

(A) 13.2.3 Extensions of Johnson's Rule

In this section we develop several other models for two and/or three machines where Johnson-type rules are optimal. (Later, in Section 13.3.3, we present several heuristics for the *m*-machine case that use the Johnson insights.)

Three-Machine Problems. Exact results cannot usually be found for the static makespan problem with more than two machines. For the case $m = 3$ we have seen that permutation schedules are optimal, yet exact results are usually not obtainable. One exception to this is (as Johnson showed in his original paper) that exact results are possible in certain cases where the second machine has uniformly shorter processing times than the first machine (or else the third machine).

Proposition 5. Consider a three-machine flow shop with makespan criterion, dynamic availability, and static arrivals.

(a) If $\min_j\{t_{j1}\} \geq \max_j\{t_{j2}\}$ or

(b) If $\min_j\{t_{j3}\} \geq \max_j \{t_{j2}\}$

then the problem may be solved by Johnson's algorithm by solving an auxiliary two-machine problem with times

$$P_{i1} = p_{i1} + p_{i2} \; ; \quad P_{i2} = p_{i2} + p_{i3}.$$

Proof. See Johnson (1954). ∎

Proposition 6 was also proved by Johnson for the three-machine problem. It is spiritually similar to Proposition 5.

Proposition 6. Consider a three-machine flow shop with makespan criterion, dynamic availability, and static arrivals. If solving the first two machines alone by Johnson's rule yields the same sequence as solving the last two machines by Johnson's rule, then the sequence is optimal for the full three-machine problem.

Proof. Baker (1974, Section 3.2). ∎

Next, we consider the addition of start-lags and stop-lags to the two-machine problem, as analyzed by Mitten (1959). In many practical situations, it may be possible to begin operation 2 before operation 1 is entirely complete. This might happen, for example, where jobs consist of large lots that may be split into sublots, and where completed sublots may proceed to the next machine without waiting for the full lot to be processed. This overlapping structure, often called *lap-phasing*, may also arise where the flow of work is conveyer driven. With lap-phasing there is a specific interval p_{ja} called a start-lag, such that operation $j2$ can be started p_{ja} after $j1$ begins instead of p_{j1}. In the usual flow shop model $p_{ja} = p_{j1}$, but lap-phasing allows $p_{ja} < p_{j1}$. (At the other extreme, significant transportation time or inspection between operations might require $p_{ja} > p_{j1}$.) Analogously, Mitten allowed for an interval p_{jb} called a stop-lag such that operation $j2$ must not complete any earlier than p_{jb} time units after operation $j1$ completes. Allowing for $p_{ja} \geq 0$ and for all p_{jb}, Mitten gave the following extension of Johnson's rule.

Proposition 7. Consider Johnson's problem with the addition of start-lags and stop-lags, that is, with p_{ja} and p_{jb} such that $j2$ can be started p_{ja} after $j1$ starts; $j2$ must finish at least p_{jb} after $j1$. Suppose also that only permutation schedules are to be allowed.

(a) Define U and V as in Johnson's problem.
(b) Define $y_j = \max\{p_{ja} - p_{j1}, p_{jb} - p_{j2}\}$.
(c) Define an auxiliary set of times $P_{j1} = (p_{j1} + y_j)$ and $P_{j2} = (p_{j2} + y_j)$.
(d) Order U by increasing P_{j1} followed by V with decreasing P_{j2}.

Proof. See Mitten (1959). ∎

This result can be extended to the m-machine flow shop as long as only machine 1 and machine m are bottleneck machines. If the intervening operations on each job can be performed independently of all other jobs due to excess capacity, they could be treated as part of a start-lag and Mitten's result employed.

13.2.4 Heuristics for Static and Dynamic Johnson's Problems

Although Johnson's algorithm is fast, easy to use, and exact for the static two-machine problem, it does not always allow easy incorporation into more complicated models for two machines, or for the m-machine problem. Here we define a family of possible heuristics for the general (e.g., dynamic) two-machine makespan problem. We concentrate especially on two special members of that family—the ratio heuristic and the difference heuristic. We prove that these heuristics are optimal for the unit-preemptive versions of Johnson's problem and show by example that they both often give the optimal solution. All heuristics here use the intuition that processing short jobs first on the first machine will allow more jobs to be available quickly on the second machine to keep it from having gaps.

Finally, we discuss the dynamic case briefly.

Definition. The *ratio heuristic* is a dispatch rule for dynamic arrival, dynamic availability, two-machine makespan problems, which schedules next the job with lowest (p_{j1}/p_{j2}).

Definition. The *difference (Palmer) heuristic* uses $(p_{j1} - p_{j2})$ instead.

Definition. The *generalized difference heuristic* uses $(f(p_{j1}) - f(p_{j2}))$, where $f(x)$ is any monotonically increasing function.

Note that the ratio heuristic is a special case of the latter, since minimizing (p_{j1}/p_{j2}) gives the same sequence as minimizing $(\ln(p_{j1}) - \ln(p_{j2}))$. It is obvious also that the difference heuristic is a special case with f the identity function.

None of these heuristics can be guaranteed optimal for Johnson's problem; the exercises ask for counterexamples. However, they are optimal for preemptive cases, and an example shows they are often optimal in given problems.

Proposition 8. Any generalized difference heuristic is optimal for the two-machine flow shop makespan problem with dynamic availability and static job arrivals if jobs are of unit length on one machine or the other.

Proof. Follows easily from Proposition 4, left as an exercise. ∎

To illustrate that the ratio heuristic and difference heuristics are quite accurate, we apply them to the same example (see Table 13.3B).

Note that the ratio heuristic does not give the same sequence as Johnson's rule, but that its sequence is also optimal. The student will be asked to verify in the exercises that the difference heuristic gives yet a third sequence, which is also

TABLE 13.3B Ratio Heuristic for Problem from Table 13.3A

				Ratio Heuristic Solution				
Job j:	2	6	3	5	4	1	7	8
p_{j1}:	2	3	1	6	7	5	7	5
p_{j2}:	6	7	2	6	5	2	2	1
Ratio:	0.33	0.43	0.50	1.00	1.40	2.50	3.50	5.00
Finish on M1	2	5	6	12	19	24	31	36
Finish on M2	8	15	17	23	28	30	33	37

MAKESPAN = 37 (optimal)

optimal. This is very common. Makespan problems often have large numbers of alternate optima. In fact, there is a simple bound on the makespan for Johnson's problem, which can often verify whether the solution is optimal.

Proposition 9. A lower bound on the makespan in Johnson's problem is given by

$$C_{\max \text{ lb}} = \max[(\Sigma p_{j1}) + \min(p_{j2}), \min(p_{j1}) + (\Sigma p_{j2})]$$

Proof. The first half of the formula simply says the last job on machine 2 cannot be completed until all jobs are done on machine 1 (even if they have no gaps) followed by its own time. The second half of the formula simply says that after the first job is done on machine 1, at least all of the times on machine 2 must follow. ■

We leave as an exercise the generalization of Proposition 9 to the *m*-machine case (Baker 1974).

Generalization of the ratio and difference heuristics to the case of dynamic arrivals is quite simple. Simply employ the priority index on jobs currently available, and choose the next with highest value. That is, use the difference heuristic as a dispatch heuristic.

13.2A Numerical Exercises

13.2A.1 Apply the difference heuristic to the two-machine data of Table 13.3A. Is the heuristic optimal for this problem?

13.2A.2 Find an example where the ratio heuristic is optimal but the difference heuristic is not. (Verify optimality via attainment of the lower bound or by applying Johnson's algorithm as convenient.)

13.2A.3 Find an example where the difference heuristic is optimal but the ratio is not.

13.2A.4 Find an example where both heuristics miss optimality by a large amount.

13.2A.5 A manufacturer of charm bracelets has five jobs to schedule for a leading customer. Each job requires a stamping operation followed by a finishing operation, which can begin on an item immediately after its stamping is complete. The table below shows operation times per item in minutes for each job. In addition, preparation for each job at the stamping facility requires a setup before processing begins, as described in the table. Find a schedule that completes all five jobs as soon as possible.

		Operation Time per Item		
Job	Number in Lot	Stamp	Finish	Setup
1	20	2	8	100
2	25	2	5	250
3	100	1	2	60
4	50	4	2.5	60
5	40	3	6	80

13.2B Software Exercises

13.2B.1 Write a program to simulate the two-machine makespan problem with dynamic arrivals. Generate a set of test problems. Compare various ratio and difference heuristics and Johnson's algorithm, in each case applying the rule only to jobs currently available. Limit your attention to permutation schedules.

13.2B.2 Test the ratio versus the difference heuristic as dispatch heuristics for makespan dynamic arrival problems. How did you design your experiment?

13.2C Thinkers

13.2C.1 Prove Proposition 8.

13.2C.2 Generalize Proposition 9 to obtain lower bounds for the m-machine static makespan problem.

13.2C.3 In the lag problem of Proposition 7, suppose $p_{ja} = p_{jb}$ for all jobs. Show that an equivalent rule is: j precedes $j + 1$ provided $\min\{p_{j1} + y_j , p_{j+1,2} + y_{j+1}\} \leq \{p_{j2} + y_j, p_{j+1,1} + y_{j+1}\}$.

13.3 GENERAL MAKESPAN

13.3.1 Overview

Makespan problems become very difficult when there are more than two machines. Except for the very special cases discussed in Propositions 5 and 6 of the previous section, a large makespan problem with only three machines cannot be solved exactly in a reasonable amount of computation, even using the fact that attention may be restricted to permutation schedules.

Branch-and-bound has had some success in exactly solving smaller problems, perhaps with three or four machines and 10 to 15 jobs for the makespan criterion. There is an extensive literature on this subject; we present a brief summary of some of it in Section 13.3.2.

Johnson-type heuristics, which are based on the generalized difference heuristics for the two-machine case, have been studied extensively. They are discussed in Section 13.3.3 along with several newer bottleneck heuristics. More recently, a number of high computation heuristics for flow shop problems have become available, including the shifting bottleneck method, beam search, tabu search, simulated annealing, and bottleneck dynamics. These methods are deferred to Section 13.5, where we discuss general objective functions. (Bottleneck dynamics will be discussed in Chapter 14 when we develop the economic makespan criterion.) The shifting bottleneck method is especially interesting for makespan problems in flow shops and/or job shops. This is because the fact that a given heuristic solution is optimal can often be verified by identifying this solution with a known lower bound.

13.3.2 Branch-and-Bound

The branch-and-bound procedure as applied to the m-machine flow shop problem, with makespan objective, and restricted to *permutation schedules* was developed by Ignall and Schrage (1965), and independently by Lomnicki (1965). It is commonly called the Ignall–Schrage algorithm. The integer programming formulation given later in Propositions 13 and 14, however, is not restricted to permutation schedules.

The branching tree obtained by Ignall and Schrage looks very similar to the graphical worksheet shown in Section 5.5, except that we branch on the first job to be sequenced, and then on the second one, and so on, so that the partial permutation σ represents the *first* k jobs sequenced rather than the last k. At any node on the tree, a lower bound on the makespan can be obtained by getting a good lower bound on the makespan on each machine and taking a maximum. In turn, one lower bound on any machine can be obtained by considering work already assigned to it, plus all remaining work assuming no wasted time, plus the time to complete the last job on the remaining machines.

We illustrate the bounds for $m = 3$; extension to the general case will be left for the exercises.

Dynamic Machine Availability, Static Job Availability. For the moment we assume dynamic availability of machines but static availability of jobs. We consider generalization to dynamic arrivals later. For a given beginning partial sequence σ and remainder set σ', first assign the jobs in σ one at a time, making sure no operation is placed on a machine before its preceding operation finishes the previous machine and until the machine is ready for it. Suppose this partial sequence of k jobs leads machine 1 to be available at q_1, machine 2 at q_2, and machine 3 at q_3. The amount of processing time still required on machine 1 is

$$\sum_{j \in \sigma} p_{j1}$$

Moreover, there must be some (unknown) last job i that is the very last one on machine 1. After it is completed, job i must be done on machines 2 and 3, which takes at least $(p_{i2} + p_{i3})$ as shown in Figure 13.5a. The very best that could happen is:

(a) There is no idle time in assigning the remaining jobs on 1. (This is actually true on machine 1, but not when we repeat this argument for machines 2 and 3.)

(b) There is no idle time in assigning any of the operations of the last job i.

(c) i has the minimum possible sum $(p_{j2} + p_{j3})$ among the jobs in σ'.

Thus a lower bound on the makespan, considering primarily processing on machine 1, is

$$b_1 = q_1 + \sum_{j \in \sigma'} p_{j1} + \min_{j \in \sigma'} \{p_{j2} + p_{j3}\}$$

We may do a similar lower bound for each machine, and the overall makespan must be at least as large as any of the lower bounds, leading to the following proposition.

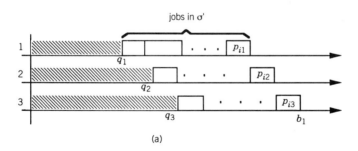

(a)

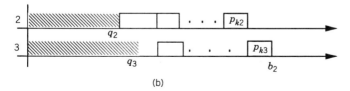

(b)

Figure 13.5

Proposition 10. For the three-machine flow shop makespan problem with dynamic machine availability and static job availability, a lower bound at a node represented by a partial forward permutation σ is (permutation schedules)

$$B = \max\{b_1, b_2, b_3\} \quad \text{where}$$

$$b_1 = q_1 + \sum_{j \in \sigma'} p_{j1} + \min_{j \in \sigma'}\{p_{j2} + p_{j3}\}$$

$$b_2 = q_2 + \sum_{j \in \sigma'} p_{j2} + \min_{j \in \sigma'}\{p_{j3}\}$$

$$b_3 = q_3 + \sum_{j \in \sigma'} p_{j3}$$

Proof. Given above. ■

Extension of Proposition 10 to the m-machine case is left as an exercise. The branch-and-bound procedure employed by Ignall and Schrage is essentially best-first search (see Section 5.5), together with the dominance property given in Proposition 11, which allows some branches to be eliminated.

Proposition 11. Suppose partial sequences $\sigma^{(1)}$ and $\sigma^{(2)}$ contain the same jobs in different order. If $q_k^{(1)} \le q_k^{(2)}$ for every machine k, then $\sigma^{(1)}$ dominates $\sigma^{(2)}$ so that $\sigma^{(2)}$ need not be considered any further in the search for the optimum.

Proof. Suppose in contradiction that $\sigma^{(2)}$ were the first part of an optimal solution. Then the $\sigma^{(2)}$ part of the solution could be reordered to be $\sigma^{(1)}$, which would save makespan on every machine and hence give an optimal solution starting with $\sigma^{(1)}$. Hence $\sigma^{(2)}$ is unnecessary. ■

We illustrate the calculations by considering the four-job three-machine flow shop problem described in Table 13.4. Suppose the first node generated by the branch-and-bound algorithm corresponds to the subproblem P_{11}, for which job 1 is assigned first in the sequence and $\sigma' = \{2, 3, 4\}$. For this partial sequence $q_1 = p_{11} = 3$, $q_2 = p_{11} + p_{12} = 7$, and $q_3 = p_{11} + p_{12} + p_{13} = 17$. The lower bound calculations are

$$b_1 = 3 + 28 + 6 = 37$$
$$b_2 = 7 + 22 + 2 = 31$$
$$b_3 = 17 + 20 + 0 = 37$$
$$B = \max\{37, 31, 37\} = 37$$

The computations to obtain the optimum are given in Table 13.5. It is interesting to note that the conditions for Johnson's rule to hold for the three-machine problem fail. (That is, the second machine is not completely dominated by either machine 1

TABLE 13.4

Job j	1	2	3	4
p_{j1}	3	11	7	10
p_{j2}	4	1	9	12
p_{j3}	10	5	13	2

Note: Ignall and Schrage did not address whether the time spent in making the dominance check was more than paid for by branching time or storage space saved.

or machine 3.) Nevertheless the three-machine version of Johnson does give the optimal sequence 1–3–4–2.

Note also that the problem shown in Table 13.4 exhibits a case where the best-first search algorithm creates the minimum possible number of nodes on the way to an optimum. Ignall and Schrage often found this to be the case (on small test problems such as ($m = 3$, $n \leq 10$). Brown and Lomnicki (1966) did not find this to be the case, however, for larger problems.

There are a number of ways to strengthen these bounds, all of which add complexity. These include refining the Ignall–Schrage bounds, developing job-based rather than machine-based bounds, utilizing dominant machines, and developing more complex dominance properties. See Baker (1974). We shall not develop these here. We turn instead to a more general model: the m-machine makespan problem with dynamic machine availability and dynamic arrivals.

Dynamic Machine Availability, Dynamic Job Availability. We first give bounds for this problem via a method that is an extension of the Ignall–Schrage method. Then we give much stronger lower bounds for large problems via linear programming. (Thus we are, of course, also giving stronger lower bounds for the static problem discussed above.)

We now add arrival times $r_1, \ldots r_j, \ldots, r_n$ for the jobs to the static problem. First,

TABLE 13.5

Partial Sequence	(q_1, q_2, q_3)	(b_1, b_2, b_3)	B
1	(3, 7, 17)	(37, 31, 37)	37
2	(11, 12, 17)	(45, 39, 42)	45
3	(7, 16, 29)	(37, 35, 46)	46
4	(10, 22, 24)	(37, 41, 52)	52
12	(14, 15, 20)	(45, 38, 35)	45
13	(10, 19, 30)	(37, 34, 37)	37
14	(13, 25, 27)	(37, 40, 45)	45
132	(21, 22, 35)	(45, 36, 37)	45
134	(20, 32, 34)	(36, 38, 39)	39

what is the earliest possible time r_{jk} for which job j could arrive to machine k, considering only its own processing time? Clearly,

$$r_{j1} = r_j \, ; \, r_{j2} = r_{j1} + p_{j1} \, ; \, r_{j3} = r_{j2} + p_{j2}; \, \cdots \, r_{jk} = r_{j,k-1} + p_{j,k-1}$$

As before in the Ignall–Schrage algorithm, let $q_1, q_2, \ldots, q_k$ be the completion times on the machines from the partial permutation σ already scheduled. Now instead of the remaining jobs on a given machine producing an additional completion time of simply $\Sigma_{j\in\sigma'} p_{jk}$ as for the static case, we must face the question of there being idle time due to there being no jobs available to schedule at some points in time. This is not difficult; we know that the minimum makespan with dynamic arrivals on a machine k starting at a point q_k can be obtained simply by scheduling any available job whenever the machine becomes free. Suppose the incremental makespan found this way is T_k. Then the jobs in σ require q_k on machine k, the remaining jobs on k require at least T_k more, and the last job must also be processed on all machines after k. Thus in analogy to Proposition 10 we have the following proposition.

(A) Proposition 12. For the m-machine flow shop makespan problem with dynamic machine availability and dynamic job arrivals r_k, and q_k and T_k developed as above (permutation schedules),

$$
\begin{aligned}
B &= \max\{b_1, b_2, b_3, \ldots, b_k, \ldots, b_m\} \quad \text{where} \\
b_m &= q_m + T_m \\
b_{m-1} &= q_{m-1} + T_{m-1} + \min_{j\in\sigma'} \{p_{jm}\} \\
b_{m-2} &= q_{m-2} + T_{m-2} + \min_{j\in\sigma'} \{p_{j,m-1} + p_{j,m}\} \\
&\quad \vdots \\
b_k &= q_k + T_k + \min_{j\in\sigma'} \{p_{j,k+1} + p_{j,k+2} + \cdots + p_{j,m}\} \\
&\quad \vdots \\
b_1 &= q_1 + T_1 + \min_{j\in\sigma'} \{p_{j,2} + p_{j,3} + \cdots + p_{j,m}\}
\end{aligned}
$$

Proof. After verifying that T_k is the proper replacement for $\Sigma_{j\in\sigma'} p_{jk}$ in Proposition 10, the proof is identical. ∎

We turn next to developing much stronger lower bounds for large problems via linear programming. We first give an integer programming model for solving the

m-machine dynamic availability dynamic arrivals flow shop problem for the make-span problem. (A similar but much more general integer programming formulation for makespan or other objectives can be given for job shops, project shops, and resource-constrained multi-resource problems. However, it is useful to give a special case now to develop insight.)

Proposition 13. The m-machine flow shop with makespan objective, dynamic availability, and dynamic arrivals can be solved by the following integer program, where (general case):

C_m is the completion time of all jobs on m.

C_{jk} is the completion time of job j on machine k.

C_{jkt} is 1 if $C_{jk} = t$; 0 otherwise.

x_{jkt} is 1 if job j is scheduled on machine k in period t; 0 otherwise.

p_{jk} is the processing time of job j on machine k.

T is the total number of time periods in the model (T may be estimated by any heuristic solution to the problem).

r_j is the arrival time of job j.

a_k is the initial availability time of machine k.

$$\min \ C_m$$

$$\text{s.t} \quad C_{jk} = \sum_{t=1,T} t C_{jkt} \quad \text{for all } j, k \tag{1}$$

$$x_{jkt} = \sum_{u=t,t+p_{jk}-1} C_{jku} \quad \text{for all } j, k, t \tag{2}$$

$$C_m \geq C_{jm} \quad \text{for all } j \tag{3}$$

$$C_{j1} \geq r_j + p_{j,1} \quad \text{for all } j \tag{4}$$

$$C_{jk} \geq C_{j,k-1} + p_{j,k} \quad \text{for all } j, k \tag{5}$$

$$\sum_{j=1,n} x_{jkt} \leq 1 \quad \text{for all } k, t \tag{6}$$

$$x_{jkt} = 0 \quad t < a_k, \text{ all } j, k \tag{7}$$

$$C_{jk} \in \{0, 1\} \quad \text{all } j, k, t \tag{8}$$

Proof. We go through each equation. We are to minimize the makespan or completion time on machine m. In equation (1) precisely one term of the sum will be nonzero, yielding t, where t is the actual completion time of job j on machine k. Equation (2) says that job j will be processing on machine k in time t if and only if it completes somewhere between t and $t + p_{jk} - 1$. Constraint (3) says that the overall makespan must be at least as large as the completion time for each job. Constraint (4) says that no job can start until it arrives. Constraint (5) says that an operation cannot complete until the previous one does plus the current operation process time. Constraint (6) says that at most one operation may be using machine k at any time period t. Equation (7) says that no job may process until the machine initially becomes available. Constraint (8) says that part of operation jk cannot finish on machine k in period t; either none or all will.   ■

Note that the number of variables in the above formulation can be reduced almost in half by substituting in equations (1) and (2) for C_{jk} and x_{jkt} as they appear in later equations. It is not clear whether this would speed computations or not, since the summations would then need to be calculated repeatedly. In any event, the version we have presented here is much simpler for expository reasons (Baker, 1974).

Of course, this is a typically too large of an integer program to solve directly. However, if we allow the x_{jkt} to be fractional, we obtain the *free-preemptive flow* version. The free-preemptive flow case means that the operation jk can freely be split into several parts, with each started at different times. Several such parts from different jobs might be processed concurrently, provided the total resource usage does not exceed 1.0.

Proposition 14. If the integer program in Proposition 12 is relaxed to allow x_{jkt} to be fractional, then the resulting linear program can be used for lower bounds at a node in any branch-and-bound, where now a_k is interpreted as scheduling to date on the machine and the jobs in the IP as those not yet scheduled.

Proof. Straightforward.   ■

13.3.3   Heuristics for *m* Machines

Johnson's exact algorithm for the two-machine makespan problem with static arrivals and the various associated difference heuristics for the same problem are based on finding a job with the strongest tendency to progress from short times to long times in the sequence of operations. We might hope to exploit this same idea for *m*-machine static arrival problems; in this section we study three heuristics that do exactly this. In addition, we study three more recent OPT-like bottleneck type heuristics (which can handle dynamic arrivals also) for comparison. (The use of the heuristics discussed will be illustrated in an example at the end of this section.) Other methods for makespan such as iterated myopic, shifting bottleneck, and bottleneck dynamics are mentioned briefly; they are covered elsewhere in the book.

Classic Heuristics. The difference heuristic for $m = 2$ calculates an index $\pi_j = -p_{j1} + p_{j2}$ and sequences the job with the largest π_j first. This suggests for $m = 3$ an index $\pi_j = -2p_{j1} + 0p_{j2} + 2p_{j3}$, which leads finally to the following heuristic.

Palmer's Heuristic. (Palmer, 1965). For an *m*-machine flow shop with makespan criterion and static arrivals, calculate the priorities $\pi_j = -(m - 1)p_{j1} - (m - 3)p_{j2} - (m - 5)p_{j3} + \cdots + (m - 3)p_{j,m-1} + (m - 1)p_{jm}$ and sequence the highest priority job first. (Permutation)

(An example comparing Palmer's heuristic with a number of other heuristics for a sample problem is given at the end of the chapter.)

Note that an analogue to the ratio heuristic can easily be constructed by substituting $\log(p_{jk})$ for p_{jk} in Palmer's heuristic.

Gupta (1972) sought a priority rule similar to Palmer's except based directly on Johnson's rule, so that it would produce perfect schedules for $m = 2$, near optimal ones for $m = 3$, and outstanding ones in general. He noted that when Johnson's rule is optimal in the three-machine case, it can be cast as a priority scheme where

$$\pi_j = [e_j] \, / \, [\min\{p_{j1} + p_{j2}, p_{j2} + p_{j3}\}]$$

and where we define $e_j = 1$ if $p_{j1} < p_{j2}$, and $e_j = -1$ if $p_{j1} \geq p_{j2}$. (It is left as an exercise to show that this priority rule formulation is equivalent to Proposition 5).

Generalizing from this structure Gupta obtained the following heuristic.

Gupta's Heuristic. (Gupta, 1972). For an m-machine flow shop with makespan criterion and static arrivals, calculate the priorities

$$\pi_j = [e_j]/ \left[\min_{k=1,m-1} \{p_{jk} + p_{j,k+1}\} \right]$$

where we define $e_j = 1$ if $p_{j1} < p_{jm}$ and $e_j = -1$ if $p_{j1} \geq p_jm$ and sequence the highest priority job first. (Permutation)

(An example comparing Gupta's heuristic with a number of other heuristics for a sample problem is given at the end of the chapter.)

In a computational study, Gupta found his heuristic to generate better makespans than Palmer's in a substantial majority of cases. (For most of the small problems we have already investigated, Gupta's heuristic gives the optimum.)

Perhaps the most accurate extension of Johnson's method to the m-machine static arrival problem is the method of Campbell, Dudek, and Smith (CDS) (1970). It is also somewhat more like our general purpose heuristic methods for flow shop problems, in that it generates a number of solutions and picks the best (multi-pass heuristic).

Remember again from Proposition 5 that in cases where Johnson's rule was optimal for three-machine problems, an auxiliary Johnson's problem is solved, with the auxiliary first machine having the sum of times on machines 1 and 2, and the auxiliary second machine the sum of times on machines 2 and 3. Then the auxiliary problem is solved via Johnson's rule. CDS extend this idea to more machines.

CDS Heuristic. (Campbell et al., 1970). Solve for m (possibly) different schedules in m different iterations. For iteration i set

$$P_{j1} = \sum_{k=1,i} p_{jk} \quad \text{and} \quad P_{j2} = \sum_{k=1,i} p_{j,m-k+1}$$

(i.e., P_{j1} is the sum of the times for job j on the first i machines and P_{j2} is the sum for the last i machines). Solve the resulting two-machine Johnson's problem, and save the corresponding sequence and makespan M_i. Finally, take the makespan as $M = \min\{M_i\}$ and use the corresponding schedule from iteration i^*. (Permutation)

(CDS have fancy rules for tiebreaking that we do not report.) Campbell et al. (1970) tested their heuristic extensively versus Palmer's heuristic. They found that the CDS heuristic was generally more effective for both small and large problems, although it has a somewhat larger computation time for large problems.

OPT-like Heuristics. We have considered the general idea of OPT-like bottleneck methods in Chapter 10. It is instructive to construct such heuristics for the m-machine flow shop makespan problem, some of which are suitable for hand calculation.

Botflow-1 Heuristic for Makespan

Step 1: Choose a "bottleneck" machine (e.g., by total process time).

Step 2: For each job calculate the "head" h_j = job's total processing time before the bottleneck machine. (We are assuming earlier machine capacities do not limit arrival times.)

Step 3: Similarly, calculate the tail t_j = job total processing time after the bottleneck machine (again, assuming no machine limitations).

Step 4: Take a value of the makespan M'' that is desirable to achieve (best known makespan, or perhaps best less 1.0 to try for a better one).

Step 5: Calculate derived due dates off the bottleneck machine as $d_j = M'' - t_j$.

Step 6: Solve the one-machine (bottleneck) problem with arrival times $r_j = h_j$, due dates d_j, and bottleneck (k) processing times p_{jk}. (A good, simple heuristic is dynamic due date; that is, schedule the available job with the earliest due date.) (Permutation)   ∎

As we shall see, Botflow-1 competes well with the earlier heuristics and is suitable for hand calculations. However, it is easily improved. Botflow-1 makes a very crude estimate of the heads and tails for the problem. We may simply use the solution from Botflow-1 to generate some real heads and tails, and rerun.

Botflow-2 Heuristic for Makespan

Step 1: Run Botflow-1.

Step 2: Use the updated heads and tails.

Step 3: Rerun Botflow-1.

Step 4: If the makespan improves, go to step 2.

Step 5: Output the best solution to date. (Permutation)

Botflow-2 basically combines the best features of OPT-like procedures and lead time iteration. In Botflow-3, we allow for the fact that our idea of the bottleneck may be incorrect, utilizing the shifting bottleneck ideas.

TABLE 13.6 Makespan Example

Job j:	1	2	3	4	5
p_{j1}	6	4	3	9	5
p_{j2}	8	1	9	5	6
p_{j3}	2	1	5	8	6

Botflow-3 Heuristic for Makespan

Step 1: Run Botflow-2 m times, using each machine in turn as the bottleneck.

Step 2: Output the best solution. (Permutation)

Since Botflow-2 and Botflow-3 need heads and tails for any job at some machine k given that a particular permutation sequence is being employed before and after k, we briefly give these procedures.

Finding Heads

Step 1: For the m-machine problem, for the given permutation sequence, simulate the entire solution forward.

Step 2: The starting times at machine k are the required heads for k.

Finding Tails

Step 1: Reverse the order of machines and their associated processing times.

Step 2: Simulate the entire reverse solution.

Step 3: "Starting" times at machine k are the "ending" times for the original problem, and thus the required tails for k.

To compare the three classic heuristic methods and the three OPT-like heuristic methods for static flow shop makespan problems, consider the five-job three-machine problem shown in Table 13.6. Palmer's method for $m = 3$ sets the priorities to $\pi_j = -2p_{j1} + 2p_{j3}$. This yields priorities for the jobs in the order given of $-8, -6, 4, -2, 2$ so that the heuristic yields a sequence of 3–5–4–2–1, for which the makespan $M = 37$.

With Gupta's heuristic the priorities are $-1/16, -1/2, 1/12, -1/13, 1/11$, yielding the sequence 5–3–4–1–2 for which $M = 36$.

The CDS heuristic yields the job ordering 3–5–4–1–2 (at both stages!) and for this sequence $M = 35$.

To implement Botflow-1 we choose machine 2 as the bottleneck, since total processing times on the machines are 27, 29, 22. (Using machine 1 gives the same answer in this case.)

The modified one-machine problem is as follows:

Job j:	1	2	3	4	5
h_j	6	4	3	9	5
p_j	8	1	9	5	6
d_j	32	33	29	26	28

(The due dates are based on $M'' = 34$, which is a unit improvement of the known makespan above of 35.) Next, we schedule the one-machine problem, scheduling EDD among jobs that have arrived at time t. Job 3 arrives first at $t = 3$; since we are using a dispatch policy it must be scheduled. It finishes at $t = 12$, at which time all other jobs have arrived, and will be scheduled in EDD order. Hence the schedule is 3–4–5–1–2, achieving $M = 34$, which is better than any of the three older heuristics.

We may use the output of Botflow-1 to generate the modified one-machine problem as input to Botflow-2, trying to improve our result by setting $M'' = 33$:

Job j:	3	4	5	1	2
h_j	3	12	17	23	27
p_j	9	5	6	8	1
d_j	16	24	30	32	33

Scheduling the earliest due date among jobs to arrive begins as follows. Job 3 arrives at $t = 3$ and must be scheduled, finishing at $t = 12$, when again there is only one job to schedule, and so on, yielding again 3–4–5–1–2. Thus Botflow-2 terminates, giving the same sequence with makespan $M = 34$.

To do Botflow-3, we use the output of Botflow-1 to generate problems for machines 1, 2, or 3 as the bottleneck. Without giving the details, for machine 1 as bottleneck we generate ($M'' = 33$) the following:

Job j:	3	4	5	1	2
h_j	0	0	0	0	0
p_j	3	9	5	6	4
d_j	3	12	17	23	29

this again yields the sequence 3–4–5–1–2 with $M = 34$. Machine 2 as the bottleneck for this sequence has already been done. For machine 3 as the bottleneck we generate the following:

Job j:	3	4	5	1	2
h_j	12	17	23	31	32
p_j	9	5	6	2	1
d_j	33	33	33	33	33

Thus Botflow-3 terminates, since there is no improved sequence with which to proceed. It is very tempting at this point to guess that $M = 34$ is the optimal makespan. The student will be asked to investigate this question in the exercises.

Iterated Myopic Heuristics/Bottleneck Dynamics. Since the one-machine problem in a makespan problem is to minimize maximum lateness, for which the heuristic is DEDD, resource usage does not enter, and myopic and bottleneck dynamics heuristics are identical. The procedure is analogous to Botflow-2 except that in each iteration we construct sequentially a solution to the problem by simulating through the problem and iterate that, rather than iterating the solution to a bottleneck machine. These issues will be explored for other objective functions in Chapter 14. It may also be considered a special case of resource-constrained project scheduling, discussed in detail in Chapter 18.

Shifting Bottleneck Algorithm. The shifting bottleneck algorithm is developed in Chapter 10 for makespan for the more general case of a job shop. It is basically a sophistication of Botflow-3. Since it has been successfully tested for the job shop, it should work well here.

13.3A Numerical Exercises

13.3A.1 Consider the following four-job three-machine static makespan problem:

Job j:	1	2	3	4
p_{j1}	13	7	26	2
p_{j2}	3	12	9	6
p_{j3}	12	16	7	1

Find the minimum makespan using the Ignall–Schrage method. Count the nodes generated by the branching process.

13.3A.2 Consider the following six-job three-machine static makespan problem:

Job j:	1	2	3	4	5	6
p_{j1}	6	12	4	3	6	2
p_{j2}	7	2	6	11	8	4
p_{j3}	3	3	8	7	10	2

Compute a good lower bound on the makespan for this problem. Can you find a "commonsense" solution that attains this perfect lower bound?

13.3A.3 Solve the problem in 13.1A.2 exactly by the Ignall–Schrage method. Comment on the relative amount of work involved in 13.3A.2 and 13.3A.3.

13.3A.4 Add dynamic arrival times to the problem in 13.3A.1 and set up the resulting problem as an integer program.

13.3A.5 Relax integrality in the previous problem, and run the resulting LP on an LP package. Of what use is the answer you obtain?

13.3A.6 Find the makespan generated by each of the first three heuristics for m-machine static arrivals for the following problem:

Job j:	1	2	3	4
p_{j1}	4	3	1	3
p_{j2}	3	7	2	4
p_{j3}	7	2	4	3
p_{j4}	8	5	7	2

13.3A.7 Find the makespan for the problem of 13.3A.6 by each of the three Botflow heuristics.

13.3A.8 What is the optimal makespan for the example problem in the chapter comparing the four m-machine static arrival heuristics?

13.3B Software/Computer Exercises

13.3B.1 Modify an earlier best-first branch-and-bound routine that you might have available, to handle the flow shop makespan problem.

13.3B.2 Validate your program by running exercises either with the tables in the chapter or with the numerical exercises above.

13.3B.3 Write a computer program to test the first three heuristics for the m-machine makespan problem with static arrivals. Conduct some tests to try to choose the most accurate heuristic.

13.3B.4 Add the Botflow heuristics to your program in 13.3B.3, and do a more thorough testing.

13.3C Thinkers

13.3C.1 Give equations for finding the Ignall–Schrage bounds for general m; explain your result.

13.3C.2 Add dynamic arrivals to the problem, and again find the analogous bounds for general *m*; explain your result.

13.3C.3 Why is it helpful when the value of a priority does not depend on the position in the sequence, as for the Palmer, Gupta, and CDS heuristics?

13.3C.4 Try to generalize the OPT-like algorithm for the case when two different machines are equally the bottleneck and a compromise must be struck.

____14
SCHEDULING FLOW SHOPS: OTHER OBJECTIVES AND EXTENSIONS

14.1 INTRODUCTION

In this chapter we continue our discussion of flow shops by investigating other objective functions, developing model extensions, and presenting a concrete case of a complex flow shop: a nuclear fuel tube shop.

We argued previously, in the discussion on parallel machines, that direct measurement of resource usage—economic makespan—is superior to makespan as a measure of shop utilization in a rolling horizon environment. In Section 14.2 this methodology is developed for flow shops, and the results of a computational study are reported.

In Section 14.3 we turn to investigating such objectives as weighted flow and weighted tardiness, comparing integer programming and branch-and-bound with heuristic algorithms of various complexity and computational difficulty: OPT-like, shifting bottleneck, myopic, and bottleneck dynamics, among others. Several studies giving computational experience are presented. In Section 14.4 we consider heuristics for model extensions, such as skip shops, reentrant flow shops, and finite queue shops. Finally, in Section 14.5 a real-world example—a nuclear fuel tube shop—is modeled, both to illustrate actual complex flow shops and to sketch appropriate heuristic scheduling systems that may be designed for them.

14.2 ECONOMIC MAKESPAN

14.2.1 Overview

As we pointed out in Chapters 11 and 12 on parallel machine problems, classical scheduling models that attempted to optimize dynamic job sequencing and/or routing typically used branch-and-bound to minimize makespan on a fixed finite horizon, say, from a current period t out to period $t + N$. In practice, however, only the first-period part of the solution of this very expensive integer programming procedure is utilized. Then the problem moves to consider periods $t + 1$ to $t + N + 1$ and is re-solved, using *updated forecast information*, which will be available in period $t + 1$. In effect, an expensive optimal procedure (branch-and-bound) is embedded in a larger heuristic procedure (rolling horizon), with relatively little regard as to whether the new procedures mesh well.

For parallel single-stage machines, we developed a new objective function called economic makespan. Computational results indicated that the newer procedure actually performed somewhat better on average than the embedded branch-and-bound procedure, at a small fraction of the computational cost. The heuristic was basically a bottleneck dynamics procedure drawing insights from the linear relaxation of the conventional makespan approach.

Here we develop a similar approach for the flow shop problem, giving methods for obtaining heuristic prices for each machine and heuristics for sequencing the job on the machines. Computational studies on rolling horizon problems are in process. It is expected that again the new procedure will be somewhat superior in accuracy to makespan, at a small fraction of the computational cost.

14.2.2 Economic Makespan Objective, Heuristics for Sequencing and Pricing

Objective Function. Suppose that, via methods to be discussed later, we have estimated some (static) heuristic prices R_k for machine k. As for the parallel machine problem, we argue that, in a heavily loaded shop, the objective should be to minimize resource usage, where the resource usage for each machine is the machine time used up by the job set, multiplied by the heuristic price for that machine.

Definition. The economic makespan objective of a highly utilized flow shop is $\min E = \Sigma_{k=1,m} R_k(C_k - a_k)$, where R_k is the heuristic price of machine k, C_k is the completion time for machine k, and a_k is the first availability of machine k.

A little thought makes it clear that this is equivalent to minimizing internal slack (as opposed to ending slack) in the problem.

Proposition 1. An equivalent way of expressing the economic makespan objective is $E' = \Sigma_{k=1,m} R_k I_k$, where I_k is the total idle time on k between $t = a_k$ and $t = C_k$.

Proof. We may write $R_k(C_k - a_k) = R_k(\Sigma_{j=1,n} p_{jk} + I_k)$.

Thus we see that the nonidle resource usage is a sunk cost, leaving the idle time resource wastage. ∎

Now that we have defined economic makespan in a way consistent with our development for parallel machines, we turn to discussing pricing for flow shops, and economic makespan heuristics.

We are interested in the *m*-machine flow shop problem with high utilization, dynamic machine availability, and dynamic job arrival. (The latter two aspects are essential in a rolling horizon procedure, which is our primary long run interest.)

Pricing Methods for Flowshops. As mentioned in Chapter 10, there are many simple and more sophisticated ways to estimate prices.

    (a) *OPT-like.* Estimate the bottleneck machine by the highest expected time load. Give this machine a price of 1, the rest prices of 0. (Minimizes wastage on the bottleneck machine.)

    (b) *Myopic.* Use dispatch simulation; assume the current machine has a price of 1.0, all others 0.

    (c) *Brute Force.* Try a set of prices and estimate the best sequence and appropriate objective. Try many exhaustive combinations and take the best objective obtained. (Does not work well for more than three machines.)

    (d) *Linear Programming Relaxation.* Relax the problem to allow fractional assignments and all arrivals at time 0. Use the dual variables obtained from equation (6) in Proposition 13 of Chapter 13 as the prices.

Heuristics for Economic Makespan

Opt-Like Heuristic (Economic Makespan)

Step 1: Choose bottleneck by most utilized machine.

Step 2: Set the price for the bottleneck machine = 1; set the others = 0.

Step 3: Use Botflow-1 modified as follows: (a) same heads except dynamic availability and (b) tails = 0 (minimize makespan of machine itself).

Myopic/BD Heuristic (Economic Makespan)

Step 1: Do a dispatch simulation of shop.

Step 2: At each machine prioritize choices by solving one-machine problem with actual arrivals, and all tails = 0.

Step 3: No iteration needed, since no lead times are used.

Shifting Bottleneck Heuristic (Economic Makespan)

Similar straightforward generalization of OPT-like.

14.2A Numerical Exercises

14.2A.1 Solve exercise 13.3A.2 for economic makespan by the following heuristics: (a) OPT-like and (b) myopic.

14.2A.2 Solve the problem in 13.3A.6 for economic makespan by the following heuristics: (a) OPT-like and (b) myopic.

14.2A.3 Estimate prices in 14.2A.1 by linear programming relaxation. Estimate economic makespan for the problem in 13.3A.2 for (a) noneconomic makespan, (b) OPT-like economic makespan, and (c) myopic economic makespan.

14.2A.4 Repeat 14.2A.3 for the problem in 13.3A.6.

14.2B Software/Computer Exercises

14.2B.1 Write a software program (or utilize an existing one) to compare a number of heuristics for the m-machine flow shop with dynamic arrivals, using the makespan or economic makespan criterion. Design an experiment with a fairly large number of problems, test your design, and analyze the results.

14.2B.2 Add a new criterion, $\min(C_n + 0.1F_{wt})$ (minimize makespan plus a small consideration for weighted flow). Use the same heuristics and test problems as in 14.2B.1. How robust are your conclusions for that problem with this change in objective?

14.2C Thinkers

14.2C.1 Economic makespan as we have defined it penalizes slack in the middle of the problem fully, but gives zero penalty for slack at the end. Suppose slack at the end is penalized, say, 40%, so that we define economic makespan by $\min E = \Sigma_{k=1,m} R_k[(C_k - a_k) + 0.4(C_{max} - C_k)]$. Set up an integer programming formulation to solve this problem exactly.

14.2C.2 One type of myopic economic makespan heuristic might be defined as follows. Consider all jobs capable of being assigned next. Let I_{jk} be the inserted idle time created on machine k if j is assigned next. Choose j as the available job that minimizes $\Sigma_{k=1,m} R_k I_{jk}$. Test this heuristic using LP prices. How does it compare with the others you have looked at?

14.2C.3 Suppose that, if job j is run right after i, there are setup times S_{ijk} on every machine. Modify the myopic heuristic given in 14.2C.2 to deal with this case. Test it.

14.3 FLOW AND TARDINESS OBJECTIVES

Overview

Although a great deal of research has been devoted to flow shop scheduling, almost all of it, until the last 10 years or so, has been devoted to the makespan criterion. Even now few exact practical algorithms exist except for small makespan problems. Exact methods such as integer programming, branch-and-bound, and dynamic programming can handle only very small problems. However, a good deal of progress has been made in developing accurate heuristics for a broad class of problems, and a fair amount of testing of these heuristics has now been done.

In Section 14.3.2 we briefly discuss limited branch-and-bound results. We also present two different integer programming formulations of the m-machine flow shop problem with general objective and discuss their advantages and disadvantages. In Section 14.3.3 we discuss the weighted flow objective for a number of different heuristic methods. We also present a computational study that compares myopic, OPT-like, and bottleneck dynamic heuristics for weighted flow. In Section 14.3.4 we discuss weighted tardiness for a number of different heuristic methods and present computational studies showing that the forecast myopic approach is superior to other nonbottleneck dynamics methods. The latter method is discussed in more detail in Chapter 16 in some job shop studies. Finally, in Section 14.3.5 we briefly discuss other issues.

14.3.2   Branch-and-Bound/Integer Programming

Ignall and Schrage (1965) devised a branch-and-bound procedure for determining the best permutation schedule for (unweighted) mean flowtime problems. They tested their procedure on job sets for the two-machine problem, for which permutation schedules are dominant. Their computational results were somewhat discouraging, and they concluded that the branch-and-bound approach is less successful in the mean flowtime problem for two machines than it is in the makespan problem for three machines.

However, there are several reasons to retain interest in branch-and-bound for flow shop problems. First, computer capabilities are hundreds of times as large as when Ignall and Schrage did their work. Second, linear programming provides a strong, broadly based procedure for obtaining good lower bounds. Both of the integer programming procedures to be described next can be relaxed and used for this purpose. Finally, such approximate methods as beam search can be used to greatly speed the search process.

We turn next to integer programming formulations.

(A) Proposition 2. The m-machine flow shop with dynamic availability and arrivals and weighted flowtime objective can be solved by modifying Proposition 13 of Chapter 13.

$$\min F_{\text{wt}} = \sum_{j=1,n} w_j (C_j - r_j)$$

$$\text{s.t.} \quad C_{jk} = \sum_{t=1,T} t C_{jkt} \quad \text{for all } j, k \tag{1}$$

$$x_{jkt} = \sum_{u=t,t+p_{jk}-1} C_{jku} \quad \text{for all } j, k, t \tag{2}$$

$$C_j = C_{jm} \quad \text{for all } j \tag{3}$$

$$C_{j1} \geq r_j + p_{j1} \quad \text{for all } j \tag{4}$$

$$C_{jk} \geq C_{j,k-1} + p_{j,k} \quad \text{for all } j, k \tag{5}$$

$$\sum_{j=1,n} x_{jkt} \leq 1 \quad \text{for all } k, t \tag{6}$$

$$x_{jkt} = 0.0 \quad \text{for } t < a_k \text{ and all } j, k \tag{7}$$

$$C_{jkt} \in \{0, 1\} \quad \text{for all } j, k, t \tag{8}$$

Proof. Slight modification of Proposition 13 of Chapter 13.   ∎

(A) Proposition 3. Define d_j as the due date of job j, the tardiness T_j by $T_j = (C_j - d_j)^+$, and the earliness by $E_j = (d_j - C_j)^+$. The m-machine flow shop with dynamic availability and arrivals and weighted tardiness objective can be solved by modifying the equations in Proposition 2 as follows:

(a) Replace the objective by min $T_{wt} = \Sigma_{j=1,n} w_j T_j$.
(b) Add equation (9); that is,

$$T_j - E_j = C_j - d_j \quad \text{for all } j \tag{9}$$

Proof. The objective merely says to minimize the weighted tardiness. Since the integer programming keeps T_j and E_j nonnegative, and the objective keeps T_j as small as feasible, then if $C_j > d_j$, equation (9) forces $T_j = C_j - d_j$; similarly, if $C_j \leq d_j$, equation (9) forces $E_j = d_j - C_j$.   ∎

The integer programming formulation of Proposition 13 in Chapter 13 is really very general. It is easily modified to handle different objective functions, dynamic arrivals, and precedence constraints and does not assume restriction to a permutation schedule.

However, that formulation has one great disadvantage as well. Time is discretized into units, and there is one decision variable for each job for each machine for each time unit. Thus if there are T time units, there are at least mnT variables, which drastically restricts practical problem sizes.

Therefore we turn to consider a second formulation that is less general but requires a much smaller number of variables. Hence it is more practical for simpler makespan problems. (It is also restricted to permutation schedules.)

(A) Second Integer Programming Formulation. For the first time, it is important for us to distinguish between job i, with identification fixed in advance, and the position j in the sequence where i is finally placed. If i ends up in position j, we say $[i] = j$ (position of i is j).

Let z_{ij} be an integer variable defined in the following way:

$$z_{ij} = \begin{cases} 1 & \text{if job } i \text{ is assigned to position } j \\ 0 & \text{otherwise} \end{cases}$$

A feasible solution to the problem requires that exactly one job is assigned to a position, and a job is assigned to exactly one position; that is,

$$\sum_{i=1,n} z_{ij} = 1 \tag{10}$$

$$\sum_{j=1,n} z_{ij} = 1 \tag{11}$$

We shall have need of the identity

$$p_{[j]k} = \sum_{i=1,n} p_{ik} z_{ij} \tag{12}$$

(All z's in the summation will be zero except the one for the i actually assigned to j, whose processing time is, in fact, $p_{[j]k}$.)

We also need some variables for idle time on machines and idle times for jobs. (These are not actually decision variables; choosing the z_{ij} indirectly chooses these.)

X_{jk} = idle time on machine k just before processing job $[j]$

Y_{jk} = idle time for job $[j]$ after processing on k and before $k + 1$

Note that successive blocks of time on machine k look like

$$\cdots, p_{[j]k}, X_{j+1,k}, p_{[j+1]k}, X_{[j+1]k}, \cdots \tag{13}$$

Successive blocks of time on machine $k + 1$ look like

$$\cdots, p_{[j],k+1}, X_{j+1,k+1}, p_{[j+1],k+1}, \cdots \tag{14}$$

For successive operations of a job blocks look like

$$\cdots, p_{[j]k}, Y_{jk}, p_{[j],k+1}, \cdots \tag{15}$$

And, for the job in position $j + 1$:

$$\cdots, p_{[j+1],k}, Y_{j+1,k} \, p_{[j+1],k+1}, \cdots \tag{16}$$

Now suppose we wish to figure out the blocks of time between the job in position j on machine k and the job in position $j + 1$ on machine $k + 1$. The first way would be to stay on machine k until the next job, and then follow the next job to $k + 1$. Combining (13) and (16) we have

$$\ldots, p_{[j]k}, X_{j+1,k}, p_{[j+1],k}, Y_{j+1,k}, p_{[j+1],k+1}, \ldots \tag{17}$$

The second way would be to immediately follow $[j]$ to the next machine, and then stay on that machine until the next job. Combining (14) and (15) we have

$$\ldots, p_{[j]k}, Y_{jk}, p_{[j],k+1}, X_{j+1,k+1}, p_{[j+1],k+1}, \ldots \tag{18}$$

It is important to realize that the two methods, (17) and (18), must involve the same amounts of time, since all idleness is accounted for. Equating the two gives the fundamental relationship for consistency (feasibility):

$$X_{j+1,k} + p_{[j+1]k} + Y_{j+1,k} = Y_{j,k} + p_{[j],k+1} + X_{j+1,k+1} \tag{19}$$

We now show how to use our model to derive an integer programming model for the m-machine flow shop static arrival weighted flowtime objective. Note that

$$w_{[j]} = \sum_{i=1,n} w_i z_{ij} \tag{20}$$

and

$$C_{[j]} = \sum_{i=1,j} (X_{im} + p_{[i]m})$$

(A) Proposition 4. (Baker, 1974, Section 6.8). The m-machine static arrival flow shop problem with weighted flow objective can be solved for permutation schedules (with variables defined as above) by

$$\text{minimize} \sum_{j=1,n} w_{[j]} C_{[j]}$$

subject to

$$X_{j+1,k} + p_{[j+1]k} + Y_{j+1,k} = Y_{j,k} + p_{[j],k+1} + X_{j+1,k+1}$$
$$\text{for } 1 \le k \le m - 1, \ 1 \le j \le n - 1$$

$$\sum_{i=1,n} z_{ij} = 1 \quad \text{for all } j$$

$$\sum_{j=1,n} z_{ij} = 1 \quad \text{for all } i$$

$$p_{[j]k} = \sum_{i=1,n} p_{ik} z_{ij} \quad \text{for all } j \text{ and } k$$

$$w_{[j]} = \sum_{i=1,n} w_i z_{ij} \quad \text{for all } j$$

$$C_{[j]} = \sum_{i=1,j} (x_{im} + p_{[i]m}) \quad \text{for all } j$$

$$X_{jk} \geq 0, \; Y_{jk} \geq 0, \quad \text{and} \quad z_{ij} \in \{0, 1\}$$

Proof. Explained before the formal statement above of the proposition. ∎

This formulation has not been given with as few variables as possible. It is clear that the $p_{[j]k}$ can be substituted out. (This version was chosen for expositional reasons.)

Note that the formulation seems to be nonlinear, since in the objective function $w_{[j]}$ and $C_{[j]}$ are multiplied together, and both depend on the decision variables z_{ij}. However, the $w_{[j]}$ are actually known constants; we know everything but their order. If we write the objective function (by reordering terms) as

$$\sum_{i=1,n} w_j C_{[j]}$$

it becomes clearer that the weights may be treated as constants.

In the exercises, the student will be asked to show how to adapt this formulation to other objective functions, to dynamic arrivals, and/or to precedence constraints.

14.3.3 Weighted Flow Heuristics

Heuristics for the flow shop weighted flow objective problem fall mainly into two groups: permutation sequence and dispatch. We have considered mainly permutation sequence heuristics up to this point. Their main advantage is simplicity and stability of the policy. If a job has the highest priority on one machine, it will have that on any machine. Priorities do not change radically over time, nor with the entry of new "hot jobs" into the system. Since there are only $n!$ permutation sequences, exhaustive search procedures are more practical.

On the other hand, dispatch heuristics are much much more flexible. There is no need to wait at a machine for a very late job, just because it is officially next in the sequence. Choosing among available jobs avoids many problems of gridlock and the like. Priorities can change as deadlines approach. A run of a dispatch heuristic is typically nearly as speedy as that of a permutation sequence. We shall consider both types.

Early heuristics for the flow shop weighted flow objective include several by Gupta (1972). He tested them and considered how well such heuristics would deal with makespan. By far the best known classical dispatch heuristic for weighted flow is the dispatch myopic (choose the job with highest w_j/p_{jk} at machine k), which was tested extensively by Conway (1965). See also a good discussion in Baker (1974).

We first present several permutation sequence type heuristics, and then several dispatch type heuristics. Next, we discuss in detail a recent study by Morton and Pentico (work-in-process), which indicates that bottleneck dynamics dispatch procedures seem to be more accurate than the dispatch myopic or the dispatch OPT-like procedure.

Permutation Sequence Heuristics

Botflow-1 (Weighted Flow). Analogous to Botflow-1 for makespan. Set tails = 0. The one-machine problem is to minimize weighted flow with dynamic arrivals. When the sequence is obtained, simulate the full problem to obtain the corresponding weighted flow.

Botflow-2 (Weighted Flow). Analogous to Botflow-2 for makespan. Set tails = 0. The one-machine problem is to minimize weighted flow with dynamic arrivals.

Botflow-3 (Weighted Flow). Analogous to Botflow-3 for makespan. Set tails = 0. Try the permutation sequence from Botflow-2 with each machine in turn as the actual bottleneck. If an improved weighted flow solution is found, use this for the next solution.

Neighborhood Search. Because we are only investigating permutation sequences, pairwise interchange is straightforward, if somewhat more tedious than for one machine. (Now the interchange may affect the whole remaining schedule. Thus the whole schedule from that point must be resimulated until it can be ascertained that no further changes have been caused by the interchange.) A starting procedure can be obtained by one of the bottleneck procedures above or by using multiple starts generated by biased random sampling. Use of tabu search or simulated annealing is also straightforward.

Extended neighborhood search methods can be speeded up greatly by using an inexpensive method to establish which small part of the neighborhood is most desirable. Some studies of this sort for makespan have been performed; see Section 16.3. There are further studies underway for both makespan and other regular objectives. The work is still in process and cannot usefully be reported here.

Beam Search. Beam search is best considered in connection with a simpler "guide" heuristic. One such approach would be to choose any simpler permutation heuristic, such as the simplest bottleneck technique. From any node to be evaluated, constrain the simple heuristic to start from this point and run a full solution, which is the evaluation. Choose nodes with, say, the three highest evaluations.

Dispatch Heuristics

Dispatch Myopic. Whenever a machine becomes available, or a new job becomes available for a waiting machine k, choose that job j currently available with the highest value of $\pi_{jk} = w_j/p_{jk}$.

This heuristic has by 30 years of testing established itself as the best of the classic heuristics. In the study to follow, we have chosen it as representative of older heuristics against which to test.

Dispatch OPT-like. For any machine k, determine in advance the machine $b(k)$ considered to be its downstream bottleneck. Whenever a machine becomes available, or a new job becomes available to a waiting machine k, choose that job j currently available with the highest value of $\pi_{jk} = w_j/p_{jb(k)}$.

This heuristic is similar to Botflow-1 for weighted flowtime. Instead of maximizing the cost benefit on the immediate machine as does the dispatch myopic, it tries to maximize the cost benefit of usage of the overworked machine downstream.

Bottleneck Dynamics. Determine relative prices R_k or "urgencies" or "bottleneckness" of the machines by some method. Whenever a machine k becomes available, choose that job j currently available with the highest value of $\pi_{jk} = w_j/[\sum_{i=k,m} R_i p_{ji}]$.

One way to compare these three heuristics is that all three may be considered to form priorities by dividing the job weight by the remaining resource usage required in the shop. However, the myopic procedure crudely treats all machines except the current machine as having zero prices. Similarly, the bottleneck procedure treats all machines except a bottleneck machine downstream as having zero prices. Thus, in a sense, these are special cases of bottleneck dynamics, which has a number of possible options for estimating the price at each machine.

Computational Study for Flow Shops with Weighted Flow Objective.
Morton and Pentico (work in process) recently completed several studies comparing these heuristics in a static flow shop. We report their first study fairly completely.

Testbed

- Two- and three-machine flow shop
- Static shop with initial inventories
- Process times p_{jk}, weights w_j
- Minimize weighted flow
- Dispatch
- Bottleneck machine with highest average process time per job
- Prices R_k determined by coarse search (all combinations of 1.0, 0.5, 0.001)
- Rules: (a) myopic, (b) OPT-like, and (c) bottleneck dynamics (best over search)

Morton and Pentico stated a number of hypotheses to be tested:

Hypothesis I. OPT-like will behave well if and only if it has a strong unique bottleneck stable over time with an adequate stable queue.

Hypothesis II. Myopic behaves relatively badly when OPT-like does well. Myopic does better with early bottlenecks than late bottlenecks.

Hypothesis III. Myopic is more robust than OPT-like.

Hypothesis IV. Both will perform better with smaller variability in job weights and process times.

To test these hypotheses, a main experiment and subsidiary experiments were run. The main experiment consisted of 40 jobs, three machines, and deterministic processing times drawn from a normal distribution. There were 20 replications of 90 basic problems, giving 1800 specific examples over which to compare the three methods.

The 90 problems were generated from five initial inventory conditions, six patterns of single, double, and triple bottlenecks, and three processing time spreads. Job weights were also generated randomly.

In each of the 1800 examples, the myopic, the OPT-like, and the bottleneck dynamics (with search) were compared as percentage above the best of the three. Results compared by bottleneck condition are given in Table 14.1.

The study confirmed all four hypothesis. However, the table shows that the advantage of bottleneck dynamics over the other procedures averaged only 3–4%. Both myopic and OPT-like procedures average at most 6% above the more computational bottleneck dynamics procedure. The worst cases are about 16% for the myopic and 38% for OPT-like.

Therefore it was decided to investigate types of problems that might be especially difficult for the other procedures. Some two-machine problems were set up: one set to be especially bad for the OPT-like procedure, and one set to be especially bad for the myopic procedure. It was possible to generate classes of problems with 52% excess cost for OPT-like heuristics. (Basically, the OPT-like procedure is tricked to starve machine 2 so that machine 1 is the real bottleneck.) Similarly, it was possible to generate classes of problems with 32% excess cost for myopic. (Basically, the myopic procedure is tricked to deliver in the longest processing time order to the second bottleneck machine.)

TABLE 14.1 Myopic versus OPT-like versus BD Search by Bottleneck Type

Type of Bottleneck	Myopic		OPT-like		BD	
	Average	Maximum	Average	Maximum	Average	Maximum
Triple	4.5	16.0	5.8	28.0	0.0	1.0
Middle	2.5	12.0	1.8	21.0	0.0	1.0
End	4.2	16.0	1.8	17.0	0.0	0.0
Start	0.4	6.0	0.4	6.0	0.1	3.0
S/E	4.6	13.0	4.6	38.0	0.0	1.0
S/M	3.4	12.0	5.0	31.0	0.0	2.0
Overall	3.3	16.0	3.2	38.0	0.0	3.0

In a second part of the study, Morton and Pentico shifted their focus to the flow shop with random arrivals and investigated methods for setting prices less computationally intensive than exhaustive search and also capable of varying dynamically over time. They investigated, in particular, two formulas taken from queuing theory for approximating prices. One was static, one dynamic.

BD Static Queuing Formula (Code: Static-Q)

$$R_k = w_{k\ av}\rho_k/(1 - \rho_k)^2$$

(Here $w_{k\ av}$ is a long-term average weight on machine k, while ρ_k is the long-term stationary utilization on k.)

BD Dynamic Queuing Formula (Code: Dyn-Q)

$$R_k = \sum_{j=1,L_{kt}} w_j + w_{av}L_{kt}[\rho_k/(1 - \rho_k)]$$

(Here t is the current time, L_{kt} is the current line length on machine k, w_j are the weights of jobs currently at the machine, w_{av} is the long-term average job weight at the machine, and ρ_k is the long-term average utilization on k.)

Both formulas are based on the approximations that jobs yet to arrive follow Poisson arrivals and exponential service times. (Corrections to other service distributions are easily made but seem not to be important in practice.)

Remember that prices are proportional to the length of the expected remaining busy period. The first formula calculates this expectation based on no current knowledge of the queue, whereas the second formula calculates this expectation based on the knowledge of the jobs currently in the queue.

The static formula has the advantage of stability, ease of use, and no frequent updating. The dynamic formula has the advantage that the estimated price of the machine will go up and down roughly linearly in the current length of the queue, which is satisfying intuitively. (See Table 14.2.)

Each heuristic's performance is measured above the benchmark of coarse search bottleneck dynamics. Results in Table 14.2 represent the average of seven bottleneck types and 10 replications of 200 jobs for each type.

An analysis of this study shows that

- OPT-like did somewhat better on average than myopic.
- OPT-like had a much worse worst-case.
- OPT-like was worse at multiple bottlenecks (not shown).
- New queuing heuristics are better for both average and worst case.
- Dyn-Q is somewhat superior to Static-Q; both are very robust across the study.

TABLE 14.2 Myopic versus OPT-like versus Static-Q versus Dyn-Q for Differing Utilizations[a]

Utilization	Myopic		OPT-like		Static-Q		Dyn-Q	
	Average	Maximum	Average	Maximum	Average	Maximum	Average	Maximum
0.9	1.2	5	3.2	15	1.6	8	0.5	3
0.99	5.1	15	1.8	14	2.3	14	2.4	14
0.995	5.8	25	4.1	73	2.6	16	2.6	16

[a]Percentage error above benchmark.

337

TABLE 14.3 Sinusoidal Utilization Side Study

Myopic		OPT-like		Static-Q		Dyn-Q	
3.8	(19)	4.8	(54)	2.0	(15)	1.5	(14)

TABLE 14.4 Switching Bottleneck Side Study

Myopic		OPT-like		Static-Q		Dyn-Q	
2.7	(25)	7.6	(24)	0.9	(1)	−0.1	(1)

A side study with a seasonally varying utilization factor between 0 and 2 times the average p, seven (mild) bottleneck patterns, the same three average utilizations, and 10 replications shows the average (and worst case) results (% above search benchmark) given in Table 14.3. Thus we see that in periods of fluctuating utilization, the main results remain valid.

In another side study, the dynamic variation was in the bottleneck. Every 200 jobs, the strong bottleneck switched from machine 3 to 2, and then back again, for a total of 800 jobs and 10 replications, showing the average (and worst case) results (% above search benchmark) given in Table 14.4. (Note that Dyn-Q actually did slightly better on average than the coarse-search fixed price bottleneck dynamics benchmark.)

Summary of Conclusions for Flow Shop Dispatch Heuristics: Weighted Flow Criterion

(a) The classic myopic dispatch heuristic is robust throughout all regimes tested. (It requires only processing times.)

(b) The OPT-like dispatch heuristic is somewhat better than myopic for stable bottleneck situations but is much less robust. (It requires, in addition, knowledge of the bottleneck.)

(c) The Static-Q bottleneck dynamics dispatch heuristic has, on average, about one-half the error of the myopic and is remarkably robust even in highly dynamic scenarios. (This requires, in addition, knowledge of the long-term utilization of each machine.)

(d) The Dyn-Q bottleneck dynamics dispatch heuristic has about 70–80% of the error of the Static-Q bottleneck dynamics heuristic and is especially called for in highly dynamic situations. (It requires, in addition, the total importance of the current queue at each machine.)

14.3.4 Weighted Tardiness Heuristics

Heuristics for the flow shop weighted tardiness objective function also fall mainly into permutation sequence and dispatch heuristics. It is still the case that permutation sequence rules are more stable and easier to investigate by search. It is still also the case that dispatch heuristics are more flexible. They have been found in practice to yield good dynamic results at relatively low cost.

There are a large number of classic dispatch heuristics that have been used for the weighted tardiness problem for both flow and job shops. The newer permutation sequence heuristics presented here are analogous to those for weighted flow. They are not tested but could be useful for those wanting to investigate high-quality, high-cost heuristics.

Permutation Sequence Heuristics

Botflow-1 (Weighted Tardiness)

Step 1: Analogous to Botflow-1 for makespan.

Step 2: Heads and tails are processing times before and after the bottleneck machine, respectively.

Step 3: The resulting one-machine weighted tardiness problem, with dynamic arrivals and lead times = tails, can be solved heuristically by using R&M Procedure 9 of Chapter 7.

Step 4: Slack in that formula is (due date − tail − current processing time).

Botflow-2 (Weighted Tardiness)

Step 1: Run Botflow-1.

Step 2: Use the resulting tails as improved estimate of tails, and run again.

Step 3: Run until the objective function has not improved for a specified number of iterations; stop.

Botflow-3 (Weighted Tardiness)

Step 1: Now try the permutation sequence from Botflow-2 with each machine in turn as the actual bottleneck.

Step 2: When an improved solution is found, use this sequence to generate the next set of solutions, and so on.

Neighborhood Search

The discussion is almost identical to that for the weighted flow case. Now calculating the expense of an interchange is even somewhat more expensive, but conceptually there is no problem. Use of tabu search or simulated annealing remains straightforward.

Beam Search

Beam search is probably best considered in connection with a low-cost relatively accurate heuristic such as the dispatch version of R&M. See the discussion for the weighted flow case.

We turn next to dispatch heuristics.

Dispatch Heuristics. First, we present a few of the classic heuristics that have been considered for the weighted tardiness problem in flow shops (and job shops).

Classic Dispatch Heuristics. In all cases l_{jk} is an a priori static estimate of the total lead time of job j from arriving at machine k until finishing. Typically, this estimate was either the sum of remaining processing times or a standard multiple (such as 3.0) of remaining process time.

Random

FCFS. Process the first job to arrive at the machine. (In dynamic shops this is similar to the due date rule since jobs tend to arrive in the order they are due.)

WSPT. Use the myopic policy for the flow case; that is, $\pi_{jk} = w_j/p_{jk}$. (This is good in a heavily loaded shop; we could also do bottleneck versions.)

EDD-Global. Do the earliest due date first; that is, $\pi_{jk} = -d_j$. (This is good in a lightly loaded shop; it is easy to justify on the floor.)

S/OP. Do the job with the least slack per remaining operation; that is,

$$\pi_{jk} = (d_j - l_{jk} - t)/(m + 1 - k)$$

(This tries to decide whether or not the job has enough slack.)

Critical Ratio. Process first that job whose slack is the smallest percentage of the remaining time until the due date:

$$\pi_{jk} = 1.0 - (d_j - l_{jk} - t)/(d_j - t)$$

COVERT-Weighted. If the slack is much greater than the lead time, the priority is 0; it rises linearly to w_j/p_{jk} as the slack goes to zero:

$$\pi_{jk} = (w_j/p_{jk})[\ 1.0 - (d_j - l_{jk} - t)^+/hl_{jk}\ ]^+$$

R&M–One Pass. This is somewhat similar to weighted COVERT, except that we have (a) exponential die off and (b) normalization of a multiple of process time instead of lead time.

$$\pi_{jk} = (w_j/p_{jk})[\exp\{-(d_j - l_{jk} - t)^+/k'p_{k\ av}\}\ ]$$

Newer Dispatch Heuristics. Several of these heuristics can be improved in two different ways:

(a) Lead time iteration (discussed in Chapter 10). For example, in the R&M heuristic above we estimated the lead time (tail) as a constant times the remaining process time. Just as in Botflow-2 for weighted tardiness, however, we can use the output of the simple R&M to give us better estimates of the lead times, and run the problem an additional number of times, until there is no improvement.

(b) Bottleneck dynamics. Estimate machine prices R_i by some method (such as our queuing approximations). In formulas with the term w_j/p_{jk}, replace this with $w_j/\Sigma_{i=k,m} R_i p_{ji}$).

Computational Study for Flow Shops with Weighted Tardy Criterion.

Vepsalainen and Morton (1987) did a study comparing many of these heuristics in a flow shop. Since bottleneck dynamics had not been fully developed at the time these studies were undertaken, there is no report on bottleneck dynamics versions of the heuristic. However, lead time iteration was investigated and performed well consistently.

In a preliminary study comparing the classic rules without lead time iteration, COVERT and R&M–one pass consistently outperformed other classic rules by a large margin. Since R&M–one pass also consistently outperformed COVERT, although by a smaller margin, it was decided to add R&M–iterated to the set of heuristic rules and perform the main study. (COVERT had originally been developed in unweighted form for the mean tardiness problem; the authors made the obvious conversion to weighted form to give it a fair footing.)

Testbed. Given the computational complexity, the authors did not attempt to compute optimal solutions or even lower bounds. They tested the following heuristics: FCFS, WSPT-local, EDD-global, S/OP, COVERT-weighted, R&M–one pass, and R&M–iterated.

Shops could have four or eight machines; increasing, decreasing, constant, or alternating speeds in going through the shop; 20 or 60 jobs; and low or high variation in processing time. The "tardiness factor" τ was set at 0.3 or 0.6 (roughly 30% were tardy or 60% were). The range of due dates, R, was set at 0.6 or 1.6; the due dates were correlated with the tardiness factor. Job weights were uniform between 1.0 and 2.0. There were 10 replications with a factorial design, giving 1280 problems. Since optimal solutions or lower bounds are not available, we report percent deviations from the best performer in each problem. Table 14.5 summarizes these results.

To summarize these results: FCFS, EDD, and S/OP all average more than twice the weighted tardiness of the best policy and hence should not really be considered viable for a tardiness based shop. At a 75% penalty for highly loaded, highly variable shops, the substantial simplicity and acceptance of WSPT seems hardly worth the price. The difference between COVERT and R&M–iterated is about 16% for this case. About half of this seems due to the superiority of R&M over COVERT, and about half seems due to the extra accuracy (and complexity) of iteration. Thus, if a fast heuristic is desired, one might choose R&M–one pass; R&M–iterated remains the most sophisticated dispatch heuristic now tested for high computational use. We repeat that the even more sophisticated bottleneck dynamics heuristics have not yet been tested for the weighted tardiness flow shop. However,

TABLE 14.5 Comparison of Heuristics for a Tardiness Based Flow Shop

Heuristic	$\tau = 0.3$ $R = 0.6$	$\tau = 0.3$ $R = 1.6$	$\tau = 0.6$ $R = 0.6$	$\tau = 0.6$ $R = 1.6$	Total
FCFS	627.0	241.1	148.0	246.0	281.0
WSPT	137.0	110.5	12.1	75.2	75.3
EDD	189.0	87.4	112.0	137.0	126.0
S/OP	227.0	138.0	110.0	133.0	127.0
COVERT	54.9	104.0	10.1	12.5	16.3
R&M–one pass	17.8	34.0	7.9	6.9	8.7
R&M–iterated	0.0	0.0	0.0	0.0	0.0

Note: Table gives average (%) deviation of each heuristic above the best performance for each problem.

there is computational experience for the job shop and for project management, which will be reported in later chapters.

(Table 14.5 exhibits the difficulty of presenting results in terms of percent errors. For example, the absolute differences in tardiness between COVERT and the myopic rules are much more constant between high and low tardiness cases than the percent differences.)

Vepsalainen and Morton also made a second study that investigated the robustness of their results to misspecification of the criterion. They tested the same set of problems and heuristics against three tardiness measures. Overall rankings were very similar for the criteria of (a) weighted tardiness, (b) maximum tardiness, and (c) percentage of jobs tardy. However, the two inventory measures, work in process and work in system, were rather different in ranking from the tardiness rankings and different from each other. (Work in system charges inventory at least until the due date and thus does not reward earliness as does work in process.) For work in process, WSPT comes in first, followed by the two R&M rules, then EDD. For work in system, EDD comes in first, followed by S/OP, and the two R&M rules. (Note that the R&M rules seem more robust than COVERT.)

Another useful reference for this section is Ow (1985).

14.3A Numerical Exercises

14.3A.1 Consider the following three-machine, five-job flow shop problem:

Job j:	1	2	3	4	5
p_{j1}	6	3	2	5	1
p_{j2}	3	6	4	5	2
p_{j3}	2	4	4	6	2
w_j	4	1	1	1	1
d_j	8	22	15	18	26

(a) Give an integer programming formulation for flowtime objective. (Use Proposition 2.)

(b) Show how to modify your answer for a tardiness measure.

14.3A.2 Solve the LP relaxation to obtain lower bounds for your problem in 14.3A.1 for (a) the flowtime objective and (b) the tardiness objective.

14.3A.3 For the same data in 14.3A.1, for the weighted flow objective, use existing software or write a spreadsheet to solve the problem by the following heuristics: (a) Botflow-1, (b) dispatch myopic, and (c) dispatch OPT-like.

14.3A.4 For the same data in 14.3A.1, for weighted flow objective, solve the problem by the bottleneck dynamics heuristic using machine prices equal to the square of the total load (total processing time on that machine).

14.3A.5 For the same data in 14.3A.1, for weighted tardiness objective, use existing software or write a spreadsheet to solve the problem by the following heuristics: (a) Botflow-1 and (b) Botflow-2.

14.3A.6 For the same data in 14.3A.1, for the weighted tardiness objective, use any available method to solve the problem by the following heuristics: (a) FCFS, (b) WSPT, (c) EDD, (d) S/OP, and (e) COVERT-weighted, $h = 2$.

14.3A.7 For the same data in 14.3A.1, for weighted tardiness, solve the problem by the following heuristics: (a) R&M–one pass, $k' = 3$; and (b) R&M–iterative.

14.3B Software/Computer Exercises

14.3B.1 Write a neighborhood search routine for flow shops that allows for more than one objective function, with a tabu search option. Generate several problems and test your routine's performance against some good simpler heuristics available from the numerical exercises above.

14.3B.2 With your own software or available software, set up an interesting set of test problems, and test a number of heuristics for weighted flow objective in the flow shop.

14.3B.3 Repeat 14.3B.2 for the weighted tardiness objective.

14.3C Thinkers

14.3C.1 Discuss heuristics for the $F_{wt} + aT_{wt}$ problem, where a is a given weight (mixed weighted flow and tardiness objective).

14.3C.2 Discuss heuristics for the early/tardy problem.

14.3C.3 Discuss heuristics for the number of tardy jobs problem.

14.4 MODEL EXTENSIONS

14.4.1 Overview

There are a number of simple variations on the simple flow shop, which might be called "skip shops," "reentrant flow shops," "compound flow shops," and "finite queue flow shops." Many of these variations are difficult to treat with integer programming, branch-and-bound, or dynamic programming. Thus there is no particular reason for exact methods to classify these as being in the flow shop family. However, these variations are often relatively easy to deal with in dispatch heuristic approaches, especially bottleneck dynamics, by making appropriate minor changes to the flow shop heuristics. Hence we are led to expand the concept of the (general) "flow shop." We explore these issues in this section.

14.4.2 Skip Flow Shops and Reentrant Flow Shops

Skip Shops. Recall that, in a skip shop, some jobs may skip some machines. That is, the job goes down the same ordered list of machines but visits each machine zero or one time. In dispatch heuristics this is easily handled. Fudge the situation so that the job visits all machines. Give the job zero processing time at the machines it does not really visit. Modify the heuristic (if necessary) so that all jobs with zero processing time have a very very large priority, can preempt existing jobs, and thus are processed before any normal job. Otherwise, do not change the heuristic.

Note that this trick does not work with permutation sequence heuristics, which require the same sequence of the jobs at every machine. Also, this trick does not work if significant transportation times are involved from machine to machine. (When all jobs have transport between machines k and $k + 1$, the transport can be treated as another machine in the flow shop, either considered as very high capacity resulting in a fixed lag, or with its own queue. However, if a job skips a machine, its transport routing is likely to be affected; the appropriate model is likely to become a job shop model.)

Reentrant Flow Shops. We turn now to reentrant flow shops (e.g., with annealing furnaces) where some machines may be visited more than once. For concreteness, consider a flow shop with a single reentrant machine. Looking at its queue, call any jobs visiting the first time "first pass" jobs, the second time "second pass" jobs, and so on.

It again seems somewhat difficult to modify permutation sequence heuristics for this case. The simplest equivalent of a permutation sequence discipline would seem to be: schedule the first job in the permutation sequence on all machines, including the multi-pass machines. Next, schedule the second job, and so on, using gaps in the program wherever possible, and so on. This would allow all the properties of permutation sequences to be employed. The difficulty is that gaps are likely to be left in the scheduling, causing considerable inserted idleness.

A slight modification of the permutation sequence discipline given above would turn it into a dispatch method and hence make it amenable to the modified dispatch heuristics discussed below. The inserted idleness problem should be largely eliminated. At the beginning, assign each job priorities 1, 2, 3, and so on in order of its position in the permutation sequence. Now schedule the shop by a dispatch procedure: whenever a machine becomes available, schedule the currently available job with the highest priority. (The different pass issue will not arrive since it does not affect priority.) The disadvantage here is that a partial permutation schedule cannot be calculated; a full simulation of all jobs must be carried out to evaluate a particular permutation.

All dispatch heuristic methods are easily modified for the reentrant case. Whenever a machine becomes free, we simply calculate dynamic priorities for the available jobs and schedule as before. The same formulas for calculating dynamic priorities in all cases may be used as before. A little care must be taken in reinterpreting some of the inputs to these formulas. Such factors as job due date, job weight(s), and current operation processing time are unchanged. In assessing future processing times, each pass at a given machine will be treated as a separate operation at a separate machine. In assessing present, future short-term, or future long-term loading at a machine, all passes must be added together and totaled. OPT-like tails will be the sum of all future processing times including totaling all times for which a job visits a machine. Starting tails in other procedures will be a fixed multiple of this tail. Lead times in lead time iteration will use for this iteration the actual lead time from this machine to the end from the previous iteration.

Dispatch formulas may be modified as desired at multi-pass machines, so that, for example, early pass operations may be given higher priority, or later pass operations, and so on.

14.4.3 Compound Flow Shops

Remember that a compound flow shop is a linear sequence of m compound machines. Each compound machine k is a cluster of q_k parallel machines. (In the following we may suppress k for clarity and talk of a compound machine with q parallel machines.)

Equal/Proportional Machine Clusters. The case of equal or proportional machines is easily handled by either permutation sequence or dispatch heuristics. We review the discussion in Section 11.3.2.

Let machine i in the cluster have speed s_i, where $s_i = 1$ would be a "standard machine." Define $q' = \Sigma_{i=1,q} s_i$ as the "effective number of machines in the cluster."

(a) Proposition 1 of Chapter 11 says there is an optimal sequence for the jobs such that each can be scheduled in turn on the first available machine.

(b) We now have considerable experience suggesting that modifications of rules optimal for the preemptive case usually make good dispatch heuristics for the nonpreemptive case.

(c) Proposition 3 of Chapter 11 suggests that for the free-preemption case, clusters may each be optimally aggregated into one machine having q' times the speed of a standard machine.

Thus we are basically reduced to a simple m-machine flow shop problem, which we may solve by any heuristic, using the aggregated faster machines to estimate priorities. (See Section 11.3 for details.) Once priorities have been established, however, calculation of partial permutation sequences or dispatch solutions proceeds for the actual detailed shop, using the discipline that the highest priority job at a compound machine is always assigned to the simple machine within it which can finish that job first.

General Machine Clusters: Proportional Approximation. The case in which machines within a cluster vary considerably in their relative capacity to do certain kinds of jobs (they have comparative advantages for certain types of jobs rather than just absolute advantages) is much more difficult. There are really only two obvious choices for high quality dispatch heuristics for this case:

(a) Approximate it in some fashion by the equal/proportional case.
(b) Utilize bottleneck dynamics.

It is dangerous to approximate cases by the equal or proportional case unless a good deal of care is taken. For example, a common real-world situation is that a job's processing time is the same on all machines in the cluster, except that some machines are not capable of running it at all. It is tempting to treat this as the equal machine case, treat each cluster as one machine q_k times as fast, and proceed by any good heuristic.

It is easy to give an example to show where this works poorly. Suppose there are one new machine and one old machine in the cluster. The old machine only handles the three original types of jobs. The new machine can handle the three original types and also the seven newer types. Suppose the new machine has the capacity to handle the newer type jobs, with only a little to spare. If an original type job comes up with highest priority under the equal price approximation and the new machine is available, it will be assigned to the new machine. However, it is very likely expensive to assign the job to it. It should be held for the old machine, so that the new machine can get on with the newer type jobs.

(It is possible to construct a pathological example for which the error in using the equal machine heuristic as an approximation is arbitrarily large.)

Thus the equal or proportional machine heuristics should never be used for the "forbidden machine" cases without a rather careful study showing that these issues are not important. However, if all jobs can be run on all machines, and the deviation from proportionality is no more than 20–30%, approximating the situation as proportional may be fairly reasonable. (No decent computational studies on this question are available at this time.)

Clustered Parallel Machine Stages: Bottleneck Dynamics. An in-between case is that while each stage of the flow shop is not as simple as proportional machines, it is not as complicated as general parallel machines, so that more complex pricing techniques need not be employed. A very important case of this type is where the stage is divided into several clusters, each of which has a unique input stream and (at least approximately) proportional processing times. This problem is readily solved and will be seen to be important in the nuclear fuel tube case that follows.

I. Preliminary

Step 1: Add up speed factors on each cluster k with q_k machines to obtain q'_k, the "equivalent number of machines" for that cluster.

Step 2: At each such aggregated machine, determine effective processing times $= 1/q'_k$ times old speeds.

Step 3: Set up a simulation program for this generalized flow shop with a decision subroutine for sequencing at each queue. (At each "aggregate machine" the program assigns the highest priority job in queue to the machine that can finish it first.)

Step 4: The program should keep a detailed enough record of the simulation to provide input to the pricing and lead time program to be described.

II. Starting Iteration

Step 1: Choose any simple sequencing heuristic to run the program the first time (e.g., FCFS or WSPT).

Step 2: Run the simulation with this heuristic.

III. Post-iteration Pricing and Lead Time Estimation

Step 1: From the previous iteration trace, determine the actual lead times for each job from leaving each stage until completion. Store as a table.

Step 2: At the time of each job arrival at each appropriate cluster in the problem, determine the actual jobs in the corresponding busy period for that cluster at that point of time, and their urgencies at their projected times of processing. (Depending on the pricing method, substitute "current line" for "busy period.")

Step 3: Calculate estimated prices for each cluster at each of the job arrivals for the previous iteration.

IV. Updating Iteration

Step 1: Switch to the full bottleneck dynamics heuristic.

Step 2: Simulate.

Step 3: At any sequencing decision at any cluster, use prices and lead times for a job arriving at a cluster from the last post-iteration analysis to make the sequencing decision.

V. Termination Decision

Step 1: Terminate if the termination criterion is met (no improvement in objective function last five iterations, total number of iterations, etc.)

Step 2: Otherwise return to III.

While the structure we present here has been implemented and tested for the simple case, it needs testing/validation for the cluster case.

General Parallel Machine Stages: Bottleneck Dynamics. Employing the bottleneck dynamics approach for the most general compound flow shop always involves the following components, which will require various types of estimation procedures, some iterative.

(a) Pricing the machine cluster k.

(b) Pricing machine k_i within cluster k.

(c) Estimating actual head times (arrival times) to cluster k.

(d) Estimating actual tail times (lead times) from cluster k.

(e) Developing sequencing rules at k.

One possible way of proceeding (not tested) would be as follows.

I. Starter Heuristic

Step 1: Simulate the entire problem by an initial heuristic, such as approximating by proportional machines, where proportions are estimated by average relative times on all jobs.

Step 2: Save arrival times and derived local due dates (global due date minus lead time) for each job, at each machine:

Step 3: Calculate an equivalent aggregated machine processing time for each job, at each machine (actual processing time for a job at the machine times the machine's rough proportion).

Step 4: Solve the linear programming relaxation at each machine k to obtain estimated machine prices R_{qk}.

Step 5: Add R_{qk} to obtain R_k.

II. Iteration

Step 1: Run a dispatch simulation of the problem.

Step 2: As time progresses determine and mark the simple machine "soon to be free" anywhere in the system. (Several exact definitions could be tested.)

Step 3: At specified time points, look at each compound machine with marked simple machines and determine the next operation to assign to the first machine which can finish it.

Step 4: Repeat step 3 as feasible.

Step 5: Do for all machines.

Step 6: Stop when all jobs are complete.

III. Stopping Criterion

Step 1: Has the stopping criterion (number of iterations and/or no improvement) been reached?

Step 2: If yes, exit routine.

Step 3: If no, continue.

IV. Update

Step 1: Repeat I, using the last iteration as the heuristic.

Step 2: Repeat from II.

It is important to repeat that this is a research plan at this point and has not been tested. There are a number of variations possible, each of which would need to be evaluated. For example, instead of using step 3 of I, one might use crude estimation to get the proportion of the total price for each machine in the cluster, and then use ex post busy period analysis for each machine after each iteration to estimate the cluster price. It is difficult to know which would work better without testing.

14.4.4   Finite Queue Flow Shops

The finite queue flow shop has limited storage in front of machines other than the first. In general, this is an extremely difficult problem analytically, since if a queue fills up at machine k, then machine $k - 1$ will have to hold the currently processing operation rather than start another one. This is called "blocking." In some cases a single overworked machine could cause blocking for several stations upstream. However, we can give some treatment for this case via heuristics.

An easier and very important special case is when there is no intermediate storage at all, simply due to the layout design or the fact that delays would make the process infeasible, such as working hot metal.

No Intermediate Queues. Since there is only one queue, we actually have a single-resource, serial machine problem, which can be optimized as a permutation sequence.

More specifically, we are assuming dynamic availability a_k for machine k, dynamic arrivals r_j, and due dates d_j (if any) for job j, with processing times p_{jk} of job j on machine k. The lack of waiting simply says that the start time of a job on machine k is equal to its ending time on machine $k - 1$. If a job i starts on machine one at a certain time, then clearly it completes on machine m later by a time precisely equal to the sum of its processing times. Also, the start of the next job in sequence may have to be delayed enough so that it cannot overtake the previous job and cause a conflict. Also, if it hasn't arrived yet to the first machine by that point, there will be a further delay.

Proposition 5. (Wismer, 1972). Consider the no-delay (no queue) flow shop problem with dynamic availability, dynamic arrivals, and any regular objective. (T_{ij} is literally a start delay; however, it has the mathematical structure of a sequence-dependent setup time; hence we use that notation to recall theorems and techniques from Chapter 8 that apply.)

(a) If job i starts at time s_i, it will complete at $s_i + \Sigma_{k=1,m} p_{ik}$.

(b) If job i starts at s_i, and job j is sequenced next and arrives at $r_j \le s_i + T_{ij}$, then job j will start exactly at $s_j = s_i + T_{ij}$, where T_{ij} is independent of other jobs and may be calculated by

$$T_{ij} = \max_{k=1,m} \left[\sum_{k'=1,k} p_{ik'} - \sum_{k'=1,k-1} p_{jk'} \right]$$

(c) In general, if the job arrives at time r_j,

$$s_j = \max [s_i + T_{ij}, r_j]$$

Proof. This is left as an exercise. As a hint, the first term in the maximum in (b) is p_{i1} and represents the minimum separation needed to avoid overtaking on the first machine, and the kth term is the minimum separation to avoid overtaking on the kth machine. ∎

Proposition 6. Consider the no-delay flow shop problem with dynamic availability, dynamic arrivals, and any regular objective. Optimization of this problem is equivalent to optimization of an equivalent one-machine problem with the same objective and arrival times, due dates earlier by the total processing time for the job, an artificial initial job to represent the offset starting availability, zero processing times, and sequence dependent setup times T_{ij}. The setups are nonstandard in that they may start at the completion of the last job i rather than at the arrival of job j.

Proof. Left as an exercise, with special attention to the construction of the artificial first job. ∎

Proposition 7. Consider the no-delay flow shop problem, with makespan objective.

(a) With static availability and arrivals and cyclically repeating jobs, this is equivalent to the classic traveling salesperson problem with distances T_{ij}, considered in Section 8.2.4.

(b) With dynamic availability and arrivals, this is equivalent to a traveling salesperson problem with specified starting city and free ending city, where a city

j cannot be visited before a time r_j, which may be handled by modification of any methods in Section 8.2.

Proof. Obvious.   ■

One rather simple heuristic for dealing with a problem of the form of Proposition 7(b) would be to first solve the problem myopically (choose the job causing the smallest delay each time) and then to perform pairwise interchange.

For the problems with some other objective added, the myopic plus pairwise interchange approach should also work well. The bottleneck dynamics approach can also be adapted.

Finite Intermediate Queues. Even if we consider only permutation sequences, the existence of finite buffers between stages complicates things a great deal. It is no longer true that the lag which must be allowed between starting jobs only depends on the two jobs in question. It also depends on the state of each of the queues, which in turn depends on earlier jobs.

For simplicity, we consider only the makespan problem here. Once the queue fills up at a machine, further work cannot proceed from the previous machine and will block it. (One possibility is to have a timing discipline that any new job will be added precisely at the earliest time possible which causes no blockage during its passage.) Start with an empty system. Simulate adding a job, keeping track of the complete system state at discrete points in time. Try to add any job; determine the earliest time for which the job may be started. Try this for all possible jobs, and myopically actually schedule the job that can be entered earliest.

Once a complete myopic schedule has been formed in this fashion, pairwise interchange can be performed. (Calculation of the effect of an interchange is expensive, however. Each iteration requires simulating on average about one-half of a given full schedule.)

A useful reference for this section is Park and Steudel (1991).

14.4A Numerical Exercises

14.4A.1 Set up a skip shop as follows. Modify the problem in 14.3A.1. Let job 1 skip machine 1, job 2 skip machine 2, and job 3 skip machine 3. Let jobs 4 and 5 skip no machines. Assume a weighted flow criterion. (a) Solve the problem myopically. Let transport time from machine 1 to 2 be four; from machine 2 to 3 be one; from machine 1 to 3 be five. Argue that WSPT will give a good answer. (b) Now let transport time from any machine to any other be 4. Show the WSPT solution is not likely to be good logically. (c) Give a better solution if you can.

14.4A.2 Consider again the problem in 14.3A.1 without transportation times. Modify the shop so that all jobs must repeat machine 1 a second time after machine 3.

(a) Solve by WSPT suitably modified.

(b) Add due dates and change the objective to minimize maximum lateness. Use the following heuristic: earliest machine due date first, where machine due date is global due date less 3 times remaining processing time.

14.4A.3 Consider a shop with the shop structure of 14.3A.1 except that each machine is expanded to a cluster of two equal parallel machines. Duplicate each job in that shop three times and solve the problem myopically.

14.4A.4 Consider again the data of 14.3A.1, except change the objective to makespan, and assume no intermediate buffers.

(a) Find the myopic solution.

(b) Do at least one pairwise interchange to understand the effort level.

14.4B Software/Computer Exercises

14.4B.1 Write a simple simulation of a reentrant shop, with facilities for easily adding new input data and/or heuristics, for weighted flow and/or maximum lateness. How sensitive are your results to the exact way you estimate lead times?

14.4B.2 Write a simple simulation of a flow shop for $m = 2$, with each "machine" compoundly made up of equal parallel machines. Can you determine by experimentation how accurate the idea of aggregating clusters into an individual machine really is?

14.4B.3 Write a simulation of a finite queue flow shop that simply allows you to select the next job and specify when to start it, telling you if your choice caused blockage, or allowed too much idle time. Is it easy to make accurate guesses?

14.4C Thinkers

14.4C.1 "In a reentrant shop where the reentrant machines have low utilization, they may fairly accurately be simply treated as two machines." Formalize and justify this statement as precisely as you can.

14.4C.2 "In a parallel machine cluster where job times aren't 'too far' from proportional, machines may be treated as proportional without 'too much' loss." Set up a formal model where the deviation from proportionality is quantified, and argue that the percent error of such an approximation is no worse than linear in the deviation.

14.4C.3 Prove Proposition 5.

14.4C.4 Prove Proposition 6.

14.4C.5 Dream up and state formally three other different model extensions to the basic flow shop model.

14.4C.6 Choose one of your model extensions. Develop a simple heuristic to solve it. Develop a more complex heuristic to solve it. State your heuristics carefully and give intuitive justification for them.

14.4C.7 Do limited testing of your heuristics in 14.4C.6. Which would you use in practice, and why?

14.5 CASE: NUCLEAR FUEL TUBE SHOP

14.5.1 Overview

We have now had an exposure to a number of types of flow shops, both simple, compound of various degrees of difficulty, and otherwise complex, together with a variety of exact and heuristic methods for modeling them. It is perhaps time to present a real-world case, to see how these ideas would work in practice, and how they might need to be modified.

We are going to consider the nuclear fuel tube shop at General Metals. (Names and some of the data have been changed for proprietary reasons.) We present a very brief summary of the problem here in the overview, a longer verbal description and diagram in Section 14.5.2, followed by some analysis of the problem in Section 14.5.3.

The nuclear fuel tube shop is part of the Plattsville Plant of General Metals in southern Ohio. It produces seamless zircalloy tubes about $\frac{2}{3}$ inch in diameter and 50 feet long for nuclear reactor applications in a different division of General Metals located in northern Georgia (Atomics General). Rated at 11 million feet of tubing per year, the facility produces tubes of several diameters, sizes, and types to close specification and tight standards. The application of the product demands high levels of performance due to safety considerations.

The manufacturing process is as follows. Each incoming tube blank (1000 lb) is *extruded* or *pilgered* through a die four times to gradually reduce the diameter and is cut repeatedly to finally produce about 950 one-pound tubes. Each time the diameter is reduced is called a "pass." Each pass consists of pilgering, deburring the ends, pickling with acid to clean the tube, and annealing in a furnace to reduce the stresses of pilgering and obtain required mechanical properties. After the four passes, there are final straightening and finishing operations. At various stages of the manufacturing process, the products have to undergo tests for mechanical, chemical, metallurgical, and corrosion properties. Some of the complexities of this manufacturing process include major and minor setup times for the pilgers, ability to logically move some pilgers to different stages in the process, batch processing (annealing furnaces), and ability to handle occasional overflow with a second furnace. The cycle time for the four passes plus straightening and finishing is about 40 days.

Management is very anxious to reduce work in process (WIP); there has been a mandate from the JIT-minded central corporation to cut WIP by 30%! There are also due dates, which are somewhat soft since Atomics General maintains large lot-sizing inventories, due to the high setup cost of pilger changeover. Release of orders to the job floor is handled by a committee composed of a high-level planner from the tube shop and a high-level planner from the nuclear division downstream.

14.5.2 Details of the Case

General. This shop is basically a compound flow shop with reentrant flows, some batch machines, and a couple of other complications that will be dealt with later. The flow line is representable roughly as four passes in series and with the first three passes having three complex machines in series, the last pass having just pilgering. Each of the first three passes has the same sequence; the last pass has just the first (pilgering).

(a) *Pilger.* Extrude diameter of tube to a smaller diameter and cut. A pilger stage consists of multiple pilgers in parallel. Pilgers have sequence-dependent setup times. (It is a minor setup for the same tube family, and a major setup for a different tube family.) Some pilgers can be set up in different ways to be used in different passes.

(b) *Clean/Debur.* Clean to remove dirt; immerse in acid bath for final pickle/cleaning; remove tailings from cut. There is usually no debur bottleneck, so we may approximate this stage as parallel machines with high total capacity, leading to a fixed delay through this stage.

(c) *Annealing.* Heat to restore crystalline structure; this is a batch process. Up to 12 lots, which must be of the same family, are annealed at once. *Furnace is reentrant.* Usually a single furnace serves all three passes. Compatible lots from different passes can be annealed simultaneously.

A diagram of this flow shop can be given in two different ways, depending on how much we wish to emphasize the reentrant feature. These two approaches are shown in Figure 14.1.

First Pass Pilgers. This extrusion produces 1.95-inch diameter tubing for all products. Pilger #7 is dedicated to the first pass, and usually #12 also, each handling different families of jobs. First pass takes 3 or 4 hours processing time.

The diagram illustrates the progress of a single job through numbered machines. Machines 1, 4, 6, and 8 are pilgers, machines 2, 5, and 7 are clean/debur, and machine 3 is the annealing furnace.

Second Pass Pilgers. This extrusion reduces 1.95 to 1.20 or 1.45 depending on the product family. There are four pilgers sometimes used here, usually two will be employed. Second pass takes 5–6 hours.

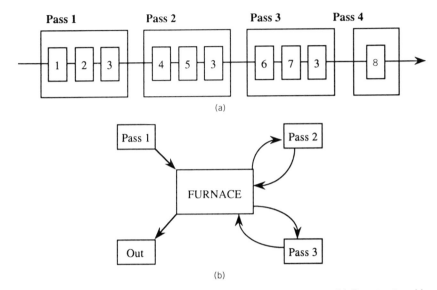

Figure 14.1. (a) Reentrant problem emphasizing flow shop structure. (b) Reentrant problem emphasizing multiple flows to furnace.

Third Pass Pilgers. This extrusion reduces 1.20 to 0.71 or 1.45 to 0.90. There are typically four pilgers here, although other pilgers can be brought in to give extra capacity as needed. Third pass takes 15 hours.

Fourth Pass Pilgers. This extrusion reduces tubes to final diameters. There are typically 12–18 fourth pass pilgers. Fourth pass takes 60 hours.

Annealing Furnace. There is one annealing furnace, #33, which usually does all the mainline products for all three passes (reentrant batch machine). It handles a 4-lot minimum and a 12-lot maximum per batch. Processing time is 9.5–15 hours per batch.

Backup Furnace. There is an entirely separate flow line for certain special products, such as tumble tubes and Matushi orders, which, for quality reasons, are processed separately. This line uses an older annealing furnace (#3) with different procedures and a longer processing time (20 hours instead of 15). This older furnace has excess capacity. Until recently, management has resisted using the backup furnace for the mainline products, citing the longer processing time, some quality questions, and difficulty in having enough product to fill the furnace.

Clean/Pickle/Debur. Initially, it was felt that this type station would never be a bottleneck, so it was modeled at each stage simply as one large capacity machine with no queue and a fixed 5-hour delay. We continue to do this in our analysis. If this area were to become a bottleneck at times, a fairly complex cell would have to be modeled.

Arrival, Timing, and Capacity Issues. To study long-term capacity issues we note that processing goes on for 24 hours/day, 6 days/week, with Sunday as backup for the furnace. (Management resists Sunday use in most cases.) Four lots per day or 24 lots per week need to be processed though the system.

Each pilger at first stage has a capacity of at least 36 lots/week, so first stage will have much overcapacity. Each pilger at second stage will have a capacity of 24 lots/week, so second stage will not likely be a bottleneck either. Third stage pilgers can handle about 10 lots/week, so utilization is up to 65% on average. *Thus third stage can be a bottleneck.* Fourth stage pilgers can handle 2.4 lots/week, so with 10–12 pilgers capacity is 24–28 lots/week. *Thus fourth stage will usually be a bottleneck.*

The main furnace takes 9.5–15 hours, averaging about 12, and holds up to 12 lots. However, each lot requires three passes in the furnace, so in the multi-pass sense the furnace holds only 4 lots. Thus its capacity would be 48 lots for 6 days/week. Suppose the furnace averages 6 jobs; it would be at 100% capacity at 6 days/week. Thus the main furnace seems a logical candidate as principal bottleneck.

Complicating Issues

(a) *Movable Pilgers.* The fact that pilgers can be used in differing parts of the process by paying a larger setup cost means that the problem is formally a job shop with complicated routing. However, such a generalist approach would be likely to reduce the flow shop insights. We choose to solve the problem for a given pilger configuration, and then consider pilger movement as an occasional ad hoc modification.

(b) *Backup Furnace.* If there are two flow shops, each with a furnace, and rules for which products go on which line, but with some complicated rules for sharing furnaces in emergencies, this again becomes a job shop problem with variable routing and rules about the use of two furnaces. We choose to solve the problem with one furnace whose capacity is varied parametrically to simulate the availability of various types of extra capacity.

(c) *Interchangeability of Lots.* Many times a customer order will require several lots of the same type, so that WIP can be reassigned from one customer to another on the floor according to urgency. This means there won't be a unique due date.

(d) *Variability of Furnace Processing.* Annealing is a complex chemical process with many dials and gauges, requiring full attention of a process engineer. The run may occasionally be recalibrated and partially redone in the middle, or occasionally aborted.

(e) *Six-Day Week.* The simulation must be modified in some way to account for one down day per week, and jobs should not be shown as being split across the gap.

(f) *Pilger and Furnace Data.* The scheduler says the pilgers rarely go down, and furnace breakdowns can be treated by increasing process times. Is this true?

(g) *Preventive Maintenance Intervals.* Preventive maintenance is not used much. This is partly due to the fact that maintenance needs are not very predictable and partly because preventive maintenance is difficult to do.

14.5.3 Analysis

First Cut Simplifications

(a) Ignore the backup furnace and the few products dedicated to it. Instead, treat furnace capacity as a parametric variable, so that the importance of the backup furnace will be captured. This avoids a more complicated model with job shop and routing structure. This seems justified for a couple of reasons: (i) schedulers have reservations about using the backup furnace and (ii) the problem structure then becomes simpler.

(b) Ignore the possibility of moving pilgers to differing stages, and ignore the possibility of making major setup changes during a simulation. This produces clusters of pilgers at a stage, each of which is composed of equal machines, a case we have seen is easy to handle. Treat the minor setups (1.0 hour between different members of a family) myopically by bottleneck dynamics. This is also justified for a couple of reasons: (i) moving pilgers and major setups seem to be handled hierarchically at a higher level by the planning committee; thus it is not clear that the simulation would have the right to make these decisions; and (ii) the problem structure then becomes simpler.

(c) Assume deterministic processing times with no breakdowns. This is required to get started with this approach. Sensitivity to the assumption can be tested.

Basic Approach. The resulting problem is a compound flow shop with one reentrant batch machine (the annealing furnace). There are 10 stages, which split 3,3,3,1 into passes. The first stage of each pass is a group of parallel pilgers, clustered into equal machines that handle a subfamily of jobs. There is a minor setup between differing members of the subfamily. The second stage of each pass is a pickling phase, treated as a single high-capacity machine. The last stage of the first three passes is the annealing furnace, which takes 4–12 jobs at a time and has rules as to which jobs are compatible at the same time. To simplify things, assume the furnace will not wait for more jobs once four or more are available and the furnace is free, but must wait when three or less are available.

The overall iteration, pricing, and lead time structure we use are exactly those we presented for the clustering case in Section 14.4.3 on compound flow shops. The appropriate objective would seem to be either weighted flow (high desire to reduce WIP) or a combination of weighted flow and weighted tardiness, both of which we have experience in dealing with in bottleneck dynamics.

There is little else to discuss except the furnace. Remember that in determining the lead time of a job, all three visits to the furnace must be determined. Given our simplifying assumption about loading the furnace, our sequencing procedure for the furnace would become:

Step 1: Prioritize all jobs as if the furnace were 12 parallel simple machines at the same speed.

Step 2: For each class of compatible jobs, pick the up-to-12 jobs with highest priority to be the potential load.

Step 3: For each potential load add up the individual priorities to determine the priority for the load.

Step 4: Schedule the load with the highest priority.

14.5A Numerical Exercise

14.5A.1 Consider a simplified version of the case, where there is only one pass, two pilgers with different subfamilies of jobs, a furnace that holds only two jobs that must be of the same subfamily, and weighted flow objective.

 (a) Make up data for a 6-job problem, and solve by WSPT.

 (b) Guess some reasonable prices (lead times not needed!) and solve by bottleneck dynamics.

14.5B Software/Computer Exercises

14.5B.1 Write a simulation for the simplified tube shop in 14.5A.1, using again some crude ad hoc prices.

14.5B.2 Add a pricing routine and iteration to your program.

14.5C Thinkers

14.5C.1 Find a real-world shop, to which you can gain some access, which has a generalized flow shop structure. Discuss its structure and problems in a 5–10-page paper.

14.5C.2 Make and justify whatever assumptions are necessary to make your shop more tractable to schedule. Make a detailed design for scheduling it.

14.5C.3 Write the software to schedule your shop, and test with artificial data.

14.5C.4 Beg, borrow, or steal real data to test and refine your approach. (Offering the software might help.)

____15

SCHEDULING
JOB SHOPS:
BASIC METHODS

15.1 INTRODUCTION

15.1.1 Overview

We see from the nuclear tube shop case of the last chapter that generalized flow shops can have many different types of machines and that the structure of the shop can depart quite a bit from linearity. It is, rather, the general flow of all the jobs in that case which are very similar (identical in the simple flow shop), especially if one aggregates the shop a bit to average out the smaller deviations from this ideal.

The distinguishing characteristic of a job shop, then, is that jobs can have quite different flows and numbers of operations. In a *classic job shop*, job j has m_j operations to be performed in sequence (serial operation precedence structure); each job/operation has been preassigned a unique machine, which may perform that operation; a job does not visit the same machine twice. These assumptions assert that there are no parallel machine clusters, no routing decisions, and no reentrant machines. (Some authors go further and restrict every job to visit every machine in the shop exactly once. This simplification does not simplify exact/heuristic methods appreciably and will not be assumed here without specific statement.)

While we largely confine ourselves in the next two chapters to the classic case, many extensions are relatively easy to make (especially for heuristics) as we have seen for flow shops in Section 14.4. The question of allowing jobs to have a more complex precedence structure is a larger undertaking: we discuss such issues for job shops in Chapter 18 (project job shops) and give the case of a custom optical equipment manufacturer.

We first introduced the classic job shop model in Section 9.2, and we repeat this development in Section 15.1.2 for convenience. In Section 15.1.3 we make a broad classification of the types of schedules one might be interested in. In Section 15.1.4 we briefly discuss how all the schedules of one type might be generated by a procedure. Finally, in Section 15.1.5 we briefly summarize the rest of the chapter.

15.1.2 The Classic Job Shop Model

A job shop has m ($k = 1, \ldots, m$) machines, each of which can do several types of operations but can process only one activity at a time. For the static case, n ($j = 1, \ldots, n$) jobs are available for processing at time zero. Job j consists of q_j ($i = 1, \ldots, q_j$) operations in series. An input table must be given of p_{ji}, the processing time of operation i of job j (Table 15.1). A second input table must be given of k_{ji}, the machine k on which operation i of job j is to be run (Table 15.2).

The pair ji represents operation i of job j, and thus represents a particular activity. If we wish to specify a priority ordering for activities on machine k, we could specify $ji(k, 1)$ as the first activity to go on k, $ji(k, 2)$ as the second, and so on. In Table 15.2, for example, we see that the activities to be scheduled on machine 3 are 31, 23, 13, and 42. A particular fixed sequencing rule might be $ji(3, 1) = 13; ji(3, 2) = 23; ji(3, 3) = 31; ji(3, 4) = 42$.

Anticipating our discussion in Section 15.1.3, if each machine has priorities giving a rigid ordering for jobs to be processed on the machine as for machine 3 above, we say machine 3 is being scheduled by a *permutation sequence* and the shop as a whole is being scheduled by a *multi-permutation sequence*. Note that insisting on following a permutation sequence set in advance could possibly be quite wasteful. In our example, once 13 is completed, suppose 31 and 42 have arrived, but the required 23 does not arrive until a good deal later. The strict rule would require holding the machine idle until 23 finally arrives. In extreme cases, gridlock could even occur. See Proposition 1 of Section 9.2.

One way to make schedules more compact (although not necessarily easier to analyze mathematically) is to simply change the rule to "schedule the highest priority job available when the machine becomes available." This would be called a *dispatch rule*. (Remember that we also discussed the possibility of *extended dispatch* rules in Chapter 10.) Also, given the fact that the priorities were calculated one time in advance, it would be called a *static dispatch rule*. If, on the other hand, the priority is calculated when the sequence choice is made using currently available information, it would be called a *dynamic dispatch rule*. All these kinds of heuristics are used in job shops; we shall discuss this issue further later.

The basic constraints for assigning an activity to start are as follows:

(a) It cannot start until the job's previous operation is finished.
(b) It cannot start until the machine is free.
(c) It cannot start until it is chosen by the sequencing rule.

Going back to our original numerical example, it is useful to realize that small- to medium-sized problems can be understood (and even solved) using Gantt charts, of which there are several kinds. The horizontal axis always represents time, and there

TABLE 15.1   Example: Processing Times

Job	Operation		
	1	2	3
1	4	3	2
2	1	4	4
3	3	2	3
4	3	3	1

TABLE 15.2   Example: Machine Routing

Job	Operation		
	1	2	3
1	1	2	3
2	2	1	3
3	3	2	1
4	2	3	1

are several bars. In the standard machine Gantt chart, each bar represents a machine, and the jobs shown occupy the machine at each point in time. In a job Gantt chart, each bar represents a job, and the machines shown occupy the job at each point in time.

We show four charts for our example problem in Tables 15.1 through 15.4. First we show machine and job Gantt charts, which take no account of constraints (a) and (b) above (predecessors must be finished, only one on a machine at a time). In these figures, the code on an operation is as follows: the first digit is the job number, the second digit is the operation number within the job, the third digit is the machine on which that operation must process.

It is easily seen that Figure 15.1 does not represent a feasible schedule. Predecessor operations for any job are done before the operation itself is scheduled; that is, 111, 122, and 133 schedule operations 11, 12, and 13 in correct sequence. How-

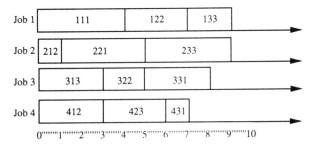

Figure 15.1. Job Gantt chart (infeasible).

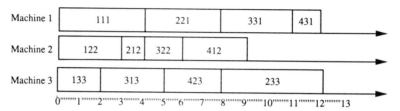

Figure 15.2. Machine Gantt chart (infeasible).

ever, Figure 15.1 repeatedly schedules two or more operations on a machine simultaneously. For example, 212 and 412 are scheduled on machine 2 simultaneously; 221 is scheduled on machine 1 before 111 has finished.

By contrast, in Figure 15.2 machines have been scheduled consistently; there are never two jobs on a machine at one time. However, here many operations have been scheduled before their predecessors have been finished. For example, all operations of job 1 have been scheduled simultaneously; also, the second operation of job 3 starts before the first is finished.

We next illustrate what a feasible schedule for this problem might look like (the student will be asked to show the feasibility of this schedule in the exercises): Figure 15-3 shows this schedule from the perspective of what time the various jobs are on each machine, while Figure 15.4 is from the perspective of when the operations of a given job are processed.

Of course, what we have illustrated is only one feasible solution to the problem. To do better, we would have to specify an objective function for the problem. Then we have to specify procedures to get very good solutions among feasible ones, either exactly or by heuristics.

15.1.3 Types of Schedules

Remember that in the simple flow shop problem there were about $(n!)^m$ schedules to examine in any search for the optimum (total multi-permutation schedules). By various types of arguments and experimentation we were able to focus attention on policies in which the sequence was the same on all machines, giving just $n!$ se-

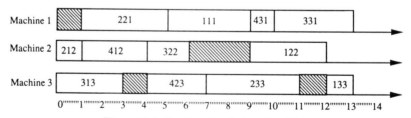

Figure 15.3. Machine Gantt chart (feasible).

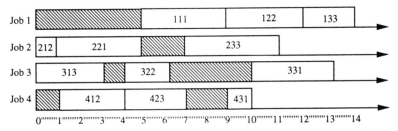

Figure 15.4. Job Gantt chart (feasible).

quences. This at least made it possible to consider branch-and-bound, for example, for small problems.

One problem for exact approaches to the job shop, then, is to find a similar subset of feasible subsets for more detailed examination. If there are on average q operations per job, there may be as many as $[(nq/m)!]^m$ feasible policies to examine, and we should, therefore, concentrate attention on some special types. This is exactly what we do in this section. This is relatively unsuccessful for exact procedures; the resulting number of choices is still much much larger than in the flow case; it is, however, quite useful for many heuristics.

In principle, there are an infinite number of feasible schedules for any job shop problem, because arbitrary amounts of idle time can be inserted. However, this is not useful for a given known sequence and regular objectives.

Definition. A *local left shift* occurs in a feasible schedule if one operation can be moved left to start earlier and keep feasibility.

Definition. A schedule is *semi-active* if no local left shifts are available.

Proposition 1. In a classic job shop problem

(a) Given a multi-permutation sequence, there is only one corresponding semi-active schedule.
(b) At least one semi-active schedule is optimal for a regular objective.
(c) There are a finite number of them.

Proof. Left as an exercise. ∎

It is possible to further improve semi-active schedules in many cases. Although an activity may not be moved immediately to the left, it may be possible to "jump" over obstructions to the left and obtain a better schedule.

Definition. A *global left shift* occurs if one operation can be shifted into a hole earlier in the schedule and preserve feasibility.

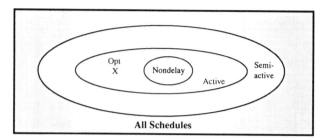

Figure 15.5. Venn diagram of schedule relationships.

Definition. The set of all schedules in which no global left shift can be made is called the set of *active schedules*.

The number of active schedules is still much too large for all but the smallest problems when using search. Thus attention often focuses on an even smaller subset called *dispatch* schedules (Figure 15.5) (also called *nondelay* schedules).

Definition. A *dispatch* or *nondelay* schedule is one in which no machine is kept idle at a time when it could begin processing some operation.

Proposition 2. The set of schedules may be classified as follows:

(a) The set of all nondelay (dispatch) schedules is part of the set of all active schedules, which is part of the set of all semi-active schedules, which is part of the set of all schedules.

(b) There is always an active schedule that is optimal for a regular objective.

(c) The best nondelay schedule may not be optimal.

Proof. Left as an exercise. ∎

In summary, the active schedules are generally the smallest dominant set in the job shop problem. The dispatch schedules are smaller in number but are not dominant. Nevertheless, the best dispatch schedule is often a very good, if not optimal, solution. In a sense, the role of dispatch schedules in job shops is similar to that of permutation schedules in flow shops.

Search algorithms, which seek to generate all active schedules, must go through a fairly complex procedure to do so (Baker, 1974, Section 7.3). Fortunately, dispatch search procedures only depend on working forward in time and, when a machine becomes free, choose which available job to assign or, if a job needing an idle machine becomes free, schedule it immediately. Thus branch-and-bound and similar methods are straightforward. We shall not discuss other generation procedures besides dispatch.

15.1.4 Chapter Summary

In Section 15.2 we discuss exact methods, including branch-and-bound (restricting attention to dispatch procedures) and two integer programming formulations. In Section 15.3 we consider one-pass heuristics of several varieties: predetermined multi-permutation sequences, simple dispatch procedures, simple bottleneck procedures, and simple bottleneck dynamics procedures. In Section 15.4 we first consider high-quality search heuristics, such as beam search, neighborhood search, tabu search, and simulated annealing. Finally, we consider iterative heuristic methods such as dispatch, bottleneck, and bottleneck dynamics.

Another useful reference for this section is Wein and Chevelier (1992).

15.1A Numerical Exercises

15.1A.1 Generate all active schedules and all nondelay schedules for the following problem with two machines, three jobs, and two operations per job:

Processing Times			**Routings**		
	Operation			Operation	
Job	1	2	Job	1	2
1	1	3	1	1	3
2	2	1	2	2	1
3	2	2	3	1	2

(While $(n!)^m = 36$, the problem is actually much easier.)

15.1A.2 Pick one of the schedules from 15.1A.1 and show the job and the machine Gantt charts for it.

15.1A.3 Choose a couple of nonactive schedules in 15.1A.1 and illustrate how to do local/global shifts to produce an active schedule.

15.1A.4 Which of the active schedules in 15.1A.3 has the minimum makespan? Minimum unweighted flow?

15.1C Thinkers

15.1C.1 Prove Proposition 1.

15.1C.2 Prove Proposition 2.

15.1C.3 Propose a general rule appropriate in makespan problems for generating a schedule using a job-at-a-time procedure. Test your idea on the example in the text; test it on 15.1A.1.

15.1C.4 Propose a job-at-a-time rule that is suited to weighted mean flowtime problems. Test your idea on the example in the text; test it on 15.1A.1.

15.2 EXACT METHODS

15.2.1 Overview

Even very small job shops (say, 10 machines and 20 jobs) tend to be too combinatorial to be solved exactly by integer programming approaches or branch-and-bound. Since the problem is NP-complete, even ultrafast computers 10–20 years from now will not make much of a dent.

Why then do we study integer programming models at all? First, there will always be certain special structure cases that are solvable. If we understand the general approaches, we may recognize these. Second, there are often various partial relaxations of the equations which can be solved and may be useful. There are two integer programming versions we shall study in particular. In the "disjunctive constraints" version, there is a model easily solved if we ignore resource conflicts, which gradually gets more and more difficult as resource constraints are added, leading to a type of branch-and-bound model. Since we know various approximate ways of solving branch-and-bound models, this leads to interesting high-computation heuristics for scheduling job shops.

The other integer programming version we shall study is a generalization of the formulations we studied for flow shops. We studied a makespan objective in Proposition 13 of Chapter 13, and a weighted flow objective in Proposition 2 of Chapter 14. Adapting these to the classic job shop is really very easy. If job j has an operation on machine k, we can no longer assume that the previous operation was on machine $k - 1$. But other than some notation and bookkeeping to keep track of the actual previous machine, the formulation is unchanged. The relaxation used for this formulation keeps track of all resource constraints and requires that once a part of an operation starts a machine, it cannot be interrupted. However, it allows several operation parts to start at different points of time.

There are a number of ways to use this in a heuristic. One is to use the relaxation as a lower bound in branch-and-bound or approximate branch-and-bound. Another way is to look at the solution and then add constraints to prevent some of the fractional starts. If this is done in a clever way, a good heuristic solution may be obtained.

In Section 15.2.2 we discuss these two integer programming methods. In Section 15.2.3 we discuss several approaches to branch-and-bound.

15.2.2 Integer Programming

Disjunctive Constraint Formulation. The disjunctive constraint integer programming formulation to the classic job shop problem is discussed in Baker (1974) and is discussed more completely by Greenberg (1968).

This formulation relies on indicator variables to specify operation sequence. As usual, let C_{jk} represent the completion time of job j on machine k (assuming that it uses machine k). Unlike the flow shop, we no longer expect that this will be the kth operation, in fact, we do not keep track of the operation number in this formulation.

It will be important to have notation for the previous machine that job j was on before k. This previous machine will be a function both of j and k. Denote the previous machine by h, so that formally $h = h(j,k)$. (We will usually just write h to avoid clumsy notation.) If jk is the first operation of j, we let h be a dummy machine that finishes at the arrival of the job, so that $C_{j0} = r_j$. The C_{jk} are decision variables and together effectively determine the schedule.

Let $K(j)$ represent the machine for the last operation of each job; we have already set the zeroth machine for all of them as a dummy machine.

As usual, we will need constraints of the form

$$C_{jk} \geq C_{jh} + p_{jk}, \quad h = h(j, k)$$

to ensure that an operation cannot complete until the previous operation is complete and the new operation is complete.

However, in this formulation we use a new method to ensure that resource constraints are not exceeded. (Review Proposition 13 in Chapter 13 for another way in which to do this.)

Note that for two jobs i and j, both needing processing on machine k, one must come before the other. If i comes first, then clearly

$$C_{jk} \geq C_{ik} + p_{jk} \tag{1}$$

If, on the other hand, j comes first, then clearly

$$C_{ik} \geq C_{jk} + p_{ik} \tag{2}$$

Since one or the other of these must hold, they are called *disjunctive constraints*. It takes a bit of cleverness to figure out how to represent disjunctive constraints in an ordinary integer program. It is useful to define indicator variables y_{ijk} as 1 if i comes before j on machine k, and 0 otherwise. Then consider the following two inequalities, where H is a large number like 10^9 (H is like the classic "big M"):

$$C_{jk} - C_{ik} + H(1 - y_{ijk}) \geq p_{jk} \tag{3}$$

$$C_{ik} - C_{jk} + Hy_{ijk} \geq p_{ik} \tag{4}$$

Note that inequalities (3) and (4) very cleverly capture the disjunctive constraints (1) and (2) with no "either–or" to worry about. If i comes before j, $(1 - y_{ijk})$ is zero, giving constraint (1) as desired. On the other hand, Hy_{ijk} becomes very large, making inequality (4) true also. If, on the other hand, j comes before i, inequality (3) becomes automatically true, while inequality (4) becomes constraint (2). In considering disjunctive constraints between two operations on machine k, we must also consider the earliest time the machine is available, a_k. It is easiest to consider this as a phony operation of length a_k, which always comes first. That is, call it operation zero, and insist that $y_{0jk} = 1$ for all j,k.

Proposition 3. Consider the classic job shop with dynamic initial availability of machines, dynamic job arrival, and weighted flow objective function. A disjunctive constraint integer programming formulation can be given by

$$\text{minimize} \sum_{j=1,n} w_j C_{jK(j)}$$

$$\text{s.t.} \quad C_{jk} \geq C_{jh} + p_{jk} \quad \text{for all occurring } (j, k) \text{ pairs} \tag{5}$$

$$y_{0jk} = 1.0 \quad \text{for all occurring } (j, k) \text{ pairs} \tag{6}$$

$$C_{jk} - C_{jk} + H(1 - y_{ijk}) \geq p_{jk} \quad \text{for all occurring } (i, j, k) \text{ triples} \tag{7}$$

$$C_{ik} - C_{jk} + H y_{ijk} \geq p_{ik} \quad \text{for all occurring } (i, j, k) \text{ triples} \tag{8}$$

$$C_{jk} \geq 0; \quad y_{ijk} \in \{0, 1\} \tag{9}$$

Proof. The objective gives the weighted average of the completion time of the last operation for each job, which is the desired objective. Inequality (5) says that operations cannot start until previous operations are finished; implicit is that the first operation cannot start until the dummy zeroth operation finishes. Equation (6) says that initial unavailability of a machine can be treated as an initial dummy operation constrained to come first. Inequalities (7) and (8) say that for every possible pair of operations on machine k, one or the other must come first. The final constraints say that completion time need not be an integer, but the disjunctive indicator variables must be 0 or 1. ∎

Greenberg (1968) developed a specialized computer code for solving this integer programming problem by a type of branch-and-bound. We discuss this in Section 15.2.3 on branch-and-bound. In the meantime, we develop the other integer programming formulation and compare it with this one.

Discrete (Unit) Time Formulation. The other formulation is a slight generalization of the one we have met for flow shops in Proposition 13 of Chapter 13. While it is easy to modify the formulation for different objective functions, we present the problem for weighted flowtime to allow easy comparison with the disjunctive integer program given above.

Proposition 4. Consider a classic job shop, with dynamic machine availability, dynamic job arrivals, and weighted flowtime objective function. Proposition 13 of Chapter 13 may be generalized to give the following integer program:

$$\text{minimize} \sum_{j=1,n} w_j C_{jK(j)}$$

$$\text{s.t.} \quad C_{jk} = \sum_{t=1,T} t C_{jkt} \quad \text{for all } (j, k) \tag{10}$$

$$x_{jkt} = \sum_{u=t,t+p_{jk}-1} C_{jku} \quad \text{for all } (j, k), t \tag{11}$$

$$C_{jk} \geq C_{jh} + p_{jk} \quad \text{for all } (j, k) \tag{12}$$

$$\sum_j x_{jkt} \leq 1 \quad (j \in \{(j, k)\}) \quad \text{for all } (k, t) \tag{13}$$

$$x_{jkt} = 0, \quad t < a_k \quad \text{for all } (j, k) \tag{14}$$

$$C_{jkt} \in \{0, 1\} \quad \text{for all } (j, k), t \tag{15}$$

(If k is the first machine for j, so that h is empty, then $C_{jh} = r_j + p_{jk}$).

Proof. The proof is a straightforward adaptation of the approach to Proposition 13 of Chapter 13, with a changed representation for previous operations. ∎

It is interesting to compare the size and solution difficulty of the formulations in Proposition 3 and Proposition 4. Proposition 3 requires a disjunctive variable for every possible pair of operations using a given machine! If in a simple case each job has m operations, where m is the number of machines, then each machine will have about n operations, giving about n^2 disjunctive variables per machine, or n^2m integer variables in total. On the other hand, Proposition 4 requires an integer variable (completion period) for every operation of every job for every time period, or about nmT in total. Thus the ratio of variables required is crudely

$$(\text{Disjunctive})/(\text{Discrete}) = (n^2m)/(nmT) = n/T$$

Thus it would seem that the advantage could go either way, depending on whether there is a large number of relatively uniform jobs arriving regularly or a smaller number of wildly varying jobs arriving sporadically.

We defer further consideration of some of these issues to the exercises.

15.2.3 Branch-and-Bound

Branch-and-Bound for the Disjunctive Method. We turn first to branch-and-bound procedures related to the disjunctive integer programming method given in Proposition 3. Greenberg (1968) developed a specialized computer code for using the disjunctive method in branch-and-bound. He first omitted the disjunctive constraints entirely ((7) and (8)). This had the effect of leaving the processing time at a machine the same, but allowing as many operations to process as desired, that is, removing all capacity constraints. The remaining problem is a relaxation, which can be solved by linear programming. If there are no machine conflicts in the resulting solution, then it represents an optimal schedule. If there is overlap between jobs i and j on machine k, then two subproblems are solved (branching). One subproblem contains the additional constraint

$$C_{jk} - p_{jk} \geq C_{ik}$$

while the other problem contains instead the constraint

$$C_{ik} - p_{ik} \geq C_{jk}$$

Both of these subproblems can then be solved as linear programs, for there are still no integer (disjunctive) variables in the formulation. The branching process continues in this way, each time branching from an infeasible relaxed solution to two subproblems containing an additional inequality taken from the appropriate pair of disjunctive constraints. Whenever an infeasible solution is obtained, the value of the objective function in the corresponding linear program can be used as a lower bound. Thus the branch-and-bound scheme solves a linear programming problem for each node in the branching tree as a means of solving the overall integer programming problem.

Contrast this branch-and-bound method with our usual ones. To date we have considered branch-and-bound procedures that successively build more and more complete partial feasible solutions. We might do LP relaxations at a node, but just to get a lower bound, not to build the solution. Here we build, at each node, a complete solution that is always better than optimal but is infeasible. As we branch we get rid of more and more infeasibility, until we finally reach the optimal solution. As we go along the infeasible solutions provide us needed lower bounds more as a by-product of the solution procedure.

Technically, our usual procedures are *primal procedures*, this disjunctive procedure is a *dual procedure*. Which approach is better for which situation seems to be a very difficult question that does not seem to have been addressed adequately either theoretically or computationally.

Dispatch Branch-and-Bound. We turn next to discussing a more conventional branch-and-bound procedure that uses the simplifications arising from focusing attention on dispatch procedures.

First we need to discuss the idea of a dispatch event simulation. (This idea has been discussed before, we present it here partly as review.) In an event simulation, there is a master list of future events sorted in order of occurrence. It starts with such events as job arrival times r_j and machine initial availabilities a_k. Time is not incremented in units, but from one event to the next. Start at the first event, say, the arrival of job 2. Add job 2 to the queue of the appropriate machine, say, 3. If machine 3 is free, start the operation and enter the finish time for the operation and the finish time (which will be equal for this simple version of the model) for the machine into the event list. If machine 3 is not free, do nothing. Now increment to the next event.

When the event is that a machine just becomes free and there are several available operations that could be performed, the choice could be made in several ways:

(a) *Branch-and-Bound.* Form a subproblem for each possible choice. Find a lower bound on the cost of each subproblem. Perhaps start working on the subproblem with the largest lower bound, and so on. We will come back to this.

(b) *Dispatch Heuristics.* Evaluate each of the possible operations to be performed by a heuristic giving its priority. Choose the operation with the highest priority and proceed.

(c) *Beam Search.* Using the same dispatch heuristic, choose the operations with, say, the three highest priorities. After a path has involved making several choices, evaluate the path as having the sum of the priorities along it. Now make the next choices of the three best paths to date.

Returning to branch-and-bound, all that is needed to turn a dispatch heuristic into branch-and-bound is a way to set lower bounds.

Lower Bounds for Dispatch Branch-and-Bound

(a) Complete the simulation from the partial problem as though there were no resource constraints; that is, lead times are simply the tails of the jobs. This is quick but typically won't give a very good lower bound.

(b) Form the discrete time linear programming relaxation of the partial problem. This is quite expensive but typically will give a very good lower bound and thus be more effective in trimming the search tree.

Overall, the question of which is better is an empirical question, which does not seem to have been explored very thoroughly, partly because job shop problems are so large and difficult to deal with.

Other useful references for this section include Wagner (1959) and Story and Wagner (1963).

15.2A Numerical Exercises

15.2A.1 Consider the job shop data given in Tables 15.1 and 15.2; use weighted flow criterion, with jobs 1 to 4 having weights 1, 1, 2, 4, respectively.

(a) Express the problem completely as a discrete time integer program.

(b) Relax the integrality constraints, and solve as an LP.

15.2A.2 Using the same input and objective function:

(a) Express the problem completely as a disjunctive integer program.

(b) Relax the disjunctive constraints, and solve in the easiest way possible.

(c) How does the solution compare with 15.2A.1(b)? Why?

15.2A.3 (a) In 15.2A.2 determine a resource constraint violated by the relaxed solution, and choose a good disjunctive constraint to fix it. Find the new solution.

(b) Repeat this procedure until the problem is feasible, or you get tired. How does the solution now compare with 15.2A.1(b)?

15.2B Software/Computer Exercises

15.2B.1 Program and test a dispatch branch-and-bound approach for the job shop problem, using whatever help you have from software developed earlier in the book. Use a very simple lower bounding procedure.

15.2B.2 Add a linear programming lower bounding procedure to 15.2B.1. Does it seem to be worthwhile?

15.2C Thinkers

15.2C.1 Specialize the disjunctive integer programming formulation given here to produce an integer programming formulation of the flow shop problem (a) for general sequencing and (b) for permutation sequencing.

15.2C.2 **(a)** If the disjunctive branch-and-bound scheme is used to solve a job shop problem, what is the maximum number of levels that can occur in the branching tree?

(b) Same question for the discrete time dispatch branch-and-bound scheme.

15.2C.3 One way of reducing computation in the disjunctive branch-and-bound scheme would be to add all "almost certain" disjunctive constraints in advance. Give some examples of this; comment.

15.3 ONE-PASS HEURISTICS

15.3.1 Overview

After seeing the very large size integer programs that arise from even very small job shop problems, it should be evident that heuristics will be necessary in most cases. In the rest of this chapter and in Chapter 16 we restrict our consideration to job shop heuristics.

In this chapter we discuss job shop heuristics from a very general point of view. First, in Section 15.3, we look at *one-pass heuristics:* the procedure simply builds up a single complete solution a step at a time. These are very useful on their own, since they represent an inexpensive solution to the problem. In addition, one-pass heuristics may be used repeatedly (iteratively) to build more sophisticated (and more expensive) multi-pass heuristics or *search heuristics*, as we shall see in Section 15.4. In Chapter 16 we look at dispatch heuristics in more detail, for differing objective functions, and report computational experience where available.

15.3.2 Multi-Permutation Sequence Heuristics

Perhaps the simplest heuristic conceptually is simply to assign in advance a sequencing rule for the operations at each machine. If there are, say, n_k operations to be performed at machine k, then the total number of such multi-permutation sequences is

$$(n_1)!(n_2)!\cdots(n_k)!\cdots(n_m)!$$

As an example, if job i entered the system before job j, then at any machine job i must be processed before job j.

One major difficulty here is that since jobs have differing paths, job j might arrive at machine k a very long time before job i. Then by our rule k would sit idle when j could have been processed.

In general, it is difficult to set the rules at some earlier time and know the complete effect. In particular, it is possible that i may never arrive because it is waiting on a machine insisting on j first, while j may never arrive because it is waiting at a machine that insists that i be done first. Such situations are called gridlock.

One easy way out is to modify the procedure to *dispatch multi-permutation sequence*. Now, if the next job in the specified sequence is not available, we do not wait. We simply put on the first job in the sequence that is currently available. That is, this is just a dispatch heuristic with a particular rule for each machine chosen in advance.

15.3.3 Simple Dispatch Heuristics

With simple dispatch heuristics, when a job arrives at an empty machine, we schedule it. When a machine becomes available, we simply schedule the highest priority job currently available at the machine.

Static priorities may be calculated one time in advance. Examples include (a) choose the job due out of the shop first, (b) choose the job arriving in the shop first, and (c) choose the job with the highest weight.

Dynamic priorities change over time. They must typically be recalculated each time a job is to be chosen. Examples include R&M, COVERT, and least slack.

Global priorities do not depend on position in the system. Release date and due date are good examples.

Local priorities depend only on the situation at the local machine. Shortest process time is a good example.

Forecast priorities depend on both the situation at the local machine and on a forecast of the job's remaining experience. Rules such as Critical Ratio, COVERT, and R&M are good examples.

Although many simple dispatch rules have been presented elsewhere, and many of these rules have now mostly historical interest, it may be useful to list in one place a large number of dispatch heuristics for the job shop. (See Table 15.3.)

TABLE 15.3   A Listing of Job Shop Dispatch Heuristics

Rule	Description
WSPT	Choose the operation with the highest weight divided by processing time.
LWKR	(Least work remaining) Choose the job with the shortest tail.
WLWKR	Largest value of (weight divided by least work remaining). (This is like bottleneck dynamics but assumes remaining machine prices are equal.)
AWINQ	Lowest value of (anticipated work in next queue). Highest priority if next machine is not busy. (This is an early bottleneck dynamics type idea.)
FOFO	(First off first on) Choose the machine that can finish the job first even if you must wait for that machine. (This is not dispatch.)
FASFS	(First arrival at the shop first served) Choose the smallest r_j first.
TWORK	Choose the job with least total work in the shop.
WTWORK	Largest value of (weight divided by total remaining work). (Again, this is analogous to bottleneck dynamics.)
EDD	Earliest due date.
FCFS	First to the machine served first.
MST	(Minimum slack time) Do the job first that has the least slack = due date − lead time − operation time.
OPNDD	Do the job first with the earliest operation due date = due date − lead time.
S/OPN	Choose the job with lowest slack per remaining operation.
TSPT	(Truncated SPT) Use SPT, except expedite a job that has waited a very long time.
Critical Ratio	Choose the job with highest total remaining process time over total time until the due date.
COVERT, R&M	Weighted tardiness heuristics; see Section 14.3.4.

15.3.4 One-Pass Myopic Dispatch Heuristics

Myopic procedures are more sophisticated one-pass dispatch heuristics that make the decision at a machine in order to optimize (or approximately optimize) that as a one-machine problem. To do that requires whatever parameters would be needed to optimize the one-machine problem. For example, the weight w_j and the local processing time p_{jk} are available. However, there will not initially be a due date available for finishing at the local machine. If the one-machine algorithm requires a due date, then an estimate of the lead time LT_{jk} for job j from machine k to its overall completion will be needed. This might be estimated simply as remaining processing time or, better, a historical multiple of the remaining processing time. (In more recent multi-pass dispatch procedures, it might be estimated from the previous iteration.) In any event, given LT_{jk}, we estimate the local (derived) due date by $dd_{jk} = d_j - LT_{jk}$. (The makespan objective is slightly more complicated. We must express the makespan problem as a network and find the late finish time (LFT) for a pair jk and set $dd_{jk} = LFT_{jk}$. This will be discussed more carefully in Chapter 16.)

**TABLE 15.4   Dispatch Myopic Heuristics for Several
Common Objective Functions**

Objective	Heuristic
Makespan	EODD (earliest operation derived due date) All global due dates set $= 0$; also called required finish time or similar names.
Maximum lateness	EODD, but with natural due dates
Weighted flow	WSPT
Weighted tardiness	R&M $\pi_{jk} = (w_j/p_{jk})\exp[-(dd_{jk} - p_{jk})^+/Kp_{\text{av}\,k}]$

For completeness, we give the myopic dispatch heuristic for several common objectives in Table 15.4. Working out dispatch myopic heuristics for some less common objectives will be left as an exercise.

Myopic dispatch heuristics are quite robust and almost always perform well in empirical studies, as we shall see in Chapter 16. However, those requiring due dates can be significantly improved by better lead time estimates. We shall also see in Chapter 16 that iterative myopic methods give much better results. In turn, myopic results can be further improved by taking into account bottlenecks downstream by using bottleneck methods, in particular, bottleneck dynamics. This assertion will also be supported by empirical evidence in Chapter 16.

15.3.5   Simple Bottleneck Heuristics

For the simple flow shop we developed a basic one-pass bottleneck procedure that was easy to apply and appears to be quite adequate. We first developed "Botflow-1 (Makespan)" quite carefully in Section 13.3.3. Then, in Section 14.3.3, we developed the straightforward modification for the weighted flow criterion, which we called "Botflow-1 (Weighted Flow)." Similarly, we developed the modification for the weighted tardiness criterion, called "Botflow-1 (Weighted Tardiness)," in Section 14.3.4.

It is useful to review the basic procedure.

One-Pass Bottleneck Heuristic Procedure

Step 1: Choose a bottleneck machine that divides the shop into three disjoint sets: (a) pre-bottleneck, (b) bottleneck, and (c) post-bottleneck.

Step 2: For each job estimate the head time h_j before the bottleneck and the tail time t_j afterward. (The simplest estimate is the actual processing time.)

Step 3: Due date d_j less t_j is the derived due date dd_j off the machine. (For makespan use any common due date for all jobs.)

Step 4: Solve the one-machine bottleneck problem with arrival times $r_j = h_j$, due dates dd_j, and bottleneck (k) processing times p_{jk}.

Step 5: The resulting sequence is then employed at each machine to complete the heuristic.

This heuristic procedure cannot be used directly for the job shop for two reasons:

(a) The same machine is unlikely to be the bottleneck for all jobs.
(b) Job shop heuristics do not typically employ the permutation sequence assumption. (Thus knowledge of the schedule on the bottleneck or bottlenecks does not directly produce the sequencing rules on nonbottleneck machines.)

We solve the first problem by defining a bottleneck cut (when it exists).

Definition. In a job shop a *bottleneck cut* $B = \{k_1, \ldots, k_b\}$ is a set of machines, such that each job visits B exactly once and each machine in the cut is the unique bottleneck for the jobs that visit it. Furthermore, the bottleneck cut divides the machines into three disjoint sets. Jobs visit set A only before their bottleneck operation, B (bottleneck cut) at their bottleneck operation, and C after their bottleneck operation.

Of course, there is no guarantee that a job will possess a bottleneck cut. If the job processes only in A or only in C, then a phony bottleneck machine with 0 processing time can be defined. However, if such a cut cannot be defined, the method cannot be used.

Botjob-1 Heuristic

Step 1: Find a bottleneck cut (if available).

Step 2: Estimate heads h_j and tails t_j from the bottleneck for j.

Step 3: Calculate derived arrival times and derived due dates for each machine in the bottleneck cut.

Step 4: Solve a one-machine problem for each of the bottleneck machines, recording actual start times for each job on a machine.

Step 5: Solve the full job shop via a one-pass dispatch heuristic for the appropriate objective. For machines in A the objective is to minimize maximum lateness in meeting the needed start times in step 4; that is, the heuristic is earliest derived due date.

Step 6: For machines in B or C use a dispatch heuristic appropriate for the problem objective (like R&M for weighted tardiness).

Step 7: For machines in A calculate any needed slack estimates as the time the job will start on its bottleneck machine less a standard lead time.

Step 8: For machines in B or C calculate slack estimates as the final due date less a standard lead time.

Several points need to be mentioned. First, how does one determine a bottleneck cut? It may be functionally obvious. If a shop has two annealing furnaces, not in parallel, which have long lines, and all jobs must be annealed, this may determine a bottleneck cut in a rather obvious fashion. A different approach would be to list, for each job, the machine that has the highest historical utilization. If these machines for all jobs cluster nicely into a cut, this would probably be sufficient.

If a one-pass procedure is to be used, estimating head lead times and tail lead times simply by the heads and tails (i.e., actual processing time) may be too crude. We may be able to determine from historical records that head lead times are typically, say, 2.3 times the head and tail lead times are typically 1.7 times the tail. Even crude historical data can often improve results greatly.

A more sophisticated procedure would be to use extensive historical data and regression packages to estimate these factors as a function of the load on the shop, the part of the shop, and so on. Of course, iterative estimation of the lead times, as we discuss in the next section, improves these estimates by the current simulation itself, at the cost of multiplying running times by a factor of 3.0–5.0.

15.3A Numerical Exercises

15.3A.1 Create a small job shop problem with makespan objective, three machines, four jobs, and two operations per job.

15.3A.2 For the problem in 15.3A.1 choose a single ordering of the jobs. Try to use this as a multi-permutation sequence. Are there any problems?

15.3A.3 In 15.3A.2 modify the heuristic to a dispatch multi-permutation sequence. Is the makespan improved? If so, by how much?

15.3A.4 Choose three of the simple dispatch heuristics from Section 15.3.3 and try them on your problem. Explain which heuristic did the best on intuitive grounds.

15.3A.5 Change the processing times on one of the machines to dominate. Run Botjob-1 on the problem. How well does it do?

15.3B Software/Computer Exercises

15.3B.1 Write a spreadsheet or simple software to make it easy to generate small job shop problems and run simple dispatch rules on them.

15.3B.2 Write software to run the Botjob-1 heuristic for makespan on small job shop problems. Make sure it is easy to change a problem slightly to run another one.

15.3B.3 Run your program on several problems, and compare with the best solution you can guess by inspection. How good is this heuristic?

15.3C Thinkers

15.3C.1 Consider a compound job shop in which each machine is replaced by a cluster of parallel machines that are equal/proportional. Modify Botjob-1 for this case; modify simple dispatch heuristics for this case.

15.3C.2 Consider a job shop in which each machine has queue room only for a certain fixed number of jobs. In variation 1 excess jobs are forced to sit and block the previous machine. In variation 2 excess jobs can be removed and returned as needed from bulk storage for a high price. Can you modify any one-pass heuristics to fit either or both of these situations?

15.3C.3 Consider a job shop where certain jobs may visit certain machines more than once (the reentrant case). Can you modify one-pass heuristics for this case?

15.3C.4 Work out myopic dispatch heuristics for the following objectives, and argue that they are appropriate:

(a) Weighted number of tardy jobs.

(b) Weighted early/tardy.

(c) Weighted flow/weighted tardy.

15.4 SEARCH HEURISTICS

15.4.1 Overview

While one-pass heuristics limit themselves to constructing a single solution, search heuristics (sometimes called multi-pass heuristics) try to get much better solutions by generating many of them, usually at the expense of a much higher computation time. Techniques like branch-and-bound and dynamic programming search until they can guarantee an optimal solution, but these techniques are simply not practical for large problems.

Randomized heuristics are an early attempt to provide more accurate solutions. While they can provide some improvements at high cost, they are probably dominated now by other techniques in this section. We discuss improved versions in Section 15.4.2. As we have learned, beam search is an approximate method for branch-and-bound that curtails the search by avoiding unpromising areas of the search space. In Section 15.4.3 we develop a type of beam search for job shops that uses dispatch heuristics to guide the search. The method generates the solution that the original dispatch heuristic would have, together with an exploration of "nearby solutions."

Neighborhood search methods, including tabu search and simulated annealing, start with a solution and try to change it in a simple way to get a better solution and so on. In Section 15.4.4 we discuss methods for applying neighborhood search to job shop problems. Neighborhood search methods are generally expensive in the job shop, since a full simulation of a job shop problem is necessary for each interchange, and evaluating the possible interchanges in a full neighborhood is also demanding. An important exception is for the makespan criterion since, in this case only, each interchange "only" requires solving a longest path problem through the shop, and prefiltering methods exist to focus search on a small part of the neighborhood cheaply. These reduction methods are now also being applied to other objective functions, but no results are available as yet.

Finally, iterative methods try to improve the values of parameter estimates, such as lead times and/or prices that are needed to make certain types of heuristics work well. In Section 15.4.5 we discuss iterative lead time methods for improving dispatch and simple bottleneck methods in job shops. We also discuss the shifting bottleneck method, which is sort of a cross between lead time methods and neighborhood search methods. Finally, we discuss methods that iterate both lead times and prices to provide bottleneck dynamic methods for job shops.

15.4.2 Guided Randomized Dispatch

Randomized dispatch was widely used 20–30 years ago (Baker, 1974). The idea is to start with a group of possible dispatch heuristics. At each selection of an operation to run, choose the dispatch heuristic randomly, repeated throughout an entire simulation of the shop. Repeat the entire process, say, 30 times and choose the best result.

The technique provided a viable alternative before other methods in this section were available. It might typically cut errors in half, for example, at a factor of 30 or 40 in computation times. However, it is clumsy and shotgun in nature when compared with newer available techniques such as beam search, iterated lead time methods, or the shifting bottleneck method. The basic problem is that there is no valuation function, such as in beam search, to estimate how far the current decision is from "likely" solutions. If the solution is already fairly unlikely, randomly choosing a new unlikely choice is likely to ruin the entire iteration, with no recovery possible.

Various researchers tried to improve on the randomization approach. One change is to have a learning process so that more successful heuristics will have higher chances of being selected in the future. However, all such procedures are still limited by the accuracy of the heuristics being used.

Guided random dispatch tries a somewhat different approach. Here *guided* means that an excellent heuristic is needed first, to "explore" the problem and provide good guidance as to where to search. (We have already seen such a procedure in beam search.)

We start with a strong guide heuristic, which estimates priorities numerically for operations at each machine choice. Rather than the random choice being over different heuristics, the random choice is over operations with priorities close to the best. (There is a strong analogy here with simulated annealing.) One might set the probability of choosing an operation as its priority divided by the sum of all priorities of jobs currently available. Alternatively, the probability could fall off exponentially in the deviation of a priority from the highest priority. One then makes a number of runs as in any randomized method.

This method attempts basically to stay close to the heuristic policy, but to check out small perturbations from it. (It is relatively untested at this point.)

15.4.3 Guided Beam Search

At least two types of guided beam search exist for dispatch job shops. The beam search we have already discussed with full probes to the bottom of the tree via a good heuristic is clearly guided, we might call it "full guided beam search." A simpler version, which looks more at local priorities, might be called "local guided beam search."

Full guided beam search would expand every operation choice at a machine into full probes to the end of the problem with, say, the dispatch myopic to give evaluation functions. These in turn would be used to trim the decision tree and carry forward a fixed number of active paths. The problem with this is that perhaps hundreds or even thousands of myopic dispatch simulations of the job shop would be required.

If this is deemed too expensive, we might use a faster, less accurate method, *local guided beam search*, which keeps track of the unlikeliness of a current path, and keeps only the M most likely paths.

As a concrete example of the way local guided beam search might work in a job shop, consider a job shop with weighted flow criterion. Suppose the base dispatch heuristic that we wish to improve via beam search is the myopic one, that is, whenever a machine is free, calculate priorities among waiting jobs as w_j/p_{jk}, where j is the job and k the local machine. Choose the job with the highest priority and schedule that operation. Repeat.

In local guided beam search we take account of the fact that the priorities may not be perfectly accurate. Suppose we choose a beam width of four. Suppose also that, at the first decision point in the simulation, the candidate jobs have priorities of 2.5, 2.4, 2.1, 1.8, 1.6, and 1.2. We first normalize the priorities so that the highest priority will be 0, by subtracting 2.5 from everything, giving normalized priorities of 0.0, -0.1. -0.4, -0.7, -0.9, and -1.3. Since the beam width is four, we keep track of four simulations, one starting with each of the four best choices for the first operation on the first machine. This is shown in the top two levels of the beam search diagram in Figure 15.6. (The double-circled nodes are those among the top four at each level.)

The first simulation, based on choosing the top rated job in the first choice, produces four new choices at the second choice with normalized priorities of 0.0, -0.1, -0.6, and -1.0. As the valuation function for these third-level choices, we take the sum of the normalized priority for the first choice and the normalized priority for the second choice. That is, these first four choices at the third level have valuation function values of 0.0, -0.1, -0.6, and -1.0. The second choice at the second level had priority -0.1, and its offspring had priorities 0.0, -0.5, and -0.9, giving these next three choices at the third level valuation function values of -0.1, -0.6, and -1.0. The third choice at the second level had priority -0.4, and its offspring had priorities 0.0 and -0.05, giving third level valuation function values of -0.4 and -0.45. The fourth choice at the second level had priority -0.7, and its offspring had priorities of 0.0, -0.2, and -0.5, giving third level valuation function values of -0.7, -0.9, and -1.2. Now, at the third level, we pick the four nodes with the best valuation function values, which are 0.0, -0.1, -0.1, and -0.4. Note that, in this example, the dispatch

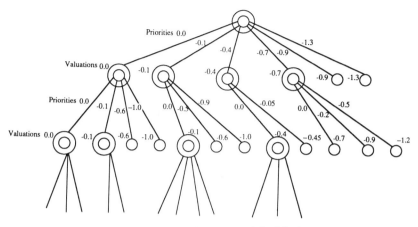

Figure 15.6. Dispatch beam search for job shops.

heuristic will always give the path with highest valuation function value of 0.0, so that the dispatch heuristic will always be produced as one of the choices. Note that the other three choices either used the dispatch choice at the first level or at the second level.

This method of obtaining the evaluation function has the advantage of being quite fast. Even if we choose the beam width at 20, we only have to basically run 20 simultaneous simulations.

15.4.4 Neighborhood/Tabu/Simulated Annealing Search

Neighborhood search methods (tabu search and simulating annealing are really just variations, and so the remarks still apply) are somewhat less practical for job shops than for flow shops. The reason is that neighborhood search works with complete solutions. From a given complete solution, we compute the cost of moving to a number of nearby complete solutions. If any improves the solution, we take that as our new base case, and so on. This procedure is inherently more complex for job shops, although we will present a special case with difficulty more like the flow shop.

Neighborhood search is reasonably practical for the classic flow shop if we restrict ourselves to permutation sequences. Given a particular permutation sequence and associated solution we can think of adjacent pairwise interchange of two jobs. This will not affect the schedule of earlier jobs, but will affect later jobs. The cost of a permutation in computing a revised solution is $O(mn)$. We have to do this for about n choices of jobs to interchange. Suppose also the algorithm required M stages to converge. Total cost would then be $O(Mmn^2)$. Part of the question would be to make M small by starting with a very good heuristic to provide a starting solution.

The most analogous case for the job shop would be to specify the *dispatch permutation sequence* discipline. In the dispatch permutation sequence, a permutation sequence determines a global priority rule: no matter what the machine, job a has priority over b has priority over c. However, "dispatch" adds a practical note in that if the highest priority job is not available, then the highest priority job currently waiting at the machine will be scheduled. Now a permutation of two jobs in the priority will require typically a complete new simulation. There will be mn operation choices to be made in the simulation, n choices to permute at m nodes, and M stages to converge. Thus the total cost of the algorithm would still seem to be $O(Mmn^2)$. However, there is a great deal of bookkeeping every time an operation is scheduled here, compared to the flow shop case, so that this is probably not an adequate analysis.

For more general sequencing disciplines, the practicality of extended neighborhood search methods depends on finding ways to reduce the effective size of the neighborhood by prefiltering and to evaluate interchanges that are actually performed. See Section 16.3 for several studies of extended neighborhood search in the classic job shop with makespan criterion.

It is not difficult to modify a straight neighborhood search procedure into a tabu search approach or a simulated annealing approach. We leave this for the exercises.

15.4.5 Iterative Dispatch, Bottleneck, and Bottleneck Dynamics Methods

In contrast to neighborhood search methods, iterative methods that improve estimates of lead times and/or heuristic prices are quite practical, typically because the iterative procedure only needs to be run perhaps three to five times. These procedures approximate the underlying process as being roughly convex *in the aggregate*. How good, in fact, this assumption is, is an empirical question. Here we describe iterative dispatch, iterative bottleneck, shifting bottleneck, and iterative bottleneck dynamics methods briefly in the abstract. Chapter 16 will be devoted primarily to presenting results of computational studies on these methods.

Iterative Dispatch Methods. Iterative simple dispatch methods solve a job shop problem by making local dispatch decisions at each machine as if it were a one-machine problem. If the objective function involves due dates, then local due dates must be estimated for jobs at each machine by subtracting an estimate of the lead time after the machine from the final due date. In turn, the lead times are not known until after the problem is solved. This simultaneous estimation problem is handled by solving the problem several times, generating successive improvements in lead time estimation. While this procedure is treated in Section 10.2, we repeat it for convenience here. Recall from Chapter 10 that this procedure can be improved both by using extended dispatch and advanced lead time iteration. We cannot discuss these issues here.

Dispatch Myopic Heuristic with Lead Time Iteration

Step 1: Obtain initial lead time estimates by some method:
   (a) Initial simulation by a one-pass heuristic.

(b) Remaining process time.

(c) Historical multiple of (b).

Step 2: Simulate by the appropriate dispatch myopic heuristic, using LT estimates.

Step 3: Update lead times; save the new cost.

Step 4: If the stopping criterion is met, stop.

Step 5: If it is not met, return to step 2.

Vepsalainen and Morton (1987) found this procedure to be excellent for the weighted tardiness problem in the job shop. This study will be presented in detail in Chapter 16. Other studies comparing simple dispatch methods, myopic dispatch methods, and bottleneck dynamics dispatch methods for several objective functions will also be presented in Chapter 16.

Iterated Bottleneck Cut Method. In the previous section we presented Botjob-1, a heuristic that produces a bottleneck solution to a job shop, assuming the bottlenecks form a bottleneck cut. Botjob-1 depends on crude estimates of lead times throughout. It is easily updated to use lead time iteration.

Botjob-2 Heuristic

Step 1: Run Botjob-1.

Step 2: Use the resulting lead times to improve lead time estimates required throughout the algorithm

Step 3: Is the stopping criterion met? If so, exit.

Step 4: Otherwise, return to step 1 with improved lead time estimates.

(While a Botjob-3 could be envisioned that iterates the bottleneck cut itself, it would probably be less complicated to use the shifting bottleneck algorithm for somewhat the same effect.)

The Shifting Bottleneck Algorithm. While the shifting bottleneck algorithm for makespan is quite well stated in Section 10.4.2 for makespan, for other objectives it is somewhat more complicated, since the embedded problem (all other machines with fixed policies, optimize on embedded machine) must be solved iterating lead times both prior to and after the machines. Thus the embedded machine cannot typically be solved optimally, but only heuristically for such objectives.

Heuristic Procedure for Optimizing Embedded Machine

Step 1: Initially obtain all lead time estimates from output of the previous embedded machine problem.

Step 2: Estimate derived arrival times and due dates for the embedded problem using the current lead time estimates.

Step 3: Optimize the resulting one-machine problem, either exactly, or heuristically.

Step 4: The stopping criterion is met if either:

(a) The solution has repeated for two successive iterations.

(b) There is no objective function improvement for x iterations.

Step 5: If criterion is met, exit.

Step 6: If criterion is not met, use new lead times and return to step 2.

Note that, since lead time estimation is heuristic, it is less likely worthwhile to solve the embedded problem exactly. Note also that it is less likely to be worthwhile to solve problems exactly early in the larger solution procedure than later in the procedure.

For completeness, we repeat the full shifting bottleneck algorithm. (Remember that for a simple embedded problem all other machines are simply replaced by lead times, which are simply total prior or posterior processing time, with no queue times considered.).

Shifting Bottleneck Algorithm

Step 1: Solve the simple embedded problem for each machine in turn.

Step 2: The machine in the problem with the highest cost is designated the *primary bottleneck.*

Step 3: Hold the primary machine at the policy just determined; otherwise repeat steps 1 and 2.

Step 4: The new machine with the highest cost is the *secondary bottleneck.*

Step 5: Balance the primary and secondary bottlenecks by alternately holding one fixed and optimizing the other.

Step 6: Then find the *tertiary bottleneck* in the same way.

Step 7: Balance over the first three.

Step 8: Similarly for the first four, first five, and so on.

Iterative Dispatch Bottleneck Dynamics Methods. Finally, iterative bottleneck dynamics models are very similar to iterative dispatch myopic models, except that prices, as well as lead times, must be estimated iteratively. Extensive studies for the job shop have been carried out by Morton et al. (1988). These also will be reported in Chapter 16 (Actually, in all these iterative procedures, there are usually dynamic but noniterative methods for estimating prices and lead times. These will be discussed in the next chapter as well.)

For completeness we show in Table 15.5 how the myopic dispatch heuristics generalize for bottleneck dynamics.

A useful reference for this section is Lozinski and Glassey (1988).

15.4A Numerical Exercises

15.4A.1 Create a small job shop problem with weighted flow objective, three machines, five jobs, and two operations per job.

TABLE 15.5   Iterated Bottleneck Dynamics Dispatch Heuristics

Objective	Heuristic
Makespan	EODD Unchanged since the formula does not involve resource usage
Maximum lateness	Also unchanged
Weighted flow	Bottleneck dynamics–WSPT $w_j/\Sigma R_u p_{ju}$] (summation over remaining operations)
Weighted tardiness	Bottleneck dynamics–R&M $\{w_j/[\Sigma R_u p_{ju}]\}\exp[(dd_{jk} - p_{jk})^+/Kp_{avk}]$ (summation over remaining operations)

15.4A.2 Solve this problem by the myopic dispatch heuristic.

15.4A.3 Solve this problem by myopic dispatch beam search with beam width of 2.

15.4A.4 Solve this problem by neighborhood search, starting from a clumsy initial solution.

15.4A.5 Solve this problem by neighborhood search, starting from a global dispatch priority solution where the priority is the weight divided by the sum of processing times.

15.4B Software/Computer Exercises

15.4B.1 Write a neighborhood search routine based on a dispatch permutation sequence idea for a classic job shop.

15.4B.2 Add tabu search and simulated annealing capability to your program from 15.4B.1.

15.4B.3 Add the ability to generate job shop problem input to your routine.

15.4B.4 Design an experiment and do some testing. What size problems are feasible? Do the fancier procedures help much?

15.4C Thinkers

15.4C.1 Iterative procedures for lead times and/or heuristic prices typically improve the objective function rapidly for a few iterations and then bounce back and forth without much improvement. What sort of stopping rule might you invent for such a procedure?

15.4C.2 For what sorts of shops might a bottleneck procedure work well? Poorly? When would iterating to improve lead times work well? Be unnecessary?

15.4C.3 For what sorts of shops might a myopic procedure work well? Poorly? When would iterating to improve lead times work well? Be unnecessary?

____16
SCHEDULING JOB SHOPS: HEURISTIC TESTING

16.1 INTRODUCTION

16.1.1 Overview

In Chapter 15, we talked about the classic job shop problem in general and developed generic, exact, one-pass heuristic and multi-pass search heuristic procedures for solving it. But we made no attempt to talk about which specific heuristics are best for which specific objective function under various conditions of shop loading, due date tightness, and the like. In this chapter we discuss simulation studies of utilization (makespan/economic makespan), weighted flow, and weighted tardiness, as well as several other objectives including a complex one. We also present and analyze studies testing a number of different job shop objectives and heuristics. While insufficient testing has been done to answer all important questions, useful answers are beginning to emerge. In Section 16.2 we present classic simple dispatch studies due to Conway (1965a,b) and Carroll (1965) for the average flow and average tardiness objectives. We also present more recent iterated myopic dispatch studies due to Vepsalainen and Morton (1987, 1988) for the weighted average tardiness problem. In Section 16.3 we review several new search studies for the makespan objective: first, a study of the shifting bottleneck method by Adams et al. (1988); second, a study of simulated annealing by Van Laarhoven et al. (1992); third, a tabu search study by Dell'Amico and Trubian (1991); and finally, a genetic algorithm study by Della Croce et al. (1992). We conclude with a short comparison of the methods.

In Section 16.4 we present recent bottleneck dynamics studies for job shops: a study of bottleneck dynamics for a combination release/inventory/tardy objective by Morton et al. (1988) and an integrative dispatch/bottleneck dynamics study by

Morton and Lawrence (work in process) comparing a number of dispatch methods (one-pass, iterated myopic, bottleneck dynamics) and a number of objective functions. Finally, in Section 16.5 we discuss model extensions such as reentrant machines, complex queues, transport, compound machines, and aggregate shops. We also briefly present a more realistic version of the nuclear fuel tube shop requiring a job shop model and contrast its advantages and disadvantages relative to the original flow shop model.

16.1.2 Job Shop Simulation Study Design

The literature on classic dynamic job shops includes simulations of both actual and hypothetical systems. Actual systems are greatly to be desired due to their realism. Unfortunately, a great variety of actual problems must be simulated to test a system properly. This is typically not possible with real systems, so hypothetical systems are also tested, where many different input systems can be created randomly and the resulting problems investigated. (One mixed procedure occasionally employed is to randomize around a real problem.)

Although simulation can accommodate almost any shop structure and assumptions, to a certain extent it is desirable for the model to be somewhat simplified in order to focus on the effects of scheduling and also to permit generalizations of the experimental results. On the other hand, if the model is too simple it may be difficult to argue persuasively that the conclusions will apply under other, more realistic conditions. All the simulation studies discussed in this chapter share the following assumptions:

(a) Operations in a job have a fixed ordering.
(b) An operation can be performed by only one machine.
(c) There are no parallel machines.
(d) Process times, routings, due dates, and the objective are known in advance (deterministic).
(e) There are no explicit setup times.
(f) There is no job preemption.
(g) An operation cannot start until its machine is finished.
(h) An operation cannot start until the preceding operation for that job is complete.
(i) The machines have no breakdowns or planned downtime.

(None of these assumptions are necessary, but they facilitate comparing studies.)

The input to the simulation model is a job file that describes the entire set of jobs in terms of arrival time, due date, routing, process times, and so on. While these factors are usually generated using random variables, there is a great deal of design work in exactly how these factors are used. For example, due dates are typically set with high or low variability and/or high or low average shop tardiness before a random number is ever drawn. Since it is known that arrival times are often positively correlated with due dates and negatively with shop load, these factors are taken into account in generating an arrival time. Operation times will tend to be longer for a long job and also for a bottleneck machine. These are taken into account in generating operation process times.

Using a large number of routings is often unrealistic, so a few routing patterns may be designed by hand and the job routings randomly selected from these. (Note that this is a far cry from older procedures, which tended to randomly equalize processing throughout the shop.)

The output of the simulation is the value of the objective function for the simulation and any side information desired. The simulation starts with the shop unloaded, which is not a normal state of affairs. Therefore the simulation will often have a *warmup* period during which the system fills up throughout with jobs and no measurements are made. At some point, the system is declared warmed up and data collection begins. Near the end, we declare *shutdown*; that is, no new jobs arrive, but those on the floor are allowed to finish. The measuring period is from the end of warmup until the last job finishes. It is desirable that shutdown to finish be a small part of the total time in the simulation.

Probably the most important feature of the experimentation is the maintenance of a reproducible job file. That way the simulation can be repeated several times using the same input each time, varying only the scheduling rules in order to focus on their differences. Nevertheless, it is still important to deal with the probabilistic nature of such results, and investigators have employed various forms of statistical analysis in an effort to support their conclusions as convincingly as possible.

16.1A Numerical Exercises

16.1A.1 Use the manual-statistical input mode in the software provided to generate a 3-machine job shop with about 10 dynamically arriving jobs. Let the criterion be weighted flow.

  (a) Using a dispatch myopic heuristic and the full trace option, trace the evolution of the shop.

  (b) When do you estimate the warmup period was over?

  (c) When should shutdown have been declared?

16.1A.2 Generate a similar problem as in 16.1A.1. Try a number of other heuristics, and compare results.

16.1A.3 Generate by manual-statistical input mode 10 similar problems as in 16.A.1 Try a number of other heuristics, and compare average results.

16.1C Thinkers

16.1C.1 Design simple analytical procedures for deciding when warmup is complete and when shutdown should start. These procedures should be implementable in a simple way in a dispatch simulation.

16.1C.2   Consider a basic job shop simulation. It is desired to add a multiple routing capability for jobs (including parallel machines as a special case).

(a) What modifications would be necessary to the simulation structure?

(b) What modifications would be necessary for the input file?

16.2   EARLY DISPATCH AND ITERATED DISPATCH STUDIES

16.2.1   Overview

In this section we cover simple dispatch methods popular in the 1960s, looking at studies for average flow and average tardiness methods. We then cover iterated myopic policies developed in the 1980s. In particular, we present a study on weighted tardiness using the R&M heuristic in conjunction with lead time iteration.

16.2.2   Classic Dispatch Methods

Makespan.   The objective of minimizing makespan in a job shop by dispatch rules has not received much attention until recently. Most classic and even fairly recent simulation studies concerned themselves with weighted flowtime or weighted tardiness. However, minimizing makespan has been more studied in project management; studies have been done to develop good myopic dispatch heuristics for resource-constrained project management with a makespan criterion. Since the job shop may be considered a special case of resource-constrained project management, those heuristics are applicable. There are a great many dispatch heuristics that classically have been employed on resource-constrained project management. Here we simplify them as appropriate for the special case of the job shop.

First, there must be a method for estimating lead times, due dates, and slacks. The older approach is to estimate the remaining lead time for each job as simply the sum of remaining processing times. That is, lead times are estimated as if there were no resource constraints. Next, the due date for all jobs to be finished at a common point is defined as the maximum lead time (the length of the "critical path" in project terminology). (For simplicity we omit the subscript k for the current machine being considered in all the following.) The slack S_j of any job is the common due date d minus the full lead time, minus the current time:

$$S_j = d - p_j - \text{LT}_j - t$$

The slack is thus how much the job could be delayed without increasing the makespan. The lateness of a job is how much it finishes before or after the due date:

$$L_j = C_j - d$$

Then there are various estimates as to when the operation for job j should start or finish at machine k.

Late finish time (LFT) is the time by which the job must leave the machine to make the overall due date and is clearly

$$\text{LFT}_j = d - \text{LT}_j$$

That is, it is the due date less remaining process time. What we are here calling LFT_j is identical to what we have previously called the derived due date dd_j.

Similarly, *late start time* (LST) is the time by which the job must start on the machine:

$$\text{LST}_j = d - \text{LT}_j - p_j$$

Clearly, the late start time less the current time is identical to what we have been calling the *slack*. Thus rules depending on minimum LST or minimum slack will produce the same heuristic.

Early start time (EST) is the time by which job j could start on the current machine if it encountered no previous waiting at earlier machines. Thus if h_j is the sum of processing times for job j before this machine, then

$$\text{EST}_j = h_j$$

Early finish time (EFT) is the first time that job could finish on the current machine; that is,

$$\text{EFT}_j = \text{EST}_j + p_j = h_j + p_j$$

For the job shop, minimum EFT is the same as shortest processing time SPT.

(All these definitions are more complicated when considering advanced lead time estimation.)

Makespan is an important exception to the above definitions. It must be treated like a project problem, as in Chapter 18. Each job has precedence constraints between successive operations. For a given set of sequencing priorities, there are precedence constraints at each machine specifying priority order for the jobs. In addition, there are final makespan constraints connecting each job completion to the dummy node representing makespan. To obtain the LFT_{jk} (perhaps using data from the previous iteration), this full (CPM/PERT) network must be solved and all LFT times derived.

The simple (noniterative) dispatch rules used classically for resource-constrained project management (which thus may be used for the classic makespan job shop) all simply simulate the problem once and apply a priority rule to available operations whenever a machine is free. The heuristics most often tested were just defined above and include the following:

Simple Job Shop Heuristics for Makespan

(a) LFT (equivalent to earliest derived due date).
(b) LST (equivalent to minimum slack).

(c) EFT (equivalent to shortest processing time).

(d) LPT.

Average Flow Studies. Flow objectives were studied extensively in the 1960s and 1970s, due to the fact that myopic dispatch policies are so simple (SPT or WSPT). Among simple dispatch priority methods, myopic policies were found to be accurate and robust. Modern lead time iteration methods do not add much here since the flowtime objective does not depend on lead times.

However, most of these early studies used rather bland shops with a nearly equal and static load on every machine, random movement from machine to machine, and so on. Thus the conditions that make bottleneck methods and bottleneck dynamics interesting—strong bottlenecks that shift over time and time-varying load and job types—simply were not tested.

Most classic studies restricted attention on flow problems to the unweighted case, where SPT is the known myopic dispatch rule. It is therefore not surprising that early major comparative studies found that SPT minimizes average flow (in rather uniform static shops) among the dozen or so simple dispatching rules that we have previously considered.

Conway (1965a) performed a more elaborate study with the unweighted flow objective. He used the simple job shop model we have already discussed, with random arrivals, exponential time between arrivals, and exponential processing times. Each job goes to every machine at random. All machines, jobs, and operations are equal except for random deviation.

In this study there were nine machines, and about 9000 jobs were simulated over 30 different rules. Table 16.1 shows the relative performance of some of these rules. (The table reports average number of jobs in the system, rather than weighted flow; however, these differ only by a constant.)

AWINQ has some of the characteristics of a bottleneck rule and does better than other rules except SPT. A true bottleneck rule would take into account work in remaining queues (machine prices) as well as work content on each machine.

Conway was able to get slightly better rules than SPT by taking elaborate combinations. For example, (0.985)SPT + (0.015)LWKR achieved a value of 23.0, an improvement of about 1%. Also, a rule (0.96)SPT + (0.04)AWINQ achieved a value of 22.7, an improvement of about 2.5%. This was the best value Conway achieved.

The effectiveness of combination rules of this sort has limited practical value for several reasons. It takes considerable effort to find the optimal weights, and there is

TABLE 16.1 Performance of Simple Dispatch Rules (Conway Flow Study)

FASFS (first arrival in shop)	57.5
FCFS (first to machine)	58.9
SPT	23.2
LWKR (least work remaining)	47.5
AWINQ (least work in next queue)	34.0

no guarantee these weights would be the right ones for a new problem with new job utilizations and parameters. Moreover, the added benefits of using a combination rule are marginal (at savings of 1 or 2%) compared with the simple robustness of SPT. (However, BD provides much better rules for the flow objective.)

Conway also did a number of side studies. In one, he let the processing times become uncertain. Even when jobs were allowed to take from 0 to 200% of their expected time, the performance of the rule was not much affected. For the deterministic case the flow was 23.2; for the 0 to 200% case, the expected flow went up only to 27.1. (This is actually to be expected since weighted flow is a linear objective; linear objectives average over uncertainty very nicely.)

Long jobs are expensive in minimizing flow, since processing them keeps every job in the system waiting a long time. However, in a busy system SPT may keep long jobs waiting an inordinately long time. Early attempts to deal with this produced some rather ad hoc rules. Truncated SPT uses SPT as the normal mode, but when the line gets longer than a preestablished cutoff (say, 30), FCFS is used until the line drops below the cutoff again. Conway found this rule behaved very badly. The reason is obvious. Right when the system is the busiest, we start processing long, low-priority jobs!

On the other hand, Relief SPT is the opposite. When the line length is above five it processes by SPT to keep things moving. When the line goes below five, it uses FCFS, trying to get the overdue but low-value jobs through while the system is not too busy. Conway found that this rule behaved much better, as might be expected.

Bottleneck dynamics would advocate a quite different approach. If very tardy jobs should be expedited, it must be that their priorities have gone up. Use an objective function where the priority increases appropriately over time (e.g., weighted tardy) and develop dynamic priorities by calculating cost/benefit ratios. When the long job achieves the highest priority, it will go through irrespective of how busy the line is. Note that deviations from the SPT rule are not central to this viewpoint, since flow is not really the correct objective.

Average Tardiness Study. An important study of the mean tardiness criterion was carried out by Carroll (1965). Carroll's heuristic is COVERT, a slack-based heuristic that has been described in Section 14.3.4. In the unweighted form that Carroll used, the priority function is

$$\pi_{jk} = [1/p_{jk}][1.0 - (d_j - l_{jk} - t)^+/hl_{jk}]$$

Here p_{jk} is the processing time, d_j the due date, l_{jk} a lead time estimate, and h is a parameter to be chosen.

Carroll's experiments compared the COVERT rule with FCFS, FASFS, SPT, TSPT, and S/OPN. Table 16.2 exhibits a set of results for a job shop containing eight machines operating at a utilization of 80%; from the data it is clear that the COVERT rule was superior to the others tested. Carroll also tested values of h other than $h = 1.0$ in the main study and found $h = 0.5$ to give somewhat better results.

TABLE 16.2   COVERT versus Other Simple Dispatch Rules for Average Tardiness

Rule	Average Tardiness
FCFS	36.6
FASFS	24.7
S/OPN	16.2
SPT	11.3
TSPT	4.6
COVERT ($h = 1.0$)	2.5
COVERT ($h = 0.5$)	1.4

(While we lack the details of the study design, we will present COVERT results in several more complete studies in any event.)

16.2.3   Iterated Myopic Dispatch Methods

Weighted Tardiness. Vepsalainen and Morton (1987, 1988) completed a job shop myopic dispatch study for weighted tardiness (with percentage of jobs tardy also reported for comparison purposes). (They also present a version of the flow shop results covered in Chapter 14. While we have separated these results for pedagogic purposes, it may be useful to study the complete paper as a unit.)

Vepsalainen and Morton consider two candidates for the myopic heuristic: (weighted) COVERT and R&M. Both rules need good lead time estimates. For each they consider three possible lead time estimation methods:

(a) STD—a fixed multiple of the remaining process time.
(b) FORM—a simple fitted formula based on queuing theory and regression, based on line length and the job priority (see the paper).
(c) ITER—lead time iteration, with no smoothing between iterations.

Four benchmark rules were also used: FCFS, EDD, S/OP, and WSPT.

The simulated dynamic job shop had 10 machines with jobs arriving continuously according to a Poisson process. The process is observed until the completion of 2000 jobs. Each job had 1–10 operations, with a randomly assigned routing and with processing time uniformly distributed around the size of a job. Three types of jobs were tested: a shop with equal machine speeds and constant job size, a shop with different size jobs (operation processing times correlated), and a shop with fast and slow machines (bottlenecks).

The load on the shop was determined by the arrival rate to yield five levels of utilization: 80%, 85%, 90%, 95%, and 97% of full capacity. Due dates were assigned randomly over a full range of flow allowance with an average of three or six times the average job processing time for relatively tight or loose due date setting. The full factorial design consisted of 30 different parameter sets (three types of shops, two types of due date settings, and five utilization levels). Each set was run for 2000 jobs. Both COVERT and R&M were adjusted somewhat differently for the different load settings.

In Table 16.3 we present the weighted tardy result, by load and due date allowance. In studying the results, overall the iterative method outperforms the multiple of remaining processing time by about 8–9% for both COVERT and R&M. However, the formula method differs: it saves over half of this total for COVERT, and only about one-sixth of this total for R&M. Thus the study seems a bit inconclusive about the effectiveness of the formula method for weighted tardiness problems.

In Table 16.4 we present the same runs, checking for robustness to the objective function, by calculating percentage of jobs tardy. Here R&M remains better than COVERT, but the effects of different lead time methods are much more random, with the formula method doing the best overall. (Since neither method is trying to minimize percentage of jobs tardy, this result may be spurious.)

TABLE 16.3 Myopic Methods for (Normalized) Weighted Tardiness[a] in a Job Shop

	Load				
Rule	80%	85%	90%	95%	97%
	A. Loose Due Dates				
FCFS	0.28	0.55	1.17	2.69	3.39
EDD	0.02	0.07	0.20	1.22	1.90
S/OP	0.02	0.03	0.08	0.92	1.50
WSPT	0.11	0.21	0.35	0.62	0.71
COV-STD	0.02	0.03	0.06	0.20	0.29
COV-FORM	0.02	0.03	0.04	0.20	0.27
COV-ITER	0.01	0.02	0.05	0.20	0.29
R&M-STD	0.02	0.03	0.05	0.19	0.29
R&M-FORM	0.02	0.03	0.05	0.19	0.28
R&M-ITER	0.01	0.02	0.04	0.18	0.27
	B. Tight Due Dates				
FCFS	0.76	0.92	2.33	3.03	5.98
EDD	0.35	0.44	1.66	2.36	4.47
S/OP	0.27	0.34	1.59	2.06	4.06
WSPT	0.25	0.30	0.56	0.67	1.08
COV-STD	0.10	0.12	0.34	0.43	0.78
COV-FORM	0.09	0.10	0.33	0.42	0.74
COV-ITER	0.09	0.10	0.31	0.40	0.73
R&M-STD	0.10	0.11	0.33	0.42	0.75
R&M-FORM	0.10	0.11	0.33	0.40	0.75
R&M-ITER	0.08	0.10	0.29	0.38	0.73

[a]Weighted tardiness normalized by total work times average weight.

TABLE 16.4 Percentage of Tardy Jobs for the Same Job Shop Study

Rule	Load				
	80%	85%	90%	95%	97%
A. Loose Due Dates					
FCFS	16.5	23.0	34.3	51.2	52.3
EDD	5.1	8.7	20.4	45.3	51.0
S/OP	5.3	9.6	19.7	48.6	54.2
WSPT	12.7	16.1	20.1	24.3	24.1
COV-STD	6.7	10.2	15.2	23.1	25.7
COV-FORM	4.8	6.6	8.7	16.8	17.5
COV-ITER	5.0	8.7	12.9	22.5	24.8
R&M-STD	4.3	6.8	8.6	16.8	18.8
R&M-FORM	4.3	5.7	7.6	15.2	17.4
R&M-ITER	3.7	6.2	8.2	15.9	17.8
B. Tight Due Dates					
FCFS	36.8	42.1	60.9	64.5	76.4
EDD	31.4	38.2	67.2	69.7	80.5
S/OP	36.1	43.0	74.5	74.9	85.1
WSPT	24.3	26.3	32.1	34.1	36.8
COV-STD	22.4	26.4	40.0	38.7	44.3
COV-FORM	17.8	20.6	29.1	30.2	34.5
COV-ITER	22.4	22.3	35.4	35.1	37.8
R&M-STD	18.4	20.7	30.2	33.1	35.4
R&M-FORM	17.0	19.0	29.0	30.1	34.6
R&M-ITER	17.2	19.8	29.0	31.2	35.1

Other useful references for this section include Baker and Dzielinski (1960), Conway et al. (1960), and Dayhoff and Atherton (1986).

16.2A Numerical Exercises

16.2A.1 Use the provided software in manual-statistical mode to generate 10 five-job job shop problems differing only in random variation in job input. Using the makespan criterion, compare a number of single-pass dispatch heuristics for average performance.

16.2A.2 Repeat 16.2A.1 for the weighted flow criterion.

16.2A.3 Repeat 16.2A.1 for the maximum lateness criterion.

16.2A.4 Repeat 16.2A.1 for the weighted tardiness criterion.

16.2A.5 Repeat 16.2A.1, adding the iterated myopic dispatch method.

16.2A.6 Repeat 16.2A.3, adding the iterated myopic dispatch method.

16.2A.7 Repeat 16.2A.4, adding the iterated myopic dispatch method.

16.2C Thinkers

16.2C.1 Truncated SPT rules in effect try to implicitly add a secondary objective function (tardiness) and make some attempt to primarily use the main rule while keeping the secondary objective in mind. Consider the objective of minimizing the number of tardy jobs. Can you come up with a procedure that also tries to minimize excessive flowtime?

16.2C.2 Compare and contrast, in 16.2C.1, the way in which bottleneck dynamics might handle the same mixed objective.

16.3   SEARCH MAKESPAN STUDIES

16.3.1   Overview

Besides the iterative myopic method, there are a number of other more recent methods for job shop scheduling that have recently begun to appear. In Section 16.3.2 we present a study applying the shifting bottleneck algorithm to the job shop for the makespan criterion. In Section 16.3.3 we present a study that solves the same problem using simulated annealing. In Section 16.3.4 we present a study for the same problem using hybrid tabu search. Finally, in Section 16.3.5 we give a study using genetic algorithms.

It is no accident that all of the high cost intensive search studies for the job shop are restricted to the makespan criterion. This special structure allows much faster computation in several ways.

(a) The dynamic myopic one-machine problem (minimize maximum lateness) may be solved extremely fast by Carlier's algorithm (Carlier, 1982), which depends directly on the criterion.

(b) Simulations of a single policy cost for the job shop may be reduced to solving a single longest path problem.

(c) Bottleneck machines and bottleneck jobs may be more sharply defined.

16.3.2   Shifting Bottleneck Algorithm

We turn now to discussion of a paper by Adams, Balas, and Zawack (ABZ) (1988), comparing the performance of the shifting bottleneck method for the makespan job shop problem with classic dispatch heuristics. Since we have discussed the shifting bottleneck method for job shops extensively in Chapter 10 and elsewhere, we do not repeat the complete discussion here, but give only a general outline.

Basically, a large number of embedded one-machine problems are each solved by integer programming (which is fast for makespan). Which embedded problem to solve next is determined by a search procedure depending on current best knowledge of the real bottleneck in the problem. An early embedded problem gives a useful lower bound. If the final solution should happen to be equal to the lower bound, the solution can be guaranteed optimal. This procedure is called SBI (Shifting Bottleneck I). If SBI cannot be guaranteed to be optimal, the user may choose to enter SBII, which applies a high-cost beam search-like improvement procedure.

As a benchmark, ABZ ran eight dispatch heuristics of the types discussed above as a composite heuristic, that is, reporting the best result of the eight answers and also recording the total computation time for them.

As a somewhat more advanced benchmark, ABZ also randomized each dispatch heuristic, again choosing the best of the eight in the same fashion. The randomized rule is to select one of the available operations at random from a probability distribution that makes the odds of being selected proportional to the priority assigned to each operation by the given dispatching rule. The run is then repeated until 10 consecutive runs produce no improvement, and the best result obtained is reported.

For each problem tested then, ABZ:

(a) Ran best-of-eight dispatch (8DISP).
(b) Ran randomized best-of-eight dispatch (RDISP).
(c) Ran SBI and noted whether optimality was guaranteed.
(d) If not, ran SBII in addition, with cutoff time not to exceed actual time for RDISP.

Forty test problems, the results of which are reported in Tables 16.5 and 16.6, were generated as follows:

(a) Choose a number of machines m and jobs n from the following cases: (m = 5; n = 10, 15, 20), (m = 10; n = 10, 15, 20, 30), (m = 15, n = 15).
(b) All jobs must be processed on all machines.
(c) The next machine is chosen uniformly from those remaining.
(d) Operation processing times are all drawn uniformly from the interval [5, 99].
(e) There are static arrivals and static machine availability.
(f) Each case is replicated five times, giving 40 problems in total.

TABLE 16.5   ABZ Makespan Study (Ratio of Cost to SBII Cost): Average of Five Replications

(m, n)	8DISP	RDISP	SBI	SBII
(5, 10)	1.091	1.042	1.024	1.000
(5, 15)	1.016	1.008	1.000	1.000
(5, 20)	1.020	1.019	1.000	1.000
(10, 10)	1.072	1.052	1.039	1.000
(10, 15)	1.160	1.099	1.053	1.000
(10, 20)	1.173	1.113	1.035	1.000
(10, 30)	1.093	1.048	1.000	1.000
(15, 15)	1.150	1.080	1.036	1.000
Average	1.097	1.058	1.023	1.000

**TABLE 16.6   ABZ Makespan Study
(Ratio of Computational Time to 8DISP Time):
Average of Five Replications**

(m, n)	8DISP	RDISP	SBI	SBII
(5, 10)	1.0	31.0	0.4	4.5
(5, 15)	1.0	28.0	0.2	0.2
(5, 20)	1.0	48.0	0.1	0.1
(10, 10)	1.0	32.0	1.1	3.2
(10, 15)	1.0	29.0	1.7	26.0
(10, 20)	1.0	33.0	1.6	31.0
(10, 30)	1.0	29.0	0.5	0.5
(15, 15)	1.0	33.0	2.5	33.0
Average	1.0	33.0	1.0	16.0

Nine other problems from the literature were also solved using only SBI and SBII. We do not report them here. However, it is worth noting that for the very difficult problem by Muth and Thompson (1963) for which an optimal solution has only recently been found at intense effort, SBII found that solution, without verifying optimality, in only 5 minutes on a VAX 780/711.

Note that, since SBII is a sophisticated and rather exhaustive search procedure, we take it as our standard of comparison in estimating "errors" of various methods. Similarly, since 8DISP is the simplest procedure, we take it as the standard for computational time.

We see that, for the 40 problems studied, the multi-pass randomized dispatch heuristic has about 60% of the "error" of the simple procedure, but at 33 times the computational cost. In stark contrast, SBI has only about 25% of the "error" of the simple procedure at about the same computational cost as the simple procedure! (However, note its cost goes up much more rapidly with the number of machines.) Also, it costs about half as much on average to run SBII as RDISP, again for these small problems.

Furthermore, an extremely remarkable property of SBI is that for any problem where there are many more jobs than machines, the initial estimate of the bottleneck machine is stable and accurate, so that SBI often achieves the (verified) optimal solution.

While the shifting bottleneck method is clearly an important one for makespan problems, there are some limitations to this particular study which should be pointed out.

(a) A better dispatch rule to use as a benchmark would be iterated LFT, which requires about half the computational time of 8DISP and is known from project work to be much more accurate.

(b) Randomized dispatch is not state of the art for high computation extensions of dispatch rules.

(c) Dispatch beam search with a beam width of 5–10 would compete better with SBII.

(d) The problem set chosen is extremely smooth and relatively unlikely to provide surprises.

    (i) All operations of a job have essentially the same length.

    (ii) All jobs have essentially the same length.

    (iii) All machines are nearly equally loaded.

    (iv) For a single operation the standard deviation is only one-third the mean.

    (v) Arrivals and availabilities are smooth.

    (vi) Only with a large number of jobs will the largest load be noticeably different than other loads.

    (vii) Machine load is quite stationary over time.

(e) It is hypothesized that the exact results obtained for large numbers of jobs accrue from the resulting more consistent bottleneck, and would not obtain in even a mildly nonstationary environment.

16.3.3 Simulated Annealing

Van Laarhoven, Aarts and Lenstra (VLAL) (1992) have looked into the issue of applying simulated annealing to the job shop problem. Their paper considers a number of issues, such as proofs of mathematical conditions under which simulated annealing converges asymptotically in probability to an optimum, computational efficiencies based on the special problem structure to make extended neighborhood search practical, as well as discussing other versions of simulated annealing and approaches to the scheduling problem. Other related research includes that by Matsuo et al. (1989).

Here we are interested primarily in explaining how simulated annealing was used by VLAL and what results they obtained by comparison with ABZ.

Their representation (which goes back to earlier authors) of a particular solution is that of a rectangular graph, with nodes being operations. Operations are connected horizontally to the right, representing precedence constraints, which are not under the decision-maker's control. On a given machine, operations are connected with arrows pointing vertically, showing which operations precede which, which is under the decision-maker's control.

At first blush any graph is in the neighborhood if it can be obtained by reversing a single arrow (pairwise interchange on a machine). This is an extremely unwieldy neighborhood; roughly mn pairwise interchanges are possible for any given schedule. However, VLAL show that only pairwise interchange of jobs on the current critical path need be considered. (After all, the critical path determines the makespan for the job, interchanging jobs with a lot of slack will have little direct effect.)

Also at first blush, a full simulation of the problem must be undertaken to evaluate the effect of any pairwise interchange. However, due to the analytic simplicity of the makespan criterion, many of these choices may be screened out by an approximate evaluation, and the rest require only a complete longest path evaluation of the problem, for which special algorithms exist. Thus neighborhood search, simulated annealing, and tabu search become practical for the job shop problem with makespan criterion, while other criteria remain elusive.

Ordinary neighborhood search will typically choose the permutation giving the best improvement. But remember that simulated annealing will pick this interchange with the highest probability, second best with next highest, and so on. If the probability falls off very rapidly, the "best" will usually be chosen; the procedure will not experiment very much, and we say we have chosen a "low temperature." If the probability falls off very slowly, some new hopeful may be chosen; the procedure will experiment quite a bit, and we say we have chosen a "high temperature."

High-temperature processes do not need to be run very long; they bounce around too much for fine searching. Low-temperature processes are good for fine searching and greater accuracy but may have to be run for a very long time. This is the trade-off.

VLAL compare their procedure with SBII on the same 40 ABZ problems discussed above. The procedure is repeated five times, and the best result is chosen. This procedure is then tried for three different temperatures, say, High, Medium, and Low. VLAL summarize their results:

> For those instances for which ABZ do not find [an optimal] solution, the running times of simulated annealing (with Medium temperature) and [ABZ] are of the same order of magnitude. [The ABZ] solution is better than the average solution of simulated annealing, and about as good as the best found in five runs of simulated annealing. Putting in [a low temperature] makes simulated annealing much slower but five runs of simulated annealing can now improve ABZ by typically between 1 and 3% [when ABZ is not already optimal].

Thus we see that if simulated annealing is run about five times as long as SBII, VLAL can obtain significant improvement. Thus simulated annealing would appear to qualify as an intensive high-quality heuristic.

It is an open question as to whether simulated annealing would remain computationally tractable for really large problems such as 30 machines and 150 jobs.

16.3.4 Tabu Search

Recently, Dell'Amico and Trubian (1991) have successfully solved the makespan job shop problem to obtain higher accuracy results at high computation times than earlier tabu search results (Taillard, 1990; Barnes and Chambers 1992) or other research methods (Adams et al. 1988; Van Laarhoven et al. 1992).

Any large-scale tabu search method must solve a number of problems:

(a) How to choose the initial heuristic to produce the starting point.
(b) How to modify the starting point into several starting points, to allow multiple search attempts.

(c) How to find a small enough definition of a neighborhood to make computation tractable, yet find excellent solutions.

(d) How to filter out most choices in the neighborhood by an approximation evaluation of the interchange.

(e) How to calculate the remaining interchanges efficiently.

(f) How to deal with various technical issues such as tabu list size, aspiration criteria, and tabu blockage.

We discuss and evaluate Dell'Amico and Trubian's solutions to each of these problems.

Initial Heuristic. Alternate scheduling an operation with maximum priority using forward simulation with scheduling an operation with maximum priority using backward simulation. That is, build up chains of partially scheduled jobs working both forward and backward. Priority is determined by using some standard heuristic such as LFT.

This heuristic can be randomized by looking at the c highest priority operations, and choosing one at random.

(The bidirectional scheduling is an attempt to compensate for the poor estimation of lead times in conventional dispatch algorithms. It would probably benefit from lead time iteration. The randomization is crude; using simulated annealing style randomization would probably provide improvement.)

Randomized Starting Points. Dell'Amico and Trubian have a procedure for breaking deadlocks when the whole neighborhood becomes tabu, which is to choose a (tabu) interchange randomly. Thus rerunning the same starting point will give a different result, which they use as their variational runs.

(Beam-seeding and indeed several other methods seem better than randomizing around deadlocks.)

Small Neighborhoods, Efficient Interchanges. Dell'Amico and Trubian use all the basic methodology of Van Laarhoven et al. (1992) here. In addition, they produce several new possible neighborhoods and other sharpenings and refinements, which we do not discuss here.

Technical Issues. The tabu list length was determined as follows:

(a) If we currently have the best objective ever, forget the tabu list.

(b) Otherwise, there are minimum and maximum lengths to the list.

(c) While we are improving, gradually decrease list length.

(d) While we are not improving, gradually increase list length.

(e) Other similar rules.

A tabu will be removed (aspiration criteria) when the estimation of the move is better than the current best value of the objective. A move is critical if all moves are tabu and none satisfies the aspiration criteria. In this case their program selects one of the moves at random.

Periodically (say, every 800 moves), the algorithm checks if a new best solution has been found in 800 moves (call this a block). If not, it goes back to the last best solution and makes a new start from there.

The program terminates if no better solutions have been found in the last block and if some maximum number of blocks have been executed (say, 15 blocks).

Results. The results of this tabu search are excellent. We show the results only for five problems for which comparisons could be made between tabu search and the shifting bottleneck algorithm. SBI represents the first shifting bottleneck version, SBII the partial enumeration extension, TS-1 the one search iteration results, and TS-5 the five iteration results. The benchmark given is the known optimal solution for the first two easier problems and the best known lower bound for the final three harder problems. For each procedure and problem Table 16.7 presents the percentage above the optimum or lower bound.

Note that for the two easy problems SBII and TS both have good accuracy; but for the difficult problems tabu search is much superior. Running times for SBII and TS-1 were of the same order of magnitude. (Note also that it is difficult to compare the accuracy of TS-1 and TS-5 from Table 16.7, since the lower bounds for the last three problems may be considerably below optimal.)

16.3.5 Genetic Algorithms

Della Croce et al. (1992) have successfully used a genetic algorithm for the job shop problem with makespan criterion. Remember that we discussed the general features and approach of genetic algorithms in Section 6.3.5. We now briefly discuss the basic characteristics of a genetic algorithm as implemented here for the job shop with makespan criterion.

Chromosome. The chromosome (really here identical to the individual) is a particular recipe for solving the job shop problem. Here it is a multi-permutation sequence, that is, a permutation sequence for each machine. The permutation sequence for machine k gives the priority of jobs (fixed) whenever a decision must be made. When a machine is free, all jobs currently available, or *to become available*

TABLE 16.7 Dell'Amico and Trubian Tabu Search Makespan Job Shop Study

Problem	n	m	Opt(LB)	SBI	SBII	TS-1	TS-5
ABZ5	10	10	1234	5.8	0.4	0.3	0.2
ABZ6	10	10	943	2.0	0.0	0.1	0.0
ABZ7	20	15	(651)	12.1	9.1	3.8	2.5
ABZ8	20	15	(627)	23.4	14.2	9.1	8.1
ABZ9	20	15	(650)	15.5	13.1	7.7	6.0

soon, are candidates for scheduling. The highest priority job is chosen. This extension of the dispatch discipline might be called x-dispatch. It allows some inserted idleness if there is a "hot" job soon to arrive.

Initial Population. In the study, a number of individuals were chosen initially at random. The population was maintained in steady state by a birth/death process described below. (Note that results could probably be dramatically improved by using a strong heuristic for an initial individual, and then beam-seeding.)

Fitness. Fitness of an individual is calculated by simply running a complete simulation using the individual's chromosome for instructions. Note that the time to estimate the fitness is similar to evaluating a pairwise interchange in the simulated annealing and tabu search methods described above.

(However, a full simulation including extended dispatch considerations will take considerably longer than a straight longest path algorithm.)

Reproduction Probability. The exact method for choosing parents was not specified, but probability is proportional to fitness of the parent.

Operators to Produce Children. The chromosomes of two parents are matched. Identical genes are assumed superior and not changed. The rest of the chromosome is swapped in pieces in various ways to produce the two children. (This was not explained too well.)

Death Probability. Probability of death is proportional to inverse fitness of the parent. Births and deaths are made equal.

Completion of Generation. Fitness of new children is calculated and they are added to the general population. This ends the generation.

Parameters of the Study

Population size $= 300$.
Births/deaths per generation $= 10$.
Number of generations $= 100$.
Total individuals $= 30,000$.
Number of runs $= 5$ (best chosen).

Della Croce et al. have run a number of studies comparing, among others, the shifting bottleneck, simulated annealing, tabu search, and genetic algorithms. We will quote these results in the next section when we give an overall comparison of these methods.

However, the gist of the story can be given here. Genetic algorithms give results that are somewhat comparable to the other methods but at computational costs often 100–200 times as high as, say, tabu search.

TABLE 16.8 Comparison of Shifting Bottleneck, Tabu Search, Simulated Annealing, and Genetic Algorithms for the Makespan Job Shop

Problem	n	m	Opt(UB)	SB	SA	TS	GA
MT06	6	6	55	0.0	0.0	0.0	0.0
MT10	10	10	930	0.0	0.0	0.5	1.7
MT20	20	5	1165	1.1	0.0	0.0	1.1
LA01	10	5	666	0.0	0.0	0.0	0.0
LA06	15	5	926	0.0	0.0	0.0	0.0
LA11	20	5	1222	0.0	0.0	0.0	0.0
LA16	10	10	945	3.5	1.2	0.0	3.6
LA21	15	10	(1048)	3.4	1.4	0.0	4.7
LA26	20	10	1218	0.5	0.0	0.0	1.1
LA31	30	10	1784	0.0	0.0	0.0	0.0
LA36	15	15	1268	2.9	2.0	0.8	2.9
Maximum				3.5	2.0	0.8	4.7
Average				1.13	0.42	0.18	1.37

Source: Della Croce et al. (1992).

The reasons are not difficult to see. Della Croce et al. have made no attempt to design for speed, or to make use of the characteristics of the problem in other ways. (Their method may very well be adaptable to other objectives without loss.) In particular, they have generated the initial population as very unfit, when methods exist to generate a very strong one. They have made little attempt to roughly prejudge the effect of a given reproduction to avoid mostly bad children being generated, which is quite time consuming. It seems that a serious effort should be made to focus this algorithm on makespan and on speed.

We turn to a general comparison of the various methods' computational experience. Table 16.8 is basically Table 2 of Della Croce et al. (1992). Here MT represents Muth and Thompson (1963) problems, and LA represents a group of problems produced by Steve Lawrence of Washington University in St. Louis, which have widely been used in these studies. SB represents shifting bottleneck (SBII), SA is simulated annealing, TS is tabu search (five repetitions), and GA is genetic algorithms. Tabulated are the percentage the given algorithm scored on a given problem above the known optimum (or above a known upper bound for LA21).

Note that tabu search is the very clear top performer, with simulated annealing the very clear second performer. Genetic algorithms and the shifting bottleneck perform fairly similarly. Note by comparison with the difficult problems in Table 16.7 that relatively easy problems are used here for which optimal solutions are mostly known. In 8 of the 11 problems at least half of the methods found the exact optimal answer, further indicating the relative easiness of these problems.

As far as running times go, the shifting bottleneck method (SBII) and tabu search have roughly comparable computation times. Simulated annealing requires perhaps

5–10 times the computation, while genetic algorithms on some problems may require 100 or 200 times as much computation.

Another useful reference for this section is Kubiak et al. (1990).

16.3.6 Guided, Forward Job Shop Methods

Morton and Ramnath (1992) argue that most high-accuracy methods of the extended neighborhood type, such as simple neighborhood search, tabu search, or simulated annealling, can be improved greatly in accuracy and speed, and thus used for much larger problems, provided two types of improvements are made:

(a) Use of a very high-quality guide heuristic to "aim" the main heuristic.
(b) Use of planning horizon methods to drastically reduce neighborhood search costs and solution evaluation costs.

They report a preliminary study of very large one-machine problems (up to 3000 jobs) with weighted tardiness cost. Their guide heuristic was the x-dispatch improvement of the R&M procedure. Besides verifying that their methods allow tabu search to be effective for very large problems, they developed a new method combining strengths of beam search and neighborhood search, which they term "iterated beam search", which seems to be computationally only $O(n^{1.2})$ in problem size.

Together with Narayan, Morton and Ramnath are in the process of extending these results to the makespan objective and the full job shop. The makespan heuristic chosen is a rolling horizon approximation to Carlier's algorithm (Carlier, 1982). No results of this work are available to report at this time.

16.4 BOTTLENECK DYNAMICS STUDIES

16.4.1 Introduction

Bottleneck dynamics studies are beginning to appear more frequently. We reported a study for the flow shop, two here for the job shop, and we will report another one in Chapter 18 for the multiple project case. We first present a rather early study entitled SCHED-STAR, and then we discuss a rather comprehensive dispatch study comparing simple dispatch policies, myopic policies, iterated myopic policies, and a bottleneck dynamics policy. (Work is currently under way on advanced dispatch policies called x-dispatch policies, which allow certain amounts of inserted idleness at little extra computation cost.

16.4.2 Bottleneck Dynamics with Release Times

Morton et al. (1988) first developed an early version of bottleneck dynamics, which they called SCHED-STAR. It iterated lead times and prices (by a post-iteration analysis). However, the method used to calculate priorities was more complicated and less easy to generalize than current bottleneck dynamics.

NPV Objective. Morton et al. included a rather large job shop study in their paper. The objective function consists of a number of revenues and costs discounted back to the present time by continuous interest:

(a) Revenue—paid at shipping date (the later of the due date and completion date).
(b) Tardiness—paid at the completion date, at tardiness weight times length of tardiness.
(c) Direct costs—paid when an operation is completed.
(d) Holding costs—paid from operation completion until shipment.

It is not worthwhile to fully develop this objective function here, but we shall attempt to give an intuitive analysis. The revenue assumption means there is not any incentive to finish a job before the due date and that lost interest on the revenue when tardy in effect adds to the tardiness cost. The tardiness cost is standard. Both the direct one-time costs and the holding costs from an operation to the due date make it preferable to schedule each operation as late as possible. If jobs are released at the start and machines are scheduled by dispatch, then operations may be performed earlier than desired.

Thus the decision process for this problem is modeled as follows:

(a) Choose a release time for the job.
(b) After release only, the job will be scheduled normally by a dispatch procedure through the shop.

The NPV objective function was normalized by dividing the NPV actually obtained in a run by the relaxed NPV if every job could be finished exactly at its due date, with no waiting at any machine.

Experimental Design. The intent of the experimental design was to test the efficacy of SCHED-STAR in a number of scheduling environments, over a broad range of resource loadings, and against a number of release and dispatch heuristics that have been reported to do well in the literature. The independent variables in this experiment were therefore chosen to be:

(a) Shop structure.
(b) Shop load.
(c) Scheduling heuristic.

A full factorial design with five shop types and six shop loads was replicated 40 times, resulting in a total of 1200 problems. Each problem was then scheduled using nine heuristics, for a total of 10,800 schedules.

Shop Structure. The study focused on five job shop types:

(a) Single machine
(b) Proportional flow shop
(c) General flow shop
(d) Ordinary job shop
(e) Bottleneck job shop

The single-machine problem required scheduling 50 jobs on a single dynamic arrival machine. The other problem types required the scheduling of 10 jobs, each of which visited five resources. Thus every type required scheduling 50 operations.

Both the proportional and general flow shops require that all jobs visit all machines in the same order. The proportional flow shop models required that the processing times of jobs on one machine be a constant multiple of those times on another machine. Both the ordinary and bottleneck job shops require that jobs visit every machine exactly once. In the bottleneck job shop, every job visited the bottleneck machine fourth. The bottleneck machine also had the longest average processing time.

Load Factor. The difficulty in meeting due dates of a problem was set by adjusting its load factor S, which is an approximate measure of the fraction of jobs that can be achieved by a benchmark. The authors used load factors of 0.5, 0.75, 1.0, 1.25, 1.5, and 2.0. (The higher the factor, the higher the load.)

Dispatch Heuristics. The basic dispatch heuristics tested in the study include:

(a) Critical ratio—CR.
(b) (Weighted) COVERT—COV.
(c) Ow/Morton early/tardy—E/T.
(d) SCHED-STAR.

Critical ratio, COVERT, and the Ow/Morton early/tardy rule were all discussed in Chapter 7. SCHED-STAR is somewhat similar to the R&M rule modified for bottleneck dynamics.

Release Heuristics. Critical ratio and COVERT were given one of three types of release heuristics:

(a) Immediate release—IR.
(b) Average queue time release—AQT.
(c) Queue length release—QLR.

Releasing jobs immediately as they came into the shop (IR) provided a benchmark. AQT takes the release time as the due date minus a fixed multiple of remaining

TABLE 16.9   Release/Tardy Job Shop Study (Reporting Normalized NPV)

Heuristic	Single	Shop Type			
		Proportional Flow	General Flow	Ordinary Job	Bottleneck Job
COV-AQT	0.09	0.39	0.23	0.29	0.17
COV-IR	−0.06	0.38	0.20	0.31	0.23
COV-QLR	−0.01	0.33	0.16	0.11	−0.03
CR-AQT	−0.75	0.16	−0.14	−0.08	−0.18
CR-IR	−0.73	0.18	−0.13	−0.07	−0.19
CR-QLR	−0.87	0.04	−0.34	−0.43	−0.49
E/T	0.08	0.36	0.22	0.21	0.11
SCHED-STAR	0.16	0.47	0.31	0.36	0.25
SCHED-STAR*	0.16	0.50	0.34	0.39	0.32

process time. QLR uses a formula to estimate the lead time in terms of the current queue lengths in the shop.

Both early/tardy and SCHED-STAR release a job when its priority crosses from negative to positive.

Two SCHED-STAR heuristics were tested in the study. SCHED-STAR* included iterated pricing and lead times. SCHED-STAR was the same heuristic, but without iteration.

We omit such details as preliminary testing to set parameters and results by load factor, reporting only the average over load factors in Table 16.9.

Summary of Results. Averaged over all runs, the performance of SCHED-STAR* was the best, with a NPV performance normalized at 100%. The noniterated version, SCHED-STAR, came in second at 91%. COVERT-AQT came in third at 68%. COVERT-IR came in fourth at 62%; early/tardy came in fifth at 56%.

Since SCHED-STAR is particularly designed as a bottleneck procedure, it is interesting also to see its performance for the bottleneck shop. Here again SCHED-STAR* came out on top with 100%. The noniterated version, SCHED-STAR, came in second at 78%. COVERT-IR came in third at 62%; COVERT-AQT came in fourth at 53%; early/tardy came in fifth at 34%.

Some conclusions may be drawn: SCHED-STAR is superior to COVERT, which in turn is superior to early/tardy and critical ratio. Iteration is worth 10–20% depending on the bottleneck condition. QLR is an inferior release heuristic. AQT and IR are less clearly differentiated.

16.4.3   Full Comparative Dispatch Study

Recently, Lawrence and Morton (work in process) have set out to demonstrate a unified set of hypotheses about dispatch methods in job shops. In the following, "myopic" means the dispatch policy that would be optimal (or very nearly optimal) for the dynamic one-machine problem; "iterated myopic" means the myopic dispatch policy with lead time iteration; "bottleneck dynamics" means the iterated myopic policy with dynamic price iteration.

Hypotheses

(a) The myopic policy performs better than other simple dispatch policies.
(b) The iterated myopic policy performs better than the myopic policy.
(c) Bottleneck dynamics performs better than the iterated myopic policy.

The study was carried out for five objective functions: weighted flowtime, weighted tardiness, maximum tardiness, weighted fraction tardy, and makespan. For these the myopic policies were taken to be WSPT, Rachamadugu and Morton (R&M), operation due date with true due dates (TODD), weighted modified Hodgson's rule (modified for dynamic case), and operation due date with identical due dates (IODD). There were also a pool of competing heuristics.

Three types of shops were tested. We do not describe them in detail; the reader is referred to Lawrence and Morton (work in process). In the *bottleneck job shop* each job visited each of 10 workstations in random order; one machine was the bottleneck in that operations required twice the time on it as on other machines. In the *limited resources job shop* there were three resources, of bottleneck strengths 1.0, 2.0, and 3.0. Jobs had a highly variable number of operations, leading to highly variable resource usage and prices over time. In the *industry example*, we recreated the 10-workstation, 13-product manufacturing network of Bitran and Tirupati (1988b), which models an actual production facility in the semiconductor industry.

Each cell in Table 16.10 is the average of 20 replications. One heuristic here may be unfamiliar: weighted modified Hodgson's rule. Basically, in Hodgson's rule one orders jobs by operation due date (ODD), finds the first tardy job, removes the longest job not later in the sequence, and iterates the procedure. In the weighted version one removes the one with the smallest (weight/processing time). In the modified version one in addition corrects for the dynamic arrival of the jobs.

Note that in 14 of the 15 objective/shop type conditions the hypothesis is fully confirmed. Note also that the relative strengths of these effects vary considerably over these 15 different treatments. While the hypothesis is robust, clearly much more experimentation would be necessary to predict the relative importance of myopic policies, lead time iteration, and bottleneck dynamics.

16.4A Numerical Exercises

16.4A.1 Using the manual-statistical input generator for the provided software, generate 10 input problems differing only in random variations in the jobs. For the makespan objective, test and compare all dispatch heuristics, including single pass, dispatch myopic, and bottleneck dynamics. Compare the average results.

16.4A.2 Repeat 16.4A.1 for the maximum lateness criterion.

TABLE 16.10   Myopic, Iterated Myopic, and Bottleneck Performance by Objective and Shop Types

Objective	Heuristic Name	Bottleneck Shop	Limited Resources Shop	Industry Example
Weighted Flowtime				
#2 Dispatch	CR	1025	607	1605
Myopic dispatch	WSPT	751	430	1244
Iterated myopic	WSPT	751	430	1244
Bottleneck dynamics	WSPT-P	747	373	1215
Weighted Tardiness				
#2 Dispatch	COV	173	91	276
Myopic dispatch	R&M	169	89	273
Iterated myopic	R&M-I	158	86	241
Bottleneck dynamics	R&M-I,P	157	76	235
Maximum Tardiness				
#2 Dispatch	SLACK	106	70	181
Myopic dispatch	TODD	104	69	162
Iterated myopic	TODD-I	101	69	144
Bottleneck dynamics	TODD-I	101	69	144
Weighted Fraction Tardy				
#2 Dispatch	WSPT	0.54	0.44	0.59[a]
Myopic dispatch	WMHR	0.43	0.31	0.66[a]
Iterated myopic	WMHR-I	0.36	0.26	0.47
Bottleneck dynamics	WMHR-I	0.36	0.26	0.47
Makespan				
#2 Dispatch	SLACK	290	220	462
Myopic dispatch	IODD	290	218	413
Iterated myopic	IODD-I	289	218	395
Bottleneck dynamics	IODD-I	289	218	395

[a]Reversal of hypothesis.

16.4A.3 Repeat 16.4A.1 for the weighted flow criterion.

16.4A.4 Repeat 16.4A.1 for the weighted tardiness criterion.

16.4B Software Project

16.4B.1 Do selected parts of the following software package:

    (a) Design a simulation package such as that needed for the Lawrence study of Section 16.4, except that in addition simple routing choices may be incorporated.

(b) Design several heuristics (including bottleneck dynamics) to be used for routing, in addition to the usual sequencing heuristics.

(c) Write and debug your software.

(d) Test for various combinations of objective functions, routing heuristics, and sequencing heuristics.

16.5 MODEL EXTENSIONS: MINI-CASE

16.5.1 Overview

The classic job shop is a long way from possessing all the features and complexity faced in a real shop. In Section 16.5.2 we discuss a few of these missing features. We have discussed reentrant machines for flow shops in Section 14.4. Dispatch methods handle this situation well. Complex queues include the special case of finite queues and no queues discussed in Section 14.4. However, it also includes a number of other cases, such as when each machine of a cluster has a small queue, served in common by a larger queue of items on a transport loop. In the simplest case, a transport system causes a fixed delay for a job going to the next work center and may therefore be considered an infinite capacity machine. However, a transport system may have its own queues and other variable delays, which we discuss briefly.

Given the number of complications that can make a shop more difficult to analyze, it is useful to consider methods for decreasing complexity in Section 16.5.3.

Compound machines are simply groups of machines that comprise a single resource; that is, there is a single queue to be sequenced and timed. We have already met some examples, such as parallel machines of all types, no-intermediate-queue flow shops, and permutation flow shops. We will analyze further examples such as FMS cells and permutation job shops. While these compound machines can simplify the shop considerably when it is possible to treat them as a single machine, the simulation of such compound machines is still often expensive and difficult. We suggest that functional approximations to lead times could be estimated, allowing us to truly deal with simplified aggregate shops for many purposes.

Finally, we take a second look at the nuclear fuel tube case introduced in Section 14.5. There we simplified the shop structure somewhat to allow it to be analyzed as a flow shop. Here we look at the full structure and show how to analyze the problem as a job shop. Finally, we discuss the advantages and disadvantages of the two types of formulations in practice.

16.5.2 Reentrant Machines, Finite Queues, and Transport

Reentrant Machines

Dispatch Methods. Reentrant situations for the job shop are similar to those for the flow shop for all dispatch methods, including classic, myopic, OPT-like, and bottleneck dynamics. For example, for bottleneck dynamics, each job arriving at such a machine needs only to know its projected lead time, and projected total future resource usage to be able to estimate its current priority. Each future visit to the machine will be counted separately; the job doesn't really care whether two visits are to the same machine or not.

In a similar fashion, in order to estimate a machine's busy period and thus its price, one need only count operations to be done and their urgency. (It matters little whether some operations are from the same job or not.)

One difference between a flow shop and a job shop, however, is that jobs in the flow shop are relatively similar, while in the job shop they may differ wildly. Thus machine prices are likely to fluctuate more strongly and bottlenecks shift more often. Thus the job shop is likely to require more sophisticated bottleneck methods.

Exact Methods. Some care is needed in using branch-and-bound to solve reentrant job shop problems. For example, both integer programming formulations in Chapter 15 (Propositions 3 and 4) use the symbol C_{jk} to refer to the completion time of job j on machine k. This is not well defined for the reentrant case and is not easily fixed without extra notation.

However, the new formulations for project job shops, given by Propositions 1 and 2 of Section 18.2.3, treat each activity as a separate job with precedence constraints. Thus these formulations allow treating the reentrant case without modification, simply by treating each classic job as a series of one-activity jobs with sequential precedence constraints.

A final branch-and-bound method is to branch on assignment choices each time a machine becomes available. This method is, in a sense, a dispatch method and is not affected by the reentrant situation.

Other Methods. The OPT-like job shop method we described in Section 15.3.4 requires the existence of a bottleneck cut, which divides the shop into "before the cut," "the cut," and "after the cut." Since, for reentrant problems, a single job is likely to pass through each of these three parts several times, that method cannot be used. (We should repeat here that we do not have access to the actual OPT procedure, which may be able to handle reentrant situations. Remember also that the OPT-like dispatch procedure, like other dispatch procedures, can handle this situation easily.)

The shifting bottleneck algorithm for reentrant machines is basically a special case of the job shop with precedence constraints (i.e., a project job shop). Solving the one-machine problem for makespan simply involves dealing with a one-machine problem with lags, as we have discussed elsewhere.

Neighborhood search methods, including tabu search and simulated annealing, do not seem well suited for the job shop, unless a global priority sequence is employed with a dispatch discipline, making pairwise interchange practical. Some rule must probably be specified for dealing with jobs at the machine in a different stage of processing, since jobs nearly finished should (intuitively) have higher priority. One such composite rule would be:

(a) Always process jobs with the least number of stages remaining before any others.

(b) Within jobs of the same number of stages, process the job with highest specified initial priority first.

Finally, dispatch beam search, like dispatch branch-and-bound, shares the robustness of other dispatch methods and is capable of handling reentrant machines directly.

Finite Queues

Review of Flow Shop Results. We already met two cases of complex queues for the flow shop in Section 14.4. For flow shops with no intermediate queues allowed (such as a steel slab rolling line), we found the following:

(a) The initial sequence cannot be changed in the processing.

(b) A fairly complex formula is necessary to tell when to release the next job, to avoid conflict on later machines.

(c) Finding the optimal sequence was related to solving a given traveling salesperson problem.

For flow shops with intermediate queues restricted in size, we suggested the following:

(a) Use the makespan criterion.
(b) Simulate the effect of adding a job.
(c) Don't do it if the queue is violated.
(d) Find the earliest that the job can be added.
(e) Find, over all jobs, the one that can be added earliest.
(f) Pairwise interchange on resulting schedule?

Dispatch Methods. Dispatch methods do not handle hard internal constraints very well because the method simply schedules forward, assuming that current feasibility will produce future feasibility. The way that dispatch methods tend to handle such situations is to use large but finite penalties for violating the constraint, for example, tardiness penalties. Of the dispatch methods, bottleneck dynamics makes the most use of quantitative economics and penalties, so we turn to looking at how bottleneck dynamics might handle finite queues.

Bottleneck Dynamics

(a) Run bottleneck dynamics as if there were no queue limitations.

(b) At machines and times with serious queuing problems, tentatively raise prices 25–50%.

(c) Increased resource costs will cause some jobs to have negative priorities and therefore to wait until the machines are less crowded.

(d) If there is still too much crowding, raise prices again.

(e) If there is now machine wastage, on the other hand, lower prices.

(f) Proper iterative search for a good set of prices remains an open research question.

Exact Methods. The discrete time formulation for classic job shops given by Proposition 4 of Section 15.2.2 can easily be modified to keep track of queue lengths and to limit their sizes on an individual machine basis. For simplicity, we do not show how to include initial conditions such as starting queues, dynamic arrival times, or dynamic machine availability. Adding these features will be left to the exercises.

Definitions.

$h(jk)$ Previous machine to k for job j

q_{kt} Queue of machine k at time t

$q_{k \, \max}$ Largest queue length allowed for machine k

Proposition 1. Consider a job shop with static machine availability, static arrivals, maximum allowable queues for machines, and weighted flow objective function. (A potentially overflowing queue does not back up a previous machine; scheduling a job that could cause such a backup is simply not allowed.) Proposition 4 of Section 15.2.2 may be generalized to give the following integer program:

$$\text{minimize} \sum_{j=1,n} w_j C_{jK(j)}$$

$$\text{s.t.} \quad C_{jk} = \sum_{t=1,T} t C_{jkt} \quad \text{for all } (j, k) \tag{1}$$

$$x_{jkt} = \sum_{u=t,t+p_{jk}-1} C_{jku} \quad \text{for all } (j, k), t \tag{2}$$

$$C_{jk} \geq C_{jh(jk)} + p_{jk} \quad \text{for all } (j, k) \tag{3}$$

$$\sum_j x_{jkt} \leq 1 \quad (j \in \{(j, k)\}) \quad \text{for all } (k, t) \tag{4}$$

$$q_{kt} = q_{k(t-1)} + \sum_j C_{jh(jk)(t-1)} - \sum_j C_{jk(t+p_{jk})} \quad \text{for all } (k, t) \tag{5}$$

$$q_{kt} \leq q_{k \max} \quad \text{for all } (k, t) \tag{6}$$

$$C_{jkt} \in \{0, 1\} \quad \text{for all } (j, k), t \tag{7}$$

Proof. Left as an exercise. [*Hint*: Equation (5) says that the new queue length is the old one, plus arrivals, less departures.] ∎

Other Methods. Finite queues tend to cause the bottleneck to be wherever the queue is currently most binding. Hence it is not obvious how to modify OPT-like methods to allow the bottleneck to move.

The shifting bottleneck method does not deal with machine queues at all, and it is difficult to see how to incorporate them directly.

Given a set of priorities for the jobs, neighborhood search will evaluate a particular sequence and will be able to say whether it violates queue restrictions or not. Unfortunately, the correct solution may be to intersperse jobs with idleness to avoid overtaxing the queues. It is not obvious how neighborhood search could put in varying amounts of idle time.

Finally, dispatch beam search cannot be considered for this problem before ordinary dispatch methods have a fairly effective heuristic method for dealing with it since it is more computationally intensive than dispatch.

Other Queuing Issues. Many modern cells, lines, and systems of machines have more complicated queuing arrangements. For example, we discuss flexible manufacturing cells in Chapter 19 as an example of multiple machines that form a single resource. Very briefly, a number of machines stand around in a ring, each with a maximum queue of length one. Jobs are loaded onto a circular belt, which serves all machines. When a machine finishes an activity, it returns it to the belt to be unloaded at the load/unload place. Basically, we have a situation of a *compound* queue. Each machine has a short queue, supplemented by the circular belt, which acts as main queue for all the machines. This is a complex situation; we will discuss it somewhat further in Chapter 19.

Transport. Transport takes many forms, such as:

(a) Overhead cranes.
(b) Forklift trucks.
(c) Worker transport.
(d) Automatic guided vehicles.
(e) Conveyor belts.
(f) In-house railroad cars.

We are primarily concerned about transportation times from a job's completion on one machine until its start on another, although transportation is also important between a machine and bulk storage or between a machine and a machine cluster, and so on. (Bulk storage is in effect an aggregate queue taking pressure off individual queues; see the discussion of two-stage queues in Chapter 19.)

Model 1. Fixed Transport Times t_{hk}. Actual transport between machines may be quite complicated, for example, via a feeder line to a central transporter and then via automatic guided vehicle to the other machine. However, if there is adequate transporter capacity so that transportation is rarely a bottleneck, we may approximate the transportation time from any machine h to any machine k as a constant, t_{hk}.

If there are no routing issues, all scheduling methods may easily be modified to handle these fixed delays. For example, a particular dispatch simulation method may encounter a job operation completion on machine h at time t. It will then enter availability at successor machine k at time $t + t_{hk}$. These transportation times will be added to lead time estimates in a simple way and into times that future resources will be required. (Use of fixed transport times in routing problems was first developed in Chapter 12.)

For branch-and-bound methods, instead of requiring that completion on k is greater than completion on h plus processing on k, it must be greater than completion on h plus transportation time plus processing on k. Shifting bottleneck methods will simply enter the new delays appropriately in the network.

Model 2. Transporter as Resource Between h and k.   If the transporter is at times a moderate to heavy bottleneck, it will be necessary to treat the transporter as a resource with a queue. The resource may be of many forms: single machine (fork lift), parallel machines (dedicated automatic guided vehicles), batch machines (truck), or permutation no-queue flow shop (conveyor line).

Obviously, any of the scheduling methods we have studied for the job shop can accommodate inserting a machine between machine h and k. The problem is that very often transport is more general purpose and may serve several machines on the same trip.

Model 3. Transporter as Multi-machine Server.   Suppose now that the transporter is a truck that transports jobs from several different machines in plant A to several different machines in plant B one-half mile away. Now we have a strange combination of a batch resource transporter and a problem something like a traveling salesperson problem in optimizing the order of pickup and delivery, and thus the transportation times as seen by individual jobs. In many cases it may not be too great a distortion just to use some sort of average transportation time for each job type from past experience. However, in some cases the route transportation time minimization could be quite important. We do not consider this issue further here.

16.5.3   Compound Machines, Aggregate Shops

Compound Machines.   By a compound machine (or machine cluster), we mean the same thing as a problem with a single resource, represented by a single input queue, and all internal decisions prespecified. The single resource is comprised of a group of machines, together with specified transport, routing, priorities, and so on.

Given arriving jobs with their known arrival times, compound machines can be scheduled by specifying the sequence in which jobs are released to the compound machine and, in many cases, by also specifying the lengths of time between releasing two successive jobs. Job release sequence has to do primarily with benefit/cost ratios in terms of greatest benefit per unit of resource used. Release spacing has to do with not overloading internal queues and/or just-in-time type objectives.

Solving detailed compound machines can be quite complicated, involving extensive simulation to evaluate a particular suggested solution. However, remember that a single permutation sequence together with inserted idleness is sufficient for optimal solutions. Thus all the techniques available for solving the one-machine problem are applicable. This is especially useful for neighborhood search, tabu search, and simulated annealing.

So far, the compound machines that we have studied include equal, proportional, clustered parallel machines, general parallel machines, permutation flow shops, and no-queue flow shops. In Chapter 19 we shall discuss several other compound machines. Perhaps the most interesting is the flexible manufacturing cell with a single loading and unloading station. (We will need extensive simulation to solve this problem even approximately.)

Once each compound machine has specified a heuristic procedure for sequencing and timing, it is instructive to ask how the various scheduling methods can cope with solving a job shop made up of such compound machines.

Dispatch methods would simply perform a detailed simulation of each compound machine as a part of a larger simulation for the entire shop. A single master list would be kept of all future events in the shop, whether external to a compound machine or internal. Whenever a dispatch decision must be made, the dispatch heuristics for that compound machine would invoke the correct sequencing and timing decision. If, as for iterative myopic or bottleneck dynamics methods, estimates of lead times and/or pricing economics were necessary, iterative solution of the shop could provide these.

It seems unlikely that exact methods such as integer programming could provide the solution to a shop made up of compound machines. It could possibly solve an aggregation of the problem, to be discussed below. OPT-like and shifting bottleneck methods might be able to solve small versions, but, again, solving an aggregate version seems more likely. Neighborhood search methods and beam search methods also seem limited without aggregation.

Aggregate Shops. In aggregate shops, a machine cluster is approximated by a somewhat simpler one-machine problem. Remember that m equal parallel machines may be approximated by one machine m times as fast, but whose jobs experience delays afterward sufficient to restore the original processing time. Again, a number of proportional parallel machines may be replaced by one machine m' times as fast, where m' is the sum of the speed factors, with each job delayed to its original processing time. Also, a number of parallel machines grouped into m proportional clusters may be aggregated into m independent machines with separate input streams, giving m completely independent one-machine problems.

Aggregation of other compound machines is somewhat more speculative. One idea is to simulate and solve the compound machine a number of times, work out a sequencing rule, and estimate lead times through the compound machine as a function of load, job priority, average priority, and so on.

What is the advantage of aggregating compound machines? At the loss of some accuracy, one is able to efficiently find solutions to much larger and more complex problems. Instead of a shop of 150 machines, for example, one might be able to concentrate on a smaller problem of 25 compound machines. Such a problem is relatively easily solved by single- and multiple-pass dispatch procedures. Branch-and-bound would be able to handle aggregate parallel machines that have no particular nonlinearities. However, it would probably not handle nonlinearities in estimating cluster lead time as a function of load. Other methods would probably be intermediate.

Procedures for aggregate shops, combined with procedures for solving compound machines, hold out promise for two-level systems where the aggregate shop is solved at the planning level and information is sent down to each cluster, allowing it to be solved in detail. This issue will be dealt with further in Section 21.4.

16.5.4 Nuclear Fuel Tube Shop (Reprise)

Review of the Flow Shop Model. In modeling scheduling problems, there is usually a trade-off between (a) a complicated, accurate model that is difficult to solve and difficult to modify and study intuitively; and (b) a simpler, less accurate model that is easier to solve and less difficult to modify and study intuitively.

To make this point clear, we are now in a position to create a more accurate job shop version of the nuclear fuel tube shop of Section 14.5, and to compare its advantages and disadvantages with the original flow shop approximation. The reader should review Section 14.5 briefly, before proceeding.

Figure 16.1 gives an overview of the flow shop version for the nuclear fuel tube shop. Tubes have their diameter reduced by a compound machine of parallel pilgers as the first operation in each of four passes, represented by compound machines 1, 4, 6, and 8. Next, in the first three passes, they are cleaned and deburred in a small compound cell modeled by a fixed delay, represented by compound machines 2, 5, and 7. Last, in the first three passes, they are annealed in a batch furnace, the "main" furnace, marked 3 in each occurrence.

There were three features, in particular, that violated the strict flow shop format:

(a) After pilgering, lots sometimes go to a testing facility before going on to clean/deburr.

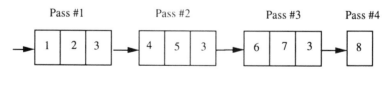

3 = Main Furnace   1,4,6,8 = Pilgers

2,5,7 = Clean/Deburr

Figure 16.1. Nuclear fuel tube shop: flow shop version.

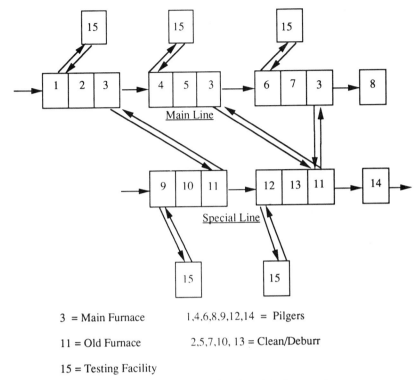

3 = Main Furnace   1,4,6,8,9,12,14 = Pilgers

11 = Old Furnace   2,5,7,10, 13 = Clean/Deburr

15 = Testing Facility

Figure 16.2. Nuclear fuel tube shop: job shop version.

(b) There is a second line using the old furnace, which is usually independent.

(c) Sometimes, however, the main furnace is overloaded, and, despite objections that the old furnace does not do quality work, jobs from the main line may be annealed there.

There were a number of other issues, including movability of pilgers in the main line and major setup costs on the pilgers for changing the major product type. However, we focus on the test facility and partial substitutability of the furnaces.

Extension to the Job Shop. Figure 16.2 shows the nuclear fuel tube shop in a job shop version that addresses these two extensions. First, we show the two pilgering lines as separate flow shop lines. Second, we connect the main furnace (3) and the old furnace (11) to show the possibility of interchange between them. (Actually, we should show the preceding clean/deburr as going to the alternate furnace and then back to the next pass pilger to be precise. However, showing alternatives by connecting them causes little confusion and simplifies the diagram a great deal.)

Also, since each pilger may send lots to the testing facility (#15, several machines in series with single queue), we show possible interconnections to and from each pilger to the testing facility. The resulting model still looks almost like two flow shop lines, with the exception of a routing choice between two furnaces, and a common testing facility.

Advantages and Disadvantages of the Two Methods. In our original discussion we gave an intuitive justification for ignoring the special line:

(a) Personnel do not like to use the second line.

(b) The second line is not needed very often.

(c) We can parameterize the capacity of the main furnace to find out how extra capacity from the old furnace would affect things.

We might justify fuller modeling by the following:

(a) Distaste for the other line can be quantified by charging the job a penalty for using the other furnace in the benefit/cost formulas.

(b) If simulation suggests cross usage even with penalties, perhaps the personnel need to reconsider the situation.

(c) Simulation can help personnel see advantages and disadvantages more clearly.

(d) Parameterizing capacity does not capture the full flavor of the situation.

Turning to the question of the testing facility, we did not justify our omission of it in the original case. However, if we had, we would have argued the following:

(a) Testing is not under the scheduler's control. It comes in three versions: (i) occasional and random; (ii) occasional and known; and (iii) response to equipment problems (random).

(b) Because it is rare, and relatively known in duration, it should simply be added to the pilgering process time, giving a distribution of processing times.

(c) The average can be used; or, possibly, simulations could use random drawings from the distribution.

As answer to this we point out that the testing facility is hardly a fixed delay. It is important to model its queue so that lead times can be more accurately modeled in the main shop.

There are no easy answers to this type of trade-off. Experience and judgment must make these tough calls. Our intention here was to clarify the trade-offs involved.

16.5A Numerical Exercises

16.5A.1 Using manual-statistical input for the provided software, create a small weighted flow objective job shop problem with a reentrant machine, and solve it with several different heuristics.

16.5A.2 Study a job shop problem with fixed transport times, using the provided job shop software. (Consider individualized transport for any single job to be a one-time machine with known processing time.)

16.5A.3 Study a job shop problem via the provided software, where transport from h to j actually requires a single machine with its own queue and fixed process time.

16.5B Software and Computer Exercises

16.5B.1 Set up a simulation of a permutation flow shop with finite queues, which allows adding a new job to an existing partial feasible set at differing possible starting times. The user manipulates the starting time to achieve feasibility. The end result is a statement of the objective value for the permutation, for any input objective function.

16.5B.2 Write a job shop simulation that handles fixed transport times simply by adding lags at appropriate places, both to adjust starting times at the next machine and to increase lead time estimates. Do limited testing.

16.5B.3 Write a compound job shop simulation where each compound machine is a set of equal parallel simple machines. You should be able to simulate the effect of using standard heuristics for parallel machines. Do limited testing.

16.5B.4 Repeat 16.5B.3, except treat each compound machine as an aggregate single machine with m times the speed and lags for each job to restore the original processing time. For a few problems, compare the effect of dealing directly with compound machines and with aggregate machines.

16.5C Thinkers

16.5C.1 Prove Proposition 1.

16.5C.2 Argue that an "aggregate machine" is something like a relaxation, in that it typically will provide lower costs than the detailed compound machine. Provide both examples and counterexamples.

PART V
PROJECT SCHEDULING AND MANAGEMENT

Project scheduling and project management are often treated as synonyms. However, project scheduling tends to be used more in the narrow sense (as in this book) of optimizing an objective function given a number of known activities, resources, and relationships to be satisfied. It is thus a *part* of project management, which is a broad hierarchical activity involving managing supervisors, workers, suppliers, customers, slippage, infeasibility, cost overruns, and so on, as well as the important subtask of project scheduling. In the next two chapters we focus on project scheduling.

A *simple project* is just a task or job with a number of activities and precedence constraints, for which the objective is to complete the entire job. A *job project* is a smaller project among many projects being completed in parallel in a job shop. An example is the manufacture of large custom complex electronic equipment. If a number of the job projects are identical, we might call this a *repetitive project* job shop. Another example might be your desk area, which probably has a number of partially completed projects on or around it at any one time. A *large project* is defined as a task of considerable magnitude in both time and money. It may be carried out once (space shuttle development) or be repetitive (shuttle manufacture). Examples include the building of a hospital complex, the withdrawal of troops from the Gulf war, development of the Saturn automobile, development of a large mainframe computer, or manufacture of the same computer.

The importance of project scheduling is seen in the fact that there are over 100 commercial software packages for the PC alone, which are dedicated to project scheduling. Project scheduling is the third most used software by operations researchers, after only simulation and linear programming. (Word processing and spreadsheets are not being counted here.) Almost all project management packages have sophisticated input, manipulation of data, output, and updating of data. Many have sophisticated ability to interact with the user. However, the scheduling heuristics used are typically not very advanced.

Project scheduling itself is an extremely complex problem. Historically, researchers have worked up from simpler problems by one of two approaches. One approach is to start from methods that work for the job shop and then to add precedence constraints and possibly complex resource usage. The only changes needed, for example, in the basic dispatch idea are to require precedence constraints to be satisfied and sufficient resources to be available before scheduling an activity. This method is discussed thoroughly in Chapter 18.

A complementary approach discussed in Section 17.5 is to start from techniques that assume there are no resource constraints (PERT, CPM, or other critical path methods) and then to add resource constraints and possibly complex resource usage. The pattern of resource usage thus obtained might then be corrected to be feasible by resource leveling, which is discussed at the end of Chapter 17. Alternatively, the lead time estimates available from the relaxed problem can be used to develop priority rules (such as minimum slack) useful in the dispatch simulation mentioned above. This approach is discussed extensively in Chapter 18.

Actually, both approaches are useful, and most current sophisticated techniques will combine elements of both, as will also be apparent in Chapter 18.

____17

NETWORK PROJECT SCHEDULING

17.1 INTRODUCTION TO NETWORK IDEAS

17.1.1 Constraints and Networks

[This chapter uses much material from Baker (1974).] The activities of a project are subject to certain logical constraints, which restrict activity scheduling to certain feasible sequences. Within a feasible sequence, however, activities may be started and stopped independently of each other, as long as the logical constraints are not violated. (This property rules out, for example, a conveyor-driven process in which stopping one activity—by shutting down the conveyor—also stops other activities along the line.)

The particular network model employed in previous chapters essentially represented activities as nodes in the network and represented direct precedence relations as directed arcs. This type of network is referred to as an activity-on-node (AON) network because of its structure. An alternative model is an activity-on-arc (AOA) network. (The AOA representation has principally been used for many years in project scheduling, but AON has become more popular recently. It is clearly useful to know both.) Since the student has already been introduced to the AON representation, we shall use principally AOA here to provide familiarity with it.

Networks are made up of nodes and directed arcs. In an AOA network, the arcs represent activities, and the nodes represent events. The distinction between activities and events in AOA networks is subtle but important. Activities are processes and are associated with intervals of time over which they are performed; events are stages of accomplishment and are associated with points in time. For example, in the development of a prototype of an automobile emissions control device, "testing cold-weather performance" might be an activity, while "test completed" would be an event.

Definition. A < B means "A is a predecessor of B."

In a network, the direction of arcs indicates precedence relations. Thus if A < B, an appropriate network representation would be that given in Figure 17.1. Event 1 (node 1) represents the start of activity A and event 3 represents the completion of activity B. Event 2 has two interpretations; it can be considered the completion of activity A or the start of activity B. The network structure indicates that these two events are not logically distinct. In other words, while they may occur *temporally* at different points, they occur *logically* at the same point. If two activities, C and D, are allowed to be concurrent but C < E and D < E, the network representation is given in Figure 17.2. The interpretation of node 6 in logical terms is the completion of both activities C and D (or, equivalently, the start of activity E, which requires that both C and D be complete). Similarly, if F < G and F < H, where G and H can be concurrent, then the network representation is given in Figure 17.3. The interpretation of node 9 in logical terms is the completion of activity F or, equivalently, the potential start of either activity G or activity H, or both.

Several conventions are usually prescribed for the construction of AOA networks. The principal rules are:

Rule 1: The network should have a unique starting event (node).

Rule 2: The network should have a unique completion event.

Rule 3: The ending event of an activity should have a larger number than its starting event.

Rule 4: Only one arc is allowed per activity.

Rule 5: No two activities have the same starting and ending event.

Rule 5 may create a problem for the basic AOA network. For example, consider the following simple project (planning and holding a fund-raising concert) and the network representation in Figure 17.4.

Activity	ID	Predecessors
Plan concert	A	—
Advertise	B	A
Sell tickets	C	A
Hold concert	D	B, C

For informal purposes, or hand calculations, this network diagram is sufficient. However, large networks often identify an arc by its starting and ending event, and here (2, 3) would be ambiguous. This is why rule 5 is included. To avoid this

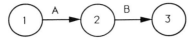

Figure 17.1

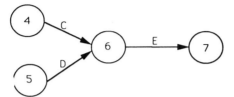

Figure 17.2

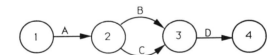

Figure 17.3

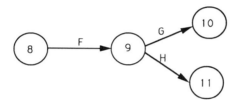

Figure 17.4. Network for planning and holding a concert.

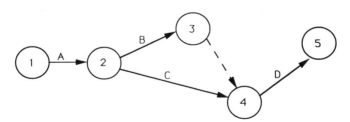

Figure 17.5. Adding a dummy activity to standardize the network.

problem it is necessary to include a dummy activity of 0 duration, shown as a dashed arc in Figure 17.5. The dummy activity allows the same logical relationships to be shown without violating rule 5.

Given these conventions, the task of constructing a suitable network requires two types of input: a detailed list of the individual activities and a specification of their precedence relations. To help provide the latter information, the following questions might be answered for each activity:

Which activities precede it? (What controls its start?)
Which activities follow it? (What are its consequences?)
Which activities may be concurrent with it?

With this information available, the next step is to draw an intelligible network diagram. Often, this will be a trial-and-error process, and many textbooks caution the novice to approach this task armed with a good eraser. Battersby (1967) offers the following guidelines for drawing good networks. (Teaching the computer to draw good networks can be harder!)

Avoid drawing arrows that cross.
Keep all arrows as straight lines.
Avoid too wide a variation in arc lengths.
Avoid small angles between arcs.
Avoid arcs going right to left.

The use of AON networks leads to a different approach to constructing network diagrams for project scheduling. Recall that in an AON network, the nodes represent activities, and the arcs represent the logical constraints. Since each arc corresponds to a direct precedence relation between two activities, there is never any need to use dummy activities. For example, Figure 17.6 shows the AON network for the fund-raising concert.

The direct correspondence of arcs with precedence information, and the fact that dummy activities are not necessary for expressing logical constraints, make AON networks somewhat easier to construct than AOA networks. (An arc can simply be labeled (A, B) as a shorthand for the relationship A < B.) For this reason, when complex scheduling problems are formulated either for project job shops or for large constrained multi-project problems, the AON type of network is usually preferred.

Nevertheless, in practical applications without resource constraints such as PERT

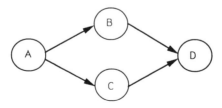

Figure 17.6. AON network for concert example.

and CPM, there are good reasons for using AOA networks. There are some computational efficiency questions we shall not go into here. In addition, showing events as arrows with lengths proportional to times can help management in visualizing the project better. In particular, events in the networks can represent milestones—points in time when project status can conveniently be updated, prospects can be reevaluated, and plans can be revised with new decisions. (This point should not be overemphasized, because there are only a few milestones. In the AON formulation milestones can be represented as dummy activity nodes requiring no time or resources, giving the appropriate review points.)

17.1.2 Chapter Summary

Section 17.2 presents the basic network approach for the deterministic scheduling problem with no resource limitations: calculation of early and late start and finish times for activities; determination of the minimum project duration, the critical path and any alternate critical paths, and identification of slack. While there are few project scheduling problems with no resource constraints, this analysis may be adequate when there are no serious resource limitations; moreover, it provides a foundation for more complete analyses. It plays much the same role as solving the embedded one-machine problem did in the shifting bottleneck method (Section 10.4).

In Section 17.3 we add the idea that critical activities may be performed faster in some instances by expending more cash. The overall trade-off between project duration and project cost then becomes of interest. These time/cost trade-off models have become known as CPM models (critical path method) although all the models of this chapter are properly termed critical path.

In Section 17.4 we go back to the simple model and add the complication that activity durations are not really exactly known but are assumed to be probabilistic (known distribution). The main method used here is called PERT (program evaluation and review technique). It employs a particular method to get managers to estimate a probability distribution for their particular activity and uses this to obtain a simple estimate of the distribution of the overall project duration. In addition, a number of more modern probabilistic issues are discussed briefly, such as: unknown network structure; estimating activity durations; activity correlation; branching, failure, loopback; loopforward, pruning; and black box subproject duration.

Finally, in Section 17.5, we consider the simple project network with added resource constraints and find heuristic solutions by a straightforward classic method called resource leveling. The idea is to plot resource usage over time and see where it is too high; then move some activities forward from the peak or back from the peak in a feasible manner, until the load is more level. In Chapter 18 we concentrate on recent more sophisticated developments in resource-constrained scheduling.

Other references for this section include Miller (1963), Archibald and Villeria (1967), Steiner and Ryan (1968), Cleland (1969), Moder and Phillips (1970), Cleland and King (1975), Wiest and Levy (1977), and Kerzner (1982).

17.1A Numerical Exercises

17.1A.1 Suppose that the tasks in changing a flat tire are as follows:

Task		Predecessors
A	Remove flat tire from wheel	—
B	Repair puncture on flat tire	A
C	Remove spare tire from trunk	—
D	Put spare on wheel	A, C
E	Place repaired tire in trunk	B, C

(a) Find an error in the AOA network diagram (Figure 17.7) for the above project.

(b) Draw a correct AOA network diagram for the project of changing a flat tire.

17.1A.2 Consider the following set of precedence restrictions among five activities: A < D, B < D, B < E, C < E.

(a) Draw an AOA network diagram for these logical constraints.

(b) Draw an AON network diagram for these logical constraints.

17.1A.3 Consider the following set of precedence restrictions among six activities: B < D, B < F, A < D, A < E, A < F, C < E, C < F.

(a) Draw an AOA network diagram for these logical constraints.

(b) Draw an AON network diagram for these logical constraints.

17.1A.4 Consider the following set of precedence restrictions among six activities: A < D, A < E, C < F, B < E, B < F. Given these logical constraints, someone in the project control staff proposes the AOA network diagram shown in Figure 17.8.

(a) Identify the error in the given network.

(b) Construct a correct AOA network diagram for the project.

(c) Construct a correct AON network diagram for the project.

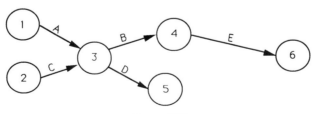

Figure 17.7

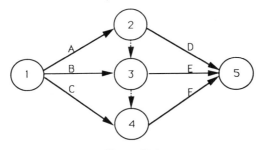

Figure 17.8

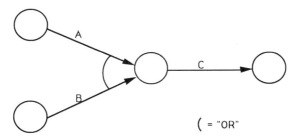

Figure 17.9

17.1B Software/Computer Exercise

17.1B.1   Suppose that an input set of event nodes and activities is such that each node is numbered, and each activity is represented as an ordered pair by its start node and ending node, for example, (6, 9). Develop a software program that renumbers the nodes in such a way that the beginning node for any activity has a lower number than the ending activity.

17.1C Thinker

17.1C.1   When we think about an AON network with relationships $B < D$, $A < D$, $C < E$, $B < F$, $A < E$, $C < F$, $A < F$, we mean an implicit "and" between precedence constraints: $(B < D)$ and $(A < D)$, $(C < E)$ and $(A < E)$, and so on. Often precedences are rather of the form "or." For example: "Before submitting your tax form, either calculate your tax *or* have an accountant do it." If A is *calculate the tax*, B is *have an accountant do it*, and C is *submit the form*, this "or" relationship could be shown as in Figure 17.9. Think of a fairly complex real-world example (like the flat tire example) with both "ands" and "ors" in it. Explain the example. Diagram as both an AOA and an AON network.

17.2   DETERMINISTIC CRITICAL PATH METHODS

17.2.1   Overview

We consider a single project (which is mathematically the same thing as a job with precedence constraints) that has no resource constraints to complicate our lives. Suppose the object is to minimize any regular function of the completion time. This turns out to be equivalent to minimizing the makespan objective, since minimizing the completion time will minimize tardiness, and so on, as well.

In this section we limit our attention to the classic basic network method, with deterministic times and no ability to change activity duration by spending extra money. This leads to the basic critical path method for minimizing project duration and to the classic ideas of early time, late time, floats, and so on.

17.2.2   Basic Analysis

We first consider a project that has no resource constraints. We also consider the most typical case, where our objective function is some regular function of the completion time of the project. No matter what the objective function, this can clearly be optimized by minimizing the completion time of the project, that is, by minimizing makespan.

Proposition 1. To minimize any regular function of the completion time of a single project, it suffices to minimize the makespan of the project. If the project has no resource constraints, this is equivalent to finding the longest path through the network, termed the critical path.

Proof. To complete the project certainly takes as long as the sum of times on the critical path, since these must be performed sequentially. On the other hand, if there were a longer makespan, there would have to be at least one sequence of serial activities taking that long. ∎

Next we turn to a procedure to calculate the minimum length of the project and, in the process, determine the critical path. To begin with, "time" refers to a point in time and is associated with the occurrence of an event; "duration" refers to an interval in time, such as is associated with an activity. Associated with each event in the network are two time values: an *early event time* (ET), which is the earliest point in time at which the event could possibly occur (i.e., the early event time of the start plus the longest path length from the start to the event), and a *late event time* (LT), which is the latest point in time at which the event could occur without delaying the completion time of the project (i.e., the makespan of the project minus the longest path length from the event to the project finish). These are obviously complementary definitions and suggest complementary ways of calculating their values.

Proposition 2. Early and late event times for every event may be calculated as follows:

A. Number nodes so the beginning node of an activity is always lower than its ending node.
B. Calculation of early event times:
   1. ET of start node is 0.
   2. For each node in turn by increasing number:
      (a) To ET of each directly preceding event add the duration of the connecting activity.
      (b) Select the maximum of these sums as the new ET.
   3. Last ET will be the completion time of the project.
C. Calculation of late event times:
   1. LT of last node is its ET (or due date if there is one).
   2. For each node in turn by decreasing number:
      (a) From the LT of each directly succeeding event subtract the duration of the connecting activity.
      (b) Select the minimum of these differences as the new LT.

Proof. Left as an exercise. ∎

Thus a forward pass calculates ET values and a backward pass calculates LT values. Once all ET and LT values are completed, attention shifts to activity information.

Definition. For each activity define the following:

(a) *Earliest start time* (ES)—same as ET of starting event.
(b) *Earliest finish time* (EF)—ES plus activity duration.
(c) *Latest finish time* (LF)—same as LT of finish event.
(d) *Latest start time* (LS)—LF minus activity duration.

To illustrate these calculations, consider the following problem:

Activity	Direct Predecessors	Length
A	—	5
B	—	4
C	—	3
D	A	1
E	C	2
F	C	9
G	C	5
H	B, D, E	4
I	G	2

We first graph (Figure 17.10) an AOA version of the problem, numbering the nodes properly, and labeling each arc with both its identification and its duration. It

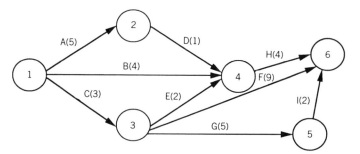

Figure 17.10. Network model for the example project.

is possible to keep track of the forward and backward passes to produce the ET and LT values shown at each node, in the calculation version of the network shown in Figure 17.11. (Since no project due date was given, the late event time for event 6 is taken to be 12, its early event time.) With these data it is possible to calculate the activity information displayed in Table 17.1.

Activities that contribute directly to the duration of the project are called *critical;* any delay in a critical activity will ultimately cause a delay in the completion of the project. The chain of arcs formed by the critical activities is called the *critical path;* it is the longest path from the origin event to the terminal event. (It may not be unique.) In the network of Figures 17.10 and 17.11, the critical path is CF. Because the logical constraints require that activities on the critical path be carried out sequentially, there is no way that event 6 can be realized prior to time 12. In general, if the project is to be completed by the ET of the terminal event, there is no room for any delay along the critical path.

For noncritical activities, however, there is some scheduling flexibility. Consider the scheduling of activities G and I in Figure 17.11. Activity G can start no earlier than time 3 and, to avoid delaying the project, activity I must be completed by time 12. Since an interval of length 9 is available and only seven units of time are

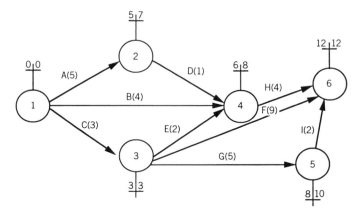

Figure 17.11. Labeled network for the example project.

TABLE 17.1 Activity Information

Activity	Duration	ES	EF	LF	LS
A	5	0	5	7	2
B	4	0	4	8	4
C	3	0	3	3	0
D	1	5	6	8	7
E	2	3	5	8	6
F	9	3	12	12	3
G	5	3	8	10	5
H	4	6	10	12	8
I	2	8	10	12	10

required to carry out activities G and I in sequence, there is some flexibility in scheduling. The two extra units of time can be absorbed before or after either activity, or perhaps in some combination, as shown in Figure 17.12. This kind of flexibility is called *float*. Along the critical path (or critical paths if there happen to be several), there is, by definition, no float, while along other paths there is some amount of float. There are a variety of ways to quantify this measure of scheduling flexibility with respect to individual activities in the network.

Definition. Suppose activity $j = (i, k)$ has length p_j; we define

(a) *Total float* (TF) $= \mathrm{LF}_j - \mathrm{EF}_j$.
(b) *Safety float* (SF) $= \mathrm{LS}_j - \mathrm{LT}_i$.
(c) *Free float* (FF) $= \mathrm{ET}_k - \mathrm{EF}_j$.
(d) *Independent float* (IF) $= (\mathrm{ET}_k - (\mathrm{LT}_i + p_j))^+$.

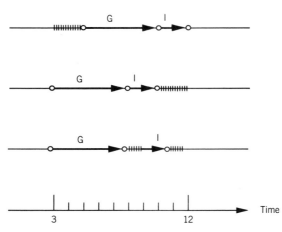

Figure 17.12. Three different ways of scheduling activities G and I.

TABLE 17.2 Activity Floats in Example

Activity	TF	SF	FF	IF
A	2	2	0	0
B	4	4	2	2
C	0	0	0	0
D	2	0	0	0
E	3	3	1	1
F	0	0	0	0
G	2	2	0	0
H	2	0	2	0
I	2	0	2	0

Of these measures, the most frequently used is total float, which actually measures float along a path. The manager of an activity could delay that activity by up to TF as long as there is no delay on any other activity on that path. Thus it is not very conservative since its assumes no other activity on the path has any problems.

The safety float is similar but more conservative: it assumes predecessors have already used up their float and asks what would still be left for this activity. Free float is sort of the opposite. Assuming all activities downstream may have trouble and shouldn't start after their early start, how much extra would be left for this activity? Finally, independent float is the most conservative of all. If upstream activities have used up their float, and we cannot use up the float of any activity downstream, how much, if any, flexibility would remain? Various types of float for our example are shown in Table 17.2. Note that critical activities are identified by the condition TF = 0.

17.2.3 Practical Considerations

The deterministic, no-resource-constraints network model we have described in this section is extremely easy to use and in fact is used very widely. Yet a little reflection will convince us that it is difficult to think of a real project planning problem where resource limitations may really be ignored! How can this be?

The answer is that the planner making up the initial diagram will usually add a number of artificial constraints to make sure that resource constraints are not violated. For example, in building a house, if two activities that have no logical precedence both require the unique cement mixer, say, "pouring the basement" and "pouring the sidewalks," the planner may decide the basement is more urgent to the project and may add a constraint that pouring the basement *must* come before pouring the sidewalks. Thus this is a judgmental constraint rather than a logical constraint. A practical difficulty in doing a more sophisticated analysis is that the decision-maker may not bother to specify which are judgmental constraints and which are truly logical constraints, or may not even understand the issue very well.

If it is not clear which is more "urgent," the planner is likely to follow common practice in the industry or rules of thumb developed over 25 years, or simply to guess. With a more experimental frame of mind, the planner might work out network analyses and project durations for each of the two choices. Of course, this kind of experimenting could get very costly if there were 10 or 15 choices. If the planner is lucky enough to have project planning software available, it would tell him or her which to do, using very crude dispatch heuristics. In Chapter 18 we shall develop better methods for doing resource-constrained project scheduling.

A very similar question from Chapter 16 is why the classic job shop model is so widely used? After all, it assumes a fixed serial order to operations, when in fact there is usually a fair amount of flexibility in the order, and even the possibility of doing some parallel processing of the job. The answer is basically the same. There is a process control engineer who artificially puts in precedence constraints to force a serial structure and to obtain an estimate of the most efficient order in which to run the jobs to suit the shop needs. These process sequences can be changed every month or two as conditions warrant. Once again, this situation can be dealt with more adequately by resource-constrained project job shop models, to be discussed in the next chapter.

Another issue is that the network logic itself is too simplified and limited for many problems. For example, reconsider exercise 17.1A.1 on changing a flat tire. In each case *every* predecessor activity must be accomplished before doing a given activity. We call these "and" constraints.

For example, "remove flat tire" and "get spare tire" both must precede "put spare on wheel." But in most real-life examples there are "or" predecessors involved as well. For example, in the flat tire example, we could remove the flat tire or we could drive on the flat to the service station and remove it. In AOA notation, alternate activities are often shown with a circular connection between those activities, as displayed in Figure 17.13.

Or in AON representation, one may similarly show alternative precedence constraints with a similar connection, as in Figure 17.14.

For a non-resource-constrained network, "ands" and "ors" are not too difficult to handle to determine project duration. In an AOA representation, if incoming activities to an event are all "and" we take the maximum completion time of these activities as the early start for the current activity. If the incoming activities are all

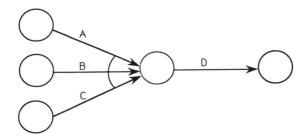

Figure 17.13. Illustration of alternative activities ("or" constraints).

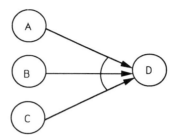

Figure 17.14. AON representation of "or" constraints.

"or" we take the minimum completion time of these activities. We create enough dummy activities/events so that there will not be a mixture of "ands" and "ors" incoming to an event. (Networks with "or" choices are often circumvented in practice by predetermining all the "or" choices in a rule of thumb fashion. Again, this often causes confusion for more sophisticated analyses.)

Another kind of change in representation now in wide use are leads and lags. Instead of the activity being able to start at the completion of the previous activity, it may be able to start somewhat before the finish or somewhat after the finish. While these can usually be modeled with extra activity pieces and dummies, it is more efficient in practice to model these issues directly. Again, for the simple network model, it is easy to see how to modify the project duration, critical path, and slack calculations. This is left for the exercises.

Activity durations can often be affected in a number of different ways: by choosing different designs, by expending larger funds, or by using other resources. Once again, these issues are often decided in advance of the network analysis. We shall discuss time/cost trade-offs in the next section and more complex trade-offs in the next chapter.

There are also a large number of issues having to do with uncertainty in projects, not only with duration, but in terms of activity rework and technological uncertainties. These will be discussed after we discuss probabilistic networks.

17.2A Numerical Exercises

17.2A.1 For the project presented in Figure 17.15:
  (a) Determine the critical path and its length.
  (b) Calculate ES, EF, LS, and LF for each activity.
  (c) Calculate TF, SF, FF, and IF for each activity in the project.

17.2A.2 Repeat 17.2A.1 for the project presented in Figure 17.16.

17.2A.3 Repeat 17.2A.1 for the project presented in Figure 17.17.

17.2A.4 Repeat 17.2A.3, assuming activities coming into node 6 are "or" related and activities coming into node 10 are "or" related.

17.2A.5 Repeat 17.2A.1, assuming the activities entering node 6 use the same

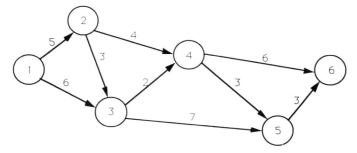

Figure 17.15

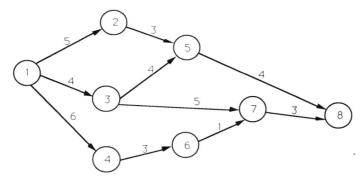

Figure 17.16

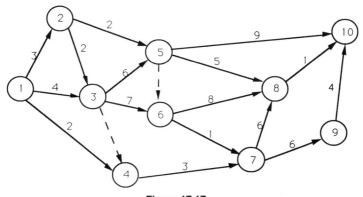

Figure 17.17

unique resource and cannot be done concurrently. (*Hint:* Solve two different problems each with an extra precedence constraint, and choose the one with the smaller duration.)

17.2A.6 Consider the project in 17.2A.3 Suppose that a special piece of equipment is required to carry out activities (5, 8) and (6, 7). Answer each part with the least effort possible, using floats and so on from the original problem if possible.

(a) Does the new requirement lengthen the project duration?

(b) Which early start times are unaffected?

17.2B Software/Computer Exercises

17.2B.1 Write a software program that accepts as inputs the activities in node pair form (such as (2, 8)) and activity duration, assuming input has been numbered appropriately with unique activities between nodes. The software program should calculate early and late times, starts, finishes, the critical path, and various floats.

17.2B.2 Upgrade the program in 17.2B.1 to handle "ands" and "ors."

17.2B.3 Upgrade the program to handle one or more pairs or triples of jobs requiring a common resource, basically by brute force addition of all possible combinations of necessary precedence constraints. If more than N critical path solutions would be necessary, the program should give an error message.

17.2C Thinkers

17.2C.1 Prove Proposition 2.

17.2C.2 Derive rules for dealing with leads and lags, and prove them. Specifically, assume for any pair of activities, one a direct predecessor of the other, that the second activity may start a specified amount before the finish of the other, or else must wait a specified amount after. (Note that this is more complicated than simply allowing the event node to have a time duration.)

17.2C.3 It is often proposed that networks should be "hierarchical;" that is, groups of activities in one part of the network could be aggregated into an aggregated activity, which would be part of a simpler aggregated network. Draw a very complex network and try this out. Do you run into any problems?

17.3 TIME/COST TRADE-OFFS (CPM)

17.3.1 Overview

One obvious generalization of the simple network model is to treat the activity durations as variable, depending on the amount of, say, labor or capital or both expended. Said simply, this means that the expenditure of more money can reduce the duration of an activity, which will be especially important if it is on the critical path. Thus there is a time/cost trade-off for each activity in the project and an overall trade-off involving project duration and project expense.

In the next subsection we develop a simple marginal technique for solving the linear version of this model and then show how to solve the same problem by linear programming. In the last subsection we discuss a few practical issues involved with CPM.

17.3.2 Methodology

The most common structure of such a model, and the one we shall study here, satisfies the following properties:

Each activity's duration is a linear function of the cost expended.

Allowable durations are bounded below and above.

Under these conditions, the time/cost trade-off for a given activity can be represented by the graph shown in Figure 17.18. (We suppress the subscript j since only one activity is being discussed.)

Definition.

m	Minimum feasible duration
M	Maximum feasible duration
t	Activity duration
C_0	Total cost of activity at minimum duration
c	Cost per unit time duration expedited
K	Total cost of activity at actual duration
D	penalty for each extra day until project completion

We first give an example, to show that a small problem can be solved optimally by a simple hand procedure, which we term the *cost–benefit time/cost method*. Then we give an LP formulation for solving larger problems.

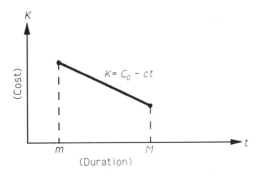

Figure 17.18. The time/cost trade-off.

TABLE 17.3A   Time/Cost Example

Activity	Predecessors	M_j	m_j	c_j	C_{0_j}
A	—	3 days	1 day	$40	$140
B	A	7 days	3 days	$10	$110
C	A	4 days	2 days	$40	$180
D	C	5 days	2 days	$20	$130

D = project penalty cost per day = $45.

TABLE 17.3B   Initial Solution

Activity	Duration	Cost
A	3	$60
B	7	$70
C	4	$100
D	5	$70
	Penalty cost	$540
	Total cost	$840

First, suppose that all activities are scheduled at their maximum durations. Table 17.3A gives the problem input, and Table 17.3B shows a breakdown of the initial cost of the project. First we need to verify that the critical path of the initial network is A–C–D, but that B will also become critical after 2 days are taken off the critical path. Among the critical activities, D is the least expensive to expedite. A 2-day reduction in its duration (costing $40) achieves a reduction in penalty project costs of $90. Thus total costs are reduced from $840 to $790. Now B also becomes critical, so that the A–B and A–C–D paths must be reduced simultaneously for further savings. The alternatives for reducing the length of the project are:

(a) Expedite B and D at a cost of $30/day, netting $15/day.
(b) Expedite A at $40/day, netting $5/day.
(c) Expedite B and C at $50/day, netting −$5/day.

Clearly alternative (a) is most desirable, but a reduction of only 1 day is possible, since activity D must be at least 2 days in length. This reduces our total costs from $790 to $775. Now alternative (b), expedite A, can be applied to reduce A by 2 days to its minimum duration, and the resulting total cost is lowered to $765. At this stage the only feasible alternative left on the list is (c), but it is not cost effective. *Therefore the cost of $765 is optimal.*

We leave it as an exercise to generalize the procedure sketched in the example above and prove that it indeed finds the optimal solution.

As the simple example illustrates, when costs of shortening activities are traded off with benefits of shortening the critical path in a simple network situation, it can be expected that total costs will be a U-shaped function of project length, with a unique optimum.

For larger projects this solution method will seldom be practical. With a great many more activities present there will be more stages and more and more paths will become simultaneously critical. The identification of all alternatives for reducing project length in such cases can be a formidable task, not to mention that there might be a very large number of stages. Therefore solutions to large-scale time/cost problems rely heavily on computer-oriented techniques.

When the cost functions are linear, as in the situation just described (or even piecewise linear if the problem is still convex), the problem can easily be solved by linear programming.

Proposition 3. Given the time/cost definitions immediately above, and representing activity j in event representation by subscript ik (i.e., i is the starting node and k the ending node of activity j), ET_k represents the earliest completion time of node k and ET_N the completion of the project (node N). The problem of minimizing total project cost may be solved as

$$\text{minimize } D(ET_N) - \Sigma c_{ik} p_{ik}$$

subject to

$$ET_k - p_{ik} - ET_i \geq 0 \quad \text{for all } ik$$

$$p_{ik} \leq M_{ik} \quad \text{for all } ik$$

$$p_{ik} > m_{ik} \quad \text{for all } ik$$

$$ET_i, p_{ik} > 0$$

Proof. Left as an exercise.   ■

This linear programming formulation is important for solving very large problems. It is also significant because it is a simple, direct way of stating the problem, and because it can accommodate a number of important variations of the problem described here. (See the exercises.)

An important generalization of the linear time/cost problem clearly involves nonlinear time/cost functions. Exact solutions may not be possible for large projects; accurate and practical heuristic approaches are likely to be available.

17.3.3 Practical Issues

One important issue in time/cost trade-offs is that crashing one activity often changes the cost of crashing another. For example, if a second cement mixer is brought in for a day to crash the time required to pour the basement, it will provide extra capacity for that day and make it quite cheap to do the driveway and walks as well. Once again, these issues are often solved in practice by putting in precedence constraints or by combining activities in such a manner as to "force" activities to be done in a certain way. This issue is best represented on the computer by evaluating a large number of individual scenarios. It is not easy to exactly or even approximately optimize such a problem.

Another issue is that there may be a number of different ways to crash the same activity. In our simple situation, when all ways may be reduced to a cost, this is not too difficult. But if there are several resources required for the activity, which may be crashed in different ways, it can become combinatorial. We will discuss some approaches to this problem in the next chapter.

Doing time/cost trade-offs for networks with leads and lags involves relatively little modification of the linear program, since the problem remains convex. However, adding "or" precedence constraints destroys the convex nature of the problem, and so linear programming cannot be employed directly. Even heuristic search procedures are unlikely to converge to a local optimum. Since "or" constraints are very common in real problems, this presents a challenging research direction.

Projects involving heavy uncertainty in either activity duration or project structure itself, or in the technology for performing activities, make any attempt to make time/cost trade-offs especially challenging. The most successful procedure is probably simply to devise a number of alternative deterministic scenarios and to analyze them. (Complex stochastic models often are likely to obscure more than they edify.)

17.3.A Numerical Exercises

17.3A.1   A small maintenance project consists of the jobs in the following table. With each job is listed its normal time (maximum) and its minimum (crash) time.

Job	Normal Time (days)	Minimum Time (days)	Expediting Cost/Day
(1, 2)	9	6	$20
(1, 3)	8	5	$25
(1, 4)	15	10	$30
(2, 4)	5	3	$10
(3, 4)	10	6	$15
(4, 5)	2	1	$40

(a) What is the normal project length and the minimum project length?

(b) If L is the length of the normal schedule, find the minimum costs of schedules of length L, $L - 1$, $L - 2$, and so on.

(c) Project duration penalty costs are $60/day. What is the optimal length schedule, trading off both kinds of cost?

(d) List the scheduled durations of each job for your solution in (c).

17.3A.2 Change the data in 17.3A.1 so that after an activity is shortened by 2 days, the cost of expediting goes up by another $10/day for further expediting. (Maximum and minimum times still apply.) Now re-solve 17.3A.1 using the new data. (*Hint:* Divide each activity into two activities in series. The second activity starts at length 2 days and can be shortened to zero at the lower cost. The first activity starts 2 days shorter than originally and can be shortened to the original length, but at the higher cost. Now use the original methodology.)

17.3A.3 Formulate a linear program for the following variation of the time/cost model:

  (a) There is no overall penalty cost.

  (b) There are standard time/cost trade-offs.

  (c) The entire project must be completed within time *T*.

  (d) We desire to minimize the cost of the entire project.

17.3A.4 Formulate a linear program for piecewise linear crashing costs. (See 17.3A.2 as an example.)

17.3B Software/Computer Exercises

17.3B.1 Write a computer code to solve simple time/cost trade-off problems by the benefit/cost method outlined in the chapter. Verify the program by solving a small problem both by hand and by computer.

17.3B.2 Generalize your program to allow piecewise linear increasing marginal costs of crashing and increasing marginal costs of longer projects. Validate your program, and investigate a problem data set.

17.3B.3 Generalize your program to allow nonlinear increasing marginal costs of crashing and longer projects. Do you obtain optimal solutions? Justify your answer.

17.3C Thinkers

17.3C.1 Prove Proposition 3.

17.3C.2 Specify formally the exact benefit/cost procedure illustrated in the text for the time/cost problem. Pay special attention to problems as more and more paths become critical. Argue that the procedure is optimal.

17.3C.3 Suppose there are only a finite number of ways to do any activity, each with a cost and a duration. Specify a heuristic for the problem, and state under what conditions it is likely to work well.

17.3C.4 Specify an integer programming formulation for 17.3C.3.

17.3C.5 Specify a branch-and-bound formulation for 17.3C.3.

17.4 PROBABILISTIC ANALYSIS

17.4.1 Overview

Deterministic network models are useful in providing foundations from which to build for both complex job shops and large-scale project management. Deterministic models are much less adequate for the latter, since projects may run months, years, or even decades. Under such circumstances uncertainties (and even vagueness) of many types will play a heavy role.

The simplest way to incorporate uncertainty into a network model with length-based objective is to assume that duration distributions can be estimated for each activity and that these distributions are independent. If the critical path is notably longer than other paths, a distribution for project length is then relatively easily estimated. This allows float distributions for each activity to be estimated and subsequent cost/time trade-offs or other redesign to be investigated. This classic model (Baker, 1974) developed in the 1950s by the Booz Allen consulting company is known as PERT for "program evaluation review technique" and is explained in Section 17.4.2. There are a number of theoretical limitations of PERT discussed in Section 17.4.3. Other practical probabilistic issues will be presented in Section 17.4.4.

17.4.2 Probabilistic Durations (PERT)

Now we allow activities to vary in duration with known distributions and ask: "What is the distribution of time until the project can be completed?"

The simple PERT model requires three basic assumptions.

Simple PERT Assumptions

(a) The activities are probabilistically independent.
(b) The critical path is "considerably longer" than any "near-critical" path.
(c) The critical path can be analyzed approximately by the central limit theorem.

Both (b) and (c) can be stated more carefully; but this is not central to our purpose here. (Roughly, the condition for the central limit theorem is that no single activity contribute more than a "small" share of the critical path variance.)

Proposition 4. Given the simple PERT assumptions, let p_j be the (random variable) duration of activity j with mean μ_j and standard deviation σ_j. Determine the associated deterministic critical path by using durations μ_j. Then if the length of the (deterministically determined) critical path q is $L_q = \Sigma_{j \in q} p_j$ with mean μ_q and standard deviation σ_q, L_q has the following distribution:

(a) $\mu_q = \Sigma_{j \in q} \mu_j$.
(b) $\sigma_q^2 = \Sigma_{j \in q} \sigma_j^2$.
(c) Approximately normally distributed.

Proof. (a) is true for any sum of random variables; (b) is true for variables that are independent; and (c) is true by the central limit theorem holding approximately. ∎

Note that for any particular realization of all activity durations, q may not be the true critical path; q is simply the longest path *on average*. We will return to this point in the next subsection.

Given the mean and standard deviation of the deterministic critical path, we assume the reader is familiar with using the unit normal table to determine the probability of meeting a project deadline of d:

$$\Phi((d - \mu_q)/\sigma_q)$$

The unit normal table is available in any introductory probability/statistics text. The problem set will assume that such skills and the table are available.

One of the strengths of PERT is to provide a way to estimate μ_j and σ_j for each activity. The manager for each activity j is interviewed and asked to provide an optimistic duration O_j, a most likely duration M_j, and a pessimistic duration P_j for that activity. From these inputs, PERT estimates the mean and standard deviation by

$$\mu_j = \tfrac{1}{6}O_j + \tfrac{4}{6}M_j + \tfrac{1}{6}P_j$$
$$\sigma_j = \tfrac{1}{6}(P_j - O_j)$$

While these formulas are often explained in terms of properties of beta distributions, a more fundamental analysis seems also useful. The formula for the mean suggests a weighted average of the three estimates, which is intuitive. It further suggests that the manager is unbiased and weights the (subjectively) most likely value at four times the weight for either end point. The formula for the standard deviation suggests that the manager for some reason sets a range for the estimate of six standard deviations. These assumptions can be verified and/or modified by statistical analysis of past manager behavior. We discuss this practical issue further in Section 17.5.3.

As an example, consider the project described in Table 17.4 in which μ_j and σ_j^2

TABLE 17.4 PERT Example

Activity	Predecessors	O	M	P	μ	σ^2
A	—	2	4	12	5	2.78
B	—	3	6	9	6	1.00
C	A	1	2	9	3	1.78
D	A	3	3	9	4	1.00
E	B	1	2	3	2	0.11
F	B	2	8	8	7	1.00
G	C	1	2	9	3	1.78
H	D, E	4	5	12	6	1.78
I	F	1	3	5	3	0.44

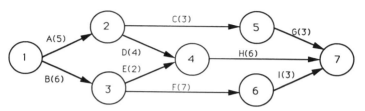

Figure 17.19. Network of the project in Table 17.4.

are first calculated from O_j, M_j, and P_j. The next step is to construct the network diagram and to label activity j with its mean duration, μ_j, as shown in Figure 17.19. Then, by using these mean values, we can identify the PERT critical path as B–F–I.

From these calculations and Proposition 4 it can be seen that the length of the path B–F–I has a mean of 16, a variance of 2.44, and a standard deviation (the square root of the variance) of 1.56. The probability that path B–F–I can be completed by, say, time 15 is then given approximately by $\Phi((15 - 16)/1.56) = 0.26$. Thus PERT would estimate that the probability of completing the project by time 15 is 26%. Actually, for this particular problem, the true mean completion time is larger than 16, and the true probability of completion by time 15 is considerably less than 26%, due to the fact that there is a "near-critical path." We analyze this and other difficulties in the next subsection. Nevertheless, PERT is widely used in practice because it is easy to use and apply.

17.4.3 Theoretical Aspects of PERT

Four main types of objections have been raised about PERT over the years:

(a) Its estimate of the mean project duration is always too small.
(b) It gives inadequate attention to other paths, which will sometimes be critical.
(c) It assumes activity durations are independent.
(d) The estimation process for activity durations may be inaccurate.

We discuss each of these in turn and give some remedies in the next subsection. We first show that the PERT estimate of project mean duration is biased downward. (In the following, q represents the deterministic critical path, p an arbitrary path, and L_p the random length of path p.)

Proposition 5. Consider a PERT network. The true mean project duration is greater than or equal to that of the deterministic critical path.

Proof.

$$\text{True mean} = E[\max_p\{L_p\}] \geq E[L_q]$$

$$= \max_p\{E[L_p]\} = \text{PERT mean.} \qquad \blacksquare$$

(In the exercises the student will be asked to explain this proof more completely.)

For example, notice in our example just above that while we focused on B–F–I with length 16 and standard deviation of 1.56, near-critical path A–D–H has a mean of 15, a variance of 5.56, and a standard deviation of 2.36. Thus there is a substantial probability (it turns out to be 36%) that A–D–H will be the actual critical path rather than B–F–I. Since both paths must be completed before the project can be, this means that our estimate of 16 for the project's mean length is low, and our estimate of 24% for the probability of finishing by time 15 is too optimistic. What is worse, even B–E–H at length 14 may sometimes be the critical path. These effects, which are noticeable in the example, would have been much less significant had the second longest path been less "near critical." This issue has been studied by a number of authors including Baker (1974).

The second difficulty is related: locating a single critical path, when in actuality different paths may be critical ex post depending on the particular realization of the durations. Remember that the usefulness of the critical path lies in knowing where there is no float: where managerial attention should be paid and where time/cost trade-offs are worthwhile.

One simple idea is to change the critical path idea to *critical path probability*. For each likely path, estimate the probability it will be critical (such as by the Monte Carlo simulation methods to be discussed below). In our example, the critical probability of B–F–I might be 59%, A–D–H 33%, B–E–H 6%, and miscellaneous 2% (not calculated). High-probability paths would be given the most attention.

A second simple idea is to focus on the *critical activity probability*. The idea here is that an activity that is on several "likely" critical paths is more important. Clearly, the critical activity probability is the sum of the "likely" critical path probabilities for which the activity is a part. In our example, activity B would have a "likely" critical activity probability of 59% + 6% = 65%.

A more precise idea, which would require somewhat more calculation, would be to determine the distribution of, say, total float for each activity. Small float times are probably worth managerial attention almost as much as zero float times.

A practical, if somewhat expensive, alternative for practical implementation is Monte Carlo simulation. A set of durations is chosen randomly, and the deterministic network resulting is analyzed. This can be repeated, say, 1000 times. Any set of statistics about critical paths, activities, slacks, and so on can be generalized and studied. See Baker (1974).

A third criticism of PERT is that activities are often very correlated in practice. To see why this is so, remember that we have said that non-resource-constrained project networks often really deal with resource constraints implicitly by adding artificial precedence constraints to make sure both activities needing the resource do not operate simultaneously. Clearly, if this resource turns out to be inadequate, for example, and operates at only partial efficiency, both activities will be affected. Or several activities may require the same late raw material shipment. Or it may rain off and on for the entire project.

This can be handled relatively well again by Monte Carlo simulation as follows. Besides creating the usual random variables for each of the two correlated variables, create a common "fudge factor" random variable, which adds to or multiplies the basic independent variables. A single fudge factor is chosen at random and applied to both events.

Finally, there may be considerable practical difficulty in estimating the mean and standard deviation of activity durations. We will discuss this question briefly in the next subsection.

17.4.4   Other Probabilistic Issues

There are a number of serious issues that come up in dealing with large, long-duration, probabilistic projects. These include:

(a) Estimating network structure.

(b) Estimating activity durations.

(c) Activity correlation.

(d) Branching, failure, loopback.

(e) Loopforward, pruning.

(f) Black box subproject duration.

These are rich topics, with much anecdotal evidence on which little formal research has been done. We can only discuss them very briefly here.

Estimating Network Structure. Consider a large R&D project over a 5-year horizon, such as designing, prototyping, testing, and protomanufacturing of a futuristic mainframe computer. The network structure of the later part of the project can only be guessed at. If certain key planned activities turn out to be impossible, there may be planned alternatives, or they may be generated "on the fly." Can several alternative conventional network diagrams treated as scenarios do the job, or should the diagram somehow represent "ands", "ors" and "?". What are good rolling horizon procedures for updating the network structure and durations without losing the planning feature? The answers chosen to these are varied in practice. We will indicate some of them below.

Estimating Activity Durations. For our mainframe design problem, suppose there are 500 aggregate activities. (Each might be composed of a subnetwork of 10–20 micro-activities; ignore this here.) Suppose the project involves 10 department heads, each supervising four or five managers, each in charge of perhaps 10 of the aggregate activities at differing points in time.

Getting decent optimistic, most likely, and pessimistic estimates from the line managers and department heads is no easy matter. There may be a desire to give high time estimates, so that set goals are easy to meet. There may be a desire to lowball the time estimates to be certain of being assigned the activity. Optimistic and pessimistic mean different things to different people. Also, there is always a lot of turnover among personnel, so that the relevant future manager may not even be available. Times depend on resource allocation, which is often a political process (and a dynamic and ongoing one) involving several departments heads, a number of managers, the project head, and often top management.

It may be possible to correct somewhat for manager idiosyncrasy by analyzing goals and performances of managers for past projects by econometric means. This is complicated though, and the past does not always predict the future. This is especially true if the manager learns the biases expected of him, and reverse games the system.

Activity Correlation. As indicated in the previous subsection, activity correlation is usually related to common resources used (in parallel and/or at different times) by the activities in question, where the effectiveness of the resource is itself stochastic. If the situation can be well modeled, extra common random variables can be included across activities, and the resulting model can be solved by simulation.

Unfortunately, common "resources" are often quite subtle, such as common technological assumptions, common business cycle, common cost breakthroughs, common equipment test reliability, common esprit de corps, and common management pressure. A likely possibility is not to model all these factors explicitly, but simply to pick pairs and triplets of activities with assumed correlation factors. These can be translated into common additive or multiplicative "fudge factors" with standard deviations chosen appropriately.

Branching, Failure, Loopback. Many times in a long-duration research and development project, it is not clear which of several possible avenues of investigation will actually be followed. An external random event (state of nature) may occur at some point in the project, and with certain probabilities this will determine which avenue to follow. The resulting diagram looks something like a mix between a usual project diagram and a decision tree. The random event might be shown as a circle with a cross in it, and possible branches marked with their probability as shown in Figure 17.20.

The diagram example shows that after the finish of A, B, and C, we will know the result of the marked probabilistic outcome. With 60% probability D, E, and F

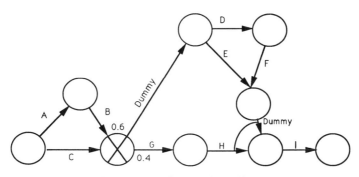

Figure 17.20. Random branching.

will be next; with 40% probability G and H will be next. Then the two branches join again with an "or" connection, and the project concludes with I. To analyze such a project for duration, take two projects. One contains D, E, and F; suppose the overall project duration is 42 in this case. One contains G and H and has an overall project duration of 38. Then the actual expected duration of the project is 0.6(42) + 0.4(38) = 40.4 (One could determine the expected cost or expected resource usage of the project in the same way.)

Note that random branching is similar to an ordinary "or" except that here we do not choose the branch. Rather we analyze two problems and weight them together.

Failure is a particular kind of branching. If testing at a certain point indicates "failure" of the procedure, then something must be done to "fix" the failing activities. The most common way this is shown is by "loopback." If the test shows that a group of activities fails, then the arrow loops back to force those activities to be completed again, as shown in Figure 17.21. If the probability of failure is 60%, then conventionally it is assumed that the repeated activities take the same amount of time to be completed the second time, and that the probability of failure is equal and independent each time. Then we come to the test again with again a 60% chance of failure, and so on.

Thus if the subproject within the loop requires time T each iteration, with probability of failure Q each time, the expected time to successfully complete this part of the project would be

$$E = T(1 + Q + Q^2 + Q^3 + \cdots) = T/(1 - Q)$$

This analysis of loopback is very common, yet it is highly inaccurate. Failure is often due to inadequate understanding of the process, especially in R&D. Doing the activity once helps (through learning) in doing that activity the second time. On the other hand, failure the first time suggests scrambling to do something different the second time, which may require much more time. Similarly, the failure probability the second time could be argued to be much higher or somewhat lower, and so on.

If we know how the times T and probability Q change with iteration k and assume

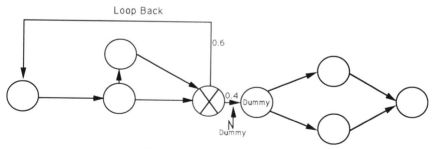

Figure 17.21. Simple loopback.

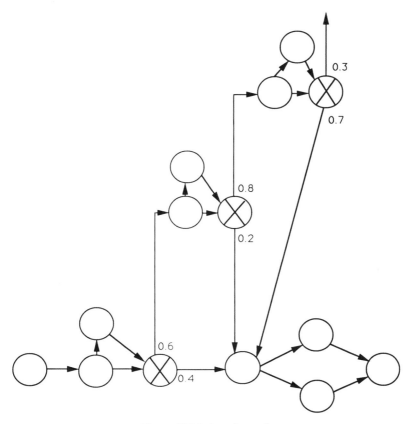

Figure 17.22. Loopforward.

independence, it is still fairly easy to calculate an expected total time for success. This is left as an exercise. However, if there is no rough independence, or if the looping process is more complex, an alternative method called loopforward may be useful, which we turn to next.

Loopforward, Pruning. The main advantage of showing failure situations in terms of loopback is to take advantage of symmetries, such as constant subproject duration time, or constant probability of loopback, or constant subproject structure. If none of these really holds in a research project (which is usual), it may be more useful to simply represent the failure possibilities in a forward direction, which we might dub *loopforward,* shown in Figure 17.22.

This does not emphasize a false repetition of times, structure, or failure. The main difficulty with loopforward, of course, is that far too many branches will be generated. A reasonable way out of this difficulty is to remove (prune) branches of less than a certain minimum probability and to fudge their consequences roughly into their more important neighbors. This procedure is necessarily judgmental in

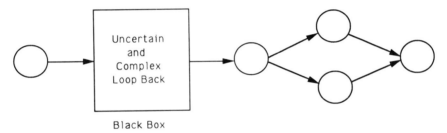

Figure 17.23. Black box.

nature but has the advantage of being the way in which managers seem to think about an uncertain future.

Black Box Subproject Duration. Highly uncertain subprojects (e.g., R&D planned for 5 years in the future) may be felt to represent extensive failures, repeats, or redesign, without much ability to specify what every individual activity will be, where failures will come, and what to do. This very fuzzy inner structure does not lend itself well to detailed project structuring, even with failure probabilities (see Figure 17.23).

Yet it may be important to estimate, say, a subproject duration before the start of the main project, and subsequently to estimate remaining subproject duration with only such indicators as various resources used to date. Some sort of econometric analysis of "similar" research subprojects performed in the past might provide a regression equation or equations for estimating these durations.

17.4A Numerical Exercises

17.4A.1   What is the PERT estimate of the probability that the following project will be completed in 20 days? Justify whether or not the estimate is good.

Activity	Predecessors	O (days)	M (days)	P (days)
A	—	1	4	7
B	—	1	5	9
C	A	3	6	9
D	B	1	2	3
E	A	1	2	9
F	C, D	2	4	6
G	C, D, E	2	9	10
H	F	2	2	2

17.4A.2   What is the PERT estimate of the probability that the following project will be completed in 18 days? Justify whether or not the estimate is good.

Activity	Predecessors	O (days)	M (days)	P (days)
A	—	3	5	7
B	—	4	6	20
C	A	1	2	3
D	A	3	3	3
E	A	1	2	9
F	B, C	0	6	6
G	B, C	2	5	14
H	D	1	8	9
J	E, F	1	3	11
K	G	2	2	2

17.4A.3 In the project shown in Figure 17.24, all activities have independent distributions with mean μ and standard deviation σ as given. What is the probability that A–C–E–G will be the longest path in the network? (Use the normal approximation for the sums of random variables.)

Activity	μ	σ
A	30	4
B	50	20
C	40	30
D	20	10
E	80	8
F	60	60
G	40	25
H	40	15

17.4A.4 To the example of 17.4A.3 add an activity J of known length 20 between the B–D node and the C–E node. Suppose C and J are related by "or" rather than "and." What is the deterministic solution? What would PERT give as the distribution of project duration?

17.4A.5 In 17.4A.4 discuss as carefully as you can the bias of the PERT solution.

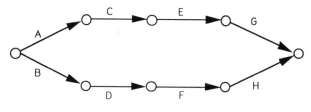

Figure 17.24

17.4A.6   To 17.4A.4 add the same activity J, but assume rather that J and D are probabilistic branches with probabilities 0.7 and 0.3, respectively. Find the expected project duration if activity durations are deterministic.

17.4A.7   To 17.4A.4 add the same activity J, with activities having random durations. Assume that either D–F–H or J–E–G must be chosen after observing the actual durations of A, B, and C. How much can you say about the proper way to make this choice?

17.4B Software/Computer Exercises

17.4B.1   Given an input project network with each activity represented as a pair of numbers, and each activity given a mean time and a standard deviation, set up a Monte Carlo simulation to simulate all durations for the network N times. Each time calculate and save the critical path, the project duration, and the total slack for each activity. Tabulate the frequency of the 10 most common critical paths, the mean and distribution of project duration, and individual slacks.

17.4B.2   Using the software from 17.4B.1, set up an experimental design, and test the adequacy of the PERT solution under various conditions.

17.4B.3   Add the capability of probabilistic branching of various types to your program in 17.4B.1. Again check the adequacy of PERT solutions.

17.4B.4   Add "or" capability to your program in 17.4A.3:
  (a) If the choice of which activity to do can be made after the durations are known.
  (b) If the choice of which activity to do must be made when the activities can be started but their durations are unknown.

17.4C Thinkers

17.4C.1   Formulate a correlated probabilistic network model as follows. Let the duration of any activity be the mean plus several normal random variables with mean zero. One of these variables is unique to the activity, but the others will be in common across one or more activities, representing common problems. Estimate the mean and standard deviation for the PERT critical path.

17.4C.2   Consider loopback, where the subproject activity times (deterministic) and the probabilities of loopback vary with k, the number of times through the loop; that is, we have $T(k)$ and $Q(k)$. Assume the probabilities are independent. Give a formula for the expected total duration until the loopback is finished.

17.4C.3   Suppose that in a certain project structure a "preparation" activity precedes each loopback subproject. There is a quality/time/cost trade-off on the preparation activities. A higher quality preparation requires more preparation time and higher cost but results in a lower probability of failure Q in the next loopback. Assume some kind of linear relationship, like $Q = Q_0 - c_1 t - c_2 d$, with limits on the range of t (the preparation time) and d (the preparation dollars). Discuss this model qualitatively, including what happens on or off the critical path, and relative magnitudes of Q_0, t, and d.

17.5 SIMPLE RESOURCE LEVELING

17.5.1 Overview

So far we have considered only the important special case of project scheduling in which:

(a) Activities need not compete for limited resources.
(b) There is a single project.
(c) We wish to minimize its duration.

This basically led us into analyzing the longest path in a network, a critical path analysis. Although we looked at related models such as time/cost trade-offs and probabilistic networks, these were still founded conceptually on the simple critical path analysis, performed in the same fashion.

In Chapter 18, by contrast, we will develop methodologies for more complex project scheduling in which:

(a) Activities compete actively for limited resources.
(b) There are multiple projects, with various types of possible objective functions.

This will lead us, as one example, into combining the bottleneck dynamics ideas of Chapters 15 and 16 with a much more complex notion of critical path. If we dub this "project bottleneck dynamics," we will find that the same basic ideas can be extended into multiple resource/cost trade-off models, interrelated project models, net present value models, and so on.

There is one resource-constrained approach, however, which requires only the critical path methods of this chapter for implementation. Because it is basically a simple critical path method, and because it provides a good introduction to the more complex methods of the next chapter, we present it here, even though, in a sense, it belongs in the next chapter.

In Section 17.5.2 we first present a simple resource-constrained model for a single project for which a minimum duration is sought. We then discuss in rather general high-level terms the ideas underlying the resource leveling approach. There are many ways to implement the approach, all requiring human judgment in the process. We need human interaction with the computer (i.e., a decision support system). We choose to present a search tree approach based on adding or not adding artificial constraints to shape the schedules obtainable. The approach is quite practical but requires a great deal of human input by comparison with methods in Chapter 18. Finally, in Section 17.5.3, we discuss briefly a number of practical issues that often arise in implementing such an approach.

17.5.2 Methodology

The single-project simple resource-constrained model can be defined as follows. There are N activities, indexed $1, ..., j, ..., N$; there are K resources, indexed $1, ..., k, ..., K$. The immediate predecessor set of an activity j is denoted by $P(j)$; if $P(j)$ is empty, j is an initial activity. Start times of activities are labeled S_j, and finish times C_j. It is desired to minimize the completion time C_N of a unique completion activity. [N can be defined as belonging to no sets $P(j)$.] p_j is the deterministic processing time of j. An activity j is called active at time t, written $j \in A_t$, if and only if $S_j \leq t < C_j$. There is a fixed and invariant quantity of resource k available, termed RA_k. Any activity j uses resource k at rate r_{jk} for as long as it is active. Total usage of any resource must not exceed its availability at any point in time. This model can be stated in the form of a mathematical program.

Proposition 6. An optimal solution to the simple resource-constrained model given above can be found by solving the following mathematical program:

$$\min C_N$$

$$\text{s.t.} \quad S_j \geq 0 \quad \text{if } P(j) \text{ is empty}$$

$$S_j \geq C_i \quad i \in P(j), \text{ for all } j$$

$$C_j = S_j + p_j \quad \text{for all } j$$

$$j \in A_t \leftrightarrow S_j \leq t < C_j \quad \text{for all } j, t$$

$$\sum_{j \in A_t} r_{jk} \leq RA_k \quad \text{for all } k$$

Proof. Straightforward. ∎

This particular formulation is easy to understand intuitively as a simple extension of the unconstrained network model. We will discuss some variations in this context. It is perhaps less obvious that if this formulation is discretized into unit time intervals, that it can be reformulated as a linear integer program. We will come back to this issue in Chapter 18.

If, in the model of Proposition 6, all r_{jk} are 0 or 1, and all RA_k are 1, then we have equivalently a project job shop with only one job, with minimizing overall makespan as the objective, and possibly multiple resources per activity. If, in

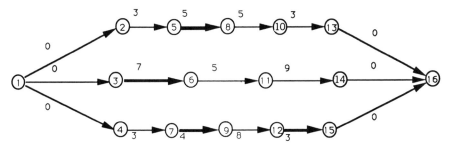

Figure 17.25. Resource leveling example.

addition, r_{jk} is 1 for only one resource, we have a project job shop with machines as the only resources. We might also term this a job shop with makespan objective and precedence constraints on each job. If, instead, some jobs have individual objectives such as weighted flow or weighted tardiness, we need only change the objective to be the sum of functions of lateness for the completion activity for each job.

Now let us return to trying to approximately solve the formulation in Proposition 6 by resource leveling. We first ignore the resource constraints and solve for the simple unconstrained network solution. For that solution we can tabulate a time–resource profile for each resource, showing when we are using more than what is available and when we are using less.

As an example, consider the one-resource problem shown in Figure 17.25. There is a bottleneck resource with availability 1.0. All activities either use the bottleneck resource at amount 1.0 or use no resources (or resources in adequate supply). We show the example as a network, with times marked on each activity. Activities competing for the resource are shown as heavier lines.

The unconstrained solution has a duration of 21 and would produce the resource utilization chart in Figure 17.26 (starting each job at its early start time).

There is overutilization of the resource from time 3 to time 7 by the activities

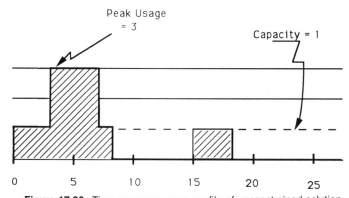

Figure 17.26. Time resource usage profile of unconstrained solution.

(5, 8), (3, 6), and (7, 9). It is fairly obvious that adding precedence constraints to force these activities not to occur simultaneously would produce the optimal solution if all six permutations were tried. A heuristic for the solution might be as follows:

Step 1: Schedule (3, 6) first to avoid major inserted idle time.

Step 2: Schedule (7, 9) second; it has the longest lead time left at that point.

The solution so generated has a duration of 24. It is easily verified that this is in fact optimal.

In general, the idea is to use the resource profile to locate the difficult peak time period(s). Next, heuristically move activities earlier/later to spread the peak out over a longer time interval and eventually to reduce it within the maximum resource usage, while delaying the project as little as possible. Note, in particular, that activities with positive total float in the unconstrained solution can be moved from their early start back to as far as their late start, with no effect on project duration. With a little computer assistance, these ideas can be formalized in a user interface oriented heuristic.

User Interface Heuristic for Resource Leveling

Step 1: Enter the problem data.

Step 2: The computer gives the non-resource-constrained solution.

Step 3: The computer gives the time resource profile and notes if solution is feasible.

Step 4: If we are done searching; go to step 7.

Step 5: The computer prompts the user to add/subtract new precedence constraints.

Step 6: Go to step 2.

Step 7: Show and/or print out the best *n* feasible solutions.

In some situations it may be useful to add constraints of the form: activity not to start until time 48. This is easily accommodated by adding a non-resource-using activity of length 48 and no precedence constraints. Then add a precedent constraint forcing the real activity to start after this time.

This heuristic may also be turned into a full optimal branch-and-bound procedure if desired.

17.5.3 Practical Issues

One important practical issue in resource leveling is preemption. It may very well happen, for example, that one would want to interrupt a long concrete pouring to do a shorter activity just available, which is on the critical path. In some cases this may be possible at basically no differential cost; in some cases this may be impossible without ruining the longer job. In many other cases the situation will be somewhat in between. The long job may be interrupted, but with some degradation of its quality, or with some wastage or other setup cost.

If there is no preemption allowed, our analysis stands unchanged. If preemption is free, the long activity may simply be represented as a number of short activities in series. The case in which preemption involves extra setup costs can basically be treated by bottleneck dynamics methods, as discussed in Chapter 8. We would need the fuller project methods of Chapter 18 as well.

In practice, there are often a number of other methods for accomplishing resource leveling than simply moving activities later. For example, it may be possible to add extra resources temporarily by renting equipment, working overtime, working at a greater pace, or having supervisors pitch in. In some cases there are alternate ways of doing the activity that require less of the critical activity or spread the use out over a longer time. A speedup of the activity by deliberately accepting a slightly lower quality may be possible. Creative overlapping of activities may be useful. Some of these issues will be discussed and/or modeled in Chapter 18.

17.5A Numerical Exercises

17.5A.1   Heuristically, resource level the problem presented in Figure 17.27 with one critical resource of amount 1.0. Resources using the critical resource are marked with heavy lines and use amount 1.0.

17.5A.2   Heuristically, resource level the problem presented in Figure 17.28 with the same assumptions as in 17.5A.1.

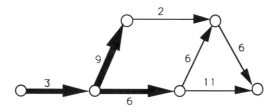

Figure 17.27

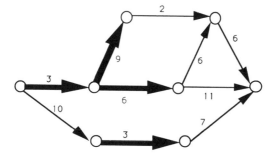

Figure 17.28

17.5B Software/Computer Exercises

17.5B.1   Write an iterative computer program for simple resource leveling as discussed in the text.

17.5B.2   Create a number of interesting problems and gain practice with your program.

17.5B.3   Add a limited ability to resource level by overtime and so on to your program.

17.5C Thinkers

17.5C.1   Create a formal procedure to make the resource leveling procedure exact by changing it to a full branch-and-bound procedure.

17.5C.2   Create a resource leveling heuristic that incorporates the possibility of preemption.

_____18

RESOURCE-CONSTRAINED PROJECT SCHEDULING

18.1 INTRODUCTION

18.1.1 Overview

As we discussed in the introduction to the last chapter, project scheduling and project management are very pervasive activities. When speaking of project scheduling, we tend to think of large one-time efforts such as developing the space shuttle, a classic example of a large single project. However, in actuality NASA develops a number of such projects that overlap in time (such as the shuttle, Mariner, Hubble, new rockets) and share various resources with each other. Here we have multiple large individual projects, each with its own objectives, sharing common resources. Or how about the assembly line putting together a number of copies of the space shuttle? Here we have multiple large identical projects, each with its own due dates, sharing common resources.

Other examples are a large construction company building several custom skyscrapers, using its own shared resources plus some subcontracting; another construction company building identical tract houses, each with different due dates, using common limited resources; a job shop (or software house) producing large complex products, one at a time; a similar custom shop producing a number of fairly complex products simultaneously, each with a due date; an assembly shop putting together rather standard products on a large scale; the corporate staff of a company carrying out a number of somewhat more diffuse projects simultaneously; or your desk, which typically has a number of partially completed smaller "projects" in process.

Thus project management and scheduling are potentially extremely broad and inclusive. An important feature of all the examples we have just mentioned is that dealing with resource limitations/resource constraints is a central part of the problem. Yet classic project scheduling methods typically deal only with the non-resource-constrained single-project model (for which the makespan objective suffices). For example, in Chapter 17 on early network methods, we discussed deterministic critical path analysis (CPM), time/cost trade-offs, and probabilistic (PERT) networks, none of which assume resource limitations.

These network models represent basically full relaxations of the resource constraints on the real problems of interest. As we have seen, relaxations are extremely useful in giving insight and helping us to produce heuristics for the full problems, which are usually very difficult or impossible to solve in their original form. A second fact is that practitioners have grown adept in adding artificial precedence constraints to the relaxed problem to avoid overuse of a scarce resource at a point in time. For example, if a simple unconstrained network solution involves using a single available crew on two different activities simultaneously, a constraint can simply be added that one activity must come before the other.

Thus in this chapter we wish to use our intuition developed from studying the relaxed problem in Chapter 17, adding to that our experience with job shops in Chapters 15 and 16. These will allow us to develop strong heuristics for project job shops, which in turn may be applied to develop good procedures for more general projects.

18.1.2 Chapter Summary

In Section 18.2 we discuss classic project job shops, which are basically classic job shops in which each job has more complicated precedence constraints. That is, each job is a (possibly smaller) project with its own objective function (not necessarily makespan). We discuss exact and heuristic procedures, especially bottleneck dynamics.

In Section 18.3 we move to the case of scheduling multiple large projects (MLP), where the emphasis is more aggregate. An activity simply uses a known overall fraction of each of a number of available resources. The modifications required to adapt exact methods or bottleneck dynamics are not difficult. Computational experience is given.

In Section 18.4 we turn to a generalization of the time/cost trade-offs of the last chapter. Now there may be a number of ways to accomplish a given activity, requiring differing amounts of various resources and times to complete. This becomes, in essence, a multi-resource routing problem. We develop some initial heuristics for this problem and again give some preliminary computational experience.

In Section 18.5 we briefly develop a number of possible model extensions. If some of the resources are used up in the process of completing activities, we have "consumable resources." If we are allowed to change the total amount of resources available at points throughout the project, we have "resource replanning" models.

For long horizon projects, the timing of rewards becomes important. This requires that these be discounted back to the present time and added together, leading to "net present value" (NPV) models. Finally, we may define a project as a special activity (final completion, or else a milestone) whose timely completion carries reward and/or punishment, together with all the predecessor activities it is necessary to accomplish. We consider solving the case where projects are interrelated by common activities.

In Section 18.6 we give a real-world case example, both to illustrate a realistic multi-project situation and to sketch appropriate heuristic scheduling systems for it. The example is producing custom optical equipment, a typical project shop.

Useful references for this section include Cleland (1969), Davis (1973), Cleland and King (1975), Kerzner (1982), and Morris (1982).

18.2 CLASSIC PROJECT JOB SHOPS

18.2.1 Overview

A classic project job shop is a classic job shop, except that rather than an individual job having a serial activity ordering, each individual job (project) may have activities with complex activity precedence constraints. For some purposes (such as integer programming) it will be useful to think of every activity as being a separate job, with most activities having zero weight, and precedence constraints relating the activities/jobs. (This formulation is actually slightly more general, since it allows some projects to depend on other projects.) Other times (dispatch methods) we treat the problem as independent projects with common resource constraints.

Actually, the ordinary job shop is usually a project job shop in disguise. Most jobs do not have a strict serial order in which they can be performed. That is, there is usually a precedence network showing which activities must precede other activities. But in a classic shop, without much flexibility, it is considered useful to preselect the ordering in which activities are to be done (i.e., to impose a serial ordering on the activities) to simplify scheduling in the job shop and avoid confusion. A *process engineer* will typically study the forecast job load over the next 3 months or so and decide on this ordering, which is set out in a *process sheet* for use on the floor.

Although making these decisions ahead of time is very convenient, it is often very expensive, since flexibility on the floor can save 20–30% on such objectives as flowtime or tardiness. Newer shop systems with higher flexibility allow making these choices on the floor and capture these savings. Of course, scheduling heuristics must then adapt well to the project schedule.

Since this section is more or less a direct extension of the classic job shop models, graphical methods, scheduling classification, and exact and heuristic methods of Chapter 15, in the same order, we make use of that chapter and present primarily differences. The reader should reread the appropriate section of Chapter 15 each time a new topic is brought up. (Presentation of the project shop material first to avoid duplication would be mathematically correct but poor exposition.)

Again, the techniques we discuss here are actually special cases of the multiple large projects models to be discussed in Section 18.3. However, the simpler case in which each activity uses up exactly one resource (machine) is often much easier to solve using the special structure. Again, understanding the more advanced models of Section 18.3 will depend on the discussion here.

In Section 18.2.2 we present two forms of the basic classic project shop method, together with graphical and charting methods. We also discuss various restrictions that might be placed on the types of schedules to consider.

In Section 18.2.3 we consider exact methods for solving this problem. We first present a disjunctive integer programming formulation and then a discrete time integer programming formulation. We then discuss some possible methods for solving these formulations using branch-and-bound.

In Section 18.2.4 we consider dispatch and search heuristics. First, we discuss classic dispatch approaches. Next, we show that dispatch beam search and the shifting bottleneck method may both be extended in a reasonably straightforward manner to this problem.

In Section 18.2.5 we discuss the iterated myopic dispatch approach and, finally, bottleneck dynamics.

18.2.2   Model, Charts, and Types of Schedules

This subsection is a direct extension of Sections 15.1.2 and 15.1.3, which should be reviewed as necessary.

Model. Two types of notation are sometimes used, depending on the situation:

Version 1. We may consider every activity a separate "job" with subscript j. There is a set of immediate predecessors $i \in P(j)$ for each job j, each of which must finish processing before j can start. There are objectives $f_j(C_j)$; the objective function is the sum of these. (Any f_j that is not identically 0 is labeled a "project completion activity.") We might call this formulation "activity oriented."

Version 2. Alternatively, we may label the projects by i, and activity j within project i by ij. There is a set of predecessors $ik \in P(ij)$ within the same project. There is an objective $f_i(C_i)$ for each project. We might call this formulation "project oriented."

Version 1 is slightly more general, since it allows projects to be interconnected. Version 1 is less complicated for representing integer programs and for graphical purposes. However, Version 2 is more insightful for certain types of heuristics that try to estimate floats within a project and thus utilize ideas from Chapter 17. We shall use either version as most useful.

The simple classic job shop has m machines indexed by k, and n operations (one-operation jobs) indexed by j, both available initially. Each operation j has an immediate predecessor set $P(j)$ and objective $f_j(C_j)$. Tables 15.1 and 15.2 providing input for the simple job shop may be replaced here by one slightly more complicated table (Table 18.1).

TABLE 18.1   Sample Input for a Classic Project Shop with T_{wt} Objective

Job	Machine	Processing Time	Weight	Due	Immediate Predecessors
1	3	1	—	—	—
2	2	1	4	5	1
3	1	4	—	—	—
4	1	2	—	—	—
5	1	2	—	—	3, 4
6	2	4	—	—	—
7	3	1	9	10	5, 6
8	3	2	—	—	—
9	3	2	—	—	—
10	2	4	—	—	—
11	1	4	3	8	8, 9, 10

The basic constraints for assigning an activity to a start time on a machine are:

(a) It cannot start until the machine is free.

(b) It cannot start until all its predecessors are finished.

(c) It cannot start until chosen by the sequencing rule.

Charts. Looking at Figure 15.1, the analogue of a job Gantt chart would have to be a project Gantt chart. It would need to have several lines per project, to avoid showing activities on top of each other timewise while on different machines. The idea of a machine chart works well, but we must check much more carefully to see whether a given schedule is feasible. Figure 18.1 represents a feasible schedule for the problem given in Table 18.1.

Types of Schedules. The discussion in Section 15.1.3 for job shops requires very little change for project shops. A *local left shift* in a feasible schedule occurs when an activity is moved earlier without affecting feasibility. There is a new condition to check as well. Does moving left violate prior completion of predecessors? A sched-

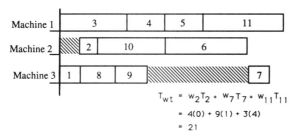

Figure 18.1. Machine Gantt chart (feasible).

ule is *semi-active* if no local left shifts are available. There are a finite number of semi-active schedules, and for a regular objective at least one is optimal.

A *global left shift* allows feasibly jumping over obstructions when moving to the left. If no global left shift can be found, we have an *active schedule*. If a schedule never leaves a machine idle when it is feasible to schedule an operation, it is called *nondelay* or, equivalently, *dispatch*. Proposition 2 of Chapter 15 obviously remains true for project shops. We restate it here for convenience.

Proposition 2 of Chapter 15 (Restated). For project shops:

(a) The set of all dispatch schedules is part of the set of all active schedules, which is part of the set of all semi-active schedules, which is part of the set of all schedules.

(b) There is always an active schedule that is optimal for a regular objective.

(c) There may be no optimal dispatch schedule.

Nevertheless, dispatch schedules are easy to work with and will usually be good if lengths of operations are not large compared with the length of the schedule. (Quantifying and testing this question might make a good research topic.) Search procedures that attempt to generate all needed active schedules for project shops are complex, difficult, and expensive to use. We do not discuss them here (Baker, 1974).

18.2.3 Exact Methods

This section is an extension of the exact methods studied for the classic job shop in Section 15.2. First we look at extensions of the two integer programming methods studied in Section 15.2.2; then we turn to branch-and-bound approaches generalizing Section 15.2.3.

Integer Programming—Disjunctive Formulation. We reformulate the disjunctive formulation for the classic job shop to be appropriate for Version 1 of the classic project shop. Remember each job is now a single operation. Let C_j be the completion time of job j and $h \in P(j)$ be any immediate predecessor of j. (For a job with no predecessors, h will be a dummy job finishing at the arrival r_j; we thus handle dynamic shops.) The C_j are decision variables and, together with ensuring feasibility, determine the schedule. As usual, we need constraints to check that predecessors are finished:

$$C_j \geq C_h + p_j \quad \text{for all } j, \text{ and } h \in P(j)$$

But we also need to ensure that when two jobs are both performed on the same machine k, both jobs cannot be performed with any time overlap. For two jobs i and

j both done on machine k, let y_{ij} be 1 if i comes before j and 0 otherwise, and, as in Section 15.2.2, let H be a very large number. Then we immediately have an extended version of Proposition 3 of Section 15.2.2.

Proposition 1. Consider the classic project shop with dynamic initial availability of machines, dynamic job arrival, and weighted flow objective (for every job/operation). A disjunctive constraint integer programming formulation can be given by

$$\text{minimize} \quad \sum_{j=1,n} w_j C_j$$

$$\text{s.t.} \quad C_j \geq C_h + p_j \quad \text{for all } j, \text{ and } h \in P(j) \tag{1}$$

$$C_j - C_i + H(1 - y_{ij}) \geq p_i \quad \text{for all } i, j \text{ on each } k \tag{2}$$

$$C_i - C_j + Hy_{ij} \geq p_j \quad \text{for all } i, j \text{ on each } k \tag{3}$$

$$C_j \geq 0; \quad y_{ij} \in \{0,1\} \tag{4}$$

Proof. Almost identical to the proof of Proposition 3 in Section 15.2.2. Since jobs and activities now correspond one to one, only a single subscript is needed per activity. The weights w_j will be zero except for activities representing the completion of a project. Initial unavailability of a machine can be represented as a dummy activity of the proper length, which must precede all other activities on that machine. Arrival times for activities (either initial or otherwise) can be treated as a dummy activity of the proper length preceding the given actual activity, which is performed on any unique dummy machine. ∎

It is easy to modify this formulation for other objective functions. For example, for weighted tardiness, create nonnegative variables for the earliness and the tardiness of each job. Change the objective function to a weighted sum of the tardiness variables. Add an equation for each job saying (completion + earliness − tardiness = due date).

The classic project shop version of this proposition is both more general and easier to write than the job shop version. Why do we bother to give both versions? Although the general version is easier to write down abstractly, it may involve up to about n^2 constraints versus about n constraints for the job shop version and is much harder to solve. Thus formulating a job shop problem with the general machinery may not be useful in practice. We discuss solution procedures for this integer program shortly.

Integer Programming—Discrete Time Formulation. We turn first to extending the other integer programming formulation of Section 15.2.2, given there as Proposition 4. Let C_{jt} be 1 if job j is completed at time t, and 0 otherwise. Let C_j be the completion time for job j. Let x_{jt} be 1 if j is processing at time t, and 0 otherwise. Let $J(k)$ represent the set of j done on machine k.

Proposition 2. Consider a classic project shop with dynamic machine availability, dynamic job arrivals, and weighted flowtime objective function. Proposition 4 of Section 15.2.2 may be generalized as follows:

$$\text{minimize} \quad \sum_{j=1,n} w_j C_j$$

$$\text{s.t.} \quad C_j = \sum_{t=1,T} t C_{jt} \quad \text{for all activities } j \tag{5}$$

$$x_{jt} = \sum_{u=t,t+p_j-1} C_{ju} \quad \text{for all } (j, t) \tag{6}$$

$$C_j \geq C_h + p_j \quad \text{for all } j, h \in P(j) \tag{7}$$

$$\sum_{j \in J(k)} x_{jt} \leq 1 \quad \text{for all } (k, t) \tag{8}$$

$$C_{jt} \in \{0,1\} \quad \text{for all } (j, t) \tag{9}$$

Proof. Identical to Proposition 4 of Section 15.2.2. Again, use artificial activities and machines to represent initial conditions, as in Proposition 1 of this chapter. ∎

Branch-and-Bound. Section 15.2.3 applies almost verbatim. We summarize that material briefly for the project shop case.

(a) The disjunctive formulation may be used as a dual branch-and-bound procedure (see Section 17.5) by first relaxing all disjunctive constraints and then continually branching on whether or not to add one more disjunctive constraint. Each such problem may be solved as a linear program, giving problems with fewer and fewer infeasibilities and LP lower bounds on the objective function.

(b) Either formulation may be used as a regular branch-and-bound procedure to find optimal dispatch procedures. Begin to work a dispatch simulation forward in time. Each time there is a choice on which activity to start, form a branch for each possibility.

LP relaxations on either integer programming formulation may be used to provide good lower bounds. As an alternative to a full branch-and-bound procedure in the dispatch simulation, one can always use dispatch procedures or beam search.

18.2.4   Simple Dispatch Methods/Search Methods

Dispatch methods, including dispatch beam search, myopic dispatch, and bottleneck dynamics, all use the Version 2 formulation, which treats each project as a unit, and uses ideas from Chapter 17 to heuristically analyze it, as well as job shop ideas from Chapter 15. We assume there are I projects, with project i having weight w_i, due date d_i, and objective function $f_i(C_i)$. Activity j within project i will be labeled by the double subscript ij, and have processing time p_{ij}. The predecessor set to ij is given by $ik \in P(ij)$. The predecessor set contains no activities belonging to other projects. (Unlike the integer programming formulation, projects must be independent.)

Estimation of Slack. All simple dispatch methods estimate slack-related variables such as slack, operation due date, early start time, or early finish time, by one of two closely related methods:

Aggregated Project Method

(a) Combine all projects into a single project by connecting individual completions to a single grand completion, using dummy activities with no resources or time requirement.
(b) Consider the relaxed version with no resource constraints.
(c) Calculate late finish time for each activity by the standard critical path method of Section 17.2.

Individual Project Method

(a) Treat each project separately.
(b) Consider the relaxed version of each project with no resource constraints.
(c) For each activity ij calculate the longest path from the activity to the end of that project.
(d) The late finish time (operation due date) for ij is the due date of the project, less that longest path.

The aggregate project method is clearly correct if the objective is to minimize the overall makespan. Since most commercial software is limited to makespan, this method has dominated in the past. However, the individual project method is much superior if, for example, the objective is the sum of weighted tardiness for individual projects, and weights and due dates vary greatly.

Dispatch Procedure. Given an appropriate priority rule (we discuss this in detail below and in the next section), the dispatch procedure is straightforward:

Step 1: Keep an event clock.
Step 2: When the next machine becomes free, find all activities that use this machine and have all predecessors and other constraints satisfied.
Step 3: Find the activity with highest priority, if any, and schedule it.
Step 4: Add any new event such as an activity finish to event clock list.
Step 5: Advance event clock to next event.

Priority Rules. There are a great number of simple priority rules that have been proposed. We discuss some of them here. Others are discussed in Lawrence and

Morton (1989). (l_{ij} is the late start time of activity i in project j; e_{ij} is the early start time. Either may be calculated by the aggregated project method or the individual project method given just above.)

FCFS	First to machine, first served	
EDD	Earliest project due date	d_i
LFS	Least float per successor	$(l_{ij} - e_{ij})/(\#)$
LFT	Late finish time (operation due date)	$l_{ij} + p_{ij}$
MAXPEN	Maximum penalty	w_i
MINSLK	Minimum operation slack	$l_{ij} - t$
MTS	Most total successors	$(\#)$
SPT	Shortest process time (minimum)	p_{ij}
WSPT	Weighted SPT (maximum)	w_i/p_{ij}
R&M	Slack corrected WSPT	$(w_i/p_{ij}) * \exp[-(l_{ij} - t)^+/kp_{av}]$

Remember that operation due date is the myopic rule for minimizing maximum lateness and hence is of special interest in a project, where completion times need to be coordinated. We come back to this point later.

These rules and many others have been tested extensively in the Lawrence–Morton (1989) study on the more general multiple large project problem. We report on this study extensively in Section 18.3. The comparison between simple dispatch and more advanced methods given there may be expected to hold in this special case as well.

Guided Dispatch Beam Search. The basic guided dispatch beam search method is very similar to that discussed for the job shop. Run a basic dispatch simulation, with some guide heuristic that might either be a simple or more complex heuristic. At the first choice of operation, calculate the indexes corresponding to the heuristic and normalize them in terms of absolute deviation from the index of the best one. Choose the b best choices, and continue with b problems, one for each choice. At the second choice point for each problem, add the deviation from the best on the second round to the deviation from the first, to get an overall evaluation function. Over all second-round choice functions on all first-round choices, pick the b subproblems to continue with lowest *total* deviation from the highest cumulative index. Continue.

If the problem involves lead time iteration, things become a little more complex. Continue the entire beam search process with original lead time estimates. At the end, choose the best problem, and utilize its lead time estimates as an input to a full second beam search, and possibly a third.

Shifting Bottleneck Algorithm. For definiteness look at a single project makespan problem in a job shop. The only issue that changes is how to define and solve the simple embedded problem. For the job shop, the simple embedded problem is to minimize the maximum lateness on a single machine with dynamic arrivals and local due dates for independent jobs. This problem has been solved for large numbers of jobs, and so the overall algorithm is efficient.

For a single-project project shop, the simple embedded problem is to minimize the maximum lateness on a single machine with dynamic arrivals and local due dates with delays interspersed on the machine. This problem has been worked on by Balas and Vazacopolus (work in process).

18.2.5 Myopic Dispatch/Bottleneck Dynamics

Iterated myopic dispatch and bottleneck dynamics are similar in three respects (the first two are true in job shops also):

(a) They both use lead time iteration to provide improved estimates of lead times for slack oriented dispatch heuristics.

(b) They both treat the current machine as a dynamic one-machine problem to be optimized (or approximately optimized) and use the same formulas.

(c) They both use two-stage dispatch procedures for selecting the next activity. (See the discussion of two-stage dispatch procedures below.)

However, in myopic dispatch the processing time used in the formulas is that of the operation under consideration; in bottleneck dynamics it is the weighted sum of all remaining operations to finish the project. The weights are the current estimated prices of the machines under consideration. Thus the operation process time is replaced by the estimated value of the total remaining resource usage needed for the project.

Lead Time Iteration. The simple dispatch procedure estimates lead times (slacks) by relaxing all constraints and solving the resulting critical path problems to compute early and late start times for the activity. This is not particularly good, since it is basically being assumed that there is no waiting for any resource.

An improved method would be to assume that the total time for processing at a machine plus waiting in queue at the machine for an activity is a fixed multiple of processing time. Then processing times at the machine could temporarily be increased and critical path calculations could be made with these adjusted processing times. However, lead time iteration for project scheduling leads to even better results. (See Section 18.3.4 below.)

Vepsalainen and Morton (1987) showed that iterative estimation of operation lead times can greatly increase weighted tardiness performance in flow shop scheduling. Morton et al. (1988) demonstrated that iterative updating of estimated machine prices significantly improved discounted cash flows in flow shop and job shop scheduling. In the study discussed in Section 18.3.4, Lawrence and Morton (1989) modify lead time iteration (LTI) for project shops and multiple large projects. (The statement we give here applies to either the model of this subsection or that of the next.)

With LTI updating, the time q_{ij} an activity ij spends in queue waiting for resource(s) is smoothed from iteration to iteration:

$$(qf)_{ij}^n = (qf)_{ij}^{n-1} + \alpha(q_{ij}^{n-1} - (qf)_{ij}^{n-1}) \tag{10}$$

where q_{ij}^{n-1} is the actual queue time experienced by activity ij in iteration $n - 1$, $(qf)_{ij}^n$ is the queue time forecast for upcoming iteration n, and $0 \le \alpha \le 1$ is an empirically determined smoothing parameter. (Note that the actual queue time in an iteration is the time from when all activity predecessors have finished until the activity actually starts processing, whether a single resource or multiple resources are involved.)

The processing time of activity ij is temporarily increased to include forecast time in queue:

$$(pf)_{ij}^n = p_{ij} + (qf)_{ij}^n \tag{11}$$

These in turn are used to compute resource-constrained adjusted activity late start times $(lf)_{ij}^n$ by

$$(lf)_{ij}^n = d_i - L_{ij}^n \tag{12}$$

where L_{ij}^n is the length of the critical (longest) path from activity ij to the end of the augmented project network [using process times $(pf)_{ij}^n$] in iteration n. (In a single project we measure the longest path from the end of the project; with multiple projects slacks need to be measured from the due dates.)

Note that the augmented processing time $(pf)_{ij}^n$ is used for critical path calculations and not for other calculations that required activity processing times p_{ij}. Iterative scheduling continues until some stopping rule is satisfied. The stopping rule in the Lawrence–Morton study presented in the next subsection was to conclude scheduling when five successive iterations failed to provide any improvement in the best value of the objective function to date.

Two-Stage Dispatch Procedure. Lawrence and Morton (1989) made the following observation about the dispatch process: candidate activities within the same project cannot be differentiated on the basis of importance, since all must be performed to obtain any reward. But they also cannot be distinguished on the basis of resource usage or process time, since, again, all must eventually be performed. Thus, for activities within the same project, the objective is to minimize maximum lateness, to coordinate completion times as much as possible. Minimizing maximum lateness is achieved by taking activities in order of their operation due dates or, equivalently, in terms of minimum late finish time (LFT) $(l_{ij} + p_{ij})$, where l_{ij} is a current estimate and p_{ij} is the original process time.

Thus the two-stage procedure is as follows:

(a) In each project, choose from among eligible activities the one with minimum LFT.

(b) For each such candidate, calculate the priority by the appropriate myopic heuristic.

We list these for a few examples next.

Myopic Heuristics by Objective Function

(a) *Makespan.* Treat the entire problem as a single project; choose activities by minimum LFT.

(b) *Maximum Lateness.* Use LFT within a project and also among projects.

(c) *Weighted Flowtime.* Set $d_i = 0$, so that all projects are "tardy"; then use the weighted tardiness heuristic.

(d) *Weighted Tardiness.* Set d_i at actual project due date. Use LFT within a project, and then R&M between projects.

Perhaps the weighted flowtime heuristic requires some explanation. If a project has no available activities that are on the critical path for completing the project, then w_i/p_{ij} is too high an estimate of doing that activity. It should be discounted somewhat depending on its slack.

Summary of Myopic Dispatch Versus Bottleneck Dynamics

(a) Both use lead time iteration and obtain the same slack forecasts.

(b) Both choose candidates within a project by LFT.

(c) Both choose among resulting candidates by a one-machine heuristic.

(d) They are identical for makespan and maximum lateness.

(e) Bottleneck dynamics requires estimates of resource prices R_k. This estimation is discussed in Section 18.3.4.

(f) If formulas have a benefit/cost part, myopic dispatch uses w_i/p_{ij}, while bottleneck dynamics uses $w_i/(\sum_k R_k p_{ik})$, where the summation is over remaining activities in the project.

18.2A Numerical Exercises

18.2A.1 Consider the example given in Table 18.1. Solve this problem for the makespan objective using a simple dispatch rule: (a) LFT, (b) MINSLK, and (c) WSPT. Compare your results.

18.2A.2 Consider the example given in Table 18.1. Solve this problem for the weighted flow objective using a simple dispatch rule: (a) LFT, (b) MINSLK, and (c) WSPT. Compare your results.

18.2B Software/Computer Exercises

18.2B.1   Write a program to solve a single-project project shop by simple dispatch rules. Include a choice by the user to select some subset of perhaps 10–20 rules, and perhaps four or five objective functions.

18.2B.2   Create an appropriate data set, and test heuristics against objectives, using your program from 18.2B.1.

18.2B.3   Enrich your program from 18.2B.1 to perform lead time iteration.

18.2B.4   Repeat your tests in 18.2B.2 using lead time iteration. What can you say about the amount of improvement?

18.2C Thinkers

18.2C.1   Create a small- to medium-sized job shop example, and give two specific integer programs for it, using the two formulations of Section 18.2.3.

18.2C.2   Take each dispatch heuristic in Section 18.2.4 and show how it simplifies for the special case of a job shop.

18.3 MULTIPLE LARGE PROJECTS

18.3.1 Overview

The two main differences in formulation between multiple large projects and project shops of the previous section are (a) that activities may require more than one resource and (b) that an activity may only use part of a resource. (These are related to the fact that activities and resources may now be larger aggregated entities.)

Now we may specify that an activity ij requires resource amount r_{ijk} for each resource k. We also specify that the total availability of resource k is RA_k. Sometimes time-varying-resource versions of these large shops are modeled, where the availability is RA_{kt}, and the resource usage for an activity varies as it is processed; that is, we have $r_{ijk}(t - t_s)$, where t_s is the start time of the activity. Integer programming and/or dispatch formulations can be given. However, even dispatch is quite difficult to compute in this case, since an activity may not be started without sufficient resources throughout its processing. In any event, we shall confine ourselves to static formulations here.

In Section 18.3.2 we discuss classic exact and heuristic methods for this problem briefly. In Section 18.3.3 we also discuss search methods, iterative myopic methods, and bottleneck dynamics. These discussions lean heavily on Section 18.2. Finally, in Section 18.3.4 we discuss extensively the Lawrence–Morton (1989) study comparing simple dispatch, iterative myopic, and bottleneck methods.

18.3.2 Classic Methods

Integer Programming.   Up to this point, as we have considered more and more complex model generalizations, we have always been able to present two integer programming formulations: (a) the disjunctive formulation and (b) the discrete time formulation.

But now that any given activity may require several resources and any given resource may support more than one activity simultaneously, it is no longer particularly useful to ask whether activity i is performed before or after activity j on resource k. Thus the disjunctive formulation can no longer be usefully formulated.

However, the discrete time formulation is much more general. We now present this formulation for the multiple large project case. This formulation can even be extended to the case of time-varying uses of resources within an activity and time-varying availability of resources. These extensions are left as exercises. This model can also be generalized to extending resource availability at a cost through such devices as overtime, subcontracting, and rental, as we shall see in Section 18.5.

The model we give was proposed by Pritsker et al. (1969). Their approach is reasonably efficient in its use of decision variables. The formulation uses Version 1; that is, activities and jobs are identified; projects appear only implicitly as activities in the objective function with a non-zero weight. We define the following:

r_{jk} Amount of resources of type k required by activity j
RA_k Total resource units of type k available
C_j Completion time of activity j
$w_j(C_j)$ Cost (weight) of completing activity j at time C_j (in the simplest case 0 except for project completions)

Let the decision variables C_{jt} and x_{jt} satisfy the following:

$$C_{jt} = \begin{cases} 1 & \text{if activity } j \text{ completes in time period } t \\ 0 & \text{otherwise} \end{cases}$$

$$x_{jt} = \begin{cases} 1 & \text{if activity } j \text{ is in process in time period } t \\ 0 & \text{otherwise} \end{cases}$$

At the outset a scheduling horizon of HZ periods is chosen so that C_{jt} may be defined for $1 \le j \le n$, and $1 \le t \le HZ$.

Proposition 3. The multiple large project problem, with activities using multiple resources and resources servicing multiple simultaneous activities, may be given by the following

$$\text{minimize} \quad \sum_{j=1,n} w_j(C_j) \tag{13}$$

$$\text{subject to} \quad C_j = \sum_{t=1,HZ} t C_{jt} \quad \text{for all } j \tag{14}$$

$$C_j \ge C_i + p_j \quad \text{for all } i \in P(j), \text{ for all } j \tag{15}$$

$$x_{jt} = \sum_{u=t,t+p_j-1} C_{ju} \quad \text{for all } j, t \tag{16}$$

$$\sum_{j=1,n} r_{jk} x_{jt} \leq \text{RA}_k \quad \text{for all } k, t \tag{17}$$

$$\sum_{t=1,HZ} C_{jt} = 1 \quad \text{for all } j \tag{18}$$

$$C_{jt} \in \{0,1\} \quad \text{for all } j, t \tag{19}$$

Proof. The objective function (13) says that each activity incurs a cost (or reward) that depends on when it completes. These costs are then added together. Thus this is quite general, including any additive functions of completion time, whether regular or not. For weighted flow, for example, this would simply be $w_j C_j$. (For objectives such as makespan, we add a completion dummy activity n having all other activities as predecessors, and change the objective to minimize C_n.) Equation (19) says that an activity cannot complete in pieces. It either has no completion or a full completion in a given time period. Equation (18) says that each activity must complete exactly once in total over all possible time periods. In equation (14), since all C_{jt} are 0 except in the completion period, the summation becomes t^* times 1, where t^* is the completion time. Equation (16) says that an activity j is processing in time t precisely if it will finish in the next p_j periods including now. Inequality (17) says if we add up all usages for activities currently using k in time t, it must not exceed the total resource available.   ■

Pritsker et al. (1969) also discuss the extension of the integer programming formulation to projects with substitutable resources, concurrency requirements, and preemptable activities. These and other extensions are considered in the exercises and in Sections 18.4 and 18.5.

However, it does not seem practical to solve this integer program exactly for even middle-sized problems. Suppose there are $HZ = 20$ time periods, $n = 100$ activities, $m = 5$ resources, and $d = 5\%$ (meaning that 5% of the potential precedence constraints are actually realized). A somewhat more compact version of the above integer program can be given with about $nHZ = 2000$ variables, and $dn^2 = 500$ constraints. This is an extremely large integer program, which does not seem practical to solve now or in the foreseeable future. Thus we turn next to approximate solution methods and heuristics.

Branch-and-Bound. Of the two branch-and-bound methods discussed in Section 18.2.3, the disjunctive approach is no longer possible. Hence we turn to studying exact solutions to dispatch problems. Events in the dispatch process are now not tied to one machine. Whenever an activity finishes, resource and activity availabilities will change. Then we determine the set of activities (if any) that can start and proceed. For branch-and-bound, we would naturally need lower bounds, which would be obtainable by solving the LP relaxation of the Proposition 3 integer program given above.

However, the LP relaxation itself has now become so cumbersome that solving it hundreds or thousands of times becomes impractical. Here we suggest a faster approximate alternative to the LP relaxation. Form the unit-preemption of the problem, and solve that approximately by some dispatch heuristic such as the iterative myopic or bottleneck dynamics approach. Since the unit-preemption problem has less problems with integrality and rounding, the dispatch heuristic may be expected to be quite accurate. Since the true LP solution was only intended to be a lower bound, the accurate heuristic approximation has a very good chance to remain a lower bound in many cases. (This procedure has been used in the software to provide approximate lower bounds, with good success.)

18.3.3 Search Methods/Bottleneck Dynamics

Heuristic methods for the multiple large project case only require minor modification from classic project job shops, which we therefore discuss only briefly.

Dispatch Beam Search. As mentioned before, the generalized dispatch process is not tied to a specific machine. In general, there is now a single eligibility list of activities whose precedences have been satisfied. Whenever an activity is finished, the following are updated:

(a) Current resource availability.
(b) Current activities eligible to be processed.
(c) Activities eligible with sufficient resources available.

From the third list, an activity is selected and started, current resource availability is updated, and its finish time is entered as a new future event in the event list. As long as the list is not empty, this process is repeated. If the third list is empty, the process moves forward to the next activity finish on the event list.

When the list has more than one activity available to choose, the activities are prioritized by some guide priority rule (such as LFT). If b is the beam width, then b problems are started. After the next activity has been completed in all b problems, the larger list of candidates is prioritized by the sum of priorities over the two stages, as before. Again, b are selected, and so on.

Beam search for these complicated problems has not yet been tested.

Iterated Myopic/Bottleneck Dynamics. Here we use the Version 2 formulation, which considers projects i and activities ij. The iterated myopic heuristic and the bottleneck dynamics method both use similar heuristics (optimal or at least near optimal for one machine in the job shop) but differ in their estimate of the resources used by the decision to schedule an activity.

Let us consider the weighted tardiness problem for concreteness. Both approaches use an analogue of the R&M priority function:

$$\pi_{ij} = (w_i/(\text{RU})_{ij})\exp[-(l_{ij} - t)^+/kp_{\text{av}}]$$

where $(\text{RU})_{ij}$ is a measure of the early resource usage by scheduling ij early.

The myopic rule treats this as resources involved for ij itself, hence the term "myopic." However, bottleneck dynamics considers that an activity is moved early only to finish the job or activity early, and hence that all other remaining activities should be considered part of the activities used early.

Thus for the *myopic* point of view:

(a) In the project job shop $(RU)_{ij} = R_k p_{ij}$.
(b) In the large project case $(RU)_{ij} = [\Sigma_{k=1,m}(r_{ijk}/RA_k)R_k]p_{ij}$ (that is, for each resource that ij uses, multiply the fraction it uses by the total price to get its cost per unit time on that resource).

Note that, in the job shop and project job shop, we are comparing priorities for different jobs considering using machine k. Since the price R_k is the same for all choices and thus *relative* priorities are independent of prices, price estimation is not necessary.

To give the bottleneck dynamics point of view, we need notation. Let U_i be the set of remaining activities in project i, and let J be one of these unfinished jobs. Then for the *bottleneck dynamics* point of view:

(a) In the project shop $(RU)_{ij} = \Sigma_{J \in U_i} R_{k(J)} p_{iJ}$.
(b) In the large project case $(RU)_{ij} = \Sigma_{J \in U_i} \Sigma_{k=1,m}(r_{iJk}/RA_k)R_k p_{iJ}$.

There is a third way of estimating costs, which might be called the process time costing point of view. It approximates resource usage as simply proportional to processing time, which is an advantage in avoiding pricing entirely, but a disadvantage in terms of accuracy.

Then for *process time costing*:

(a) In the project shop $(RU)_{ij} = p_i$.
(b) In the large project case $(RU)_{ij} = \Sigma_{J \in U_i} p_J$.

A remaining question is methods for determining resource prices. We discuss this in Section 18.3.4.

We turn next to computational experience.

18.3.4 Computational Experience

Lawrence and Morton (1989) conducted large scale testing comparing simple dispatch heuristics, iterated myopic methods, OPT-like methods, and bottleneck dynamics for various lead time estimation and pricing estimation methods for multiple large projects with weighted tardiness objective.

General Definition of Priorities. A general bottleneck-oriented priority function for weighted tardiness (and similar objectives) is defined by activity priority = (project weight)(activity urgency)/(implicit remaining project cost)

The activity's *project weight* is self-explanatory. The *activity urgency* is any function ranging from 1.0 to 0.0 as total slack increases. Total slack is the sum of the activity's slack within the project and the slack of the project within its due date. The *implicit remaining project cost* represents the marginal opportunity cost of committing resources, both for the current activity and for downstream successor activities.

Resource Pricing Methods. Heuristic resource prices, used to calculate implicit remaining project cost, are rough estimates of the shadow prices of resource availability for the relaxed problem. They are then used in heuristics as if they were dual prices for the original problem. Experimentally, this procedure works quite well for job shops (see Chapter 16).

The study tested five different approximations for the heuristic prices R_k: uniform, resource load, bottleneck, busy period, and constraint relaxation.

(a) *Uniform Resource Pricing.* $R_k = 1.0$ (benchmark case).

(b) *Resource Load Pricing.* $R_k = \rho_k = \Sigma_{ij}(r_{ijk}/RA_k)p_{ij}/M$ (M is the rough makespan, ρ_k is the effective machine utilization).

(c) *Bottleneck Pricing.* b is the k giving $\max_k \rho_k$; $R_b = 1.0$, and $R_k = 0.0$ otherwise.

(d) *Busy Period Pricing.* $R_k = (WU)_{av}[\rho_k/(1 - \rho_k)^2]$. (We have seen this approximation in Section 10.5. $(WU)_{av}$ can be any relatively crude approximation to the average weight times slack factor for jobs using resource k.)

(e) *Constraint Relaxation.* Run a full simulation of the problem once with normal amounts of resource k, and once with X times as much, giving observed objective values of T_0 and T_1, respectively. $R_k = (T_0 - T_1)/X$ ($X = 10$ proved robust).

Benchmark Heuristic Study. Fifteen simple dispatch scheduling heuristics, which have been used in the literature, were tested in an initial pilot study. Many of these are given in Section 18.2.4. Three of these provided superior results in producing low weighted tardiness schedules. These were EDD, MINSLK, and LFT. Since these are unweighted in problems where weights should be important, weighted versions were developed and tested:

WEDD	Weighted EDD	$\max w_i/(d_i - t_0)$
WLFT	Weighted LFT	$\max w_i/(l_{ij} + p_{ij} - t_0)$
WMINSLK	Weighted MINSLK	$\max w_i/(l_{ij} - t_0)$

(While t_0 is a free parameter, the study used t_0 as *now*. Thus WEDD becomes "project weight over time until due date," WLFT becomes "weight over time until late finish time," and WMINSLK becomes "weight over slack remaining.")

Experimental Design

Project Parameters. Four problem sets were generated, with one, two, three, and five resources, respectively. Each problem set included 40 problems of five projects each. The number of activities per project was uniformly distributed [25, 50] and marginal project tardy penalties w_i were uniform over [1, 10]. Activity durations p_{ij} were integer and uniform over [1, 10]. The number of resources for an activity was distributed uniformly between 1 and the maximum. Resources were assigned RA_k uniformly on [0.3, 1.0]. Fractions r_{ijk} were determined by multiplying RA_k by a number uniform on [0.5, 1.0]. This provided resource requirements ranging from 0.15 to 1.00, while ensuring a wide range of resource tightness.

Project Due Dates. Three levels of project due date tightness were included in the main experiment. For the tightest due date settings an average of four projects were tardy; for the middle setting three projects were tardy; and for the loosest setting two projects were tardy.

Precedence Constraints. The number of precedence relations for a given project was measured (order strength) as a fraction of the number possible. This was uniformly distributed on [0.05, 0.15], exclusive of redundant constraints.

Experimental Factors. The main factors of the study were the number of resources (four levels), due date tightness (three levels), scheduling rule (30 policies), and random cell replications (40 replications), giving a total of 14,400 individual scheduling problems for the main experiment alone.

Larger Pilot Study. The design of the larger pilot study was identical to that of the main experiment just described above, except that there were only 20 replications of each problem. The principal objectives of the pilot study were to determine good parameter values for the R&M priority formula and the smoothing constant for lead time iteration, and to obtain preliminary results regarding the relative performance of the several resource pricing schemes.

The value of k for the R&M rule in project shops turned out to be relatively insensitive between 6 and 28, and $k = 12$ was chosen. (Note that the value of k useful for projects turns out to be larger than for single machines or job shops.) The value of α, the smoothing constant for lead time iteration, was insensitive between 0.04 and 0.2, and $\alpha = 0.1$ was chosen for subsequent experiments.

The pilot study showed that bottleneck dynamics dominated the myopic policy and the process time pricing policy for all complex dispatch rules. This result suggests that important information is embodied in the prices of downstream activities that can significantly improve scheduling performance. Therefore only simple dispatch rules and bottleneck dynamics were considered further.

A final finding of the pilot study was that the two-stage procedure first discussed in Section 18.2.5 is superior to a straight simple or complex dispatch procedure:

(a) At a choice point use LFT to choose the highest priority activity available for scheduling. Thus form a reduced candidate list.
(b) Use some dispatch priority rule to choose the activity to schedule from the reduced candidate list.

The reasoning behind this is that LFT is essentially a "minimize operation due date" (ODD) rule, which is known to minimize maximum tardiness for a single machine. On average, the LFT preselection policy improved the performance of the benchmark heuristics by 6%, with improvement ranging up to 28% for some dispatch policies. For the balance of the study, all reported results make use of the LFT preselection policy.

Computational Results. The main experiment consisted of the 14,400 problems previously described. The 30 rules previously mentioned consisted of 20 benchmark rules, five heuristic pricing rules described at the beginning of this subsection, and the heuristic pricing rules augmented by lead time iteration (LTI).

Pricing Rules. The performance levels of the five pricing rules were not statistically significantly different in this experiment. This suggests that more investigation is needed to decide whether the method of determining global resource prices is critically important, or whether computationally inexpensive methods such as bottleneck or uniform pricing policies may work well.

Lead Time Iteration. The LTI updating policy reduced the weighted tardiness of all pricing rules by over 18% on the average and thus proved highly useful. It found its best schedule on the average after 3.8 iterations.

Experimental Results. Across all problems the new rules WEDD, WLFT, and WMINSLK yielded the lowest total average weighted tardiness among the 20 benchmark rules considered. Several other rules, such as MINSLK and MAXPEN, which have been shown previously to provide good results, also performed relatively well but were not competitive with the best benchmark rules tested. Table 18.2 summarizes the performance of the seven best benchmark rules and bottleneck dynamics with and without lead time iteration. It compares them also to the composite rule, which picks the best of the 20 benchmark rules.

Note that bottleneck dynamics (BD) with lead time iteration is by far the best performer, followed by bottleneck dynamics using the relaxed network to estimate lead times. The new simple dispatch heuristics WEDD, WLFT, and WMINSLK do quite well, followed by their unweighted versions. The fall-off to the rest of the table is quite sharp; the increase from MINSLK to the adjacent MAXPEN is about 25%.

Note that the improvements of the weighted simple dispatch heuristics over their unweighted versions increase with due date tightness, as would be expected.

Note that BEST/20 is only about 10% worse than bottleneck dynamics with lead time iteration. However, the former involves running a full 20 dispatch simulations,

TABLE 18.2 Experimental Results: Top Performers[a]

Policy	Due Dates			Average
	Loose	Medium	Tight	
BD–LTI	95.6	162.7	291.8	183.4
BD–NO LTI	119.4	205.6	348.5	224.5
WEDD	140.6	218.5	356.1	238.4
WLFT	148.3	231.9	376.4	252.2
WMINSLK	149.2	236.2	385.7	257.2
EDD	143.5	247.5	439.8	276.9
LFT	149.2	255.3	459.2	287.9
MINSLK	150.0	260.1	468.9	293.0
MAXPEN	293.0	364.5	473.6	377.1
BEST/20	106.4	182.5	325.7	204.8

[a]In terms of weighted tardiness.

the latter only about four. Using beam search with a width of five would further improve the bottleneck dynamics result at about the same cost as BEST/20.

Other references for this section include Balas (1970), Davis and Heidhorn (1971), Zaloom (1971), Gorenstein (1972), Davis and Patterson (1975), and Talbot and Patterson, (1978).

18.3A Numerical Exercises

18.3A.1 Re-solve exercise 18.2A.1 using lead time iteration for the same simple dispatch rules. Two iterations are sufficient.

18.3A.2 Re-solve exercise 18.2A.2 using lead time iteration and bottleneck dynamics. Two iterations are sufficient.

18.3B Software/Computer Exercises

18.3B.1 Modify your program of 18.2B.3 to handle bottleneck dynamics.

18.3B.2 Use your data set from 18.2B.2 to test heuristics against objectives.

18.3B.3 Enrich your program to handle a multi-project project shop.

18.3B.4 Further enrich your program to handle large-scale multiple projects.

18.3C Thinkers

18.3C.1 Give an integer programming formulation for large-scale multiple projects, which allows for renting extra resources in a period or periods at an appropriate rental.

18.3C.2 Suppose that one of the resources in a large-scale multi-project formulation is a batch resource:

  (a) Multiple running activities must start and stop at the same time.

  (b) Only activities of the same "type" may be run together.

  (c) There is a sequence-dependent setup between different type batches.

  Design a bottleneck dynamics procedure to fit this situation.

(A) 18.4 RESOURCE/COST TRADE-OFF MODELS

18.4.1 Overview

In Section 17.3 we studied time/cost trade-offs (CPM) for simple unconstrained single project networks. The basic assumption was that critical activity lengths could be "crashed." Specifically, an activity length could be decreased, within maximum and minimum limits, at a cost linear in the amount of shortening. The cost of the activities and the cost of the resulting project duration could then be added together and the total cost minimized. Alternatively, a trade-off curve showing project duration versus crashing cost could be constructed to give graphical assistance to management in making decisions.

A similar analysis can be performed for more general resource-constrained project scheduling. The problem is now more interesting, since shortening the activity may increase (or in some cases decrease) the total and/or the per-unit-time usage of the various resources, as well as increase the cost.

We shall look at two different formulations. The first model, presented in Section 18.4.2, allows continuous "crashing." As for the time/cost model in Chapter 17, it assumes costs of doing an activity go up linearly as the activity length decreases. It further assumes that total use of each of the other resources goes up proportionately at the same time. The continuous linear form allows alternate iterative updating of the current "optimal" processing times and the schedule until the stopping criterion has been met. An iterative procedure for this model has been developed, and initial testing has been performed by Lawrence and Morton (work in process). Sketchy qualitative initial results are reported here.

The other approach allows only a finite number of discrete ways of doing the activity. Each activity–method pair has associated a specific usage of each resource and of money and yields an associated process time. This can be considered a routing problem, with each method for doing the activity considered a parallel method (analogous to a parallel machine). During any iteration, for an activity one can cost out each possible method by the current prices and slack times and implement the method that is cheapest. (If the cheapest method is not yet feasible, some rules would have to be developed about whether or not to wait. This approach is discussed in Section 18.4.3. It has not yet been programmed or tested.)

18.4.2 Continuous Resource/Cost Models

General Assumptions. The general assumptions are as follows:

Projects

(a) There are multiple projects.
(b) Each project has a due date.
(c) The objective is to minimize the sum of activity costs and weighted tardiness. For weighted flowtime, set due dates equal to zero.
(d) There are standard precedence constraints.

Activities

(a) Each activity has a cost and also uses one or more limited fixed resources.
(b) Use of resources is limited; use of costs is not.
(c) Cost and resource usages vary linearly in activity process time.

This time/cost model is the first to combine project crashing, dollar costs, and resource constraints. Such a model should have broad applications in practice for: engineering firms, contractors, software developers, and R&D teams.

Overview of the Heuristic Solution Procedure

Step 1: Decompose the problem into two subproblems.
Step 2: Subproblem 1 is to determine activity durations:
   (a) Calculus: trade off duration costs with anticipated tardiness costs.
   (b) Factor in imputed resource costs.
Step 3: Subproblem 2 is to schedule activities to minimize tardiness. Use multiple large project bottleneck dynamics heuristics.
Step 4: Iterate between subproblems 1 and 2.
   (a) Use information from the previous iteration and smoothing to improve the current estimates of activity slacks, resource prices, and activity durations.
   (b) Stop iterating when further improvement in the objective function value is not forthcoming.

Resources, Cost Usage Versus Activity Duration.

The following model is based on the time/cost model of Proposition 3 in Section 18.3.

We assume that both the cost A_{ij0} of completing activity ij, where ij means activity i within project j, and the total resource usages A_{ijk} for resource k are linear in activity duration p_{ij}:

$$A_{ij0} = a_{ij0} - b_{ij0}p_{ij} \tag{20}$$

$$A_{ijk} = a_{ijk} - b_{ijk}p_{ij} \tag{21}$$

Note that cost and individual resource usages are allowed to vary independently as the activity is crashed. In particular, while cost is assumed to increase as duration decreases, some resource usages might be so substitutable with cost as to actually decrease as cost increases.

As usual for crashing models, we impose limits on the duration p_{ij} allowed:

$$p_{ij\ min} \leq p_{ij} \leq p_{ij\ max} \tag{22}$$

Since processing costs are explicit, while tardiness costs are implicit, we multiply the tardiness cost by an adjustment factor ω, which can be varied to generate a tardiness/cost trade-off curve. (In the following, resource 0 is cash. Cash always has a price $R_0 = 1$, and resource limit $RA_0 = $ infinity.)

Proposition 4. The discrete approximation to the continuous resource/cost multi-project model can be represented mathematically by the following:

$$\text{minimize} \quad \sum_{ij} (a_{ij0} - b_{ij0}p_{ij}) + \omega \sum_i w_i(C_i - D_i)^+ \tag{23}$$

$$\text{s.t.} \quad C_{ij} = \sum_{t=1,HZ} tC_{ijt} \quad \text{for all } ij \tag{24}$$

$$C_{ij} \geq C_{hj} + p_{ij} \quad \text{for all } hj \in P(ij) \text{ and all } ij \tag{25}$$

$$x_{ijt} = \sum_{u=t,t+p_{ij}-1} C_{iju} \quad \text{for all } ij, t \tag{26}$$

$$\sum_{ij} (a_{ijk}/p_{ij} - b_{ijk})x_{ijt} \leq RA_k \quad \text{for all } k, t \tag{27}$$

$$\sum_{t=1,HZ} C_{ijt} = 1 \quad \text{for all } ij \tag{28}$$

$$p_{ij\,min} \leq p_{ij} \leq p_{ij\,max} \quad \text{for all } ij \tag{29}$$

$$C_{ijt} \in \{0,1\} \quad \text{for all } ij, t \tag{30}$$

(Note that p_{ij} in (26) is a decision variable, which makes solution difficult.)

Relevant Marginal Activity Duration Costs for Bottleneck Dynamics

Activity Duration, Direct Costs, and Implicit Resource Usage. For fixed expected tardiness of projects and fixed implicit resource costs, the total explicit and implicit cost A_{ij} of completing activity ij is a linear function of activity duration p_{ij}:

$$A_{ij} = \sum_{k=1,m} (a_{ijk} - b_{ijk}p_{ij})R_k \tag{31}$$

Thus the marginal cost of activity duration, holding other things constant, is given by

$$A'_{ij} = - \sum_{k=1,m} b_{ijk}R_k \tag{32}$$

Marginal Activity Tardiness Cost. If p_{ij} is shortened a marginal amount, the R&M heuristic says the saving in weighted tardiness is given approximately by

$$T'_{ij} = w_i \exp[-(d_j - LT_{ij} - p_{ij})^+ / \kappa p_{av}] \tag{33}$$

where LT_{ij} is a measure of activity lead time (thus $(d_j - LT_{ij} - p_{ij})^+$ measures the slack), κ is a planning factor, and p_{av} is an average activity duration.

First-Order Condition, Local Optimum Activity Durations for Other Fixed Factors. The first-order condition with respect to each activity time p_{ij} is given by

$$A'_{ij} + \omega T'_{ij} = 0 \tag{34}$$

If slack is positive, solving for p_{ij} yields

$$p^*_{ij} = d_j - LT_{ij} + \kappa p_{av} \ln\left[\left(\sum_{k=0,m} b_{ijk}R_k \right) \Big/ (\omega w_i) \right] \tag{35}$$

It is interesting to note that, at $\omega =$ zero, cash outlay is very important and tardiness is of no interest. Thus we expect large activity durations to be selected so as to minimize cost outlays while maintaining resource feasibility. As ω increases, shorter durations will be selected to help tardiness, and optimal costs will go up slowly. For very large ω we expect to approach the pure multi-project solution; that is, durations will stabilize at the pure tardy case, and costs will simply go up linearly in ω. This intuition can be checked in equation (35). For large ω machine prices should go up proportional to ω, and cash should become unimportant. Indeed, the logarithm term would become constant in such a case.

Iterative Solution Procedure. We use CPM for the non-resource-constrained relaxation to the fixed processing times case.

We use MLP for the multiple large projects bottleneck dynamics solution to the fixed processing times case.

Step 1: Initial Network Relaxation
  $n = 0$.

Initially solve the CPM for $p_{ij}^0 = p_{ij\ max}$.

Save slacks, average processing times, and prices (0).

Step 2: Initial Fixed Process Time Solution

   $n = 1$.

   Initially solve the MLP for $p_{ij}^0 = p_{ij\ max}$.

   Save slacks, average processing times, and prices.

Step 3: Update Iteration

   $n \leftarrow n + 1$.

   Smooth and update activity slack values.

   Smooth and update average processing times.

   Smooth and update prices.

Step 4: New Processing

   Use equation (35) to calculate apparent new optimal processing times.

   Smooth and update processing times, noting constraints.

Step 5: New Fixed Process Time Solution

   Solve the MLP for the current fixed processing times.

   Save slacks, average processing times, and prices.

Step 6: Test Stopping Rule

   If stopping rule satisfied, go to step 7.

   Otherwise go to step 3.

Step 7: Termination

   Record best schedule and other desired statistics.

   Stop.

Qualitative Results to Date. Lawrence and Morton (work in process) have written preliminary software and have run early checks. Real testing has not begun at this writing.

Hypotheses

   (a) Bottleneck dynamics provides superior solutions.
   (b) Multiple smoothing of parameters provides even better solutions.
   (c) On average, the best solution will be found in 5–15 iterations.
   (d) Problem is complex, difficult; much fine tuning will be useful.

Future Efforts

   (a) Major testing.
   (b) Refinement of iterative procedure.
   (c) Refinement of various parameters.
   (d) Develop simple dispatch benchmark heuristics.
   (e) Compare with discrete approach (see below).

18.4.3 Discrete Resource/Cost Model

Continuous Versus Discrete Model. The continuous and the discrete resource/cost models are identical, except in the way that different methods of doing an activity affect the cost and resource usages. In the continuous case we just looked at, there are an infinite number of choices. When an activity is shortened, cost and each resource usage are all affected linearly. This allowed us to use calculus to estimate optimal process times as a function of current slacks and machine prices. In actual use at a dispatch point, we must be careful because the unconstrained (in the calculus sense) "optimal" activity may be infeasible due to excess resource usage. Thus we choose the shortest activity at least as long as the "optimal" and choose this as the process time to use at any dispatch point. Then we decide among the possible activities thus defined to choose the activity to schedule at each dispatch point.

In the discrete model, there are only a finite number of methods $q_{ij\,max}$ to perform an activity, which seems less general. However, the different ways of doing the activity can be entirely different from each other; linearity is not assumed. For example, there can be several methods of doing the same activity, which involve the same processing time but different mixes of costs and resources, or there may be different methods that take different processing times and differing mixes of costs and resources.

Resources and Cost Usage Versus Activity Duration. Any method q of the $q_{ij\,max}$ methods for doing activity ij has an associated processing time p_{ijq}, total cost A_{ijq}, and resource usage per period $r_{ijq1}, r_{ijq2}, \ldots, r_{ijqk}, \ldots, r_{ijqm}$, which may be arbitrary nonnegative numbers. By the nature of things, no limits are now necessary on the p_{ijq}. In the following define $z_{ijq} = 1$ if ij is assigned to method q, 0 otherwise. Define $C_{ijt} = 1$ if ij completes in period t, 0 otherwise. Define $x_{ijt} = 1$ if ij is in process in period t, 0 otherwise. Define hj as any immediate predecessor activity to ij.

Formal Model

Proposition 5. Defining C_i to be the final activity of project i, the discrete resource/cost multi-project model can be represented mathematically by the following:

$$\text{minimize} \quad \sum_{ij} \sum_{q} A_{ijq} z_{ijq} + \omega \sum_{i} w_i (C_i - D_i)^+ \qquad (36)$$

$$C_{ij} = \sum_{t=1,HZ} tC_{ijt} \quad \text{for all } ij \qquad (37)$$

$$p_{ij}^* = \sum_{q=1,q_{ijmax}} p_{ijq} z_{ijq} \quad \text{for all } ij \qquad (38)$$

$$C_{ij} \geq C_{hj} + p_{ij}^* \quad \text{for all } hj \in P(ij), \text{ for all } ij \qquad (39)$$

$$x_{ijt} = \sum_{u=t,t+p_{ij}^*-1} C_{iju} \quad \text{for all } ij, t \tag{40}$$

$$\sum_{ij} \sum_{q=1,q_{ij\text{max}}} r_{ijqk} z_{ijq} C_{ijt} \leq RA_k \quad \text{for all } k, t \tag{41}$$

$$\sum_{q=1,q_{ij\text{max}}} z_{ijq} = 1 \quad \text{for all } ij \tag{42}$$

$$\sum_{t=1,HZ} C_{ijt} = 1 \quad \text{for all } ij \tag{43}$$

$$z_{ijq} \in \{0,1\}; \quad C_{ijt} \in \{0,1\} \quad \text{for all } ijq \tag{44}$$

Proof. Equations (44), (43), and (42) say that an activity is done by exactly one method and that it completes in exactly one particular period. Inequality (41) says that all activity–methods current at a point of time cannot have resource usages totaling more than available. (Note that this is a nonlinear equation.) Equation (40) says an activity is active in a period if its completion is between this period and p_{ij}^* − 1 periods in the future. Equation (38) selects the actual processing time of each activity; inequality (39) uses this to give the precedence constraints. Equation (37) changes zero–one variables into the actual completion time of ij. Equation (36) gives total costs as the sum of activity–method costs and tardiness costs. The double summation selects the costs only for activity–methods actually used. ■

As before, this formulation gives a precise definition to the problem but is not particularly useful for computation. We go on to the heuristic approach for the problem.

Heuristic Selection of the Best Method for an Activity. At the beginning of any iteration, given prices and lead times from the previous iteration, we first cost out every activity–method in the program by

$$(AM)_{ijq} = A_{ijq} + \sum_{k=1,m} r_{ijqk} p_{ijq} R_k$$

This is our best current estimate of the explicit plus implicit cost of doing each activity–method.

Then, at each dispatch decision point during the new iteration, we select the cheapest feasible method for each activity. Then, for this method in each case, use the multiple large projects bottleneck dynamics heuristic with current prices and lead times to select the activity with the "best" benefit/cost ratio as before.

Iterative Solution Procedure. The procedure is very similar to the continuous case. As before, we use CPM for the non-resource-constrained relaxation to the fixed processing times case. We use MLP for the multiple large projects bottleneck dynamics solution to the fixed processing times case.

Step 1: Initial Network Relaxation

Step 2: Initial Fixed Process Time Solution

Step 3: Update Iteration

Step 4: Cost Out Activity–Methods

Step 5: New Iteration. Solve MLP, using, for each activity, the cheapest feasible activity method at each decision point.

Step 6: Test Stopping Rule

Step 7: Termination

Lawrence and Morton have neither programmed nor tested the discrete model at this writing.

18.4A Numerical Exercises

18.4A.1   Create two simple examples, each with one project, three or four activities, and one resource. All but one activity should only be doable in only one way.

    (a) For the first example, let the remaining activity be continuously variable in activity length.

    (b) For the second example, let the remaining activity have three methods.

18.4A.2   From 18.4A.1 compute a heuristic solution to the continuous version, by hand, spreadsheet, or computer program, whichever seems appropriate.

18.4A.3   Solve 18.4A.2 again for the discrete version.

18.4B Software/Computer Exercises

18.4B.1   Write a spreadsheet or a program capable of solving somewhat larger discrete problems.

18.4B.2   Invent a data set, and test your program from 18.4B.1 fairly extensively.

18.4C Thinkers

18.4C.1   Suppose that we replace RA_k by RA_{kt} in the discrete formulation and make time discrete also. What problems are there in adapting the solution procedure given?

18.4C.2   Suppose, for some activities, a number of discrete choices can each be varied continuously? Discuss some formulation for this generalization, and discuss any problems that might occur.

(A) 18.5 MODEL EXTENSIONS

18.5.1 Overview

Resource-constrained multiple project scheduling is an extremely large subject. First, it essentially provides a unifying framework for all of our scheduling models in the book to date: single machine, parallel machines, complex single resources, multiple resources per activity, flow shops, job shops, classical project scheduling, project shops, ordinary multiple large projects, and resource/cost models. Second, higher level strategic situations, mid-level planning, design, and R&D can now be modeled.

However, there is a difficulty. While all these situations can be seen to have many modeling parts in common, there are so many variations that just keeping track of major model types can get to be complicated. It is almost impossible to adequately test all these types. We present four such major types in this subsection, which are clearly very important but have not yet been fully developed and are largely untested. In each case we give a basic mathematical representation and very briefly discuss possible heuristics.

In Section 18.5.2 we discuss the case in which some of the resources get used up by the project. (Cash is one example.) In Section 18.5.3 we discuss the case in which we can supplement the initial resource endowment by choosing to rent more resources in desired periods. In Section 18.5.4 we discuss the case in which the project occurs over a long time period so that discounting costs and revenues becomes important. Finally, in Section 18.5.5 we discuss the case in which a project may have several reward points (milestones) and other similar cases in which reward points may not be independent of each other.

18.5.2 Consumable Resources

We consider the standard MLP (multiple large projects) model, except that there are two classes of resource: nonconsumable (trucks, labor, machines) and consumable (cash, heating oil, iron ore), which have an initial availability and are not convertible or replenishable for the project. That is, there are m_1 nonconsumable resources indexed by $k1$, with initial availability $RA_{k1,0}$. (Here we avoid subscripting subscripts, since it will probably cause no confusion.) Activity ij uses resource $k1$ at rate r_{ijk1}. There are $m2$ consumable resources indexed by $k2$, with initial availability $RA_{k2,0}$. Activity ij uses resource $k2$ at amount r_{ijk2} per period. Leftover resource in the last period HZ of amount $RA_{k2,HZ}$ has value v_{k2} per unit. Then we may generalize Proposition 3 as follows.

Proposition 6. The MLP problem with resources split into $k1$ nonconsumable resources and $k2$ consumable resources may be given mathematically by (see Proposition 4) the following:

$$\text{minimize} \quad \omega \sum_{ij} w_{ij}(C_{ij}) - \sum_{k2=1,m2} v_{k2}RA_{k2,HZ} \qquad (45)$$

$$\text{s.t.}\quad C_{ij} = \sum_{t=1,HZ} tC_{ijt} \quad \text{for all } ij \tag{46}$$

$$C_{ij} \geq C_{hj} + p_{ij} \quad \text{for all } hj \in P(ij), \text{ for all } ij \tag{47}$$

$$x_{ijt} = \sum_{u=t,t+p_{ij}-1} C_{iju} \quad \text{for all } ij, t \tag{48}$$

$$\sum_{j=1,n} (a_{ijk1}/p_{ij} - b_{ijk1})x_{ijt} \leq RA_{k1} \quad \text{for all } k1, t \tag{49}$$

$$RA_{k2,t+1} = RA_{k2,t} - \sum_{j=1,n} (a_{ijk2}/p_{ij} - b_{ijk2})x_{ijt} \quad \text{for all } k2, t \tag{50}$$

$$p_{ij\,\min} \leq p_{ij} \leq p_{ij\,\max} \quad \text{for all } ij \tag{51}$$

$$\sum_{t=1,HZ} C_{ijt} = 1 \quad \text{for all } j \tag{52}$$

$$C_{ijt} \in \{0,1\} \quad \text{for all } ij, t \tag{53}$$

Proof. The only changes from the equations of Proposition 4 are as follows. The objective function (45) has an additional term subtracting the salvage value of the leftover consumables from the costs. Cash does not appear per se in the objective function. (However, cash is allowed to be a consumable resource with a large limit. $v_{k2} = 1$ for cash.) Constraints (49) and (50) are changed. The first gives resource limits for the nonconsumable items, which are the same as before. The second says that consumables available in period $t + 1$ are those in period t less resources consumed in t. ∎

Bottleneck Dynamics Heuristic for Consumable Resources. First, notice that while having a salvage value for leftover consumable resources seems to make solution more difficult, it can easily be handled by a trick. Add the constant $\Sigma_{k2=1,m2}v_{k2}RA_{k2,0}$ to the objective function; that is, require that initial resources be paid for at the salvage value rate. Part of this extra cost can be charged against the eventual salvage value; the rest can be used to cause the cash cost of any activity to increase by the salvage value of each of the resources used. This alternative problem has the same solution as the original up to a constant and involves no salvage values on leftover resources. We first make this transformation of the problem.

The heuristic we propose (which is currently untested) is somewhat similar to that for the continuous version of the resource/cost trade-off model. (See Section 18.4.2.)

Step 1: Initial Network Relaxation. $n = 0$.

Step 2: Initial Fixed Process Time Solution. Add resource constraints; solve MLP for $p_{ij\,\max}$.

Step 3: Update Iteration

Step 4: New Processing Times. Using resource prices, smooth and update process-ing times, noting constraints. Make only a small change each time to minimize chances of violating constraints.

Step 5: New Fixed Process Time Solution

Step 6: Test Stopping Rule

Step 7: Termination

The interesting empirical/theoretical question is estimating stationary prices on the consumable resources sufficiently useful to solve for new processing times.

18.5.3 Resource Replanning

In this model, we consider the basic MLP problem with activities using multiple resources and a resource servicing multiple simultaneous activities, as given in Proposition 3. The first main difference is that we allow the resource limits to be dynamic; that is, we replace RA_k by RA_{kt}. The other main difference is that we can, for period t, rent additional resources. If in period t we have RA_{kt} amount of resource k, we can choose to rent for that period D_{kt} more units at a cost v_{kt} per unit.

Proposition 7. The MLP problem with addition of resource replanning (renting extra resources) may be given by the following:

$$\text{minimize} \quad \sum_{j=1,n} w_j(C_j) + \sum_{t=1,HZ} \sum_{k=1,m} v_{kt} D_{kt} \tag{54}$$

$$\text{subject to} \quad C_j = \sum_{t=1,HZ} C_{jt} \quad \text{for all } j \tag{55}$$

$$C_j \geq C_i + p_j \quad \text{for all } i \in P(j), \text{ for all } j \tag{56}$$

$$x_{jt} = \sum_{u=t,t+p_j-1} C_{ju} \quad \text{for all } j, t \tag{57}$$

$$\sum_{j=1,n} r_{jk} x_{jt} \leq RA_{kt} + D_{kt} \quad \text{for all } k, t \tag{58}$$

$$\sum_{t=1,HZ} C_{jt} = 1 \quad \text{for all } j \tag{59}$$

$$C_{jt} \in \{0,1\} \quad \text{for all } j, t \tag{60}$$

Proof. The formulation differs from the MLP in Proposition 3 only in paying for extra units rented in the objective (54), and having dynamic basic resources and purchased extra resources in (58).   ∎

Sketch of Heuristic Solution Method. Now the number of resource prices needed has increased dramatically (kt instead of k). Thus it is suggested, if practical, to obtain these by solving the linear programming relaxation at each iteration and using the LP dual variables as the resource prices.

Step 1: Solve the LP relaxation above, and save the resulting rental amounts and resource prices.

Step 2: Fix the rental amounts and resource prices and solve the MLP.

Step 3: Add the current solution of the MLP as constraints into the LP, solve, and save the resulting rental amounts and resource prices.

Step 4: Repeat steps 2 and 3 until the stopping criterion is met.

18.5.4 Net Present Value

The net present value model is really just a special case of Proposition 3, the multiple large projects (MLP) problem. Continuous interest discounting is introduced for each activity completion. For example, the objective function for MLP for the weighted tardiness problem and independent projects is

$$\text{minimize} \sum_{j=1,n} w_j [C_j - D_j]^+ \tag{61}$$

where $w_j = 0$ unless j is the completion of one of the projects.

But recognizing continuous interest would change the objective to

$$\text{minimize} \sum_{j=1,n} w_j e^{-I(C_j)} [C_j - D_j]^+ \tag{62}$$

While both (61) and (62) are special cases of the general objective given by (13), and can thus be solved by analogue integer programs, remember that this IP is nonlinear and hence cannot be solved for even rather small problems. We must look to heuristics.

We discussed the full net present value sequencing analysis for bottleneck dynamics in Section 10.5.4. Under the simplifying assumption that all completion times were under a year, we showed how to modify the inputs in order to fit in the standard bottleneck dynamics methodology. We also sketched how to calculate the marginal cost of delaying an activity for longer horizons. Once this is done, the standard bottleneck dynamics methodology for the MLP problem can be used.

This is only a sketch of a possible full NPV methodology. We do not attempt to carry this further here. It awaits actual software development and testing.

18.5.5 Milestones/Dependent Projects

While Proposition 3 handles very general projects from an integer programming point of view, our heuristics to date have been limited to independent projects.

Definition. *Independent projects* represent the case where n activities $j = 1, \ldots,$ n can be divided into q disjoint projects $i = 1, \ldots, q$, with q_i activities each. The final activity in each project has a reward or penalty $w_i(C_i)$; the other activities have no reward or penalty.

Definition. In *dependent projects*, any activity plus all its predecessors may be considered a project. (The activity itself would be called the project completion.) These are of several types:

(a) If the activity has no successors and carries a major reward for finishing or a major penalty for not finishing promptly enough, we term the activity and its predecessors a (conventional) *project* and the activity itself a *project completion*.

(b) If an activity has a project as a successor, but itself carries major rewards or punishment on completion, we term the activity and its predecessors a *milestone* and the activity itself a *milestone completion*.

Figure 18.2 shows a case of independent projects. Figure 18.3 shows dependent projects with milestones. (We use activity-on-node representation.)

In Figure 18.2 the projects are quite obvious. Project 1 is comprised of activities

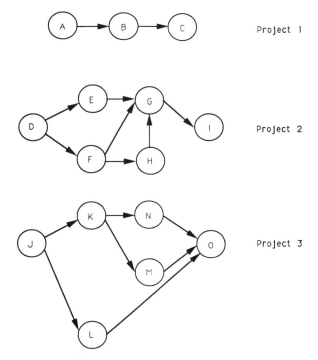

Figure 18.2. Independent projects.

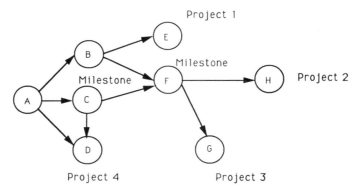

Figure 18.3. Dependent projects.

A, B, and C. Project 2 is comprised of activities D, E, F, G, H, and I. Project 3 is comprised of activities J, K, L, M, N, and O.

In Figure 18.3 we must keep track of predecessors more carefully. A, B, and E comprise Project 1. A, B, C, F, and H comprise Project 2. A, B, C, F, and G comprise Project 3. A, C, and D comprise Project 4. All overlap with each other, so that, for example, when we try to calculate the priority of doing activity C, we must decide whether we are mostly trying to benefit Project 2, Project 3, or Project 4. We will return to this question when we develop our heuristics.

It is common practice in large projects to give partial payment for completing an activity important in finishing the final project. For example, suppose in Figure 18.3 that we get payments when C is completed. A,C is the milestone, while C is the milestone completion. Similarly A,C,F is a milestone, while F is the milestone completion. Note that an activity may be a milestone for several projects at once: F is a milestone completion for both Project 2 and Project 3; C is a milestone completion for Projects 2, 3, and 4.

We turn now to solution methods for the dependent project case.

Integer Programming. The integer program given by Proposition 3 on MLP problems is very general in that it allows a general objective function and arbitrarily interconnected projects, milestones, and so on. In fact, it allows rewards and punishments at every activity. The main problem as stated then is that it is not practical to solve a very large problem, since the number of precedence constraints themselves can be of the order of n^2, and discrete time can multiply the number of variables by a factor of 20–100 and many of the constraints by a similar factor. (It may be practical to solve the LP relaxation as one method of estimating resource prices, however.) Therefore we turn to heuristic methods.

Heuristic Methods. The bottleneck dynamics method is really quite similar to the one for the MLP problem outlined in Section 18.3.3. Pricing and lead time iteration are no different.

To see where modification is needed, consider the dispatch procedure, when an activity has completed and a next activity to be scheduled must be chosen.

Step 1: Find all feasible activities to be scheduled next.

Step 2: Treat each project/milestone as an independent project.

Step 3: For each project find the highest priority activity in it by the LFT (there may be none), producing a reduced candidate list.

Step 4: If an activity is chosen by more than one project, create one candidate for each, so that a candidate will correspond to exactly one project.

Step 5: Calculate the priority of the candidate as would be done for the independent project case. For example, for weighted tardiness use the R&M rule. Calculate a benefit/cost ratio. The numerator would be the sum of the delay costs for the project and of any internal milestones. The denominator would be the sum of all resource usage remaining in the project.

This extension of the MLP heuristics has been neither programmed nor tested at this writing.

A useful reference for this section is Li and Willis (1991).

18.5A Numerical Exercises

18.5A.1 Modify your discrete example from exercise 18.4A.1. Include a consumable resource. Give a mathematical program for it in the format of Proposition 6.

(a) Is the relaxation a linear program? Why or why not?

(b) Is the relaxation of any other use?

18.5A.2 For your example in 18.5A.1, invent a crude heuristic for finding a decent solution and apply it.

18.5A.3 How much can you improve your solution by commonsense adjustment?

18.5B Software/Computer Exercises

18.5B.1 Computerize your heuristic invented in 18.5A.2, and add input routines to make it easy to create new problems. Test your heuristic fairly extensively. Look each time at any manual ways to improve your initial solution.

18.5B.2 Using the experience gained from 18.5B.1, improve your heuristic, add it to your software, and test the improved version on the problems you have already investigated, as well as some new ones.

18.5C Thinkers

18.5C.1 Create a new model variation of the type given in this section, explain the applicability by real-world example, and formulate it carefully as an integer program (if you can).

18.5C.2 Consider a standard project job shop (see Section 18.2) except that now there are allowed to be both "and" constraints and "or" constraints. Formulate an integer programming formulation for this case. (*Hint:* You have seen "disjunctive" formulations that are essentially "or" constraints.)

18.6 CASE EXAMPLE: CUSTOM OPTICAL EQUIPMENT

18.6.1 Overview

We have now had an exposure to a number of types of projects, project shops, and large-scale multiple projects with various kinds of resource constraints and other types of assumptions, together with a variety of exact and heuristic methods for modeling them. Here we present a real-world case, to see both how these ideas would work in an actual situation and to what extent they might need to be modified. We present a very brief summary of the problem here in the overview, with a longer and more careful description in Section 18.6.2, followed by a partial analysis of the total scheduling problem in Section 18.6.3.

We consider the manufacturing shop of OptiQuip, a medium-sized manufacturer of all types of large custom optical equipment. (Names and data have been changed for proprietary reasons.) OptiQuip has about 200 employees in an engineering group that designs new products to customer request, about 150 in the manufacturing group, and 150 in marketing, inventory, shipping, and so on. It produces such items as large refractive telescopes, reflective telescopes, small and large laser equipment, fiber optics, laboratory equipment of all types, and electron microscopes. In a few cases it produces subassemblies and assemblies for others to build precision equipment, but management prefers complete creative control. Many orders consist of a single large instrument worth $100,000 to a few million dollars, but a custom order can also consist of say 50 identical items worth $10,000 each. Although cost is an important consideration, the company prides itself on extremely high scientific, engineering, and manufacturing capabilities.

In the past few years customers have increasingly pressured OptiQuip to shorten lead times, both in engineering and in manufacturing. Meeting strict deadlines has become increasingly important, and in addition to rewards at the completion of milestones and the final order, there are now very often contractual penalties for tardiness.

The manufacturing process is complex and varies a great deal according to the product being produced. There is a rather standard metal fabrication shop. There is a lens grinding shop. There is a paint shop that has both a paint room and a bake room. The paint room can simultaneously accommodate a number of items with the same color. A paint room batch moves together (with no waiting allowed) to the bake room. There is an "assembly shop" that is basically one very large worktable for assembly, together with various types of clamps, jigs, and other features. The assembly shop also has several smaller worktables for assembling smaller jobs and/or subassemblies.

Personnel basically divide into four classes.

Class 1 is specialized and can operate only one machine. This includes the paint shop and many machines in the metal shop.

Class 2 can operate any remaining machine in the metal shop.

Class 3 can operate any lens grinding machine.

Class 4 represents several crews. Any crew can do any assembly job.

The production process is roughly as follows. Some items and subassemblies will be purchased elsewhere; most will be fabricated in the metal shop. All lenses are ground on the premises. Items are painted and baked. The assembly process consists of several stages of subassembly and assembly, including customized packaging for shipment. High levels of precision WIP on the floor cause many problems. If a hot job preempts another job for several weeks, for example, it may be hard to find all the parts, and there is always the danger of damage in the interim.

18.6.2 Details of Case Example

Model Type. This can be cast fairly well as a multiple large project model. ("Large" might be reworded as "medium to large.") All items in a customer order comprise a single project with a common due date, whether a single large telescope or 20 laboratory instruments. (If the project is large enough to have milestones, each would be treated by the heuristics as a separate project.)

Resources—Metal Shop. Machine clusters can be treated as an individual machine and aggregated to an equivalent aggregate machine. Attached workers can be ignored (treated as attached to machine). These machines are not considered to require a second resource. On the other hand, the flexible workforce is a resource to be assigned as needed.

Thus, overall, the metal shop becomes a classic job shop with one extra resource that can be shifted to machines as needed. This all fits the multiple large project model without change.

Resources—Lens Grinding Shop. This is basically a large group of parallel machines, with clustering. Each cluster handles one type of lens and can be considered proportional machines to be aggregated into a single composite machine. Operators may be transferred across clusters and hence become another resource. The paint and bake rooms both accept batches of items at one time. There are groups: each group has the same color, type of paint, baking time, and temperature. Since no waiting is allowed between paint and bake, each batch must wait before starting painting so that the batch currently baking can finish before this painting cycle finishes. Thus this is a single-resource multi-processor situation.

This composite resource can be treated as a continuous resource, with special rules about when a batch can start and which items can be processed together.

Resource—Assembly Shop. Each assembly table can be considered a separate resource. Small tables can handle only one activity at a time, but the large table can handle one large project or several small projects, and so on. When several activities are fitted on the large table, there is some inaccuracy in just representing each activity as requiring r_{jk} fraction of the resource. The jigs can also be taken to represent separate resources and could complicate the issue somewhat.

The small tables can clearly be aggregated to a single resource, the large tables to another, and the work crews as another. However, some activities may use either the large table or the small tables, so that there is a routing choice to be considered.

18.6.3 Analysis

General Strategy. The problem is basically a multiple large project (MLP) problem except for several difficult features: the paint room is batch; the bake room is batch; no delay is allowed between paint and bake; and there is a routing issue in the assembly shop.

Our strategy is as follows:

(a) Approximate these features so the resulting simplified problem fits the MLP assumptions.

(b) Solve the MLP by the bottleneck dynamics method in Section 18.3 to get estimated resource prices (static or dynamic as desired) and estimated lead times for each activity.

(c) Create a full simulation model with details restored, and make a single pass by bottleneck dynamics using the estimated prices and lead times, making any modifications necessary in the assembly shop/paint shop areas.

Approximating the Paint/Bake Shop

(a) Without loss of generality assume the bake shop time is the bottleneck.

(b) Each item is assumed to require a simple lag time equal to paint time, followed by a small bake time equal to the actual bake time times the percentage of the bake shop used, followed by a simple lag time equal to the rest of the actual bake time. (That is, we are eliminating batching, making baking divisible, and putting the rest in as simple lags.)

Approximating the Assembly Shop

(a) Approximate the assembly shop as a single resource. (That is, simply add up all table space, and allow any activity to require a certain amount of table space and one crew. The main inaccuracy here is that the approximation will allow larger projects requiring more than one small table to use several tables at once.)

Solving the Simplified Model. The resulting simplified model fits the MLP formulation exactly and can be solved fairly easily by bottleneck dynamics. Note, in particular, that fixed leads and lags can be treated as activities requiring only time but no resources. The resulting resource prices and lead times become inputs to the full simulation, which is run only once.

In particular, the bake shop will have a price or prices, while the paint shop will receive a price of zero, which is probably reasonable. All tables will receive the same price per square foot, which may not be so reasonable. We will return to this issue later.

The Full Simulation. The full simulation must add back an accurate representation of the paint/bake shop and the assembly shop.

The Full Simulation—Paint/Bake Shop. As activities arrive in front of the paint shops they are accumulated into compatible batches and prioritized by the methods of Section 11.4. At the appropriate start time, chosen so that the previous bake batch will finish on time, the highest priority batch is started. The determination of the highest priority batch depends on the prices and lead times in the earlier simplified solution.

The Full Simulation—Assembly Shop. As space on a small table becomes available, the highest priority item fitting on a small table can probably be assigned without too much inaccuracy. However, the large table is more difficult. As space becomes available on it, we must consider both the priority of the waiting jobs and the best fitting of jobs on the table. We may also wish to hold part of the table, so that an arriving large "hot" job will be able to start. These issues would require either some fitting heuristics to be devised, or else some human–machine interface so that a skilled operator could make some of the difficult decisions.

A possible fitting heuristic for the large table when empty is just to try a number of possible combinations of jobs to fit on the table, choosing the configuration for which the sum of priorities is the highest.

PART VI
OTHER ISSUES

One major advantage of heuristic scheduling systems, as opposed to integer programming or dynamic programming based exact systems, is that various types of model extensions are relatively easily incorporated into the basic model. In Chapter 19 we extend the basic model in which an activity is processed by exactly one resource (machine). Here we consider both the case in which it takes several resources to process one activity, and the case in which it takes several machines to comprise one resource.

In Chapter 20 we also give many examples of small- and medium-sized modifications to the basic model including open and closed shops, assembly and continuous shops, push versus pull systems, release setting, due date setting, and variable speed resources.

A major advantage of bottleneck dynamics scheduling systems is that major modifications involving interconnecting the system to human operators or to other systems, such as high-level planning, logistics, MRP, and accounting, may be made, much as tinker toys allow attaching assemblies together. Here prices and lead time estimates passed between levels and modules act as the tinker toy connectors. We discuss these issues more completely in Chapter 21.

____19
COMPLEX RESOURCES AND/OR COMPLEX ACTIVITIES

19.1 INTRODUCTION

19.1.1 Overview

Both classic and modern scheduling methods are at their best in the classic flow shop, job shop, or project job shop. In all these cases each relatively simple activity requires precisely one relatively simple machine for processing. For a one-machine shop therefore, only an optimal permutation sequence must be determined, whether by branch-and-bound related methods, neighborhood search related methods, or dispatch priority related methods.

For multi-machine shops, if activities and machines remain simple, dispatch methods remain much the same, with the added complication of estimating lead times and/or downstream resource usage in order to be able to approximately decompose the problem and solve a number of related one-machine subproblems.

However, real-world problems are becoming more complicated; complex resources and/or complex activities are becoming more the norm than the exceptions. Situations where large numbers of machines form a single resource are quite common. We have met some of these situations in earlier chapters. Examples include equal parallel machines, proportional parallel machines, clustered parallel machines, permutation flow shops, and no-delay flow shops. We will meet more examples in this chapter.

The great advantage of multi-machine problems which may be considered single resources is that they possess a single input queue. Thus scheduling requires determining only a single optimal permutation sequence. Thus, for example, one-dimensional neighborhood search is possible for a quite complex looking problem. This overstates the situation somewhat, however. Often a given sequence requires estimation of delay times for each job before entering it into the system, even if it is known to be next in sequence. These delay times require analytic or simulation estimates of the earliest time that entering the job will not jam the system. Simulation will always work but may require several repetitions for each delay to be estimated. Similarly, determining the effect of pairwise interchange on the permutation sequence requires a complete simulation of the system—several if a new delay time must be computed. Thus there is a trade-off. These types of systems may get quite complex. A single-resource complex system being solved by neighborhood search may very well require about the same number of interchanges for its solution as for a simpler one-resource system. However, the interchanges themselves require rapidly increasing effort in the increasing complexity of the system.

Turning to activities requiring multiple resources, dispatch methods such as bottleneck dynamics often allow putting a price on each resource and simply aggregating resource usage, obtaining a total dollar value of resource usage for the activity. Then the activity with highest benefit/cost ratio may be given the highest priority as before. This procedure works quite well in aggregate situations like multiple large projects.

Unfortunately, as we get into situations that involve careful coordination of several resources, interval scheduling rather than dispatch scheduling becomes increasingly important, and so bottleneck dynamics must be modified to remain useful. For example, if a specific oversize drilling machine, a specific one-of-a-kind tool, and a particular skilled operator are all necessary to perform a given operation, then it is of little interest to determine a priority for the activities simply by adding up the imputed resource usages. It is necessary to make a time schedule currently feasible for all three to try to assure all will be available. This may introduce secondary idleness for some of the resources, which must be added to the resource cost.

Thus the message of this chapter is twofold:

(a) Conventional methods can often be "stretched" to deal with more complex real-world situations.
(b) Such "stretching" is subject to decreasing returns if taken too far.

Developing methods that are more robust to increasing complexity seems to provide important future directions.

19.1.2 Summary of Chapter

In Section 19.2 we discuss techniques for dealing with multiple machines per resource. Section 19.2.1 gives a short review of such results that we have already seen for parallel machines and flow shops. Section 19.2.2 discusses machines with an external setup. Such machines allow the setup for the next job to be performed while the current job is being run. We discuss this problem when the external setup serves a single machine and when it serves equal parallel machines. Section 19.2.3 discusses complex machines with serial multi-processing, such as a simple assembly line. Finally, Section 19.2.4 discusses lines made up of finite-queue parallel machines and also flexible manufacturing cells.

In Section 19.3 we discuss ways of analyzing the case of multiple resources for an activity. In Section 19.3.1 we review the bottleneck dynamics approach for aggregating downstream resources, and we also review the pricing methods we used in scheduling multi-resource projects. In Section 19.3.2 we begin to develop aggregate joint-resource methods where dispatch methods must be augmented by various interval scheduling adaptations. In Section 19.3.3 we discuss scheduling machines that also require an operator resource. We consider both the case in which there is one operator needed per machine and more complex situations. Finally, in Section 19.3.4 we consider machines with tool magazines. We consider both the case where the machine has a fixed lead time in obtaining and changing tools and more complex situations.

In Section 19.4 we present the case of a computer card line sector, which combines both the elements of multiple machines per resource and multiple resources per activity. In Section 19.4.1 we give a description of the problem. In Section 19.4.2 we present first the base case, in which operators are not the limiting factor, then the general case, in which we must jointly deal with sequencing and operator assignment.

19.2 MULTIPLE MACHINES PER RESOURCE

19.2.1 Overview

In this section we develop the issues, advantages, and disadvantages of treating single-queue multi-machine problems as a single resource. We first turn to a short review of the parallel machine results, which we have already obtained, and then we give a review of similar flow shop results.

Review of Parallel Machine Results. In Proposition 1 of Chapter 11 we proved a fundamental result that any general parallel machine grouping may be considered to be a single resource. "Consider a general parallel machine problem with dynamic arrivals, dynamic availability, and regular objective function. There is an optimal solution of the form: sequence the n jobs in some order in a list, and schedule each in turn to the machine that can finish it first. For the dynamic arrival case, this remains true whether the discipline is dispatch or inserted idleness is allowed."

The implication of this is that any exact or heuristic method for scheduling a single machine, such as branch-and-bound (and approximations), neighborhood search (and extensions), or dispatch (and pricing extension) methods, may be used for general parallel machines. Consider simple neighborhood search, for example. The interchange of two jobs in the master list now will require up to a full simulation of the problem for a given sequence and on average a half simulation (depending on where the interchanged jobs are in the sequence.)

Proposition 1 does not, however, suggest intuitive dispatch heuristics suitable for sequencing the master queue. A good part of the rest of Chapter 11 does just that, for the special cases of equal and proportional machines. This technique is as follow:

Step 1: Aggregate the parallel machines into one machine m' times as fast, where m' is the sum of the speed factors for the machines.

Step 2: Order the list by the heuristic for this objective for the equivalent one-machine problem.

Step 3: Schedule this list to the original problem, scheduling each job to the machine finishing it first.

Obtaining heuristics for ordering the line for general parallel machines is somewhat more difficult. BD Heuristic 3 of Section 12.3.3 achieves this, using the principle of "maximum regret." We do not repeat this development here.

Review of Flow Shop Results. It has historically been popular to limit flow shop investigations to policies that preserve job sequence from machine to machine, the so called permutation sequences. In many cases permutation schedule limitations are quite reasonable. It was shown in Section 13.1.3 that permutation schedules are optimal for:

(a) All regular objectives, static input, two machines.

(b) Makespan, static input, three machines.

In addition, the permutation sequence limitation turns a flow shop into a single-resource problem, which makes such techniques as branch-and-bound reasonable for small to small–medium problems with three or four machines. Historically, investigators largely limited flow shop work to permutation flow shops with makespan objective. Six different makespan heuristics, both classical and modern, given in Chapter 13, are all permutation sequence. (However, dispatch methods do not lend themselves to the permutation sequence assumption; see Chapter 14.)

A somewhat more natural way that permutation flow shops arise is when there is no (or only inadequate) storage between stages of the shop, making it difficult or impossible to change the sequence halfway through. These model extensions are discussed in Section 14.4.

Consider first the no-intermediate-queue problem; that is, the start of a job on machine k is equal to its ending time on $k - 1$. There are actually two variants:

(a) The job can wait, blocking machine $k - 1$ until k is available.

(b) For various shop design or process reasons, the job must not finish on $k - 1$ until k is available.

Both of these situations arise in practice. However, given a deterministic (or perfect forecast) problem, it is easy to see that, with a regular objective, there is no reason to release the job early so that it blocks the system. It might as well be held until it can be released with no interference. Thus point (b) may be assumed to hold.

Proposition 5 of Section 14.4.4 asserts that for no-storage flow shops and two adjacent jobs i and j there is a lag T_{ij} such that j can be started T_{ij} after i (or later, if not arrived yet). It is important that T_{ij}, which depends on checking interference between the two jobs through the stages, depends only on jobs i and j. Thus, once the T_{ij} matrix is calculated, no simulation is needed. The effect of interchange is very easily calculated. In fact, solving such a problem for minimum makespan turns out to be equivalent to solving a traveling salesperson problem.

Turning to the problem with restricted queues and permutation sequences, delay times must typically be estimated by simulation. Again, for deterministic problems we may insist jobs never be added that would block the system. To choose the next job in a makespan situation, one might simulate delays for all possible choices and schedule the job with smallest delay.

19.2.2 Machines with External Setup

Single Machine with External Setup. Many modern machines are so expensive that it is not economical to hold them idle during the time a job is being set up. Setup is basically a lower level activity that does not utilize the entire sophistication of the processing center.

There are a number of schemes for processing one job while the next one is being set up. We illustrate just one of these; the others are similar in concept. There is a turntable, say 8 feet in diameter, in front of the main process. The turntable is split in half by a vertical wall. Each has a duplicate of the holding jigs and duplicates of any parts of the machine needed in the setup. Two jobs may be mounted at once, one on each side. One has been set up previously and can currently be processed. Simultaneously, a worker may set up the other job on the other half. Setup is scheduled to end exactly as the job being processed is finished. The old job is unloaded as part of the processing time. The turntable turns to face the next job for processing, and the following job for setup. All of this is illustrated in Figure 19.1. (Other minor variations on this model are quite possible. In some situations where the setup times are smaller than the load–unload times it might make sense to revolve the turntable *before unloading* job j and before setting up job $j + 1$. This variation is not difficult but will not be considered here.)

A little thought should make it clear that the one-machine problem with external setup may be considered to be a two-machine flow shop with no delays, where new jobs are delayed at the start so that they may be processed through both machines without waiting. It is fairly easy to derive some insights for this two-machine problem, which are not immediately obvious from the general case.

The following simple results are closely related to those in Proposition 5 of Section 14.4.4. Let r_j be the arrival time, s_j be the start time of the setup, S_j be the setup time, p_j be the main process time, and C_j be the finish time for job j. Let 0 and p_0 be initial availabilities for the setup and the machine, respectively, where p_0 is the main processing time for the previously assigned job.

Proposition 1. Consider the one-machine external setup problem with dynamic availability, dynamic arrivals, and any regular objective.

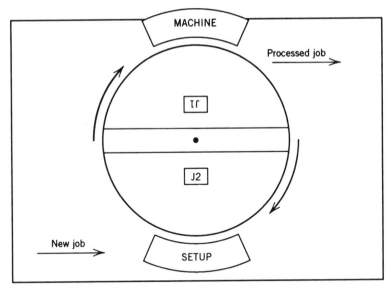

Figure 19.1. One machine, external setup.

(a) If a job starts at s_i, it will finish at $s_i + S_i + p_i$.

(b) The correct delay time between i and $i + 1$ to avoid conflict is $T_{i,i+1} = S_i + (p_i - S_{i+1})^+$, provided $i + 1$ has arrived.

(c) The start time for job 1 is $s_1 = \max((p_0 - S_1)^+, r_1)$.

(d) In general, $s_{i+1} = \max [s_i + T_{i,i+1}, r_{i+1}]$.

(e) The makespan of a given static schedule is (without loss of generality start at $t = 0$)

$$\text{makespan} = \sum_{j=0,n-1} \max(S_{j+1}, p_j) + p_n$$

(f) Provided $\max_{j=1,n}(S_j) \le \min_{j=0,n}(p_j)$ then makespan $= \Sigma_{j=0,n} p_n$ regardless of the sequence for a static problem.

Proof. Most of these properties follow directly from Proposition 5 of Section 14.4.4. The simplified makespan results follow from the fact that the setup for $j + 1$ cannot start until the processing for j starts. Thus $\max(S_{j+1}, p_j)$ breaks the interval nicely into units of analysis. ∎

Note the following facts.

(a) Setup times are usually considerably less than the process time. When true, Proposition 1 could be simplified.

(b) If (f) holds for the static case, then one may simply sequence the p_j to optimize some other objective, such as weighted tardiness, by using the usual one-machine heuristics.

(c) If (f) holds for the dynamic case, there may be some small loss due to inconvenient arrival times. However, the one-machine dynamic arrival heuristics should still be reasonably accurate.

(d) If (f) "largely" holds (with exceptions) a good procedure would be the following:

(i) First use the one-machine heuristic as before.

(ii) If no idle time on the main machine is introduced, the solution is correct.

(iii) If a small amount of idle time is introduced, use the methods of Section 7.6 to trade-off reducing idle time versus changing the objective.

(e) If makespan is the objective and there are many exceptions to (f), schedule using Johnson's algorithm, followed by pairwise interchange.

Parallel Machines with External Setup. While the use of an external setup for an expensive machine is very efficient for that machine, it is quite wasteful of the external setup facilities. If the external setup is relatively simple, such as a worker with a few hand tools, this may not matter very much. However, the setup machinery may itself be expensive and precision in nature. In such cases it makes sense to inquire if the external setup may be shared among several primary machines.

The answer is often yes, but more care must be taken than for the one-machine case. If the setup is typically 20–25% of the processing time, then probably only three machines can be accommodated. First, the multi-machine process is less accurately coordinated. Second, the variability of setup and processing times and time to availability of the next machine leads to a certain amount of loss, which should be borne by the setup rather than by the main machines. Finally, the two current jobs may no longer be left permanently mounted as in Figure 19.1. As shown in Figure 19.2, a new job may be left mounted on the revolving turret while it is being set up. However, when it is delivered to machine A, for example, it must be transferred and mounted in a position where machine A can work on it, so that the turret can move without it to service other machines. Also, the previous job on A must be returned to the turret and delivered to a dismount position before the new job is installed.

Thus we are dealing with parallel machines with a common external setup, which is a much harder problem than the one-machine problem. It is easy to demonstrate, via a counterexample, that even with the analogue of Proposition 1(f), that is, max $(S_j) \leq \min(p_j/m)$, it still does not follow that idle time on the machines can be avoided. This will be left to the exercises. However, if $m - 1$ storage slots can be left available on the turret (i.e., remove the restriction of limited storage) then it can indeed be proved that idle time *can* be avoided. Again, we leave this to the exercises.

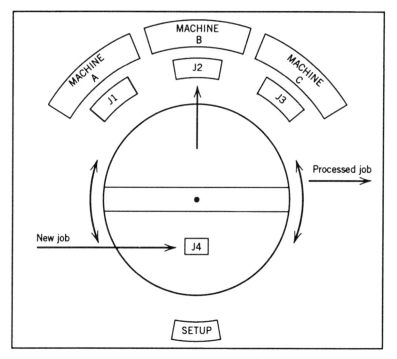

Figure 19.2. Parallel machines with external setup.

19.2.3   Complex Machines with Serial Multi-processing

Consider a continuous bread oven in an automated bakery. Dough that starts at one end appears as finished bread a time D later, the same for all loaves. The line can enter L loaves per hour. An order of Q loaves uses up Q/L hours of the overall line time and $(Q/L) + D$, including the flowthrough time. Thus the machine sees the order as requiring processing time Q/L and the order sees it as requiring $(Q/L) + D$.

In terms of our analytical work, this is best seen as equivalent to a two-machine flow shop with no delays. The first machine is a normal resource with a price; both the machine and the order see the time as Q/L. The second machine is an infinite resource with 0 price and process time of D. Equivalently, the second machine may be treated as a simple delay.

Note a similarity with other aggregate-machine problems, such as proportional parallel machines. There we aggregated the problem into a faster machine, followed by a delay, to make times come out correctly. This same situation occurs here for the one-machine serial line.

This analysis can also be pressed into service for apparently more complicated problems. The oven temperature might be programmed to change depending on current orders going through. In a chemical process, the line could split. While processing, time and/or quality could be, to some extent, random variables. The main point is that scheduling decisions do not occur except initially.

It is possible to have several such lines in parallel or in series, without changing the basic analysis of a compound machine including a delay. Situations in which new decisions can be made in the middle of the process require multi-machine formulations, to be considered next.

19.2.4   Multi-machines: Line or FMS Cell

Line of Parallel Machines. As preparation for the case to be presented in Section 19.4, we present a simplified version of part of it here. Consider a compound flow shop with a set of equal machines at each station. There are different numbers of machines at each station; the number can also vary by how many operators are at the station. The line is rather automated. All internal queues are finite; several are very short. When the scheduler enters a job into the line, this ordering remains fixed. The scheduler must also determine a delay between entering the job and entering the next. It is highly desirable for the delay to be assured of avoiding jamming the line, without wasting valuable processing time. (There are a number of stochastic issues that we do not discuss until Section 19.4.)

This is a rather classic case of a one-resource multi-machine flow shop with finite queues. One essential for this problem is a simulator that can estimate the present and future positions of jobs currently in the line and can estimate the amount of conflict (jamming of the line where queue space is limited) if any new job is entered at any point in time. Since trying all possible new jobs and all possible delays is prohibitive, some simplifications seem to be needed. One possible simplification is the following heuristic.

Possible Heuristic Delay Estimator

Step 1: Pick a job to try.

Step 2: Estimate a crude delay (such as average or maximal process time of internal jobs).

Step 3: Simulate with this job, and find the internal job most likely to be in conflict, and the number of jobs that could be added before causing a jam (plus or minus). (Simulate with aggregate stations.)

Step 4: Adjust the crude delay by the amount of conflict in step 3.

We turn next to looking at selected sequencing methods for the problem.

Neighborhood Search Sequencing

Step 1: Choose an initial sequence.

Step 2: Simulate and cost out the entire sequence, using the delay estimator to insert the proper delays.

Step 3: Choose two jobs for adjacent pairwise interchange.

Step 4: Simulate and cost out the revised sequence.

Step 5: Do this for all possible interchanges and implement the one with the best improvement.

Step 6: Repeat this process until no improvement is possible.

Note that adjacent pairwise interchange is possible but extremely expensive. We turn to developing an implementation of bottleneck dynamics for the model.

Bottleneck Dynamics Pricing and Sequencing for this Model

Step 1: For each stage add up all the processing times expected over all the jobs in the external waiting line.

Step 2: Divide this total by the number of active processors in the stage to obtain the total current load on that stage.

Step 3: Estimate the current utilization rate for each stage.

Step 4: Estimate the sum of the delay costs over all waiting jobs for each stage.

Step 5: Estimate the price of the aggregate machine at each stage by the static or the dynamic queuing formula.

Step 6: Estimate lead times initially from the recent history of similar jobs; in general, estimate lead times by lead time iteration.

Step 7: Estimate priorities for each job by the usual bottleneck dynamics priority functions. (Note that processing times must be divided by the number of active machines at each stage.)

Step 8: Simulate the entire scheduling process and record lead times and objective function value.

Step 9: Repeat to obtain lead time estimation improvement, until no further improvement in the objective function value is obtained for several iterations.

Step 10: Terminate.

Other Methods. More complex approaches such as branch-and-bound, beam search, and the like do not seem computationally attractive. Some kind of OPT-like simplification of bottleneck dynamics might be possible. Although bottlenecks in this problem often shift, choose, say, the three most likely bottleneck stages. For a job, add its processing time on these three stages and run a dispatch OPT-like approach. Let this composite processing time act as a surrogate for resource usage in the bottleneck dynamics priority formula, and let some recent lead time experience for similar jobs substitute for lead time iteration.

Flexible Manufacturing Cells. There are a number of variations of flexible manufacturing cells. We describe one in some detail as representative of the general approach. Look at Figure 19.3. The cell bears some resemblance to the parallel machines with external setup example previously presented in Figure 19.2. A flexible manufacturing cell is typically quite a bit larger than an external setup group. Thus the revolving turret is replaced by an elliptical or circular transporter with a

WORK CENTER BOUNDARY

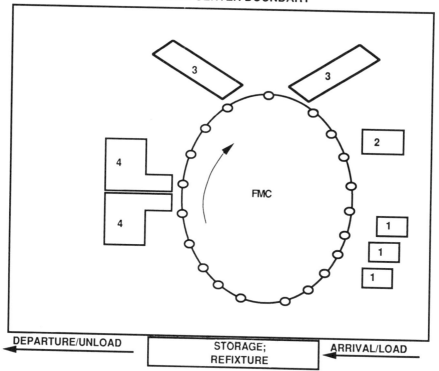

Figure 19.3. Typical flexible manufacturing cell.

number of buckets or harnesses for jobs. There are a number of machines around the perimeter of the belt. Each machine has a single position that can be used to work on a job or store that piece until the belt can take it away. There is also a load/unload position that is manned by a worker or two. It also provides supplementary storage for jobs beyond that given by the belt itself. Finally, the workers can use this area for refixturing jobs. That is, if the job is initially run on one machine, it might come back and be reconfigured with jigs and repositioning suitable for the next machine to work on it.

Perhaps it would be worthwhile to follow a hypothetical job through the flexible machine center. A job 110792 arrives at the load area and waits in storage until its turn on one of the machines (3). It is loaded on the belt to arrive just as the second (3) machine becomes free. Four operations are then performed on it by (3). Then it is returned to the belt to wait for a machine (2). However, all the machines (2) currently have high-priority work, so the job circles on the belt awhile. The belt gets too full, and the job is returned to the storage area. Eventually, it becomes top priority for a machine (2) and is placed back on the belt in time to arrive just after (2) becomes free. After completing the next operation, it is placed back on the belt and returned to storage for refixturing before going to one of the (1) machines. As it happens, the belt is now lightly loaded; one of the (1) machines is free, and the job is sent there directly. It is then sent back to the load/unload area for departure to another part of the shop.

Note that this problem is relatively easy to handle by dispatch/bottleneck dynamics methods as long as storage on the belt and in the load/unload area is in good supply. For dispatch, the problem has some features of skip shops and reentrant shops (see Section 14.4) and may be handled in much the same way.

Some fixed-time delays on the belt from one machine to the next must be added in for cases in which there is no waiting for the next machine, but these are easy to compute. If the next machine is not currently free, however, the exact path length may be unknown. On average, one might estimate half one transit of the belt once the machine becomes free.

In case storage is more limited, the problem can get much more complicated. One would have to specify fixed rules as to when jobs get returned from the belt to main storage and what to do if the job inventory in main storage becomes too large. Simulation of the system to forecast future positions would probably be required.

Bottleneck dynamics price/lead time estimation would require techniques similar to the finite-queue complex line discussed earlier.

Useful references for this section include Buzacott and Shantikumar (1980), Klahorst (1981), Dupont-Gatelmand, (1982), Iwata et al. (1982), Arbel and Seidmann (1984), Suri and Dille (1984), Suri and Hildebrant (1984), Kimemia and Gershwin (1985), Buzacott and Yao (1986), Arbib and Perugin (1989), Stecke (1989), Agnetis et al. (1990), Arbib et al. (1991), Das and Khumawala (1991), Park and Steudel (1991), Rabinowitz et al. (1991), Schweitzer et al. (1991), Blazewicz et al. (1992), Sengupta and Davis (1992), Sethi et al. (1992), Wang and Wilhelm (1992), and Younis and Mahmoud (1992).

19.2A. Numerical Exercises

19.2A.1   Consider the following one-machine with external setup problem:

Job	0	1	2	3	4	5	6	7	8
Setup	—	1	3	2	2	1	3	3	2
Process	5	7	9	4	4	6	5	9	9

   (a) For the given sequence calculate start times, delay times for each job, and the makespan.

   (b) Change the sequence (except job 0) and repeat (a).

19.2A.2   In 19.2A.1 change setup 3 to 8, and setup 6 to 10.

   (a) Re-solve 19.2A.1.

   (b) Minimize makespan for the changed problem as best you can.

19.2A.3 In 19.2A.2, suppose the objective function is weighted flow, with all weights equal 1.0.

(a) Find the best solution you can by starting with WSPT.

(b) Suppose idle time is worth 2 per unit on the main machine. Redo part (a).

19.2B Software/Computer Exercises

19.2B.1 Write a computer program to solve the one-machine with external setup problem by neighborhood and/or tabu search for any mixture of two regular objectives. Create some test problems and solve them.

19.2B.2 Write a computer program to solve the m equal parallel machines external setup problem by first solving the comparable one-machine problem and then searching around that problem by pairwise interchange. Create some test problems with "small" setups and solve them.

19.2B.3 Write a simulation of a flexible manufacturing cell, allowing user input to control decision points, including how to deal with an overloaded belt.

19.2C Thinkers

19.2C.1 Prove Proposition 1.

19.2C.2 Prove via counterexample that the analogue of Proposition 1(f) does not hold for parallel machines.

19.2C.3 Prove that, if $m - 1$ inventory slots are added to the turret, then Proposition 1(f) does indeed hold.

19.2C.4 For the complex line with finite queues and permutation schedules, list and discuss possible less computationally demanding ways of estimating the necessary delay for a starting job.

19.3 MULTIPLE RESOURCES PER ACTIVITY

19.3.1 Overview

Multiple resources per activity can be handled quite easily if the resources are fairly divisible and there is no serious problem matching the resources in order to be available at the same time. This was true in estimating the value of remaining resources needed in calculating bottleneck dynamics. It was also true in assigning fairly divisible items (work crews, cement trucks) on large aggregate projects.

However, the bottleneck dynamics method is less useful, without modification, in assigning classrooms, instructors, and students together, all of which must match schedules exactly. It is also less useful in matching operators with various skills to machines.

Review of Bottleneck Dynamics Resource Aggregation. The very basis of bottleneck dynamics is the ability to aggregate several prices to get a composite price or several different resource usages to charge against a decision to start an activity. For example, in Section 10.5.2 we developed the idea that activity urgencies over the busy period could be added up to give the resource price:

$$R_k = \sum_{i=1,N} w_j U_j(t)$$

Similarly, in Section 10.5.3 we developed the idea that an activity's priority should be its own price divided by the sum of all resource usage by the job downstream:

$$\pi_{jk(i)} = \frac{w_j U_{jk(i)}}{\displaystyle\sum_{m=i,M} R_{k(m)} P_{jk(m)}}$$

In studying large aggregated multi-project and multi-resource problems (Section 18.3) we developed the simple rule for the total resources an activity needs:

$$\left[\sum_{k=1,m} (r_{ijk}/\mathrm{RA}_k) R_k \right] p_{ij}$$

That is, if an activity ij needs a number of resources, multiply the fraction needed of each resource by the price and add, getting the total price of resources used. Then multiply this by the time used, p_{ij}.

If it is desired to know the total cost of all activities downstream, then we add such an expression up for every such downstream activity (call this U_i):

$$\sum_{J \in U_i} \sum_{k=1,m} (r_{iJk}/\mathrm{RA}_k) R_k p_{iJ}$$

19.3.2   Multiple Resource Activities: Dispatch Versus Interval Methods

First we discuss rather general conditions under which each of the following may be appropriate when an activity uses one/multiple resources: (a) pure dispatch, (b) dispatch with heuristic pricing, (c) pure interval, and (d) mixed methods.

Then we illustrate this rather abstract discussion with two fairly detailed examples: (a) scheduling meetings between executives/professionals/and so on and (b) manufacturing products that simultaneously require several raw materials.

General Discussion of Methods for Multiple Resource Activities.
Here by dispatch we simply mean scheduling the highest priority activity as soon as resources are available. (In the general dispatch case, one could also include the possibility of waiting a short interval of time for the availability of a "hot job.")

Clearly, if the job's activities each require a single resource, dispatch will tend to minimize resource usage and hence is a reasonable discipline. If there is one resource for the current activity, but several unitary resources that must be matched for a downstream successor activity, then things are a little more complicated. We could probably fudge this into a dispatch situation by using the downstream interval schedule start time as a heavy due date, and change the local objective to tardiness for the current operation.

If the situation is scheduling a large-scale project with rather divisible resources, then dispatch will work fairly well in many cases if we include the possibility of included idleness.

When we are scheduling several resources simultaneously, the time will often be determined by an interval "reservation." If no resource was inconvenienced by this reservation, heuristic resource usage may be determined simply as the sum of individual resource prices times the reservation interval length. However, if some resource has been inconvenienced by leaving gaps in the rest of its schedule or more complicated interchanges of low- and high-priority items, then simply "price multiplied by time used" will give too low an estimate of the cost of making that resource available. If the changes in the rest of the schedule are fairly simple, they may be costed out. For example, extra idle time can be costed at the extra time multiplied by resource price. Finally, if the reservation forced the resource to simply delete other activities, these may be costed by the lost revenue (contribution margin). The more complicated the resources being called upon for the activity under question, and the more they each have complicated interval scheduling with other activities that must all be scheduled, the less likely marginal pricing analyses will be adequate, and the more likely that a full costing of different possible schedules may be required.

In effect, the more the situation is one of complex, high-utilization interval scheduling requiring extensive fitting, the less likely are heuristic pricing methods to be useful without modification. An important practical question to be investigated is under what conditions a mixed approach can be employed: complex interval scheduling, but with price estimates to decide which potential changes in the overall schedule are worthwhile and which are not. Little research or empirical investigation has been done to date on this question.

Scheduling Meetings.
Some well known examples may clarify these ideas somewhat. Consider a 10-person software development group. Each person in the group has a share of several projects to do. His/her work is quite preemptible, and

the deadlines are a bit soft. Few spend much time in meetings except those called by the boss. The boss, on the other hand, spends a great deal of time in meetings, with a complicated calendar, which he/she or an executive secretary updates personally.

Both the boss and the upper level employees have pretty good ideas of the opportunity cost of their time, which is not determined so much by a simple busy period analysis as by considering activities refused due to "lack of time." Lower level employees are assigned definite work to do; busy period analysis is more appropriate for them if all assigned work must eventually be done.

When the boss calls a 2-hour review meeting, he/she makes it fit his/her own calendar and worries little about the employees' fit, since they can typically handle it with no cost but the direct lost time. The boss does not usually allow employees to skip meetings. His/her decision problem is a standard bottleneck dynamics one: Does the good from the 2-hour meeting outweigh the sum of employee costs including his/her own?

We can easily make this scenario more challenging by changing the employees to busy executives with complex calendars of their own and assuming that one of the executives calls the meeting to carry out some generic directive of the boss-executive.

The first change is that it is almost impossible to schedule a meeting. (A famous rule of thumb is that no more than six executives will *ever* all be available at the same time.) If there is heavy pressure to come to the meeting, the total cost of distorting individual schedules is likely to be too complex to estimate even approximately. Another rule of thumb is that the person calling the meeting will value it much more highly than those invited. At the same time, their schedules are more likely to be grossly violated. Those who appear may come a half-hour late, with explanations about other important meetings that preempted the time. A meeting will often be left early with similar excuses.

Clearly, analytic (even heuristic) scheduling of a large number of highly committed individuals with their own objectives is a challenging and largely unsolved problem.

Even the case of two executives dealing with each other by phone produces the notorious problem of "telephone tag." In full blown telephone tag, the executives take turns in trying to do interval scheduling to obtain 5–10 minutes over the phone. If they are unsuccessful but of goodwill, they soon resort to recorded communication: voice mail, E-mail, memos, faxes. Such devices attempt to reduce the hopeless interval scheduling communication by changing one of the two resources to an activity that can be done when the other becomes available; that is, these devices attempt to reduce interval scheduling to dispatch scheduling.

Products Needing Several Raw Materials. MRP practitioners make the point that one should not simply maintain typical stocks plus safety stock allowance for every raw material with no common planning. If an activity requires 10 raw materials, each with a 90% chance of being available, then the probability that the activity can be performed (at random) is about $(0.90)^{10}$, which is only about 40%. If we plan and coordinate we improve our odds.

But there are times when we cannot replenish a given material (or several) in time. Then we must decide how to allocate between different needy activities. If there is a single raw material, we can put a Lagrange multiplier on the current value of that raw material and do the allocation that way. However, if there are a number of materials in short supply, we will probably need to run a full (LP?) planning model to do the allocation. Simple pricing becomes somewhat less practical.

19.3.3   Machines with Operators

Machine with Unique Operator. In the special case where, on a given shift, there is a unique operator for a given machine and that operator is only assigned to that machine, the situation is relatively easy. Since a decision to use the machine is a decision to use the operator, and vice versa, there is no mathematical reason to distinguish the operator and machine. Aggregate them to a single resource, with usual pricing methods and heuristic pricing decision-making.

Now consider a slightly more complex situation where the operator is in charge of, say, three machines. An activity arrives at one of the three machines and must be done by that machine. There is no overlap in processing; the operator cannot start one machine and then proceed to the next. Under these situations the operator can be treated as the single resource, since all three machines are available whenever the operator is. The only contribution of the three machines in the analysis is to determine what the processing time of each activity is on its appropriate machine. The operator will mentally combine the three queues and use a benefit/cost analysis to assign heuristic priorities to each. The operator then goes from machine to machine, running the highest priority available job next. Note that the problem structure made the operator the limiting resource, which meant that prices for the machines were irrelevant. Note that if there is significant travel time from machine to machine, this may simply be treated as sequence-dependent setup times between the corresponding activities. In a more complicated case, the operator keeps several machines going at once and depends on the problem structure. We will not attempt to discuss it here.

Machine Groups with Operator Pools. Sometimes there may be a large group of machines to be staffed on a given shift. Some operators are highly skilled and can be assigned to almost any machine. Some operators are medium skilled and can be assigned to some of the machines. Some operators are inexperienced and can be assigned to only a few machines. Some operators may also be preferred to others on a given machine; we leave this complication for the exercises.

To analyze this assignment, we consider the computer card line sector to be analyzed in Section 19.4. Suppose we have a compound flow line with K stages and N_k identical machines at each stage k. Looking at future jobs over perhaps the next day, the load forecast for stage k (if only one machine is active at the stage) is L_k. If n_k of the N_k machines at stage k are assigned operators, the highest effective load will be the bottleneck. By assigning some stages with smaller effective load through giving them more operators, we would like to even out the load, that is, to *minimize the highest effective load.*

Minimizing the maximum L_k/n_k is a nonlinear objective since L_k is a constant and n_k is the decision variable. However, this is easily remedied. Define the "comfort" of a stage as n_k/L_k, that is, one over the effective load. Now this variable is linear in the n_k. It is also clear that the objective "maximize minimum comfort" is equivalent to the objective "minimize maximum load."

Let i be an operator and I the total number of operators. Let v_{ik} be 1 if i is competent to work at stage k, 0 otherwise. Similarly, let x_{ik} be 1 if i is assigned to stage k, 0 otherwise. Similarly, let there be a dummy stage 0 to allow some operators not to be assigned.

Proposition 2. The above described operator assignment problem can be solved using the maximin version of the generalized assignment algorithm given as follows:

$$\max \left[\min_{k=1,K} \{n_k/L_k\} \right]$$

$$\text{s.t.} \quad \sum_{i=1,I} v_{ik} x_{ik} = n_k, \quad k = 0, \ldots, K$$

$$\sum_{k=0,K} x_{ik} = 1, \quad i = 1, \ldots, I$$

$$n_k \leq N_k, \quad k = 1, \ldots, K$$

$$x_{ik}, v_{ik} \in \{0,1\}$$

Proof. Left as an exercise. ∎

We also leave as an exercise that:

(a) This can be represented as a linear program.
(b) This can be represented as an ordinary assignment problem.

19.3.4 Machines with Tool Magazines

Permanent Tool Assignment. If there are a number of interchangeable tools kept permanently in a tool magazine, the machine and tools need not be considered as separate resources. For a fixed sequence of tools in, say, a rotary magazine, there will be incurred an activity sequence-dependent setup time, which can be handled by the methods of Chapter 8.

A somewhat more interesting problem arises if arriving jobs tend to cause tools to be used preferentially in certain patterns. For example, each arriving job might consist of three to five serial activities, all to be done on the same machine, all with different tools. Then certain sequences of tool use might be more popular than others. In this case there are two sequencing activities to be determined simultaneously: tool sequencing within the circular tool holder (rotation time becomes the tool change time), and operational sequencing of the jobs, both of which will affect the total tool setup time as well as the lateness of the jobs. (Note that the tool assignment bears some resemblance to the operator assignment problem.)

If one plans over a 12-hour horizon, for example, one could first fix the tool assignment and then solve for the activity sequence. Then holding the latter constant, one could solve for a revised tool assignment, and so on.

Shared Tooling. In a more complex situation, each of three machines might have its own tool magazines, perhaps accommodating 30 tools each. In addition, there might be a centralized automatic tool handling magazine holding perhaps 100 tools. A machine is capable of drawing a new tool from the central magazine, perhaps necessitating returning a tool for which it no longer has space.

In this situation the machines and the tools all represent separate resources. The machines will have queues of activities; the tools will have queues of needy machines. Sophisticated rules are difficult; simple rules are often employed. One common set of rules is:

Rule 1: Don't transfer in a tool until it is just needed.

Rule 2: Transfer out the tool with greatest time until next use.

This set of rules works rather well if tool supply at the central magazine is high; local needs can be well forecast. It also involves sequence-dependent setups and iteration between the activity sequence and the tool sequence and transfers.

Useful references for this section include Eaves (1986), De Souza and Bell (1991), De Warra and Widmer (1991), Chan (1992), Ghosh et al. (1992), Lin and Solberg (1992), and Veermani et al. (1992).

19.3A Numerical Exercises

19.3A.1 Consider a compound flow shop with four stages, each with three equal parallel machines. There is a pool of seven operators, two of whom can handle any machine, one who can handle stages 1 and 3, and the other four are specialized to stage 1, stage 2, stage 3, and stage 4, respectively. The total loads at the four stages are 100, 150, 300, and 150, respectively.

    (a) Find an operator assignment that is feasible and evens the load well, using your own invented heuristic methodology.

    (b) Explain your heuristic principles.

19.3A.2 Program 19.3A.1 on the computer as an LP or assignment problem. (Note that it will be useful to do the appropriate thinker exercise below first.)

(a) Compare the optimal answer with your heuristic in 19.3A.1.

(b) Can you improve your heuristic in light of this exact solution?

19.3A.3 Try to make up some more interesting operator data and redo 19.3A.1 and 19.3A.2.

19.3A.4 Give the operators ratings other than 0 or 1, and try to equalize the load divided by total rating at each machine. Repeat 19.3A.1 and 19.3A.2.

19.3B Software/Computer Exercises

19.3B.1 Write a bottleneck dynamics simulation involving an operator tending two machines with separate queues and travel time.

19.3B.2 Write a simulation for a machine with a limited tool magazine, but able to get any tool from a central tool crib with a fixed access time and to remove any current tool at no delay. There is a known queue of waiting jobs and weighted flow objective. Write an iterative procedure that improves tooling and activity sequence alternatively.

19.3C Thinkers

19.3C.1 Prove Proposition 2.

19.3C.2 Prove that Proposition 2 can also be represented as a linear program or by an ordinary assignment problem. In both cases it is sufficient to exhibit the actual corresponding formulations.

19.3C.3 Consider two machines with tool magazines and a common centralized tool pool and automated service. Try to develop commonsense rules for allocating tool requests to the central server, and justify your heuristics.

19.3C.4 Discuss common situations other than scheduling meetings that illustrate variations in which dispatch scheduling (with or without joint interval scheduling) is (a) adequate and (b) inadequate.

19.4 CASE EXAMPLE: COMPUTER CARD LINE SECTOR

19.4.1 Overview

U.S. Computers is a gigantic computer company with manufacturing facilities all over the world. It makes its greatest profits from mainframe computers, although it is also heavily into PCs and software. Its plant in Seattle makes mainframe computer boards for assembly into large computers at other plants in the United States and Mexico.

Seattle is especially proud of its mainframe computer board line, which is both automated and versatile. The line is very long. It is actually a series of 18 sectors (some sectors, such as repair sectors, are reentrant). Each sector is in itself a compound flow shop with an initial queue and associated sequencing and release decisions. Each also has finite internal queues, provision to remove jobs that "jam" the system, provision to shift operators to relieve bottlenecks, internal testing and loopback for internal repairs, and sophisticated forecasting to estimate processing time for each type of board at each stage.

Since no internal resequencing is allowed (with the exception of some of the internal repairs, for which no decisions are required), a sector may be considered a single resource (although complicated) by our definition. Thus, at the aggregate level, each sector is a machine, and the appropriate model is an 18-machine reentrant flow shop with large queues at each sector (machine) and resequencing allowed. This aggregate problem can be solved by bottleneck dynamics, given some method of estimating product lead time through a sector.

Sector lead time can, in turn, be estimated by solving each sector as a subproblem, and the whole hierarchical planning and scheduling system can be solved iteratively. The issue of solving the whole system of sectors will be discussed in the continuation of this case given in Section 21.4 on the integration of planning and scheduling. Here we concentrate on a more manageable building block: sequencing, job release, and operator balancing for a single sector, with assumed known arrival dates to the sector, and derived due dates from it.

A computer board is a modular piece of a computer, 18 inches by 30 inches by 1–2 inches thick. Typically, the value added by completing the entire line of 18 sectors might be $20,000. The composition is highly variable, but it includes layers of circuit board wiring, clusters of software and hardware chips, and so on. There might be 10–20 such boards in an average mainframe. If something on a board does not work, the whole board is simply replaced, to improve uptime. It is often called a "card," which is somewhat a misnomer. A typical electronic card is much smaller in size, weight, and value and occurs in all sorts of electronic equipment.

Boards coming into a sector typically have widely different characteristics. A particular mainframe board is often manufactured one or two at a time. Furthermore, there are always many engineering boards testing various design modifications, and also boards going through again to fix remaining subtle flaws found later in the process or by an unhappy customer. The flexibility of the machinery allows such diversity to be handled; however, it is important to have a good forecast of the processing time of the board at each stage of the sector (including expected rework) to facilitate bottleneck management via sequencing, release, and operation management. Engineering personnel have worked out a fairly careful regression scheme to forecast these times from the features of the board. Time allowance for correction loopback and some allowance for probabilistic factors are also built in. Although in the deterministic case careful estimation of release delays would result in never jamming the system, in the real world uncertainties make this impossible. The size of the release delay must balance the probability of jamming, and the probability of not maintaining a peak production rate.

One rather sticky issue with this problem is how to treat the random process resulting if a board is found defective in the middle of the sector. Depending on the defect, there are two main outcomes:

Outcome 1: The item loops back and repeats a stage or two, with the same process times.

Outcome 2: The item is removed from the sector and sent to a repair sector, after which it will restart this sector.

Again, each of these two stochastic outcomes can be represented in different ways depending on the purpose.

Need 1: To estimate the distribution of time until the customer will get delivery of this board. (We assume the board is eventually fixed.)

Need 2: To estimate the expected load on any stage at any point in the future, considering both normal processing time and rework times.

Need 3: To estimate at what earliest time a new board can be released to the sector with low probability of causing future line jam at any point.

19.4.2 Case Analysis

In analyzing the case, we first discuss possible solutions for the stochastic failure problems just raised. Next, we analyze the case in which adequate operators are available (so that the problem is a complex one-resource problem). Finally, we add to the preceding analysis the real-world situation in which operators must be allocated dynamically across machines.

Dealing with Stochastic Failure. Of the three needs presented above, estimating loads on the stages at points of time in the future is relatively easy for either internal or external correction. Basically, make a loopforward branching tree, using lags for typical processing delays. For every point, for every time, for every stage in the system, there will be a probability the activity is being processed there (which may be zero). Now simply add up all these piecemeal probability loads to get the total expected load on the system at that point of time.

Another need is to calculate the distribution of lead times for each board until completion. Use a similar sort of loopforward branching tree and lags for lead times through sectors, and so on. Rather than tallying loads on machines, tally the times the board finishes and calculate a mean and variance of the total lead time.

The final need to estimate job release without causing delays is difficult, especially if computation times are to be kept low: deterministic simulation of the shop is already very expensive. One approach here would be simply to increase the effective processing time of all jobs to cover expected repair time, and treat the simulation as deterministic. Individual jobs would not be presented well in the simulation, but averaging over all jobs in the simulation may produce adequate results. More complex approaches may not be justified computationally.

One Resource Case (No Operators)

(a) Simulations are run as deterministic, with job times increased appropriately to handle expected rework.

(b) Simulate the sector for each job that could be handled next, choosing for each potential job the smallest simulated delay that does not cause a future simulated jam.

(c) Estimate relative loads on the stages by adding up spread out loads estimated from each future job.

(d) Estimate stage prices by some simple function of current estimated load, such as simple proportionality or the dynamic queuing approximation.

(e) Calculate each job's bottleneck dynamics priority, using estimated lead time through the rest of the system, myopic priority, and total resource usage in the sector.

(f) Choose the job with the highest priority, and start at the appropriate delay time.

(g) As probabilistic events occur, update all estimates affected.

(h) Do not iterate lead times at the sector level, but at the aggregate planning level. (See Section 21.4.)

(i) This procedure is for (weekly) regular scheduling; corrections for reactive rescheduling are discussed in Section 21.2.

We turn next to correcting this analysis when operators can be redistributed to alleviate suggested bottlenecks.

Operator Redistribution. So far, we have done an analysis for a given a priori load on each machine, that is, for a given assignment of operators to each stage. The resulting loading of the stages can be improved by redistributing operators to the stages via the generalized assignment method discussed in Section 19.2, with objective to minimize the maximum effective load at any stage, and so on.

Iterative Procedure for Operator Redistribution

Step 1: Choose an initial feasible operator assignment to the stages, resulting in a priori loads at each stage.

Step 2: Run the bottleneck dynamics procedure to sequence and time the various jobs through the sector.

Step 3: Using the resulting loadings of the stages, reassign the operators by the generalized assignment method.

Step 4: Adequate convergence? If so, exit.

Step 5: Return to step 2.

Note that there is also an outer loop of iteratively updating the lead time estimates, done at the higher planning level. This is discussed in Section 21.4.

A useful reference for this section is Wittrock (1989).

19.4A Numerical Exercise

19.4A.1 Consider a dynamic deterministic compound flow shop with four stages, each with three equal parallel machines. There is an infinite queue in front of the first stage; no queue is allowed in front of the other stages. The objective function is weighted flow. Two operators have been assigned to the first and second stages, one to the third, and three to the fourth. Ten jobs are to be processed with the following characteristics:

j	w_j	r_j	p_{j1}	p_{j2}	p_{j3}	p_{j4}
1	2	0	6	6	3	8
2	1	0	3	4	2	5
3	1	0	4	5	3	6
4	3	4	7	5	5	5
5	1	8	4	4	1	5
6	2	10	6	5	0	7
7	2	15	6	7	3	9
8	1	19	4	4	0	4
9	5	25	5	5	5	5
10	4	30	4	4	4	6

(a) Using prices as proportional to average loads, estimate the relative prices for each stage and each machine, using the first five jobs to estimate load for time 1 to 100 and the second five jobs to estimate load for times after that.

(b) Estimate constant prices over the whole time horizon by using all jobs to estimate the average load. Comment.

19.4B. Software/Computer Exercises

19.4B.1 For 19.4A.1 write a computer program to find a bottleneck dynamics solution, giving both sequencing and delays, using the approach suggested in Section 19.2.4.

19.4B.2 For the operator availabilities of 19.3A.1 and job input of 19.4A.1, extend your program in 19.4B.1 to alternatively and iteratively redistribute operator load via the assignment method, and sequence and set delays for jobs by bottleneck dynamics.

19.4B.3   Create new data sets for 19.4B.2 to try to ascertain when the method works well and what difficulties it may present.

19.4B.4   Revise your program from 19.4B.2 to deal with random failures in some simple but useful way. Create some data and test it.

19.4C Thinkers

19.4C.1   Suppose some operators are more preferred on a given machine than others also trained for it. Develop and justify a model for this situation.

19.4C.2   Suppose senior operators may be called in to work as needed at premium rates of pay. Develop and justify a model for this situation.

19.4C.3   Suppose the failure/repair situation is approximated by giving each board a processing time probability distribution at each station. Suppose boards can be removed from the line at jam time for a fixed cost and returned to the initial queue, with processing times revised to zero on previously finished stages, and full repeat of other stages. Discuss analytic and/or simulation methods for addressing this model.

20
MODEL EXTENSIONS

20.1 INTRODUCTION

To this point we have always assumed that a given activity is well defined and does not change its purpose over time. In particular, it has exactly one type of job for which it is used; the job is well defined and has a particular customer in mind, with given weights and due dates. These are all part of the standard job shop and/or project job shop assumptions.

In Section 20.2 we consider various types of changes to these assumptions. Open shops build partly to stock, and not order, so that completion urgency may be less well defined. In assembly shops the same part may later be used for more than one possible outstanding job, so that again priorities are less well defined. In continuous shops the activity itself may be hard to isolate. For example, a roll of paper may be cut in a number of different ways; the roll represents several possible sets of activities.

In Section 20.3 we consider push versus pull systems, and related topics such as JIT, Kanban, and Conwip. We also briefly consider methods for determining the release time of a given job into the shop, and for a salesperson and customer to jointly set the due date for a job.

A push system releases items to the floor and subsequent priorities push them along. A pull system waits until an item is sold and then asks for another to be finished, which asks for another to replace that one. Just-in-time (JIT) is a more general philosophy that tries to maintain zero inventory in the system to the extent practical. It can be implemented either as a push or pull system. Kanban is a special kind of pull system used in Japan, in which cards are passed from machine to machine to authorize the pulling process. It maintains a rather constant inventory at each machine. Conwip is a related system that tries to maintain constant inventory in each *subsystem* on the floor rather than at each machine. Jobs are withheld from the floor primarily to avoid the higher inventory costs for WIP rather than raw materials, or similarly to avoid excessive congestion. Setting of due dates by the order taker and the customer is a complex compromise between customer desires for a short lead time, customer desires for the order to be done neither early nor tardy from the due date selected, and the order taker's desire to avoid overloading the shop with short due date jobs.

Finally, in Section 20.4 we consider situations in which bottlenecks can partially be managed by speeding up the bottleneck resource. Speeding up the bottleneck resource invariably leads to other new costs: excess tool wear, worker fatigue, and lower product quality. Bottleneck dynamics is a good framework for considering these trade-offs.

20.2 OTHER TYPES OF SHOPS

20.2.1 Overview

In classic job shops or classic project shops, an activity is associated with a unique customer and final priority. Thus knowledge of the activity and its lead time to this finish gives a unique priority for the activity.

In less classic shops, the final use, lead time, and priority of the job become fuzzy for various reasons. Many times this problem can be fixed simply by keeping a list of possible uses and priorities for the item and updating them as necessary. However, this does carry a practical overhead in the system in terms of computer space and computation time. Thus in some situations cruder approximations may become necessary.

20.2.2 Open Shops

In an open shop, many customers require shorter lead times than the system can accommodate by making "to order" (making it after the order). So a final stock of inventory of each such type of item is maintained, from which such orders may be filled. A related problem is that two or more customers may be competing for the same WIP on the floor, which could be used to make either order. (For example, one customer may want 100 of an item, the second 60 of the same item, with only enough WIP for 120.) We consider both of these problems here briefly.

Final Inventory Stock Problem. For the final inventory stock problem, we modify the (existing orders)/(forecast orders) records as follows:

(a) List the known orders by increasing priority.
(b) List forecasted orders by increasing forecast due date.

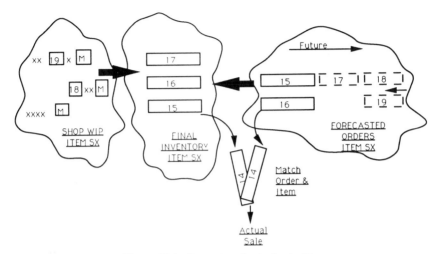

Figure 20.1. Open shop (make-to-stock).

Forecasted due dates are simply listed as spaced by 1.0/(demand rate). Because they are only forecasted, we simply prioritize them in order of due date. A cutoff giving the maximum time in the future for which we wish to produce can also be given.

We allocate the existing final inventory against these phantom demands. Any remaining phantom demands represent orders against the floor. The floor item with the least forecast lead time until finish will be matched against the phantom demand with the earliest due date, and so on. See Figure 20.1.

Whenever there is a change, all assignments' "labeling" is changed dynamically. For example, if forecast demand decreases, and phantom demands are moved further out or canceled, the corresponding labels for items are revised. In particular, if an activity had two possible uses, this might cause its use to be redirected. This is discussed below.

Competing Customer Problem. Particular WIP lots on the floor might not correspond exactly with customer needs. Suppose there is no final inventory right now, and WIP lots on the floor in order of increasing remaining lead time are 50, 100, 50, 50, and 50.

There is a single customer, who wants 175. She is assigned 50 from the first, 100 from the second, and 25 from the third. The remaining 25, 50, 50 will be assigned against the first phantom demands.

Note that the split lot does not need to be split physically, but needs double accounting on the computer record. An easy way to do this would be to consider every unit as a separate order, so that the single customer would be considered 175 orders. While simple conceptually, this blows computer storage and computation time out of the water. We need to design a relatively simple multiple labeling scheme that can be changed at frequent intervals. While this is an important issue, we only sketch it here.

Now suppose an urgent customer suddenly appears who needs 125. We would instantly need to change phantom forecasts if necessary to increase forecasts (or decrease if the customer simply came early). We need to relabel lots so that 50, 75 would be assigned to the urgent customer, 25, 50, 50 to the second customer, and so on.

Things could easily get much more complicated, if there were five or six customers, all with changing priorities for the item. Each floor lot would need to carry a record for each of several jobs currently claiming part of it. Each time there is a change of priorities of customers or a revision in lead times of various floor lots, a revision of the lot priorities and assignment splits would be necessary.

A revision of uses/priorities for a lot very often will cause part of a lot to have one priority and the rest another, while it is desired not to split the lot. In such a case, we determine an aggregate priority by summing the benefits of expediting each part and dividing by the summed costs of expediting each part.

20.2.3 Assembly Shops

An assembly shop is very closely related to the project job shop, which we studied and derived heuristics for in Section 18.2. In one sense it is a special case, since the network structure of each project or job is an assembly tree. In particular, this means that the initial sets of any two activities (see Section 8.4) are either completely independent or one set is a part of the other.

But in an assembly shop, the completion of an activity (or of the activity and its initial set) always creates a subassembly, which represents work in process. In an ordinary project shop these partially completed projects would all be distinct and visually identifiable with the job or project for which they are destined. However, in an assembly shop many activities may be equivalent and have equivalent initial sets. This will lead to an inventory of identical items in the WIP.

Now if a dispatch heuristic is considering whether a new activity (subassembly) can be performed, it checks whether or not all predecessors have been completed. But due to duplications, there may be more than one possible predecessor finished and sitting in WIP on the floor, which would allow this high-priority activity to proceed, but none of which are currently labeled as part of this project. In such a case we clearly want to relabel to allow this activity to proceed.

Another way to think of it is that there may be a subassembly X2A6 sitting on the floor, with two different projects wanting to use it. (It represents an identical initial set for both projects in our terminology.) A similar question arises. How do we know the priority for manufacturing another X2A6 if we are not sure of its final destination and, therefore, its priority?

This is basically a relabeling problem again, but we must be a little careful. We have basically to perform an iterative process of (a) lead time estimation, (b) pricing estimation, (c) labeling estimation, (d) simulation of solution, and (e) iteration.

A sketch of the process is as follows.

I. Initialization

Step 1: Set initial lead times at a given multiple of the relaxed resource network lead times.

Step 2: Set prices equal to the total load on each resource or by another simple heuristic.

Step 3: Create full networks for all jobs (projects) and label in a natural fashion.

Step 4: If there is initial WIP on the floor to be used, delete equivalent activities from the network, making nondetermined choices arbitrarily.

Step 5: Begin a standard dispatch simulation for MLP.

Step 6: When checking a high-priority activity for predecessors, if the natural predecessor is finished, use that.

Step 7: Otherwise, if a copy exists in WIP, do that. Swap labeling so it matches that of the new activity; go to the unfinished predecessor that is being replaced by the item in WIP and give it the other label.

Step 8: Complete iteration.

II. Next Iteration

Step 1: Update lead time estimates in standard fashion.

Step 2: Update prices in any standard fashion.

Step 3: Labeling has been updated from the last iteration of the simulation.

Step 4: Simulate and relabel as before.

Step 5: Termination criterion? Yes, exit.

Step 6: Next iteration.

This procedure provides a superior alternative to floor level MRP, since lead times and urgencies are being updated automatically. It may become impractical if there are large numbers of identical items in WIP, which could involve an excessively large number of jobs and projects. Here the practical solution would be to aggregate and standardize transfer lots as individual items. For example, suppose that two orders produce 80 and 120 of the identical item, and that the transfer lot is 20. We would then treat this as four units and six units (transfer lots) of the same item, which would not strain computational facilities. We return to the question of providing improvements to MRP in Chapter 21.

20.2.4 Continuous Shops

Most of the manufacturing we have described in this book is properly described as *discrete part manufacturing*; a project, job, assembly, or subassembly comes in a standard size or integer multiples. *Continuous manufacturing* has the characteristic that a large batch of soup, molten steel, or paper is manufactured and can be divided up in arbitrarily many ways into jobs or subassemblies.

It is sometimes thought that the process of making the product continuously over time is what makes it continuous. Although this issue has some interest, the primary characteristic is the difficulty of defining a job in advance. Soup and molten steel are often made in stationary batches but are continuous processes. Continuous steel casting and paper roll manufacture are continuously created over time and are similarly continuous processes.

All continuous processes share the fact that they are made in batches. This might not seem obvious for continuous casting or paper, but in fact paper ingredients are mixed in a batch and inserted on the machine with a setup cost and/or time just as for other batch machines we have studied. Similarly, when a casting batch is done, a new batch with different chemical characteristics and setup time will be put on.

This lays to rest another common misconception that a classification is "batch" versus "continuous." Such a classification would have to be "stationary batch" versus "continuous time batch."

We discussed batch machines in Section 11.4, and much of the bottleneck dynamics we developed for that case still apply. For instance, in deciding which jobs to put in the batch, one principle is to assign them in order of highest priority. However, a rather different one is to fit with as low a percentage of waste as possible. These may conflict in more difficult ways than in our earlier discussions.

If we are planning a roll of paper, say, 50,000 feet, with a certain composition, we may, for example, have total jobs for this paper of 42,000 feet of various sizes and shapes. Our simple model of Section 11.4 would simply suggest taking as many top priority jobs as possible. If this leaves us using, say, only 38,000 feet, this may be alright if the roll can be redesigned as 38,500 feet. Otherwise, we may wish to combinatorially try to fit more into the roll. The combinatorial problem itself varies from the simplicity of a one-dimensional bin packing problem, such as a batch of molten steel to be poured into different sized ingots, to a two-dimensional cutting stock problem, such as cutting different widths and lengths from a large roll of paper, to a three-dimensional problem, such as fitting odd-shaped objects into a kiln.

A different issue is that we also generally have a complex relabeling problem. This may occur when a hot job causes us to replan the uses of a given molten steel heat. Or it may occur in the iterative process of scheduling a shop, as changing estimates of lead times and prices cause lot priorities to shift and change the composition of a given batch.

This is a complex problem that, to date, has only been partially worked out. We sketch a few initial ideas, starting with the one-machine case, moving briefly to the parallel machine case, and then briefly to labeling issues for the general case.

One Machine. We first consider the case where all jobs have the same characteristics (or raw materials) and can be processed together—the one-class case. Suppose that n jobs arrive dynamically with processing time p, arrival time r_j, due dates d_j, weights w_j, and slack factors $U_j(t)$ (see Section 11.4). The machine currently has price $R(t)$. Suppose that the entire batch has a relative "volume" 1.0, and that if a subset of the n items is scheduled, it "wastes" a fraction $v(x)$ of the batch, and uses $1.0 - v(x)$ of the batch. The imputed machine cost of this waste is $v(x)R(t)p$. In addition, the wasted material in the batch (a waste ingot or part of a paper roll) is either thrown away or reprocessed at a cost Q, per batch. For each job expedited the value is $w_j U_j$ per time unit expedited, and the total interest cost on using the machine a time unit early is $IR(t)p$.

Thus the net marginal value of running batch x is given by

$$\pi_x = \sum_x w_j U_j(t) - IR(t)p - v(x)[R(t)p + Q]$$

The subset of jobs x creating the highest value of this priority should be scheduled.

Here we have assumed inserted idleness (waiting for another arriving job) is not allowed. (See Section 11.4 for the type of approximations required otherwise.)

In general, the $v(x)$ (fitting) function may be quite complex and require an integer program to exactly optimize, for example, optimal cutting of a roll of paper into individual orders. Heuristics are widely used for that purpose. In particular, if $v(x)$ is additive in volumes v_j of individual jobs, then it is possible to calculate priorities π_j for individual items, and quite a good heuristic is to assign items to the batch in order of decreasing priority per unit volume π_j/v_j.

The multi-class case is quite simple if we don't allow inserted idleness. Calculate the best lot above for each possible class, and start the one with the highest priority.

Parallel Batch Machines. For equal parallel batch machines the extension is quite easy. (Here we mean equal in processing time, the possible layouts for the individual jobs, and the corresponding amount and cost of wastage. Fortunately, this case is common.) For the batch machine that will be available next, conduct the analysis above to determine the highest priority batch, and run it.

Unfortunately, even the usual proportional case may be very difficult. A new paper machine may be three times the speed of the old one, but the number of jobs accommodated may go up by a factor of 1.5, and so on. We may wish to save the longer roll machine for an incoming order that isn't here yet, and so on. This issue is too complex to go into here.

Labeling Issues for the General Case. Actually, the labeling issues are quite similar to those for batch shops, open shops, and assembly shops. First, in continuous systems in which there is no identical WIP on the floor for different jobs, or identical final inventory for different customers, the problem of relabeling WIP or final inventory does not arise. In the process of dealing with hot jobs, or iterative solution of the scheduling problem, the composition of jobs for a given batch may change, which is more a problem of relabeling the batch than the job, so that we have really a pure batch shop situation.

On the other hand, if there are multiple copies of WIP and/or final inventory, the labeling procedures sketched above for open shops and assembly shops are fairly adequate.

Useful references for this section include Box and Herbe (1988), Bolander and Taylor (1990a, 1990b), Lefrancois and Roy (1990), Diaz et al. (1991), and Takahashi and Kateeshock (1992).

20.2A Numerical Exercises

20.2A.1   Set up a mini-open shop: one machine, two products, two final inventories, forecast sales, weighted flow, 10 jobs, and dynamic arrivals. Illustrate relabeling caused by a new hot order, or similar. You may find putting this on a spreadsheet helpful.

20.2A.2 Modify 20.2A.1 into an assembly shop: three machines, two products, two final inventories, forecast sales, weighted flow, 10 jobs, and dynamic arrivals. Machine 1 makes a wooden toy train; machine 2 paints it red; machine 3 paints it blue. (The two products are red trains and blue trains; orders may be for 14 red trains, and so on.)

20.2B Software/Computer Exercises

20.2B.1 Modify exercise 11.4B.1, in which you wrote a routine to schedule a batch machine with dynamic arrivals and a single class, to handle a steel furnace that makes steel and pours ingots. Assume each desired ingot has a certain volume. The furnace can handle only three possible volumes, so there will be wastage. Using the bin packing heuristic of selecting in order of decreasing priority per volume used, write a program to "solve" the problem for arbitrary input.

20.2B.2 Modify 20.2B.1 to handle several different possible steels to choose for a batch.

20.2C Thinkers

20.2C.1 A certain mainframe computer maker manufactures its basic "vanilla" computer to stock in order to shorten lead times to customers from 6 months down to 1 month. Fifty percent of customers typically want vanilla, 20% will want modifications added making chocolate, 20% will want raspberry, and 10% strawberry. These modifications take about 1 month. Sometimes the shop isn't very busy and the manager goes ahead and modifies appropriate numbers of computers to other flavors. If the manager guesses wrong, and an important order comes in, he/she will tear these flavors back to vanilla and remanufacture them, requiring 6–7 weeks and considerable cost. Orders are seasonal, random, and trendy. What can you say quantitatively (or otherwise) about this situation?

20.3 PUSH VERSUS PULL SYSTEMS, JUST-IN-TIME, AND RELEASE/DUE DATES

20.3.1 Overview

In this section we address several larger issues that affect the long-run future of a shop. Two major points of view center around push systems versus pull systems. The push system has been emphasized mostly in this book. It basically assumes a completely customized order that must traverse the entire shop after an order has been placed. By contrast, the pull system assumes a few standardized products, so that the order may be taken out of final inventories with little lead time. The final inventory is also quickly replenished by finishing WIP permanently stationed in the system for the product.

Just-in-time (JIT) is an idealized shop/inventory system in which raw materials, WIP, and final orders are delivered to the next stage exactly when needed. It thus implies absolutely minimal inventories and hence works best in a (standardized product)/(stable demand) situation using a pull system.

Kanban and Conwip are WIP control methods on the shop floor related to pull systems and JIT. Kanban uses a system of cards to maintain roughly constant inventory levels at each machine. It was developed at Toyota in Japan. Conwip is an American generalization for when demands in the shop are somewhat too erratic to maintain constant inventory levels at each machine. Instead, one groups the machines into sections of the shop and maintains constant inventory in each section. Push, pull, JIT, Kanban, and Conwip are discussed in Section 20.3.2.

Setting release dates in a push system is extremely important because it controls the overall inventory level in the shop, which is so important in a pull system as well. Priority rules such as early/tardy, which build high costs of finishing early into the priority function, can be used to release a job when the priority rises above 0.0. We illustrate a couple of these models in Section 20.3.3.

Finally, setting the due date is a rather delicate negotiation with the customer. The customer wants a short flowtime but, even more importantly, wants delivery not to be tardy from the agreed upon date. The scheduler prefers a longer flowtime to minimize the chance of tardiness. We discuss this issue further in Section 20.3.4.

20.3.2 Push, Pull, JIT, Kanban, and Conwip

Push Systems, Pull Systems. In a *push system* (which we have mainly studied in this book) a customer order triggers the following:

(a) Creation of paperwork to create the order from scratch.
(b) Negotiation of a due date with the customer.
(c) Decision when to release the job order to the shop floor.
(d) Dynamic calculation of priorities.
(e) Priorities push the job through the stages.
(f) Complete job goes to customer.

In a *pull system* a customer order triggers the following:

(a) Ship from final inventory.
(b) Instruct the final machine to take existing WIP and finish a new unit for final inventory.
(c) Last machine orders replacement WIP from previous machine, which in turn orders WIP from previous machine, and so on.

Note that the pull system has a number of advantages. (a) The lead time from order to delivery is short and automatic; the item is taken out of final inventory. (b) The lead time to replenish the final inventory item is short and automatic; a final machine WIP item of the right type has its final operation performed and is added to final inventory. (c) The lead time to replenish the final WIP item is equally short, and the procedure repeats automatically through earlier and earlier machines. (d) Inventories are essentially constant and stable across the floor; there is no confusion.

Note that the pull system has a number of disadvantages as well. (a) There must be only a few rather high-volume products, so that keeping permanent WIP for each product in front of each machine is practical. (b) The volumes of the products must be fairly stable over time, both for constant WIP to work and to keep demand surges from swamping the automaticity of the system. (c) The system will not tolerate even brief failures of machines; it will bring associated product production to a halt. (Actually, the Japanese would consider this an advantage in terms of focusing attention on finding and repairing the system defect quickly.)

Advantages of a push system complement those of a pull system. (a) Completely customized one-time orders are handled well. (b) The order can often be modified repeatedly during manufacture. (c) No WIP is kept except for orders currently being processed. (d) Overall volume can be quite lumpy; the priority system, lead time expandability, rerouting, and overtime will take up a great deal of slack. (e) Failure of individual machines is less serious.

Just-in-Time (JIT). The Japanese have important strategic reasons for preferring pull systems and to redesign products and processes to make pull systems more universally desirable. They argue that manufacturers should aim for zero inventories in the long run. This concept is called just-in-time (JIT), since, if there were no final inventories, the product would need to be finished just in time for delivery. If there were no raw material inventories, they would need to be delivered to the factory just in time for first-stage manufacture, and so on.

JIT motives for keeping inventories very very close to zero include: (a) immediate recognition if defective product is being hidden in WIP on the floor; (b) immediate recognition and quick tracing to worker or machine if bad product is being currently produced; (c) repeated small improvements to the production system due to ferreting out these defects; and (d) resulting consistent long-term learning that produces lower costs, higher quality, and higher flexibility.

Note that there is an intimate connection between JIT and pull systems, since pull systems produce constant easily controlled inventory levels. We look at this relationship more carefully next. The classic Japanese system for maintaining a constant inventory pull system is Kanban. A recent American generalization for more complex systems is called Conwip.

Kanban and Conwip. Kanban literally means "card" in Japanese. It is just that— a manual system of cards to control a pull system and maintain a constant WIP at each machine at all times. For ease of understanding, we present a slightly simplified version of the approach that conveys the essential idea.

Remember that JIT tries to keep inventory absolutely as low as possible at all times. Therefore we must keep a machine from producing if incoming inventory to the next machine in sequence already has "enough." Kanban initially issues one card to each stage of a product for each unit of WIP to be allowed. If the WIP falls below that, the extra card is sent to the preceding machine, which is allowed to manufacture the extra item, and return that and the card to the WIP, which is now up to snuff.

Let's see how it works. Suppose there is one card at the WIP in front of each machine, and at the start there is exactly one unit of WIP in front of each machine, and one unit of finished goods inventory. Suppose there are three stages. Now a customer comes and buys the unit of final inventory. It now has an extra card, sent back to stage 3, which takes its own unit of WIP and manufactures it, sending it and the card back to final inventory.

But now there is no WIP in front of stage 3, and an extra card, which gets sent back to stage 2. It manufactures a stage 2 WIP unit, and returns the card. In turn, the stage 1 machine manufactures a stage 1 WIP unit. In turn, the raw material unit in front of stage 1 is gone, and its card is sent to bring a new unit of raw material.

So far we have described a system running on just one unit of inventory at each stage. But suppose a stage breaks down briefly, or the operator is sick, or a WIP unit is defective. Then it would have been nice to have more than one unit of inventory. We might run on three units of inventory, for example. If a machine breaks down the inventory can be used for a short while, but the machine must get back to running quickly and, in addition, manufacture the extra units whose cards have been accumulating.

Kanban in its simplest form only works if only one product is manufactured, or at most a few very similar ones. If there are a great many different kinds of products using different machines, a pure Kanban/pull system will not be practical. However, a mixed push/pull system may be quite practical. Divide the shop into major groups (e.g., machining, fabrication, sanding, painting, packaging). Within each *group* set a maximum amount of inventory that is to be allowed (in some units, e.g., dollars of orders). One is allowed to release items into packaging only when the packaging WIP has dropped below a certain limit, for example. In particular, one cannot release new jobs to the floor in the first place until the machining WIP has dropped below its limit. This in effect gives a type of order release policy. (Order release policies will be discussed in the next subsection.)

In practice, if a WIP is at or below its limit, it could accept a new piece of any size. But above its limit, the WIP would have to wait until it is below its limit. So far we have discussed the pull side of our idea. The push side is that in order to decide which item to release to the next stage, we always choose the item with the highest priority in the conventional push sense.

20.3.3 Setting Release Dates

It is not our purpose here to develop the large and important literature on setting release dates, but simply to introduce the idea, and present a model or two that fits into our other framework in this book.

A job with a regular objective function (which depends only on wanting the job to finish as early as possible) will always prefer to be released to the shop as early as possible. After all, suppose releasing the job early increased its cost. But one could always have held it to the original start time, and used the original schedule, giving the same objective function.

Thus to find why shops are interested in setting release dates, we are led to consider more general objective functions and the corresponding motives to release. Actually the motive, in almost all cases, is that it is more expensive to store WIP on the floor than the associated raw materials in the raw material warehouse. If, as in many cases, the completed job won't be paid for until the due date in any case, then finishing it early will simply cause a longer period of the expensive storage.

The idea that it is expensive to store WIP may be explicit, in the form of a given holding cost charged from release until the job is sold and revenue received, or it may be implicit, in the idea that too much WIP on the floor exceeds physical space or simply causes too much congestion. In such a case we may develop an order release based on the existing floor WIP, as in the Conwip system described in the last subsection.

We present a model for finding the best release date for the weighted inventory/weighted tardy problem. We develop it for the one-machine problem, although the procedure is similar for the general case.

We are talking about only one job j, and so for simplicity we suppress that subscript in the formulas. Let the job have inventory cost rate 0 until we decide to release it at r, after which it has inventory cost w_1 until at least the due date d. After the due date, for a time unit of tardiness it still incurs w_1 and, in addition, the tardiness cost w_2. We write this total cost as $w_1 + w_2 = w_3$.

Clearly, there is no point in releasing the job to the single machine unless its priority is positive (or zero) and until it has the highest priority. Thus we need to calculate that priority.

The objective function for this problem is

$$f(C, r) = w_1(d - r) + w_3(C - d)^+$$

Since we start processing as soon as we release, we have the identity that $r = C - p$. Thus we may simplify the objective function to

$$f(C) = w_1(d + p) - w_1 C + w_3(C - d)^+$$

Taking the derivative of this term with respect to C, adding the usual R&M smoothing correction, and dividing by p for resource usage gives the priority function:

$$\pi = (1/p)\{-w_1 + w_3\exp[-(S)^+/kp_{av}]\}$$

(Compare Procedure 12 in Section 7.2. Compare also the early/tardy formulas in Section 7.3.)

As we have said, we cannot release the job until it has both a positive priority and the highest priority. To decide when to release, it is useful to calculate the slack factor for which this priority crosses zero:

$$(S)^+ = -kp_{av}\ln(w_1/w_3)$$

20.3.4 Setting Due Dates

There is a large and important literature on setting due dates, which we do not have time or space to develop here. We do want to illustrate the basic ideas and tie them in to the BD framework.

When the order taker and the customer sit down to negotiate a lead time, there are several considerations. The customer wants a due date that is not too far in the future. Once the due date is determined, the customer will want (typically) the finished job to be neither early nor tardy. These two costs roughly add for the customer and are somewhat analogous to "long-range costs" and "short-range costs" in economic theory. The cost of setting the due date at time T in the future might typically be rather flat, monotonically increasing, and convex, as shown in Figure 20.2. For a given due date, the cost of delivery deviating from that due date might typically be rather sharp, especially on the tardy side, U- or V-shaped, and convex, as also shown in Figure 20.2. Note that the customer may very well be more concerned about accurate completion time than early completion time.

The cost curve of the customer is not, of course, directly available. However, we assume a cooperative relationship so that neither side has any incentive to lie. Furthermore, each wishes to minimize total costs to both sides. (If that favors one side or the other, the side gaining too much can make adjustments in the job price or make other "side payments" to make it seem fair to both sides.)

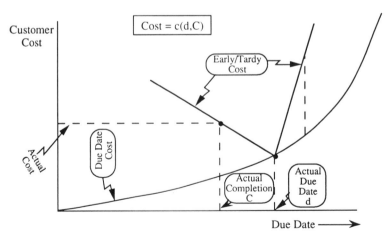

Figure 20.2. Negotiating a due date: customer cost as a function of due date and actual completion.

Suppose for simplicity that all shop costs are fixed and that the shop manager, for a given due date and early/tardy function (learned from the customer), would set a release for the job and run it together with all other jobs already in the shop to minimize early/tardy costs for all customers with WIP in the shop, using standard heuristic scheduling procedures.

Thus given the aggregate cost functions for a customer class, and given any due dates selected, one could simulate the shop and find the total cost to all orders, including the one of interest. To this we could add the due date cost from the customer's curve. This then could be repeated for four or five representative due dates, and the cheapest one chosen.

Although this seems like a very good scheme, there are several potential problems. First, simulating a full shop with good heuristics four or five times for each order and potential due date would require a great deal of computation. One way to speed this up would be not to do multiple iterations of the problem, but simply to estimate lead times by some simpler method. This might save a factor of 5 in computations. Another way would be to group customers into classes by urgency and type, and simulate a single due date time into the future to be given to each class. This might be updated only, say, six times a year.

A second potential problem is that the customer's cost functions in due date and tardiness are typically not known. Here we might assume that the order taker is skilled enough through experience to make fairly good guesses.

A third potential problem is that the supplier and each customer have separate objective functions, while we have simply added everyone's costs together and tried to minimize that. This assumption can actually be quite good if we assume long-standing cooperation, so that if one customer does better at one point, the other customer or the supplier will trust to compensating corrections being made. (That is, we are assuming a cooperative game.)

Useful references for this section include Sugimori et al. (1977), Kimura and Terada (1981), Baker (1984), Bitran and Chang (1987), Krajewski et al. (1987), Gravel and Price (1988), Glassey and Resende (1988a, 1988b), Cheng (1989), Karmakar (1989), Leachman (1989), Miltenburg (1989), Sarker and Fitzsimmons (1989), Spearman et al. (1989), Fry (1990), Mahmoodi et al. (1990), Sumichrist and Russell (1990), Berkley (1991), Berkley and Kiran (1991), Bieleck and Kumar (1991), DiMascolo et al. (1991), Inman and Bulfin (1991), Kubiak and Sethi (1991), Una et al. (1991), Wein (1991), Ahmed and Fisher (1992), Deleersnyder et al. (1992), Goyal and Deshmukh (1992), Mejabi and Wasserman (1992), Miltenburg and Sinnamon (1992), and Roderick et al. (1992).

20.3B Software/Computer Exercises

20.3B.1   Set up a small two-machine flow shop with one product. Give it a pull system, with one item of WIP allowed in front of each machine and in final inventory. Let both process times be 5.0; lead time for raw material is 5.0. Design a time-stream of demands for product that works well and one that does not.

20.3B.2 Change the system of 20.3B.1 to a simple push system (one designed to keep a unit of final inventory in place). What differences do you observe?

20.3B.3 Write a more general simulation for a flow shop with one product for either pull or push, and more general rules. Test it, and try to get a more general evaluation of push versus pull.

20.3C Thinkers

20.3C.1 What sort of a relationship can you see between the order release mechanisms into the multi-resource shop we have discussed here and the release mechanisms into a multi-machine single-resource situation we have discussed in Chapter 11? Be quantitative if possible.

20.3C.2 What other mechanisms for determining the due date of an order can you think of besides the one discussed here? Be quantitative if possible.

20.4 VARIABLE SPEED RESOURCES

20.4.1 Overview

The implicit cost of processing an operation with processing time p_j through a resource with price R is Rp_j. If the machine becomes a severe bottleneck, R may be very high, and the cost Rp_j assigned the operation may become very high. We have studied several ways to lower this cost:

(a) Do not process the job until the machine is less congested.
(b) Use larger batches to reduce effective setup time.
(c) Route the operation to other machines.
(d) Use substitute operations with different resource needs.
(e) Negotiate longer due dates with the customer.

Some other important methods we have not considered include:

(a) Employ overtime/additional shifts.
(b) Use subcontracting.
(c) Refuse the order.

Another very important and flexible method that can sometimes be used is to speed up the resource to shorten the process time. We discuss this issue in this section in two different versions: when the bottleneck resource is a metal-cutting machine and when the resource is human.

Running a resource faster, whether machine or human, involves subsidiary costs that must be figured in. In the first instance running the machine faster involves extra tool wear. Running the human faster involves stress/tiring and thus less output later. In Section 20.4.2 we develop models for this case. Here we consider the rather myopic case in which we only vary the speed for the job currently being processed; the multiple job case is considered later.

Another "cost" of higher speed is that the quality of the product is likely to go down. We add this new consideration to the model in Section 20.4.3. Finally, in Section 20.4 we consider the fact that changing the speed for a number of jobs will affect the overall price R. We sketch an iterative solution to this problem.

20.4.2 Congestion, Speed, and Resource Wear

We turn first to the congestion versus speed and tool life model. This work is discussed in much more detail by Narayan and Morton (1992).

Tool Life and Economics. Taylor (1907) proposed and studied the classical relationship between average tool life and cutting speed. Taylor fit the equation

$$L = (k/v)^{(1/n)} \tag{1}$$

where L is the tool life, v is the tool cutting speed ($v = 1$ for nominal speed), and k is an empirical constant depending on tool, job, and cutting conditions. The constant $1/n$ is usually between 0.1 and 0.5.

If there is no rush for a job to be completed one ought to take the maximum possible time to process it in order to reduce the tooling costs. On the other hand, if there are several jobs waiting, it might be worthwhile to increase the processing speed somewhat. Clearly, in the short run the optimal speed will be at the point when the marginal cost of tooling equals the marginal gain in increasing the processing speed.

Basic Machine Cutting Model. Here we assume that the instantaneous price of the machine is independent of the processing speed. (However, see Section 20.4.4.)

The actual time taken to produce job j through a bottleneck resource can be given as equal to the setup time + the speed corrected process time + [process times as a fraction of tool life] × the time to change the tool.

$$q_j = p_j/v + t_m + [(p_j/v)/L]t_c \tag{2}$$

where q_j = total time to produce j

$\quad\;\; p_j$ = nominal time to produce j

$\quad\;\; v$ = speed compared to nominal

$\quad\;\; t_m$ = setup time

$\quad\;\; L$ = tool life from equation (1)

$\quad\;\; t_c$ = time to change tool

We turn next to calculate the total cost to produce this job both explicitly and implicitly. This is equal to the explicit cost associated with the setup + explicit cost of machining the part + a prorated part of the cost of changing the tool and the tool itself + the congestion cost for other jobs kept waiting. This is then

$$TC_j = C_s + C_m(p_j/v) + C_c t_c + C_T[(p_j/v)/L] + Rq_j \qquad (3)$$

where TC_j = total cost of producing one piece

$\quad C_s$ = explicit setup cost

$\quad C_m$ = explicit cost rate of machining

$\quad C_c$ = explicit cost rate for changing tool

$\quad C_T$ = tool replacement cost

$\quad R$ = current heuristic congestion price of machine

$\quad q_j$ = total actual processing time of j

Next we substitute equations (1) and (2) into (3), verify convexity, take the derivative of (3), and finally set the derivative equal to zero to solve for the optimal speed v^*, obtaining

$$v^* = \gamma\{[C_m + R]/[t_c R + C_T]\}^n \qquad (4)$$

where $\gamma = k/(1/n - 1)^n$.

Let us try to study (4) intuitively. For low machine prices R we have asymptotically, as $R \to 0$,

$$v^* = \gamma(C_m/C_T)^n = v_c \qquad (5)$$

v_c is the "explicit minimum cost speed" defined in the classic literature. If C_m is small relative to the tool cost (normal), then as the price of the machine becomes very small, one would be inclined to slow the machine almost to a standstill: tool wear would dominate congestion.

On the other hand, for high machine prices R we have asymptotically, as $R \to$ infinity,

$$v^* = \gamma/(t_c)^n = v_t \qquad (6)$$

Again, in classic literature the same v_t is referred to as the minimum cutting time speed.

An Illustrative Example. Let us look at an illustrative example where

$\quad C_m$ = \$1.0/minute

$\quad t_c$ = 1.5 minutes

$$C_T = \$24.75$$
$$k = 600$$
$$n = 0.23$$
$$g = 454.4$$

We plot optimal values of v^* as R varies from 0 to 200 (Figure 20.3).

We see that the variation of speed with change in machine price is monotonic. It is interesting to note that the increase in the optimal speed is much faster when the machine price is low. As the machine price increases the optimal speed increases at a decreasing rate. The reason for this is that as the speed increases, the rate of tool wear increases rapidly, and the reduction in time is much lower at higher speeds.

Speed Versus Cost for Human Resources. We derive a much simpler version of the model above for dealing with human workers. The model is not meant to be directly useful in practice, but to illustrate how easily the ideas may be translated to another setting.

Assume now there are no significant setup times to the job, so that the job requires a time p_j/v. The total explicit and implicit costs of job j are now given by

$$\text{TC}_j = (p_j/v)(R + W) \tag{7}$$

where R is the implicit congestion price per unit of job time, and W is both the explicit wage rate of the human resource and the implicit added costs of stress, poor performance, and so on. We assume there to be a maximum feasible job completion speed v_m and that costs increase to infinity as this limiting speed is approached, leading to a total wage as a function of speed something like

$$W = W_0 + W_1(1 - v/v_m)^{-1} \tag{8}$$

Here W_0 is the explicit part of the wage, while the second term represents either explicit extra incentives offered by the company for faster work or later lost absen-

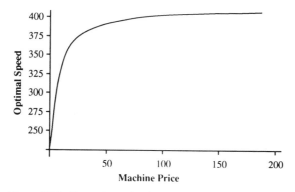

Figure 20.3. Example: optimal speed versus machine price.

teeism, slowdown, and so on, which will result from the current stress as the fastest possible speed is approached.

Thus the total cost as a function of R and v is

$$TC_j = p_j[(R + W_0)/v + (W_1/v)(1 - v/v_m)^{-1}] \tag{9}$$

By taking the first and second derivatives, it is easy to verify that the function is convex, with a unique solution $0 < v^* < v_m$. The solution is monotonically increasing in R, which is to be expected. We do not give the derivation here. In Section 20.4.4 we add the complicating fact that R will typically depend on the v we choose. The resulting equation cannot be solved in closed form but is very easily solved by iterative methods.

20.4.3 Adding Quality Considerations

We limit ourselves here to the tool wear version.

Quality Model 1. We look at two different ways to account for quality costs. In the first case we assume that the tolerance achieved in a critical dimension is associated with a cost of quality (lack of tolerance). The cost of quality is zero if the tolerance is 0 but increases with deviation either way. If m is the target dimension and y the actually achieved dimension, then the cost of quality is

$$Q(y) = Q(m + (y - m))$$

Assuming the existence of a Taylor expansion, we have

$$Q(y) = Q(m) + Q'(m)(1/1!)(y - m) + Q''(m)(1/2!)(y - m)^2 + \cdots$$

But $Q(m) = Q'(m) = 0$, so that to a quadratic approximation $Q(y) = K(y - m)^2$. Now if we assume in addition that the variation is linearly dependent on speed so that $y - m = k(v - v_0)$, then we finally obtain the approximation that the cost of quality is proportional to the square of the deviation of the actual rate from nominal:

$$Q(v) = C_q(v - v_0)^2 \tag{10}$$

If the cost of quality term (10) is added to the total cost equation (3), the resulting model is somewhat more complicated but can be solved for v^* by iterative methods. Asymptotic results are somewhat easier to find. For $R \to 0$, the asymptotic result is the solution to the simplified equation:

$$C_m = \gamma C_T(v^*)^{1/n} + C_Q(v^* - v_0)(v^*)^2 \tag{11}$$

Note that for a large enough cost of quality, and low R, we produce at v_0 to have perfect quality. Low cost of quality causes us to balance tool replacement cost against machining time costs.

For $R \rightarrow$ infinity, the asymptotic equation becomes the same as before [see equation (6)] since the price of the machine overwhelms quality considerations.

Quality Model 2. Another reasonable approach to dealing with quality is by using reject rates. Let $q(v)$ be the rate of acceptance, assumed to be a decreasing function of the material rate v. Let TC_j be the original cost from equation (3), then total production must go up by $1/q(v)$, which means that the new cost is just (3) divided by $q(v)$:

$$TCQ_j = TC_j/q(v) \tag{12}$$

We try a particular acceptance rate similar to the one due to Chan et al. (1989): $q(v) = 1/(1 + sv)$, where s is a parameter and $q(v) = 1$ at zero speed. Thus we finally have

$$TCQ_j = TC_j[1 + sv] \tag{13}$$

Differentiating with respect to v and equating to zero, we find

$$0 = TC_j'[1 + sv] + sTC_j \tag{14}$$

We may then substitute in TC_j from (3) and its derivative.

The derivative is complex, and v^* does not appear to have a closed form. However, the equation is solvable by successive approximations.

20.4.4   Considering Resource Price as a Function of Speed

In our models of this section, we have been solving for optimal machine speed as a function of a given input machine price R. In practice, however, raising the speed over all the jobs tends to lower R. Including this effect makes the model more realistic if also more complicated. We illustrate this effect with our human resource as an example.

Suppose the human resource has a given stream of jobs to be done irrespective of work speed. We approximate the price R by queuing theory, given a known current line length of L and utilization of ρ, at nominal speed of $v = 1$. By working at $v > 1$ the effective utilization rate drops to ρ/v. The queuing theory formula for machine price given current line length L, and average importance of jobs w, is

$$R = wL + w\{[\rho/v]/[1 - \rho/v]\}$$

or

$$R = wL[1 + \rho/(v - \rho)] \tag{15}$$

Substituting this back into equation (9), and taking the average job normalized to have length 1 at nominal speed, we find

$$TC = (wL)/(v - \rho) + W_0/v + (W_1 v_m)/(v[v_m - v]) \tag{16}$$

The first term is the job congestion cost; the second term is the explicit wage cost; the third term is the implicit human stress cost. Note that the congestion cost goes down like $1/v$ as speed increases fairly far above the critical speed $s\rho$. As the speed decreases toward ρ, effective utilization goes to 1, and congestion costs go to infinity.

Note that the explicit wage cost term also goes down but just as $1/v$ and does not blow up anywhere. Note also that the implicit stress cost term decreases approximately as $1/v$ far below the maximum level (mildly increasing stress but for much shorter time). As stress takes hold, the term begins to go up as $-1/v^2$; then, as v approaches the maximum possible v_m, stress goes to infinity. Differentiation confirms that the congestion term always decreases in v, the base wage term always decreases in v, and the stress term decreases up to $v = 0.5v_m$, after which it increases. Thus we see that $v^* \geq 0.5v_m$. There is clearly a global minimum greater or equal to $0.5v_m$ and less than v_m. Thus the only problem would be to prove the second derivative is positive in this region so there would not be more than one zero.

20.4A Numerical Exercises

20.4A.1   Consider the simplest speed versus human stress model summarized by equation (9). For some numerical set of parameters, compute (9) for perhaps five values of v and plot, determining the value of v^*. Repeat this process for perhaps five values of R, and plot a graph of $v^*(R)$. (A simple spreadsheet or small program may be useful.)

20.4A.2   Building on 20.4A.1, carry out a similar plan for equation (16), where R now is given by a queuing formula. You may need more points wherever the function is a little more complicated.

20.4B Software/Computer Exercises

20.4B.1   Write a program that computes total cost as a function of inputs for the (tool wear)/(quality) model 1. Plot v^* as a function of R, and any other parameters you are interested in. Pick a few data sets that may be interesting.

20.4B.2 Repeat 20.4B.1 for the (tool wear)/(quality) model 2 situation. Can you find data sets for which the two formulations are radically different?

20.4B.3 Repeat 20.4B.1 for the human stress model, where R is given by a queuing formula.

20.4C Thinkers

20.4C.1 Simplify the basic tool wear model as much as you can by eliminating setup costs or whatever. Now modify R to be given by a queuing formula. Try to solve the resulting model and/or give whatever qualitative sensitivity results you can.

20.4C.2 What other factors might limit practical machine speed? Make a relatively simple model that incorporates one or more of these.

20.4C.3 Suppose a human resource has large initial setup times for a job, which decrease radically the more times similar jobs are repeated. Can you model the relation of R, v, and experience?

___21
PLANNING, SCHEDULING, AND CONTROL

21.1 INTRODUCTION

21.1.1 Overview

Remember in Chapter 1 we classified scheduling (in the larger sense) in a company into five levels:

Long-range planning (plant expansion, layout, design).

Middle-range planning (smoothing, logistics).

Short-range planning (MRP, bidding, due date setting).

Scheduling (sequencing, routing, lot sizing).

Reactive scheduling/control (hot jobs, down machines, late activities and material).

We said then: "while all these levels can be considered scheduling in that they all have issues of sequencing, timing, routing, reconfiguration, forecasting, labeling, grouping, aggregation, and disaggregation, this formal similarity should not be pushed too far. There are a wealth of scheduling models at Level 4; these constitute the principal topic of this text. In attempting to model a different level, it is useful to recognize the similarities, but the differences are also crucial."

In this book we have concentrated thus far almost exclusively on Level 4. In this chapter we consider scheduling at other levels, drawing on the wealth of ideas we have developed already for Level 4, but showing the differences as well. We also briefly show how bottleneck dynamics principles are useful in integrating different levels, such as planning and scheduling, and for integrating the strengths of machine heuristics with the broad expertise of the human operator.

554

This chapter can only barely touch on these important issues. Neither software nor experience is available to more than sketch them at this point.

21.1.2 Chapter Summary

In a rolling horizon context, a Level 4 scheduling model with, say, a 6-week horizon might be solved perhaps once a week, with updated forecast and current shop loading as inputs. This model at best will be expensive computationally to solve, requiring perhaps 8 hours of computer time. During the week various unexpected events, from minor to major, will occur requiring revision of the schedule. It will not be feasible to rerun an 8-hour model a number of times, and so there must be provision for "reactive scheduling" Level 5, which is capable of making low-cost simple corrections to the model a number of times during the week. This topic is covered in Section 21.2.

Short-run planning is also of interest at Level 3. Existing material requirements planning (MRP) systems are quite crude in that material and release requirements for entering jobs are planned simply from working back from the due dates by fixed (inflated) lead times. Thus large WIP on average is built into the system. No lengthening of due dates with increased lead times is considered: the system merely makes some estimate of total capacity used with the unrealistic lead times to limit and shape order release. Bottleneck dynamics can achieve superior results by actually simulating the shop in aggregate fashion, thus considering resource constraints much more realistically. We discuss bottleneck dynamics versions of MRP in Section 21.3.2.

Middle-range planning such as logistics, at Level 2, covers the factory, warehouses, and distribution of product at much more aggregate levels, over a longer time horizon. Linear programming models, which do not consider congestion effects, have been successful here. A BD model is briefly developed that allows storage, routing, and production decisions to consider bottleneck issues more realistically. We discuss logistics (LOGJAM) in Section 21.3.3.

Cost (management) accounting, at Levels 1 and 2, has traditionally ignored congestion issues entirely. In effect, resources are given a heuristic price of zero when not being used to full capacity and are given a price equal to potential lost capacity when at full capacity. Simple improvements to cost accounting procedures using bottleneck dynamics are suggested. We discuss heuristic pricing improvements for management accounting in Section 21.3.4.

A large system with perhaps 200 machines might be aggregated for MRP purposes into 20 workcenters with, say, 10 machines each. At Level 4 it is probably not efficient to run a full 200-machine scheduling model, which will be poorly integrated with the MRP model. Instead, we suggest running 20 smaller Level 4 models with 10 machines each. Running the MRP model would give guidelines (prices and lead times) to each workcenter model, allowing them to run in a decentralized fashion, passing corrections back to the MRP model as necessary. We discuss these issues, and give further development of the computer card line example of Chapter 19, in Section 21.4.

Traditional manual shop floor scheduling via a human expert has a number of advantages. Humans have a very broad base of common sense; intuitive, poorly understood thinking abilities; and the ability to strategically summarize knowledge quickly and accurately. On the other hand, expertise is difficult to teach; increasingly complicated systems overload common sense. These strengths and weaknesses are complementary to those of the formal heuristic scheduling systems developed thus far in this book.

A number of methods have been proposed to try to combine the strengths of human expertise and mathematical expertise. One method is to allow the human and the machine to interact extensively with each other's point of view through a sophisticated user interface; this is the decision support system (DSS) approach. Bottleneck dynamics is particularly well suited for DSS use; we sketch this issue fairly carefully in Section 21.5.2.

A somewhat different approach is to build an expert system that copies and codifies the more routine historical expertise of the human expert. A major advantage is that this expertise can then be mechanized and mass produced. Major disadvantages include the fact that such "pure" systems cannot codify subtle expertise, create new expertise in new situations, or use broad-based commonsense knowledge. Thus there is a need to combine the best of mathematical expertise, expert systems, and the ongoing human expert. We discuss these issues in Section 21.5.3. Finally, distributed and hierarchical scheduling systems will produce the need to coordinate several user interfaces and several experts simultaneously. This issue is discussed very briefly in Section 21.5.4.

21.2   HEURISTICS FOR REACTIVE SCHEDULING

21.2.1   Overview

Although, in this book, we have primarily solved complex Level 4 scheduling problems over a fixed deterministic time horizon, we have also explained that in practice such fixed problems will be solved in a rolling horizon context, to allow updating of shop load information and forecast information. A perhaps 6-week "dynamic" model might be solved (quite expensively), the first week's part of the solution utilized, and then the 6-week model would be rolled forward a week, and solved again. Each time only 1 week of the solution would be used; the other 5 weeks of the horizon would be included simply to make that solution more accurate by taking into account interaction effects with the future.

Thus many of the long-range effects of uncertainty are handled by the rolling horizon device. However, during the week between major runs there will be dozens of small changes affecting the plan, some minor, some midsize, and some major. It will simply not be feasible to input, rerun, and analyze an 8–10-hour run many times. The system would not be real-time for one thing, and it would require massive human support for another.

One of the weaknesses of existing mathematical scheduling packages is that there is no real provision for cheaply correcting the schedule ("reactive scheduling") for small to middle-size changes. A large company reports privately that a popular large-scale scheduling package had no provision for reactive scheduling, except to make full reruns. For this reason the package was mostly ignored in-between the weekly runs, and human experts used the results of the package less and less. Because bottleneck dynamics, on the other hand, develops approximate dual prices, sensitivity analysis is rather easily performed. This allows minor reactive scheduling to be performed and allows approximations to be available for more major reactive scheduling.

It turns out that minor changes can be handled by a real-time dispatch mode, which simply leaves prices unchanged, updates the shop after each transaction, and dynamically recalculates decisions. This happens in a real-time dispatch mode; we call these "dispatch" corrections. Middle-size corrections can be handled by crudely fudging changes in some of the prices, and then continuing in real-time dispatch mode; we call these "mid-reactive" corrections. Large-size corrections must be solved by re-solving a simplified version of the full model to get revised prices, and then continuing in real-time dispatch mode; we call these "major-reactive" corrections. It is not clear as to whether the choice between "mid-reactive" response and "major-reactive" response ought to be calculated automatically or left to human intervention.

Types of Unforeseen Changes. It will be necessary to deal with a large number of all sorts of (unforeseen) changes and corrections during the week. We list a number of these, but the list is by no means exhaustive:

Dispatch Correction (Duration: hours)

- Changed activity process time
- Transport delay
- Minor machine slowdown
- Minor machine stoppage/startup
- Critical machine down/startup

Mid-reactive Correction (Duration: 1–2 days)

- Job recycles for rework
- Job scrapped, replacement started
- Job waiting for missing input
- Transport down, startup
- Machine down/startup
- Workcenter down/startup

Major-reactive Correction (Duration: 3 days or more)

- Major order canceled
- Major transport down/startup

- (Long) machine down/startup
- (Bottleneck) machine down/startup

Summary of Three Modes. (Real-time) dispatch mode is the normal operating mode during the week. Prices and lead times from the weekly solution are assumed unchanged, but minor real events change arrival times and job availabilities at a dispatch point. At a real-time dispatch decision, effectively prior priorities are applied to the jobs actually available. The dispatch mode can accept intervention from the operator to change prices or priorities. However, if more major changes occur, it will change to the mid-reactive mode. The mid-reactive mode fudges prices where feasible, or asks the operator to help in fudging them. The operator may also call the mid-reactive mode in if it judges the dispatch mode is not doing the job accurately. Either the operator or the mid-reactive mode can change to the major-reactive mode.

When the mid-reactive mode fudges prices and lead times, the mode changes back to real-time dispatch and proceeds. When the major-reactive mode is called, it solves an approximation to the full model and thus is not real-time in this period. Therefore decision modifications should be made by a lower mode for this period, switching to the new full solution when available. The "dispatch" mode is discussed in Section 21.2.2, the "mid-reactive" mode is discussed in Section 21.2.3, and the "major-reactive" mode is discussed in Section 21.2.4.

21.2.2 Dispatch Corrections

(Any of the three reactive modes have a very large number of planned variable values and partially corrected variables available. We will assume this without further comment.)

The (real-time) dispatch mode is a straightforward emulator or simulator; it keeps track of a corrected detailed plan at all times. As they become known, it enters actual times to update the plan. Whenever it is called to make a decision, such as schedule an activity to a just available machine, it selects from currently feasible activities, considering all fixed and dynamic constraints as they become known (including machine down, etc.). It then chooses the highest priority job (priority not corrected by current happenings) and schedules it, assuming that priorities have not changed much since the last major replanning. (That is, it holds priorities constant until the mode changes.)

To illustrate this, suppose a 4-hour processing time actually turns out to be 7 hours. This turns out to fall under "minor problems," and hence the dispatch mode is expected to handle it. Now a processing time change:

(a) may change the time that a given sequencing decision is made;
(b) may change which jobs are available at the next decision point;
(c) may change which resources are available then;
(d) hopefully will not noticeably change job priorities.

To summarize, dispatch keeps track of everything as it actually happens and makes dynamic choices according to the planned priorities.

The transport delay is similar to the processing time overrun problem and will not be discussed.

The remaining machine problems all involve machines with less processing than usual. This will change the machine dual prices and process times; thus it actually will affect prices and priorities somewhat. However, by the definition of this case, dual prices and process times only decrease for a few hours and then return to normal. Thus the average effect on dual prices and the average effect on process times will be fairly minimal. Hence the case is similar to the processing time variance case described above and can be handled adequately without recalculation. Note that we are only approximating prices and priorities as unchanged; the real-time dispatch procedure must keep track of actual machine availabilities and speeds as they happen.

The number of hours of longer processing time, or machine slowdown, and so on to set as the boundary between using dispatch mode and mid-reactive mode is open to study and experimentation. That is, one would need an experimental design to investigate both types of modes for different levels of disruption.

21.2.3 Mid-reactive Corrections

Remember that the dispatch mode operates by real-time simulation of the actual progress of the shop, assuming existing priorities still to be valid. The mid-reactive mode operates almost the same, except that one or more of the resource prices or job lead times have changed as a result of unforeseen mid-reactive corrections with a duration of 1–2 days. This results in a change of one or more job priorities, which is too large to comfortably remain in the dispatch mode.

Remember our basic sequence priority for an activity is calculated by

$$\pi_{jk(i)} = w_j U_{jk(i)} / \Sigma R_k p_{jk} \tag{1}$$

That is, for a current activity i that is a part of job j and currently at machine $k(i)$, its priority is the job weight times the slack factor from this machine to finishing j, all divided by the remaining resources the job requires, that is, the sum of remaining activity processing times multiplied by the appropriate machine prices. A full iterative dispatch solution of the problem from the current moment would produce the correct new prices, lead times, and processing times and allow us to recalculate this priority.

However, this would be too expensive for a medium to large number of mid-reactive corrections in a given week. Since our results are typically not too sensitive to mild errors in priorities, we make approximate corrections to process times, prices, and lead times as necessary, without re-solving the entire problem.

As making good mid-reactive corrections that are not computationally demanding is quite difficult, we start with the one-machine case in order to strengthen our intuition.

One-Machine Case. The simplest case is when the unforeseen change does not involve a different total use of the resource. Consider the case of "job waiting for missing input." In particular, jobs 1, 2, 3, . . . , n are currently waiting or will arrive in the current busy period for the single resource at time t. As job 1 would be about to start processing, it becomes apparent that job 1 will not be ready to start for some time T', due to some problem not involving the resource. For the purpose of determining the change of expected lead times, price of the machine, and priorities of the jobs, we ignore the fact that the optimal sequence may change. (When the new priorities are actually used in running the shop, however, the sequence is allowed to change. Thus we are making a "myopic" correction in lead times, prices, and priorities.)

Say that after time T' has elapsed, jobs 2, . . . , $i - 1$ have been processed and job i is in process. Thus the total delay for job 1 is $T = \Sigma_{j=2,i} p_j$; jobs 2 to i have been delayed by $- p_1$, and remaining jobs have not been affected. Therefore the lead time for job 1 is increased about T, the lead times for jobs 2 to i are reduced by p_1, and other lead times are unaffected. For regular functions, the delay cost for job 1, $w_1 U_1$, increases, while for jobs 2 to i $w_j U_j$ decreases, and for the remainder it is unaffected. The busy period of the resource is unaffected, and some delay costs increase while some decrease, so to a first-order approximation assume $R(t)$ may be left unchanged. The priority for job 1 will increase, for 2 to i it will decrease, and the others will be unaffected. (Note that during the actual processing of these jobs, the reprioritizing may move some later jobs ahead of earlier jobs. We are making a myopic correction; if it is deemed important to make a more complicated correction, one would use the major-reactive method below.)

Consider a slightly more complicated case: "job recycles for rework." Here job 1 finishes processing at time t, only to discover some difficulties in processing, so that it continues processing for some additional time Δ. Thus all n jobs in the busy period have their lead times increased by Δ, including job 1, and thus have their U_j increased appropriately as well. However, typically the extra time will allow extra jobs to arrive, lengthening the busy period as well. Suppose rather than trying to trace them, we simply use a queueing theory approximation. Remember that when a job of average length p_{av} arrives as an extra, the busy period is lengthened by $\rho/(1 - \rho)$ jobs, where ρ is the "long-term" utilization of the machine. Thus, in this case, the busy period will be lengthened on average by $(\Delta/p_{av})\rho/(1 - \rho)$. Putting all this together, if U' and R' indicate new values after the disturbance, we have:

$$R = \sum_{j=1,n} w_j U_j; \quad R' = \sum_{j=1,n} w_j U_j' + w_{av} U_{av}(\Delta/p_{av})\rho/(1 - \rho)$$

Versions of this formula could be given where we don't know the current busy period, but estimate it from the current line. We leave this as an exercise.

The case "job scrapped, replacement started" is almost the same except that the amount of time lost will be the full p_j. The case "machine down/startup" is also very similar, since the system does not distinguish whether the delay Δ is due to the machine breakdown or the failure of the first job's processing. The principal change is that the delay may be large (1–2 days) rather than small, so that the approximation may deteriorate if, for example, urgent jobs may be routed to other machines. Again, the correction to the price of the current machine gets much more complicated if future jobs may be routed away. Roughly, if our current approximation is adequate simply to correct prices and lead times, we remain in the mid-reactive correction mode. If not, we would turn to the major-reactive correction given below.

Job Shop Case. The case of "job waiting for missing input" is relatively unchanged from the one-machine case. To a first-order effect, the current machine price is unaffected. One job has an increased priority while several have reduced priorities, causing a random temporary mild lowering and raising of machine prices downstream, which to a myopic approximation may be ignored. This case is made easier by the fact that the overall load on every machine is unchanged.

An exception would be if the required waiting period for the missing part was very long and the job very important. Then the whole shop might be reorganized somewhat to expedite the job. This is clearly outside the scope of mid-reactive modeling. Either the simpler mid-reactive correction would be deemed adequate, or the major-reactive correction might be called for.

The multi-resource version of "job recycles for rework" can be much more interesting. If the recycling simply involves rework on the current processor, holding it a longer period of time, then the analysis is little changed at the current machine, which has a somewhat higher overall load. The downstream machines have first a somewhat lighter load, and then a somewhat higher load, averaging out overall. Thus we may be fairly comfortable in ignoring downstream events.

Suppose, however, that the rework involves repeating, say, the last three operations on two previous machines and the current machine. We may reasonably approximate this as three separate one-machine rework problems, again because downstream effects average out pretty well. One exception is that the delay for the delayed job at each machine will be the sum of the three process times, not just that at that machine. (This approximation, while interesting, may be a little tricky. It needs to be tested.)

Situations where a transport is down, a machine is down, or a workcenter is down are similar to the rework for a single-machine case. (Remember that transports and workcenters can be treated as a single resource if there is a single queue.)

21.2.4 Major-Reactive Corrections

Remember that the full planning mode runs once a week or so, requiring perhaps five iterations of a dispatch simulation with iteratively improving prices, lead times, and priorities. If the major-reactive correction is called in the middle of the week ($t = 0.51$ instead of $t = 0$), it is exactly as if the full planning mode were calculating the typical (6-week) plan at the beginning of the period (week), with three exceptions:

(a) The mid-reactive correction is always called first. If it is deemed inadequate, it still provides a corrected set of prices, lead times, and priorities for the first iteration of the simulation.

(b) It is run for $6.00 - 0.51 = 5.49$ weeks instead of 6.

(c) It is run for one iteration instead of five to interfere with real-time running as little as possible, followed by the appropriate update for prices, lead times, and priorities available from that run. These are then used to continue real-time running.

These points need further discussion.

The mid-reactive correction should always give an improved solution to the available dispatch mode solution. Since it is low cost, it may be preferred over the major-reactive correction, even if it is far from perfect. Even if it is necessary to continue on to the major-reactive mode, the improved prices and lead times will allow for a much improved simulation and obviate the necessity for several iterations.

The point on shortening the simulation to 5.49 weeks is that the new full plan must mesh firmly and accurately with the realized 0.51 weeks of actualized event given by the real-time dispatch mode. This is actually a relatively subtle point, since many jobs will be in process at the interruption and must be processed properly. We probably simply forecast their finish times as the actual start time plus normal processing time as usual.

One iteration instead of five is a major improvement toward making the system real-time. This saving comes because we are updating the forecast throughout the week and correcting prices, and so on, so that much of the convergence work is being done more or less continuously. This raises the possibility that if the major-reactive correction is done three or four times a week, then the weekly full replan with five iterations might simply be unnecessary. Either cut it to one iteration, or only run it if, say, more than 0.5 week has elapsed since the previous major-reactive correction.

Useful references for this section include Lehoczky and Sha (1986), Lehoczky et al. (1987), Liu et al. (1987), Rajkumar (1989), Sprunt et al. (1989), and Bai et al. (1990).

21.2C Thinkers

21.2C.1 Think of two or three reactive-scheduling problems that have not been mentioned in the book. Describe each in some detail, and try to justify whether they should be classified as minor, middle sized, or major problems.

21.2C.2 Take one of the problems you described in 21.2C.1. Model an appropriate approximate corrective procedure.

21.3 HEURISTICS FOR PLANNING LEVEL SYSTEMS

21.3.1 Overview

In this section we consider three representative systems that are needed for manufacturing:

(a) Short-run planning (such as MRP) at Level 3 is discussed in Section 21.3.2.

(b) Middle-run planning (such as logistics) at Level 2 is discussed in Section 21.3.3.

(c) Management accounting at planning Levels 1, 2, and 3 is discussed in Section 21.3.4.

In each case we first briefly discuss traditional methods for dealing with the problems to be solved. The coverage is brief partly because the reader is assumed to have some familiarity with the traditional topics. References are also given for those needing them.

After the traditional coverage of each of the three topics, we briefly discuss bottleneck dynamics methods for the same problems, drawing heavily on the scheduling material that has been developed in this book.

21.3.2 Short-Run Planning (MRP)

We first discuss traditional MRP methods for handling planning, followed by more advanced MRP. Next we discuss its advantages and disadvantages. Then a bottleneck dynamics approach is sketched.

Sketch of Traditional MRP. For each product, and for each week (representing a due date) for up to, say, 6 months in the future, a mixture of actual customer orders, forecasting, and interaction with the production floor produces a set of planned requirements called the *master schedule*. Current excess final inventories are netted against the master schedule going forward in time. To get remaining needed product at a point in time, a fixed historical lead time is assumed to do the final assembly. The need for a given quantity at a given time in the future is moved backward in time by that fixed lead time and entered as requirements for each element of the subassembly. In turn, for each of these, we net any inventory, determine lead times for the second level assemblies, move backward, and so on. Eventually, we reach the raw material level, which determines the release date into the shop. However, there may still be lead times from the raw material warehouse, or from the supplier if the materials are not in stock, to finish the process.

Note that the backward explosion process, for a due date too close in to the future, may produce release dates before the current time; that is, they may be infeasible. To correct for this, MRP simply moves the whole schedule for this forward enough so that the product may be started and accepts whatever tardiness is involved at the other end.

This MRP procedure makes no capacity checks or lead time checks to see whether assumed lead times are really too low, or too high, or possibly both. A second routine called CRP, capacity requirements planning, tries to adjust for this. After a given plan is complete, CRP plots the actual capacity being used in each workcenter as a function of time. If capacity is overused, CRP adjusts or extends various parts of the plan, until every workcenter is feasible. This, however, is not done using an actual simulation of the process but is simply based on the same historical lead times.

MRP sets lot sizes in an arbitrary way, which does not take into account whether a resource is busy or empty. Finally, it has no provision for dovetailing with a scheduling module, simply passing its crude release times and due dates to the human scheduler (dispatcher) on the shop floor.

Advantages and Disadvantages. MRP systems have a number of advantages. They provide good database support and are oriented for interacting well in a decision support system (DSS) mode. They represent the product structure adequately and deal with intermediate and final inventories well. They are relatively easy to understand.

However, MRP systems have a number of disadvantages as well. No feasible aggregate schedules are produced. Without even a minimal simulation, there is no real way to estimate capacity used as a function of time. Use of fixed historical lead times means that estimates of tardiness cannot accurately be made in light shops or heavy shops. In practice, lead times are almost always inflated to provide safety stocks of materials on the shop floor. This in turn makes the shop look even more loaded, destabilizing the associated CRP module. MRP makes no attempt to interface with the Level 4 shop floor and cannot give any guidance as to which orders can be met on time and which cannot. Other higher level planning capabilities are very minimal. While MRP-II attempts to account for interaction of manufacturing with other functions, it still does not address issues mentioned above.

A Bottleneck Dynamics Approach. We first aggregate each workcenter into an aggregate machine, and aggregate similar individual jobs with common routing and due date, where useful and feasible. We also need some initial values for the first iteration. We use the release times, lead times, due dates, and incorporation of inventory from some convenient source (such as the output of a MRP run). The aggregate machines will be treated in the Level 3 planning simulation and iteration as having a single queue. However, the "processing time" for an aggregate job is a composite idea depending on how long it takes a job to go through the aggregate machine after it leaves the initial queue and starts processing. (These processing times will typically be supplied from the Level 4 scheduling simulation for that workcenter. However, if these are not available, historical averages will do. Most of the lead time is usually spent in the initial queue that is being modeled fully.) We initially set machine prices by any crude approximation, such as by their load or a queuing approximation.

Delay costs for jobs will come from the lead times initially supplied, so that sequencing, routing, and lotting decisions may be made through the first simulation. Then prices, lead times, and so on may be updated for the next iteration, which will determine much more realistic release dates, lead times, and so on.

Advantages and Disadvantages. A disadvantage of the bottleneck approach is that aggregation of jobs, machine center, and due dates is often difficult. Too little aggregation may result in a system too large to solve; too much aggregation may result in an excess loss of accuracy.

An advantage is that aggregate capacity control, lot sizing, routing, prioritizing, and order release are extremely useful to have in a sophisticated fashion on Level 3, even if Level 4 scheduling is not being done by bottleneck dynamics. We deal with pretty much feasible schedules at all times, so that the planner really knows what is happening.

A special advantage is that after Level 3 is "solved" for each workcenter, the workcenter price, and job arrivals and local due dates can be passed down to Level 4. Then Level 4 can solve that problem in a more detailed fashion and send corrections to the master plan back to Level 3, and so on. We discuss these issues in much more detail in Section 21.4.

21.3.3   Middle-Range Planning: Logistics

Logistics plans the entire procurement, manufacture, storage, and distribution of products. This is a very aggregate analysis, at Level 2, and hence the "product" may actually be a mix of products, and the manufacturing plant is a simple entity with a capacity.

The earliest mathematical logistics model is probably the transportation algorithm. Look at Figure 21.1. There are four normal resources—plants 1, 2, 3, and 4—each with some capacity and cost of production (not shown). There are three negative resources—cities 1, 2, and 3—each with some capacity (need) and negative cost of consumption (revenue). The arrows show routing: we must choose good (and feasible) amounts to route across each.

We like to use circles for resources and arrows for routing decisions. In this sense, there are 12 missing resources on Figure 21.1: the transportation resource on each route. [We would add a circle on top of the middle of each route, with the original arc becoming two: one into the transportation route and one out. These also have costs of production (transportation). The early transportation algorithm assumed no limitation to the amount shipped on each route, which may mean no resource was considered. In our further analysis we shall always specify them as resources. This is important since a rail line may have limited capacity, congestion slowdowns, and certainly differing lead times for delivery.]

The transportation problem can be solved by a special form of LP. Because it is

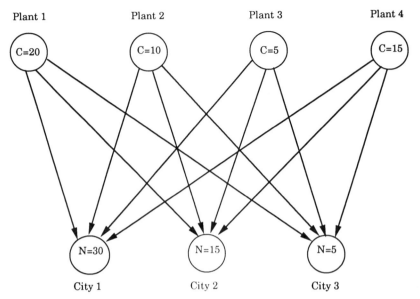

Figure 21.1. Early logistics network (transportation problem).

convex, exact dual prices are obtainable for resources, which leads us to consider it for integration into a bottleneck dynamics framework. However, this early formulation is limited in speed and in solving complex problems. Development of network algorithms, to be discussed next, gave spectacular improvement in both.

Network Algorithms and Logistics. New and powerful network algorithms (special LP structure) such as those by Klingman (1977), allowed much richer and more interesting logistics problems to be solved. See Figure 21.2.

Network methods allow us to deal with a great number and variety of raw material suppliers, vendors, manufacturing plants, warehouses, customers, and transportation providers, that is, just about any aggregate resource needed for planning purposes. Networks are quite flexible, in that a critical plant can be broken down into several resources in the network. Resources can have upper bounds and even convex increasing costs. Resources can have wastage and other losses.

Networks can, by expanding their size greatly, be made dynamic. Suppose we want to model the next 13 weeks within a rolling horizon of a one-period review, the situation shown in Figure 21.2. Make 13 copies of Figure 21.2, and label them week 1, week 2, and so on. The choice of storing in plant 1 from week 1 to week 2 is simply shown by putting an arrow from it in week 1 to it in week 2. Of course, do not show storage possibilities where they do not exist.

Another kind of complexity to add is multiple products. Simply make a full multi-leveled diagram for each product. If several products are produced in plant 2 in period 3, the total production must not exceed the capacity of the plant, giving a

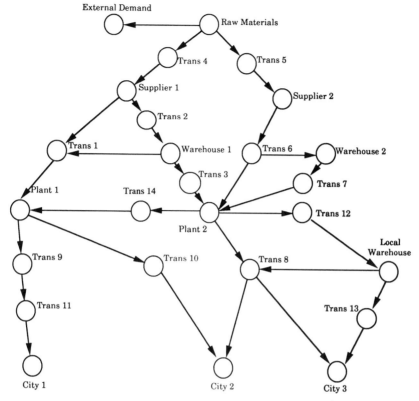

Figure 21.2. Logistics network (one-period model).

cross constraint between products involved. This makes a more complicated model to solve; however, it is still a special structure LP.

Bottleneck Dynamics Logistics Model (LOGJAM). LOGJAM (Srinivasan and Morton, work in process) is a proposed bottleneck dynamics network planning system model. It captures bottleneck congestion for resources such as vendors, plans, workcenters, transportation links, regional and local warehouses, and final customer demands.

While earlier models assume there is no congestion until full capacity is attained, LOGJAM summarizes each resource as a simple relationship with output as a concave function of input (Figure 21.3). This relationship may be estimated for each important resource by regression on historical data.

The model captures congestion effects in a very simple but reasonable way. The peak load effects that may be treated include the following:

- Seasonal/lumpy/stochastic load versus lead time.
- Smooth variation of implicit resource evaluation.
- Dynamic planning for bottleneck utilization.

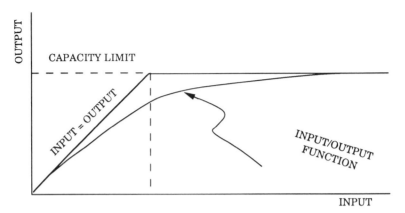

Figure 21.3. Resource congestion input/output function.

The exact prices given by LOGJAM are useful for the following:

- Dynamic resource prices.
- Dynamic shadow price of extra WIP anywhere.
- Dynamic shadow price of expediting existing WIP.

Earlier Work on Congestion in Logistics. Glover et al. (1979) found an efficient solution for the one-product, no-congestion version of the basic model for which LOGJAM is an extension. Osleeb and Cromley (1977) showed (not efficient) how to solve the one-product, no-congestion model for nonlinear costs for each resource.

Open and closed loop queueing system approaches are useful when it is desired to model rather uniform stochastic congestion effects. However, such approaches depend on stationarity and cannot model known demands, seasonality, or long-term cyclical behavior.

Single-Period LOGJAM Model. Let $1 \leq j$, $k \leq n$ be any node of a precedence logistics network. Each node represents a resource, such as plant, department, machine, rail car, truck, supplier warehouse, local warehouse, or city demands. Let m be a particular product. Let A_{jk}^{m} be an arc (routing choice) from node j to node k for product m (unique). Let $S(m, j)$ be the set of possible direct successors of j for m, $P(m, j)$ be the direct predecessors, D^{m} be the final demand nodes, and V^{m} be the initial supply (or vendor) nodes.

Let x_{1j}^{m} be starting WIP of m at j; y_{1j}^{m} be ending WIP. Let d_{1jk}^{m} be arc flow of m from j to k in period 1; let x_{2jk}^{m} be WIP in period 1 after initial inflows. Then

$$x_{1j} = \sum_{m} x_{1j}^{m} \qquad (2)$$

gives the total load on node j for period 1. We next postulate a congestion function (see Figure 21.3):

$$g_{1j} = g_{1j}(x_{1j}) \tag{3}$$

which is the total resource output given a load x_{1j}.

Next, this total output must be divided among outputs for different output nodes and products, using conversion factors a_j^m that tell how much a product uses of the throughput. Then we equate the output from equation (3) in aggregate to individual outputs:

$$g_{1j} = \sum_m \left[a_j^m \sum_{k \in S(m,j)} d_{1jk}^m \right] \tag{4}$$

The left side is the known output, while the right side equates it to the sum of outputs over products and successor nodes.

Next, we compute the ending inventory for each product and node as the starting inventory less the outflows [see equation (4)]:

$$y_{1j}^m = x_{1j}^m - a_j^m \sum_{k \in S(m,j)} d_{1jk}^m \tag{5}$$

Now we compute the next period's starting inventory as this period's inventories, plus inflows for next period:

$$x_{2j}^m = y_{1j}^m + a_j^m \sum_{k \in P(m,j)} d_{1jk}^m \tag{6}$$

(A few final equations are needed to specify all variables ≥ 0, except final demands are ≤ 0. All initial and ending boundary conditions are also specified.)

Finally, in order to completely specify the mathematical program, we must give an objective function. [Note that the mathematical program is not a linear program because of the nonlinear congestion functions from equation (3).]

Costs

c_j Per total units output from a node. Production, transportation, and so on. Aggregate, not assignable to products.

c_j^m Similar, but assignable

h_{jk}^m Arc cost for product m to be routed from j to k (not transport, which is a node).

(Arc costs are especially useful in a multi-period model, in which they can represent holding costs, stockout costs, backlogging costs, earliness costs, or tardiness costs.)

Objective Function

$$\min \sum_j c_j g_{1j} + \sum_j \sum_m c_j^m a_j^m \sum_k d_{1jk}^m + \sum_j \sum_m \sum_k h_{1jk}^m \qquad (7)$$

Remember that, for the special case in which the congestion function given by equation (3) is simply "output = minimum (input, capacity)," this model reduces to the standard multiple commodity network model, where excess capacity has a price of zero.

Multi-Period Model. Now add another dimension, with time period $1 \leq t \leq T$. We basically duplicate a single period network T times and add connective (storage) links. Node t_j means a node at time t and resource j. Arc $t_{jk}(0)$ means arc from j to k both at time t, a normal routing arc. Arc $t_{jj}(1)$ means arc from j at time t to j at time $t + 1$, a normal storage arc. Arc $t_{jj}(-1)$ means arc from j at time t to j at time $t - 1$, a backlogging arc. Arc $t_{jk}(+3)$ might mean j giving goods to a broker who contracts to give equivalent goods to k three time periods in the future. Note that predecessor and successor sets now must include temporal predecessors and successors also.

Equations (2) to (7) translate directly for the temporal case, simply by changing beginning and ending subscripts 1 and 2 to t and $t + 1$, respectively, and updating node, arc, and sets to reflect time, such as j becomes t_j.

We leave it as an exercise to generalize the objective function to the temporal case.

Modeling Lead Times

Proposition 1. Let $(g_{tj}^m)'(x)$ be the slope of the throughput function at x amount of input. Then the lead time for the marginal unit after x is $1.0/(g_{tj}^m)'(x)$.

Proof. Left as an exercise. ∎

Note that the marginal lead time is *not* the input divided by the output at x, which would be the average lead time or clearing time, although this intuition would be true for the linear input/output case.

Linearized Model. It is quite easy to make a piecewise linear approximation to the congestion functions in equation (3), so that a linear program will result. However, for very large problems, the linear program may be impractical. The single-product, noncongestion model is a network model and very fast to solve. The congestion function at each resource links these networks together. Some kind of decomposition solution may be possible.

Miscellaneous Issues. The existence of exact dual prices for resources and for inventory is useful in sensitivity analysis. It is also useful to tie logistics, MRP, and scheduling together by communicating prices between the levels.

Other extensions that can be envisioned are adding fixed minimum lead times, products that are assembled, and optimal release decisions from each stage to the next.

Computational Experience. LOGJAM has been programmed for small- to medium-sized problems as a convex nonlinear program (Srinivasan and Morton, work in process). We describe a few of the early validation runs. Further experimentation is planned but has not yet been carried out.

Seasonal Multi-plant Planning. Suppose product A can be processed only through plant 1, but product B can be processed through either plant 1 or plant 2. Plant 1 is more expensive, and A has one-third the average demand of B, but they have somewhat different seasonal patterns. As expected, B sticks with plant 2 in low demand periods and stores excess product in warehouses; in higher demand periods it shares plant 1 as well; in very high demand periods it takes product from warehouses; A and B may suffer longer lead times and tardy delivery. The congestion function helps smooth the solution over time. It also gives good dual prices for planning subcontracting and so on. Finally, it gives an easy method for estimating how lead times vary with plant load.

Seasonal Multi-mode Transportation. Suppose the train is the cheapest transportation mode for a given link but is subject to seasonal degradation due to seasonal side-loads outside the model. There is also a higher cost truck link, which is always lightly loaded. As might be expected, in seasons when the railroad link is not very side-loaded, "railroad" is preferred and gives adequate lead times. When "railroad" is congested, the truck link will be preferred, until it too becomes congested. The model tends to equalize long lead times and tardiness as much as possible. Where it makes sense, it may ship by rail in advance and store near the customer market.

21.3.4   Cost (Management) Accounting: Levels 1, 2, and 3

The reader is assumed to have some familiarity with cost accounting; however, we have attempted to make this brief account self-contained.

Financial Accounting Versus Management Accounting. Financial accounting is largely for the purpose of describing the performance of a company as a whole to outside parties: other companies, investors, auditors, and so on. It is heavily tied to rules and regulations that force presentation and computation to be in a certain prescribed way.

Management accounting, on the other hand, is concerned with helping the manager make practical economic calculations that are used only to make *decisions* within the company. It is not heavily tied to required ways of thinking. However, a company tends to resist having two ways of thinking, so very often management accounting thinking and procedures are more similar to financial accounting than they need to be in the abstract.

Management accountants tend to classify internal decisions into short run and long run. Short-run decisions are often taken as those that affect a year or less into the future. At this short horizon costs can be classified into *fixed costs and variable costs*. Fixed costs will not vary with the volume of production. These include depreciation on machinery, parts of the workforce, rent on the building, part of the utility bill, and many administrative and marketing costs. Variable costs are often assumed to vary linearly with the volume of production and may include raw materials, direct labor, some power and other utilities.

Long-run decisions are often taken as those that affect 1–10 years into the future. At this long horizon all costs become variable. Short-term variable costs are still there, but the fixed costs can be changed eventually up or down to reflect the long-term demand volume.

In practice, apparently short-run decisions often have long-run implications that are very important. However, we ignore this issue here, and simply assume we know which decisions are short-run and which are long-run. To analyze a short-run cost, we can use the equation

$$\text{SRC}(x) = F + vx \qquad (8)$$

Here x is the volume of a product to be produced in a month, SRC is the total short-run cost depending on x, F is the fixed cost allocated to this product, and v is the variable cost per unit.

The long-range version of this would allocate the fixed costs among the units:

$$\text{LRC}(x) = (F/x + v)x \qquad (9)$$

This is also called the full absorption cost.

Here we do not discuss the short-range versus long-range issue. We consider only the pure short-range equation (8) and discuss how it may be improved from the point of bottleneck dynamics.

What are some typical short-range decisions?

How much would it cost to increase production by 50 units?

How should we price the product this week?

Which work should have priority when the shop is highly loaded?

Should we make or buy the units?

Should we subcontract?

Should we work overtime?

Note that in management accounting, in the short run, the conventional wisdom is to consider the resources (machines) as sunk costs that are free for making an extra few units. Note that in the long run, the conventional wisdom is to divide the full "sunk" costs by the number of units so that the variable cost now includes a piece of the cost of the machine. (In older cost accounting, this would be done in the short run as well, the so-called full absorption costing.)

But in the bottleneck dynamics approach, there are also implicit short-run costs to be considered related to the machine. A congested machine k will have an implicit price $R_k(t)$ developed from an iterative process, which tells how much implicit costs will be added to production. When an extra unit of j is produced, it will be processed for time p_{jk} on each machine k. Thus the extra implicit variable cost, which we denote v_2, is given by

$$v_2 = \sum_k p_{jk} R_k \tag{10}$$

If we then label the explicit variable cost as v_1, we have

$$\text{SRC}(x) = F + (v_1 + v_2)x \tag{11}$$

Accountants, by ignoring this term, in effect assume that the value of extra capacity is zero, even at 80–90% utilization. This term is large at higher utilizations and needs to be taken into account. Ignoring it because it is somewhat difficult to estimate merely ignores a problem that has not gone away.

Other references for this section include Leachman (1986), Spence and Welter (1987), Sullivan and Fordyce (1990), and Shaw et al. (1992).

21.3C Thinkers

21.3C.1 Prove Proposition 1.

21.3C.2 Give an example where the fixed lead time assumptions of MRP might give reasonable results. Give another example where it might give very bad results.

21.3C.3 In the LOGJAM model, some transportation links might be shared between products in the model and external usage outside the model. Try to modify LOGJAM to treat this situation.

21.3C.4 Accountants often argue that as long as production is less than 100% of capacity, it would make sense to accept one-time orders at little more than variable cost.
  (a) Give examples where this makes sense.
  (b) Give examples where this does not make sense.

21.4 TOWARD INTEGRATION OF PLANNING AND SCHEDULING

21.4.1 Overview

It has been observed by many authors interested in planning and scheduling (whether with an operations background, a computer systems background, or an artificial intelligence background) that these functions are performed hierarchically in business. That is, there will be a planning manager and staff who build plans (say, at Level 3 or MRP), and perhaps a scheduler for each shop or department who make schedules. The planner provides a less detailed plan to each scheduler, who creates a detailed schedule roughly conforming to the plan.

There are basically two main reasons why a single planner does not simply make one very detailed plan, including detailed schedules for the entire manufacturing floor, which could be solved exactly and implemented. Those reasons might be summarized as (a) difficulties of complexity and (b) difficulties of incomplete knowledge.

First, such a model would be so large and complex that no human or group of humans could comprehend it in total. Thus the model would have be solved by computer. Even here, such a model could not be solved exactly by computer in anything like real-time. Such a model could not be solved at all accurately by approximations, since any theory of accurate heuristics for such a complex model is not available.

Second, even if the model could be solved approximately in reasonable time, humans would not understand it well enough to deal with the inadequacies of the model and to override the model recommendations when warranted.

The second reason is perhaps the most important. Humans have the ability to develop "expertise" in a special area, such as scheduling one machine center, through experience in that area and use of human intelligence. This expertise augments and complements computer modeling. Typically, a number of experts are needed, some at planning levels and some at detailed scheduling levels.

Authors recognizing this fact try to develop hierarchical models that conform to this human organization. A planning module interacts with the planner, making use of his/her expertise. It passes overall planning information down to scheduling modules, which interact with the various schedulers, making use of their expertise. If schedules seem too much at odds with the plans, such information is passed up to the planning module, to allow modification of plans.

In Section 21.4.2 we discuss a few examples of earlier hierarchical models, both those that pass production requirements from the top level to the bottom level, as well as those that pass pricing information. These models typically have a medium planning level (production smoothing) as the upper level, and a short-term planning level (MRP) as the lower level. We also discuss advantages and disadvantages of these methods.

In Section 21.4.3 we present an outline of a two-level bottleneck dynamics planning and scheduling system. The upper level is a short-range planning system (Level 3) and the lower level is a full scheduling system (Level 4). This system simulates a scheduling solution to an aggregate shop (MRP level). Prices and timing information are passed down to each workcenter. Then each workcenter is scheduled as an independent shop. In some cases, major changes in the environment of a workcenter may have to be passed back up to the planning level, the planning level module re-solved, and so on. We also discuss advantages and disadvantages of bottleneck dynamics in this context.

Finally, in Section 21.4.4 we complete our discussion of the computer board line, in order to illustrate a two-level system design in action.

21.4.2 Earlier Hierarchical Models

Newson (1975) developed integer formulations and heuristic procedures for both the detailed scheduling and aggregate-detailed scheduling. His single combined model solved only a one-period model and did not capture the actual complex job shop structure. However, he did develop the basic idea of decomposing the problem (for solution purposes) into aggregate and detailed levels, with the higher level passing down shadow prices of resources to the lower level detailed scheduling system. Moreover, he used setup times rather than setup costs, which enables the representation of the lot size–lead time–capacity interaction. The data sets used to test computational performance were very small.

Independent of Newson's work, Jaikumar (1974) decomposed the overall problem into a higher level production smoothing component and a lower level problem for deciding the number of setups and production quantities for each period. The prices and costs of the various resources are passed down from the higher level production planning solutions to determine the lower level scheduling costs. Also, the higher level plan in addition passes down information that constrains directly the lower level production quantities of individual items.

Hax and Meal (1975) suggest a hierarchical approach that partitions the production planning and scheduling problem into a hierarchy of subproblems with decisions implemented only for the immediate period. Bitran et al. (1980, 1981) have provided detailed solution procedures using this approach. These models are for the "aggregate planning," "family," and "item disaggregation" levels (and models) in single- and two-stage systems. Unlike Jaikumar's work, the shadow prices are not passed downward. Furthermore, the solution procedure has to be adapted depending on the magnitude of the setup costs. This approach is a true hierarchical one: it requires less detailed demand forecast data. It also parallels the nature of decisions, typically made at distinct organizational levels. The solution procedures do not appear to be computationally efficient.

Graves (1982) presents a hybrid approach that retains the computational advantages of the earlier single-problem approaches, while retaining the hierarchical structure, by using Lagrangian relaxation. A natural decomposition is achieved with feedback on prices/costs between both levels. However, the approach does require extended detailed demand forecasts and assumes infinite overtime capacity. The computational performance achieved here may be better than the purely hierarchical approach.

None of these hierarchical approaches consider the actual detailed shop structure: variable lead times, due dates, lot sizes, assembly structure, dynamic input, or queues. None except Newson consider setup times and hence inadequately model setup costs. It is not clear that these approaches can be extended to jobs with individual characteristics, and hence cannot easily be extended to general multi-stage systems where the capacity–lead time–lot size interactions become critical.

Finally, all these models assume one globally known "deductive" model, rather than being organized for focused inductive control.

The models discussed to date lack the ability to model accurately at each level (which might seem to require a simulation structure capability). Such structure can also allow each level to be modeled as interactive with a local expert. DSS/AI approaches, which follow up on these ideas, typically do not have strong decision-making capabilities, however. Rather than discuss DSS/AI methods explicitly here, we look at an approach that combines simulation and focused control with heuristic pricing for making decisions.

21.4.3 Two-Level Bottleneck Dynamics Systems

One of the most difficult issues for any large organization is the tension between the advantages of centralized control versus decentralized control. Businesspeople think of it as a conflict between coordinated thinking and planning versus the initiative and reactivity of local authority. Operations research people think of it as a computationally impractical global optimization versus practical but inaccurate sub-optimization. Artificial intelligence people tend to think of it as inadequate global understanding that can be combatted only by accumulated local expertise.

Overview. Here we describe a two-level bottleneck dynamics system (Morton et al., 1988), which might meet some of these concerns. It features global, aggregative control at the master scheduling (MRP) level for the entire shop, and local distributed disaggregative control for scheduling at each workcenter. The upper level module (MRP-STAR) has an aggregative scheduling capability similar to the usual Level 4 models. It does not contain such detailed features as lot sizing/enforced idleness but can add such aggregate features as overtime, extra shifts, subcontracting, and production leveling. Suppose the shop contains eight workcenters. MRP-STAR would aggregate these to eight "machines" and would also aggregate jobs with similar routings, due dates, and customer priority. It would probably allow preemption to mitigate the effects of such large "jobs" in terms of artificial integrality.

There would be eight lower level modules (SCHED-STAR), one to schedule each workcenter by the methods developed in this book. MRP-STAR and the eight SCHED-STAR modules would be coordinated in a way to be explained below. These modules might be run twice a week.

The principal advantage of such an upper level module over MRP and other similar master scheduling systems would be that a realistic schedule could be produced, which considers the actual dynamic workload and resource limitations in detail. Lead times would reflect the actual shop queues. This module might be run every week. MRP-STAR would also be capable of making decisions about rough sequencing, routing, overtime, and extra shifts.

Note that each of the eight SCHED-STAR modules can schedule independently once it has two kinds of information: (a) job characteristics and estimated arrival time to the workcenter and (b) derived due date off the workcenter and marginal penalty for violation.

The planning system would produce an aggregate schedule and aggregate prices. Parts of this information would be passed down to the individual scheduling modules. The workcenter scheduler would also react to breakdowns and other problems and pass corrected information back to the upper level module.

Resource Aggregation for MRP-STAR. After aggregation, workcenters would be treated as "black boxes" into which jobs enter, emerging one aggregated operation later. One would need parameters such as arrival time, expected process time, actual process time, start time, departure time, and average portion of the center used.

The aggregate price for a workcenter would be determined in two different ways, which would have to be reconciled.

(a) MRP-STAR treats the center directly as a machine in a scheduling/pricing procedure and hence estimates an aggregate price R directly.

(b) The local SCHED-STAR model also estimates prices R_k directly for the various resources in the center, which then may be added up to obtain R by another method.

(A simple reconciliation such as taking a weighted average of the estimates might be appropriate.)

A job coming though the center suffers three components of time loss: (a) waiting in the main center queue, (b) processing on various resources, and (c) waiting in internal queues.

Items (b) and (c) can be added together to give the job's aggregated processing times. Under the assumption that once the job leaves the main queue it is relatively high priority, item (c) might be estimated historically as a fixed percentage of direct processing time. Aggregated processing time could then be estimated in advance, while item (a) is determined explicitly in the upper level model.

A job being processed does not typically tie up the whole center. Define δ_j as the fraction of the workcenter that job j ties up on the average. One way this may be estimated is

$$\delta_j = \left(\sum_k R_k p_{jk} \right) \bigg/ \left(R \sum_k p_{jk} \right)$$

That is, multiply individual resource prices by the time job j spends on that machine to get a total center resource usage. Divide by the maximum possible usage of the whole center for the total processing time to obtain the average portion of the center used.

Now, then, $R\delta_j p_j$ is the total cost of using the center by job j; $P_j = \delta_j p_j$ is the effective equivalent time the job would tie up the whole center. In the simplified aggregate model, the main queue moves ahead as if the center were a single machine, with entering job j requiring processing time P_j on the whole center. However, the job is not allowed to leave the center until the full time p_j after beginning processing. The aggregation thus preserves gross resource and processing time usage properly, while losing some fine detail.

Job Aggregation. The most difficult thing about job aggregation is deciding which jobs to aggregate together. Clearly, the jobs should have the same routing and "similar" arrival times, process times, due dates, and tardiness costs. If interpreted strictly, few jobs could be aggregated. On the other hand, it is important that the upper level module remain manageable in size. Thus a compromise must be struck.

Aggregation of due dates is fairly straightforward. The aggregate due date should probably be somewhat earlier than the average due date (say the 25th percentile). Process time aggregation depends very much on whether the jobs are likely to run more in parallel or in series in the center. In the former case the proportions δ_j should be added; in the latter case the times p_j should be summed. Any intermediate compromise must preserve at least the estimate of total resources used (i.e., the sum of individual resources used).

Overtime, Extra Shifts, Subcontracting, and Production Leveling. If there is overtime available on a center with no extra fixed costs involved, but higher direct costs, then the aggregate module could simply model it as a second center with appropriate capacity and higher direct costs. Note that the time granularity is too coarse to worry about whether the centers should be simultaneous or not. The aggregate module's routing optimization heuristics will automatically avoid using overtime unless it is justified.

An extra shift may be treated as similar to overtime. However, there are lower direct costs, but a rather high additional fixed cost (e.g., supervisory and direct labor). To avoid combinatorial search over opening and closing shifts, the following type of heuristic might be employed. Run the problem with only overtime allowed. In weeks with overtime above a certain cutoff value, open a shift in a second run. More elaborate heuristics are easily envisioned.

Simple subcontracting, where individual jobs may be subcontracted as desired with no overall contractual arrangement and where there are well forecasted lead times, may again be modeled as an extra center with appropriate capacity and process times. Contracts, however, often involved fixed costs and/or minimum purchase amounts per year. These issues will not be considered here.

MRP-STAR will automatically consider load leveling. As previous iterations experience excessive tardiness, the new iterations will move processing further and further back into the slack period. Trade-offs between load leveling, overtime, and other choices will also automatically be made.

Passing Information Between the Two Levels. When MRP-STAR is run, it does a coarse scheduling for the machine centers. Then it passes down to each center:

(a) the aggregate price for that center (perhaps as a function of time);
(b) projected aggregate job arrival times;
(c) aggregate job derived due dates for the center.

The center uses this input as follows:

(a) The aggregate price is multiplied by the resource price share for each machine to determine prices (perhaps as a function of time) for each machine in the center.
(b) The jobs are disaggregated to produce individual jobs with arrival times, weights, and derived due dates.
(c) This establishes priorities, routing, and so on for each job at that center.
(d) Each center can then be scheduled individually in detail.
(e) At real-time dispatch the entire shop is simulated together in detail, using the prices and priorities already established.

When the workcenter experiences reactive problems such as breakdowns:

(a) if the problems are relatively small they may be handled in dispatch reactive mode or by rescheduling the machine center;
(b) if the problems are relatively large, they may be sent back up to the MRP-STAR level, to correct its own input and rerun the corrected MRP-STAR.

21.4.4 Case Example: The Computer Board Line Revisited

To illustrate how a two-level system works in practice, reconsider the computer card line problem as developed in Section 19.4. There we primarily considered scheduling an individual sector (by some version of SCHED-STAR). Here we concentrate more on how to coordinate the individual sectors with some version of MRP-STAR.

To review briefly, the full line is a flow shop with 18 sectors. Each sector is in itself a compound flow shop with an initial queue and associated sequencing and release decisions. Each also has finite internal queues and provision to rebalance by shifting operators. No internal resequencing is allowed, hence a sector may be aggregated to a single resource. At the planning level (we ignore the reentrant feature here) the upper level model is an 18-machine flow shop, with resequencing allowed and adequate queue space.

Thus, given effective processing times for each job through each sector, the aggregate problem is easily solved. These can be estimated as a multiple of the total processing time of the job in the sector, perhaps estimated historically by regression with the multiple depending on the current total load in the sector. Since there are not a great number of jobs, these need not be aggregated, which simplifies things considerably.

Thus we may immediately run MRP-STAR to get for each sector: (a) the aggregate price over time, (b) the projected arrival times of individual boards to the sector, and (c) the derived due dates of individual boards off the sector.

This then is sufficient to allow each sector to be optimized separately. Remember that this was a rather complicated process involving simulating release times and rebalancing the operators iteratively.

Again, small reactive issues can be handled at real-time dispatch with the given prices. Medium-size change and difficulties can be handled by rerunning the sector model. Large-size problems might require passing all changes to the MRP-STAR model and rerunning the master model.

Useful references for this section include Gershwin et al. (1985), Hadavi and Voight (1987), Bai et al. (1990), and Srivatsan and Gershwin (1990).

21.4C Thinkers

21.4C.1 Think of an example where aggregating jobs is not likely to work very well. Explain your reasoning carefully, with perhaps some small hand numerical support.

21.4C.2 Think of a second example where aggregating jobs can be expected to work well. Explain your reasoning carefully, with perhaps some small hand numerical support.

21.4C.3 Can you make generalizations from your findings in 21.4C.1 and 21.4C.2 to provide some sort of general guidance to someone wishing to aggregate jobs?

21.5 INTERACTIVE SYSTEMS (DSS) AND EXPERT SYSTEMS

21.5.1 Overview

Throughout most of this book we have been developing heuristic systems that might seem logical to implement by computer, ideally giving complete computer control of the planning, scheduling, and dispatch/control function. However, since the 19th century, a rather different idea has also come to the fore: a human and a machine (the computer in this case) have different strengths and weaknesses and therefore should act as a team.

Human strengths include a very broad base of common sense, intuitive but poorly understood thinking abilities, and an ability to strategically summarize knowledge. These abilities, taken together, often work to allow experts to react quickly and accurately to glitches and problems.

Human weaknesses include difficulty in teaching their expertise to others, being overloaded by increasingly complicated systems, and variability in response.

Computerized scheduling strengths include very fast computation, the capability to consider many factors, and a very reliable response.

Computerized scheduling weaknesses include difficulty in learning from experience and no broad common sense.

Thus the idea has grown that human–machine interactive systems should be able to have both sets of strengths and less of the weaknesses. This became shortened to "interactive systems" and eventually was broadened to "decision support systems."

Sketch of Interactive Scheduling Systems/(DSS). All decision support systems have at least three main parts: a database, a computational procedures base, and a user interface. (See Figure 21.4.)

For a scheduling DSS, the *database* must contain more permanent data, such as the structure of the shop, characteristics of each machine/resource, and routing information. It must also contain current data such as jobs, arrival times, processing times, routings, and due dates. It must be interfaced with the user and electronically with other parts of the facility.

The *computational procedures base* may contain one or more computational procedures for making computerized sequencing and routing decisions. It will always contain simulation procedures for tracking the system through the future, whether automatic or manual (or mixed) decisions are being made at each decision point. These procedures may be at Level 5, real-time dispatching/scheduling, or at

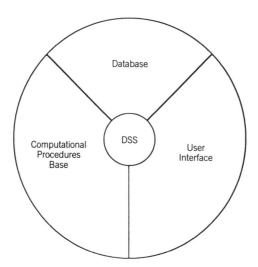

Figure 21.4. Schematic of a DSS.

Level 4, scheduling over a forecast horizon, or both. The simulation procedure will typically need some type of evaluation component for measuring the performance of schedules.

The *user interface* will have a graphics component to produce graphical aids such as Gantt charts and/or network diagrams (especially for project shops). (Note that such graphical aids may just be given at the end of the scheduling process or can be given repeatedly as a trace for more complex problems.) It will also possess a schedule editor for manually generating and/or manipulating schedules in various ways. It will also possess a database editor for similar purposes.

The *Leitstand* is such a system that has become very popular in Europe; we discuss it below. But perhaps it is worthwhile to illustrate a simple scheduling DSS in the software provided with the book, namely, the one-machine problem with the manual interchange mode. Data sets can either be manually input, generated statistically (this can be considered a computational module), or taken from a previously stored file. The user may then input an initial schedule, or enter one obtained from other automatic schedule generators, and then revise it at will, with good display of the current schedule and multiple evaluation of the schedule.

Sketch of AI/Expert Systems/Knowledge Based Scheduling Systems. AI practitioners have developed sophisticated databases (related to modern relational databases) that are very modularized and hierarchical. Instead of storing all information for each machine, much may be stored just once for a group of machines, or groups of machines themselves may store some of their data only once for the shop, and so on. Retrieval of desired data may then require going all over the database in a sophisticated way. This database is compact and allows easy additions or deletions of machines, and so on; that is, it is modular. Such a database is often renamed a "knowledge base"; the modules of the data are called "frames."

AI practitioners have modularized decision procedures also, putting modules of logic into "rules." A rule might be "don't start a job on Friday afternoon, unless overtime has previously been scheduled for Saturday or Sunday." Such rules don't need always to be quantified. This makes it possible to imitate rules that real human experts use. The rules, being modularized, can be stored in the knowledge base, instead of being fixed in the logic of the simulation. The computational base now needs computational procedures for deciding which rule to use next. Such procedures are called "inference engines." Some of these popular inference engines include PROLOGUE and OPS5. There are some AI "shells" that make it possible to build systems more easily. Knowledgecraft is an example of a shell that is frame based and allows several different inference engines.

21.5.2 Interactive Scheduling (DSS)

Earlier Systems. Godin (1978) presented an excellent survey of interactive scheduling in the 1960s and 1970s, which we summarize here. We focus here on the operational systems discussed rather than prototypes. It is striking to note that of the dozen or so systems discussed, none was successful enough to lead to commercial software, as opposed to other areas such as design.

Just one of the early systems, PRODUCE, had full interactive capability, graphics, simulation, and automatic/manual scheduling capability. We summarize the advantages and difficulties of then current interactive systems:

Advantages

(a) All system builders conclude interactive scheduling systems are a sound idea.
(b) Good CRT graphic displays are better than printed output.
(c) Full ("look ahead") simulation is better than short horizon simulation.
(d) Interactive scheduling is better than batch scheduling with a simple heuristic.
(e) New unified concept of a DSS will help scheduling applications.

Difficulties (in 1970s)

(a) Lack of good mathematical ways of solving huge combinatorial problems.
(b) Rapidly changing problems with no dynamic modeling capable of keeping up.
(c) Interactive systems still rare; people unaware of advantages.
(d) Decision-makers do not understand the advantages of computers.
(e) Cheap computer technology only very recently available.
(f) Software houses avoid the area because it is hard to prove that the system pays off, each installation needs to be almost custom, and much education is needed for both users and higher ups.
(g) Schedulers do not understand their own job; they simply build in large slacks and do not see the trade-offs.
(h) Schedulers are not motivated by customer goals; so motivation is lacking.
(i) The best current system (PRODUCE) is not being promoted.
(j) Research is very disjoint; authors do not communicate with each other.

This list of difficulties is very insightful for the 1970s. It is interesting to see which still apply in the 1990s:

Difficulties (in 1990s Perspective)

(a) There are plenty of accurate methods now available to solve these problems, as shown throughout this book.
(b) AI techniques and heuristic pricing techniques now provide easier updating of models.
(c) Interactive systems are widely available on PCs and very widely used.
(d) Decision-makers are still leery of computers (and academics).

(e) Very cheap computer technology is widely available.

(f) Difficulties of commercial software houses are very real, but see discussion of the leitstand below.

(g) Extensive retraining of schedulers is still necessary but the problem is more to motivate the CEO.

(h) Now excellent systems are being produced, for example, project management systems and the leitstand.

(i) Research is somewhat less disjoint; secrecy of commercial packages is a new problem.

Looking at the comparative lists, it seems now that the difficulties of customization and support of top management in companies are the primary remaining problems. We shall discuss these issues again below, in considering the leitstand.

Current Systems. There are many more successful commercialized scheduling interactive systems today. We shall limit our discussions to three: OPT, project management software, and the leitstand. We study each of the three types and then discuss strengths and weaknesses.

OPT (Creative Output) (Fox, 1982; Lundrigan, 1986; Meleton, 1986; Fry et al., 1992) is representative of other current finite scheduling systems, including NU-METRIX and Q-Control. OPT makes clever approximate solutions to the full mathematical programming scheduling model. In a preliminary pass, OPT identifies the bottleneck resource. Priorities are then set in the rest of the system in an attempt to "baby the bottleneck." As one example, jobs that require extensive processing in noncritical parts of the system, but use only small amounts of bottleneck time, will tend to be expedited.

Advantages of OPT include clever, intuitive, approximate solutions to large mathematical programming problems and a strong focus on understanding critical bottlenecks. Weaknesses include only a fair user interface and database. In addition, the system seems to have no capability to do reactive scheduling without re-solving the full model, which is not practical in real-time. OPT does illustrate, however, that a scheduling DSS with stronger computational procedures base is very important, since OPT was successful with basically just that.

We turn next to project management software. This is an extremely successful DSS application, as stated in Chapters 17 and 18. There are over 100 software programs for the PC alone. Since these software packages are quite similar, we simply describe a generic one.

The packages provide forward and/or backward simulation for multi-resource-constrained project scheduling. Only the makespan criterion is allowed. This allows projects to be combined and treated as a single project. Sophisticated manual editing and generation of schedules and projects are allowed.

Automated resource-constrained scheduling is available but unsophisticated: the program simply uses slacks and lead times from the unconstrained problem, and then schedules by LFT. This area is open to easy strengthening.

The success of project management is probably related to two factors, both stemming from the fact that project management is more aggregated than scheduling:

(a) The project manager can likely keep the project updated personally, on a PC, without excessive time or data input.
(b) The project manager is likely to have a clear stake in the final cost and quality of the project.

Adelsberger and Kanet (1991) describe the German leitstand, which is a generic name for several commercial software systems. These systems are rapidly becoming successful in Europe, with sales of about 400 systems. Adelsberger and Kanet (1991) provide an excellent abstract; we simply quote it in full:

> A leitstand is a model for a type of DSS for scheduling and control in manufacturing. The specification for a leitstand includes the following features: 1. a graphics component, 2. a schedule editor, 3. a database management system, 4. an evaluation component, and 5. an automation component. A leitstand provides management with a flexible, user-friendly, intuitive, logically-transparent shop floor scheduling component to help managers plan the detailed flow of production orders through a factory. The leitstand model can be positioned between a host computer in which a higher level production planning system is housed and individual machines or work centers where shop flow data collection comes into play. A major difference between the leitstand and conventional finite scheduling tools is that the leitstand approach emphasizes decision support rather than a particular scheduling algorithm or procedure.

Hypotheses often advanced as to why leitstands have been much more successful in Europe than the United States include:

(a) European managers understand the strategic nature of scheduling and are willing to support the conversion effort more fully.
(b) Europe is on the climb and more willing to innovate.

Existing and Research Leitstand Systems. Adelsberger and Kanet (1991) describe three commercial German leitstand systems—AHP, IDS, and Infor—as well as four research systems—the L1 Project at Dortmund, the Production Reservation System, the Jobplan Project at Siemens, and the Quick Response Center at Clemson.

AHP-Leitstand

Stand-alone or networked with other leitstands.

Implemented at over 300 sites.

Control individual operation or groups.

Plan forward, backward, or on bottlenecks.

Rich communication component.

IDS

Written in C, UNIX, SQL database, X-windows.

Stand-alone or as a component of production planning system.

Good user interface and help.

Infor

Part of decentralized CIM system.

CIM system is networked workstation.

Redundant data storage.

L1 Project at Dortmund

Microsoft windows.

Load leveling.

Main emphasis improved interfaces.

Production Reservation System

Cornell/Hewlett Packard collaboration.

Some distributive capability with other shops.

Several-second response times for 30 machines and 100 tasks.

Jobplan Project at Siemens

Multi-workstation, under UNIX.

INGRES as database.

MOTIF as graphical user interface.

Higher order editing functions.

Highly developed evaluation function.

Embedded neighborhood search.

Quick Response Planner at Clemson

QRP allows interactive planning of a single machine.

Immediate feedback on schedule adequacy.

Handles sequence-dependent setup times.

Good features for JIT.

Research Directions. We turn now to a consideration of how bottleneck dynamics would be especially useful in an interactive scheduling system with very high-quality automatic scheduling capability.

An example may clarify things. Suppose the nuclear fuel tube shop (Section 14.5) is being run interactively by a scheduler and a leitstand. The leitstand has advanced automatic scheduling capability in the form of, say, some nonpricing bottleneck method. The method knows the furnace should be the bottleneck and schedules items in such a way as not to overuse it. However, as it turns out, the previous resource (third-stage pilgers) was nearly a bottleneck. Careless scheduling for it causes it to be overloaded and to starve the furnace. Another way to say it is that for the moment the bottleneck has moved.

From experience, the scheduler recognizes the situation that the third-stage pilgers are starving the furnace, and knows roughly the changes she would like to make. She moves to the manual schedule editor and makes changes. Each time she makes a change, other parts of the schedule must change to preserve feasibility. She eventually becomes confused as to whether she is fixing the situation at the third-stage pilgers or fixing her mistakes in trying to get a larger and larger number of changes all to be feasible.

For the heuristic pricing version of the leitstand, the scheduler simply asks the computer to lower the price of the furnace by a small, medium, or large amount, and will receive automatically a new schedule correcting the original model error. The machine can keep the three or four feasible schedules resulting, and the scheduler can eventually make any small manual changes to the one selected. The scheduler does not get involved with the complexities of maintaining feasibility.

As another example, consider the example of reactive scheduling, which we have discussed in Section 21.2. If the weekly automatic scheduling run requires, say, 4 hours, then we will not be able to run the full run each time there are small changes, such as late arrival of jobs, correction in processing times, or machine down. Thus, without heuristic pricing, one would be driven to making numerous manual "fixes" many times during the week. While a few of these fixes might easily be made by the scheduling editor, eventually accumulated changes might make it more difficult to maintain a feasible schedule and more difficult to preserve the "good" character of the original automatic schedule. Since bottleneck dynamics for small changes allows all prices and most priorities to be unchanged, and simply requires repeating a simulation part of the original optimal schedule, the work and chance for making errors would be much smaller.

This is just a taste of how bottleneck dynamics might be used in a reactive-scheduling system.

21.5.3 Expert Systems for Scheduling

Overview. Expert systems in scheduling fit the DSS interactive scheduling model described above. They generally emphasize heavily the automatic scheduling mode. There is typically a forward chaining "inference engine," which implements rules (modularized heuristics) depending on the current conditions in the shop. There will also typically be a knowledge base (sophisticated database) based on frames.

There are a variety of ways to obtain the rules:

(a) An expert human scheduler is studied for a number of months. The rules the scheduler actually uses are formalized in rules. The rules are incorporated in the expert system. Test runs are shown to the human scheduler, who gives feedback on sharpening the system.

(b) The system builder goes to an OR expert, who suggests a system of heuristics for a given system concept. These are implemented as rules. The human scheduler again gives feedback.

(c) The system builder instead simplifies and approximates the full heuristics into rules more like the human scheduler would give. The human scheduler again gives feedback.

While classically the words "expert system" have meant rules imitating the human scheduler, in practice basing rules on heuristics is usually now also called an expert system.

Roughly speaking then, an expert system for scheduling is a complex version of an interactive system for scheduling. These complexities come in the sophistication of the data management, manipulation of decision rules by the inference engine, and sometimes in complex operations research heuristics used to form the rules.

While prototypes of a number of scheduling systems have been built, tested, and demonstrated, very few have actually been implemented successfully. Two reasons for this seem to be (a) failure of management to give the level of support necessary in getting a different system to be accepted by the users and (b) burdening an already complicated system by an advanced automatic scheduling system.

Examples of Expert Systems in Scheduling. In this section we give brief descriptions of a few of the expert system prototypes that have been built. We draw on recent surveys by Rao and Lingaraj (1988) and Kusiak and Chen (1988) and some of our own personal experience.

One of the earliest documented scheduling systems was ISIS (Fox et al., 1983), developed originally for Westinghouse. Its knowledge base (database) is based on frames, the forward chaining engine is basically OPS5, and rules are based on OR type heuristics:

(a) Multiple job objectives are modeled as constraints on the schedule.

(b) Heuristics are based on backward interval scheduling and beam search.

(c) Constraints are relaxed successively until a feasible solution can be found.

(d) A form of hierarchical planning is employed.

The popular expert system shell Knowledgecraft was an outgrowth of the ISIS research.

Another outgrowth of ISIS was a system designed for the Digital Equipment Company called CALLISTO. CALLISTO (Sathi et al., 1985) deals with project scheduling and/or project management for very large R&D projects, such as developing a new mainframe computer. CALLISTO and ISIS have similar frames, forward chaining, and OR heuristics for rules. Special features include:

(a) Multiple users in a distributed and hierarchical system.

(b) An early form of heuristic pricing that forms much of the heuristic system.

PATRIARCH (Morton et al., 1988) is a two-level planning and scheduling system. The upper level utilizes bottleneck dynamics; the lower level is an early version of OPIS. OPIS (Ow and Smith, 1987; Smith et al., 1990) builds on PATRIARCH to produce a multi-expertise system where the various sources of expertise must cooperate.

MERLE (Prietula and Thompson, work in process) is a multi-expertise scheduling system that can allow negotiation between the human user and the internal AI/OR system.

MARS (Marsh, 1985) has been developed to schedule resources for the space transportation system. It uses ART, a popular and powerful expert system shell. OPAL (Bensana et al., 1988) is closely related to ISIS. There have also been a number of similar applications to flexible manufacturing systems, such as Lecocq and Guiot (1988), Shaw (1988), and Shaw and Whinston (1989).

There are very few operational expert systems in manufacturing. By far the most successful well known systems have been developed at the Digital Equipment Corporation (Rao and Lingaraj, 1988) for in-house use. They include XSITE (an expert site planner's assistant), IMACS (to aid manufacturing), ISA (to aid scheduling), IPMA (to aid project management), and ILRPS (to aid long-range scheduling).

ISA was developed from ISIS. The inputs were customer orders, change orders, and cancellations. The output was a due date for each order, corrections for administrative errors, and alternative schedules.

Other references for this section include Adachi et al. (1988, 1989), Savell et al. (1989), and Shaw et al. (1992).

21.5C Thinkers

21.5C.1   Give and discuss a number of reasons that a leitstand with a sophisticated automatic scheduling capability is better than one without any automatic scheduling. Give and discuss a number of reasons for the opposite conclusion.

21.5C.2   Consider the one-machine scheduling module in the software provided with this book to be a primitive leitstand. Explain in some detail the correspondence between parts of this module and parts of a typical leitstand. Which parts of the software are weak and which strong in such a comparison? Be specific.

21.5C.3 Design several improvements for the one-machine scheduling module, to improve its performance as a leitstand. Give these improvements in detail.

21.5C.4 Consider students preparing for exams and term papers to be expert systems for deciding how much effort to place when and where. Do a careful study with a tape-recorder to find exactly what rules students seem to follow. Work out a set of corresponding formal rules, and describe a possible expert system for duplicating the student behavior. How accurate would your system be?

APPENDIX A

BIBLIOGRAPHY

Chapter 1

Baumgartner, K. M., and B. W. Wah (1991). "Computer Scheduling Algorithms: Past, Present, and Future," *Information Sciences* 57–58, 319–345.

Morton, T. E., S. Kekre, and T. Smunt (1990). "Predicting the Master Schedule from Partially Known Demand," *International Journal of Applied Forecasting* 6, 115–125.

Chapter 2

Adams, J., E. Balas, and D. Zawack (1988). "The Shifting Bottleneck Procedure for Job Shop Scheduling," *Management Science* 34, 391–401.

Anthony, M. T., and J. Schaffer (1990). *Computers, Chess and Cognition*. New York: Springer.

Baker, C. T., and B. P. Dzielinski (1960). "Simulation of a Simplified Job Shop," *Management Science* 6, 311–323.

Baker, K. R. (1974). *Introduction to Sequencing and Scheduling*. New York: Wiley.

Baker, K. R., and A. G. Merten (1974). "Scheduling with Parallel Processors and Linear Delay Costs," *Naval Research Logistics Quarterly* 20, 793–804.

Baker, K. R., and Z. Su (1974). "Sequencing with Due-Dates and Early Start Times to Minimize Maximum Tardiness," *Naval Research Logistics Quarterly* 21, 171–176.

Balas, E. (1965). "An Additive Algorithm for Solving Linear Programs with Zero–One Variables," *Operations Research* 13, 517–546.

Balas, E. (1967). "Discrete Programming by the Filter Method," *Operations Research* 15, 915–957.

Balas, E. (1969). "Machine Sequencing via Disjunctive Graphs: An Implicit Enumeration Algorithm," *Operations Research* 17, 1–10.

Balas, E. (1970). "Machine Sequencing: Disjunctive Graphs and Degree-Constrained Subgraphs," *Naval Research Logistics Quarterly* 17, 941–957.

Balas, E., and A. Vazacopoulos (work in process). GSIA, Carnegie Mellon University, Pittsburgh, PA.

Barret, R. T., and S. Barman (1986). "A SLAM II Simulation of a Simplified Flow Shop," *Simulation* 47, 181–189.

Baumgartner, K. M., and B. W. Wah (1991). "Computer Scheduling Algorithms: Past, Present, and Future," *Information Sciences* 57–58, 319–345.

Bensana, E., G. Bell, and D. Dubois (1988). "OPAL: A Multi-Knowledge-Based System for Industrial Job-Shop Scheduling," *International Journal of Production Research* 26, 795–819.

Brown, A., and Z. A. Lomnicki (1966). "Some Applications of the Branch-and-Bound Algorithm to the Machine Sequencing Problem," *Operational Research Quarterly* 17, 173–186.

Buxey, G. (1989). "Production Scheduling: Practice and Theory," *European Journal of Operational Research* 39, 17–31.

Byrd, T. A., and R. D. Hauser (1991). "Expert Systems in Production and Operations Management: Research Directions in Assessing Overall Impact," *International Journal of Production Research* 29, 2471–2482.

Charlton, J. M., and C. C. Death (1970). "A Generalized Machine Scheduling Algorithm," *Operational Research Quarterly* 21, 127–134.

Conway, R. W., W. L. Maxwell, and L. W. Miller (1967). *Theory of Scheduling.* Reading, MA: Addison-Wesley.

Coroyer, C., and Z. Liu (1991). "Effectiveness of Heuristics and Simulated Annealing for the Scheduling of Concurrent Tasks—An Empirical Comparison," Institute Nationale de Recherche en Informatique et en Automatique, Le Chesnay, France.

Davis, L. (1991). *Handbook of Genetic Algorithms.* New York: Van Nostrand Reinhold.

Della Croce, F., R. Tadei, and G. Volta (1992). "A Genetic Algorithm for the Job Shop Problem," D.A.I, Politecnico di Torino, Italy.

Dobson, G., U. S. Karmarkar, and J. L. Rummel (1987). "Batching to Minimize Flow Times on One Machine," *Management Science* 33, 784–799.

Dorigo, M. (1989). "Genetic Algorithms: The State of the Art and Some Research Proposals," Dipartimento di Electronica, Politecnico di Milano, Italy.

Elmaghraby, S. E. (1991). "Manufacturing Capacity and Its Measurement: A Critical Evaluation," *Computers and Operations Research* 18, 615–627.

Falkenauer, E., and S. Bouffoix (1991). "A Genetic Algorithm for Job Shop," *Proceedings of the 1991 IEEE International Conference on Robotics and Automation.*

Fisher, M. L. (1981). "The Lagrangian Relaxation Method for Solving Integer Programming Problems," *Management Science* 27, 1–18.

Fisher, M. L. (1985). "An Applications Oriented Guide to Lagrangian Relaxation," *Interfaces* 15(2), 10–21.

Fisher, M. L., and A. H. G. Rinnooy Kan (1988). "The Design, Analysis and Implementation of Heuristics," *Management Science* 34, 263–265.

Florian, M., P. Trepant, and G. McMahon (1971). "An Implicit Enumeration Algorithm for the Machine Sequencing Problem," *Management Science* 17, 782–792.

Fox, M. S., and S. F. Smith (1984). "ISIS—A Knowledge-Based System for Factory Scheduling," *Expert Systems* 1, 25–49.

Fox, R. E. (1987). "Theory of Constraints," *NAA Conference Proceedings,* 41–52.

Fry, T. D., J. F. Cox, and J. H. Blackstone, Jr. (1992). "An Analysis and Discussion of the Optimized Production Technology Software and Its Use," *Production and Operations Management* 1, 229–242.

Garey, M. R., D. S. Johnson, and R. Sethi (1976). "The Complexity of Flowshop and Jobshop Scheduling," *Mathematics of Operations Research* 1, 117–129.

Glaser, R., and M. Hottenstein (1982). "Simulation Study of a Closed-Loop Job Shop," *Journal of Operations Management* 3, 155–166.

Glover, F. (1990). "Tabu Search: A Tutorial," *Interfaces* 20(4), 74–94.

Glover, F., and M. Laguna (1989). "Target Analysis to Improve a Tabu Search Method for Machine Scheduling," Advanced Knowledge Research Group, US West Advanced Technologies, Boulder, CO.

Goldratt, E. M., and J. Cox (1984). *The Goal.* Milford, CT: North River Press.

Goldratt, E. M., and J. Cox (1986). *The Race.* Milford, CT: North River Press.

Gomory, R. E. (1965). "On the Relation Between Integer and Non-integer Solutions to Linear Programs," *Proceedings of the National Academy of Science* 53, 260–265.

Gomory, R. E. (1967). "Faces of an Integer Polyhedron," *Proceedings of the National Academy of Science* 57, 16–18.

Greenberg, H. (1968). "A Branch-and-Bound Solution to the General Scheduling Problem," *Operations Research* 16, 353–361.

Heck, H. W., and S. D. Roberts (1971). "Sequencing and Scheduling via Disjunctive Graphs," *Proceedings of the AIIE Institute National Conference.*

Holland, J. H. (1975). *Adaptation in Natural and Artificial Systems.* Ann Arbor: University of Michigan Press.

Howard, R. A. (1988). "Decision Analysis: Practice and Promise," *Management Science* 34, 679–695.

Ignall, E., and L. E. Schrage (1965). "Application of the Branch-and-Bound Technique to Some Flow Shop Scheduling Problems," *Operations Research* 13, 400–412.

Ishibuchi, H., R. Tamura, and H. Tanaka (1991). "Flow Shop Scheduling by Simulated Annealing," *Transactions of the Institute of Systems, Control and Information Engineers* 4, 111–117.

Jacobs, F. R. (1984). "OPT Uncovered: Many Production Planning and Scheduling Concepts Can Be Applied With or Without the Software," *Industrial Engineering* 16(10), 32–41.

Karmarkar U., S. Kekre, and S. Kekre (1985). "Lot-sizing in Multi-item Multi-machine Job Shops," *IIE Transactions* 17, 290–298.

Kelly, F. P. (1982). "A Remark on Search and Sequencing Problems," *Mathematics of Operations Research* 7, 154–157.

Keng, N. P., D. Y. Y. Yun, and M. Rossi (1988). "Interaction-Sensitive Planning System for Job-Shop Scheduling," *Expert Systems and Intelligent Manufacturing,* Michael D. Oliff (ed.). Amsterdam: Elsevier, 57–69.

Kirkpatrick, S., C. D. Gelatt, Jr., and M. P. Vecchi (1983). "Optimization by Simulated Annealing," *Science* 220, 671–680.

Kumar, A., and T. E. Morton (1991). "Dynamic Job Routing in a Heavily Loaded Shop," GSIA, Carnegie Mellon University, Pittsburgh, PA.

Laguna, M., J. W. Barnes, and F. Glover (1989). "Scheduling Jobs with Linear Delay Penalties and Sequence Dependent Setup Cost Using Tabu Search," Department of Mechanical Engineering, University of Texas, Austin.

Lawrence, S. R., and T. E. Morton (1989). "Resource-Constrained Multi-project Scheduling with Tardy Costs: Comparing Myopic, Bottleneck and Resource Pricing Heuristics," Working Paper #88-89-8 GSIA, Carnegie Mellon University, Pittsburgh, PA. (To appear in the *European Journal of Operational Research.*)

Lecocq, P., Q. Fang, B. Gyselynck, M. Dumollt, and P. Vandereyken (1988). "An Expert Systems Application to Increase the Flexibility and the Efficacy of Real-Time FMS Controllers," *Expert Systems and Intelligent Manufacturing*, Michael D. Oliff (ed.). Amsterdam: Elsevier, 248–263.

Lee, M. (1989). "Least-Cost Network Topology Design for New Service Using Tabu Search," *Heuristics for Combinatorial Optimization* 6, 1–18.

Lo, Z. P., and B. Bavarian (1991). "Scheduling with Neural Networks for Flexible Manufacturing Systems," *Proceedings of the 1991 IEEE International Conference on Robotics and Automation* (Cat. No. 91CH2969-4), Sacramento, CA, 9–11.

Lomnicki, Z. A. (1965). "A Branch-and-Bound Algorithm for the Exact Solution of the Three-Machine Scheduling Problem," *Operational Research Quarterly* 16, 89–100.

Lowerre, B. T. (1976). *The Harpy Speech Recognition System*. Ph.D. Thesis, Carnegie Mellon University, Pittsburgh, PA.

Lundrigan, R. (1986). "What Is This Thing They Call OPT?" *Production and Inventory Management* 27 (second quarter), 2–12.

Marquez, L., T. Hill, M. O'Conner, and W. Remus (1992). "Neural Network Models for Forecast: A Review," *Proceedings of the Hawaii International Conference on Systems Science*, IEEE.

Maxwell, W. L. (1970). "On the Generality of the Equation $L = \lambda W$," *Operations Research* 18, 172–174.

McKay, K. N., F. R. Safayeni, and J. A. Buzacott (1992). "Common Sense Realities of Planning and Scheduling in Printed Circuit Board Production," Department of Management Science, Waterloo University, Ontario, Canada.

McMahon, G. B., and P. G. Burton (1967). "Flow-Shop Scheduling with the Branch-and-Bound Method," *Operations Research* 15, 473–481.

Meleton, M. P. (1986). "OPT—Fantasy or Breakthrough?" *Production and Inventory Management* 27 (second quarter), 13–21.

Moore, J. M., and R. C. Wilson (1967). "A Review of Simulation Research in Job Shop Scheduling," *Production and Inventory Management* 8, 1–10.

Morton, T. E., and D. Pentico (1993). "Scheduling in Dynamic Flow Shops with Weighted Average Flow Time," *Proceedings, Northeast Decision Sciences Institute*, 321–323.

Morton, T. E., S. R. Lawrence, S. Rajagopolan, and S. Kekre (1986). "MRP-STAR: PATRI-ARCH's Planning Module," GSIA, Carnegie Mellon University, Pittsburgh, PA.

Morton, T. E., S. R. Lawrence, S. Rajagopolan, and S. Kekre (1988). "SCHED-STAR: A Price-Based Shop Scheduling Module," *Journal of Manufacturing and Operations Management* 1, 131–181.

Morton, T. E., and M. Singh (1988). "Implicit Costs and Prices for Resources with Busy Periods," *Journal of Manufacturing and Operations Management* 1, 305–322.

Nakano, R., and T. Yamada (1991). "Conventional Genetic Algorithm for Job Shop Problems," *Proceedings of the 4th International Conference on Genetic Algorithms,* San Diego, 474–479.

Ovacik, Irfan, and R. Uzsoy (1992). "A Shifting Bottleneck Algorithm for Scheduling Semiconductor Testing Operations," Research Memorandum No. 92-4, Purdue University, West Lafayette, IN.

Ow, P. S., and T. E. Morton (1988). "Filtered Beam Search in Scheduling," *International Journal of Production Research* 26, 297–307.

Ow, P. S., S. F. Smith, and R. Howie (1988). "A Cooperative Scheduling System," *Expert Systems and Intelligent Manufacturing,* Michael D. Oliff (ed.). Amsterdam: Elsevier, 43–56.

Pai, A. R., and K. L. McRoberts (1971). "Simulation Research in Interchangeable Part Manufacturing," *Management Science* 17, 732–743.

Potts, C. N., and L. N. Van Wassenhove (1985). "A Branch and Bound Algorithm for the Total Weighted Tardiness Problem," *Operations Research* 33, 363–377.

Prietula, M., and G. Thompson (work in process). GSIA, Carnegie Mellon University, Pittsburgh, PA.

Prietula, M. J., Wen Ling Hsu, and Peng Si Ow (1991). "A Coincident Problem Space Perspective to Scheduling Support," *Fourth International Symposium on Artifical Intelligence,* Cancun, Mexico.

Rachamadugu, R. V, and T. E. Morton (1982). "Myopic Heuristics for the Single Machine Weighted Tardiness Problem," GSIA, Carnegie Mellon University, Pittsburgh, PA.

Rubin, Steven (1978). *The Argos Image Understanding System.* Ph.D. Thesis, Carnegie Mellon University, Pittsburgh, PA.

Sathi, A., M. S. Fox, and M. Greenberg (1985). "Representation of Activity Knowledge for Project Management," *IEEE Transactions on Pattern Analysis and Machine Intelligence* 7, 531–552.

Sathi, A., T. E. Morton, and S. F. Roth (1986). "CALLISTO: An Intelligent Project Management System," *AI Magazine* 7, 34–52.

Schultz, C. R. (1989). "An Expediting Heuristic for the Shortest Processing Time Dispatching Rule," *International Journal of Production Research* 27, 31–41.

Schwimer, J. (1972). "On the *n*-Job, One-Machine, Sequence-Independent Scheduling Problem with Tardiness Penalties: A Branch-Bound Solution," *Management Science* 18, 301–313.

Shapiro, J. F. (1979). "A Survey of Lagrangian Techniques for Discrete Optimization," *Annals of Discrete Mathematics* 5, 113–138.

Shaw, M. J. P. (1988). "Knowledge-Based Scheduling in Flexible Manufacturing Systems: An Integration of Pattern-Directed Inference and Heuristic Search," *International Journal of Production Research* 26, 821–844.

Shaw, M. J. P., and A. B. Whinston (1986). "Applications of Artificial Intelligence to Planning and Scheduling in Flexible Manufacturing," *Flexible Manufacturing Systems: Methods and Studies,* A. Kusiak (ed.). Amsterdam: Elsevier, 223–242.

Shaw, M. J., S. Park, and N. Raman (1992). "Intelligent Scheduling with Machine Learning Capabilities: The Induction of Scheduling Knowledge," *IIE Transactions* 24, 156–168.

Singh, M., and T. E. Morton (1989). "Non-Preemptive Scheduling to Minimize the Maximum Lateness Based Costs on a Bottleneck Machine with Dynamic Arrivals and Precedence Constraints," GSIA, Carnegie Mellon University, Pittsburgh, PA.

Smith, G. F. (1988). "Towards a Heuristic Theory of Problem Structuring," *Management Science* 34, 1489–1505.

Smith, S. F. (1987). "A Constraint-Based Framework for Reactive Management of Factory Schedules," *Intelligent Manufacturing: Proceedings of the International Conference on Expert Systems and Leading Edge in Production Planning and Control,* Charleston, SC, M. D. Oliff (ed.). Menlo Park, CA: Benjamin–Cummings Publishing.

Smith, S. F., M. S. Fox, and P. S. Ow (1986). "Constructing and Maintaining Detailed Production Plans: Investigations into the Development of Knowledge-Based Factory Scheduling Systems," *AI Magazine* 7, 45–61.

Solberg, J. J. (1989). "Production Planning and Scheduling in CIM," *Information Processing* 89, 919–925.

Srinivasan, V. (1971). "A Hybrid Algorithm for the One-Machine Sequencing Problem to Minimize Total Tardiness," *Naval Research Logistics Quarterly* 18, 317–327.

Vaithyanathan, S., and J. P. Ignizio (1992). "A Stochastic Neural Network for Resource Constrained Scheduling," *Computers and Operations Research* 19, 241–254.

Vakharia, A. J., and Y. L. Chang (1990). "A Simulated Annealing Approach to Scheduling a Manufacturing Cell," *Naval Research Logistics* 37, 559–77.

Van De Velde, S. L. (1991). *Machine Scheduling and Lagrangian Relaxation.* Ph.D. Thesis, Technische Universiteit Eindhoven.

Van Dyke Parunak, H. (1991). "Characterizing the Manufacturing Scheduling Problem," *Journal of Manufacturing Systems* 10, 241–259.

Van Laarhoven, P. J. M., E. H. L. Aarts, and J. K. Lenstra (1992). "Job Shop Scheduling by Simulated Annealing," *Operations Research* 40, 113–125.

Vepsalainen, A., and T. E. Morton (1987). "Priority Rules and Leadtime Estimation for Job Shop Scheduling with Weighted Tardiness Costs," *Management Science* 33, 1036–1074.

Vepsalainen, A., and T. E. Morton (1988). "Improving Local Priority Rules with Global Leadtime Estimates," *Journal of Manufacturing and Operations Management* 1, 102–118.

Vollmann, T. E. (1986). "OPT as an Enhancement to MRP II," *Production and Inventory Management* 27 (second quarter) 38–47.

Widmer, M. (1989). "Job Shop Scheduling with Tooling Constraints: A Tabu Search Approach," Department of Mathematics, Ecole Polytechnique Federale de Lausanne.

Widmer, M., and A. Hertz (1989). "Tabu Search Techniques: A Tutorial and an Application to Neural Networks," *OR Spectrum* 11, 131–141.

Wilkerson, L. J., and J. D. Irwin (1971). "An Improved Algorithm for Scheduling Independent Tasks," *AIIE Transactions* 3, 239–245.

Zhou, D. N., V. Cherkassky, T. R. Baldwin, and D. W. Hong (1990). "Scaling Neural Network for Job Shop Scheduling," *IJCNN International Joint Conference on Neural Networks* (Cat. No. 90CH2879-5), San Diego, CA, 17–21.

Zhou, D. N., V. Cherkassky, T. R. Baldwin, and T. R. Olson. (1991). "A Neural Network Approach to Job Shop Scheduling," *IEEE Transactions on Neural Networks* 2, 175–179.

Chapter 3

Baker, K. R. (1974). *Introduction to Sequencing and Scheduling*. New York: Wiley.

Bector, C. R., Y. P. Gupta, and M. C. Gupta (1990). "Minimizing the Flow-time Variance in Single-machine Systems," *Journal of the Operational Research Society* 41, 767–779.

Buxey, G. (1989). "Production Scheduling: Practice and Theory," *European Journal of Operational Research* 39, 17–31.

Cheng, T. C. E. (1991). "An Improved Solution Procedure for the $n/1/\max\{(Ci)\} \to Ci$ Scheduling Problem," *Journal of the Operational Research Society* 42, 413–417.

Conway, R. W., W. L. Maxwell, and L. W. Miller (1967). *Theory of Scheduling*. Reading, MA: Addison-Wesley.

French, Simon (1982). *Sequencing and Scheduling: An Introduction to the Mathematics of the Job Shop*. New York: Wiley.

Fry, T. D., and J. H. Blackstone (1988). "Planning for Idle time: A Rationale for Underutilization of Capacity," *International Journal of Production Research* 26, 1853–1859.

Johnson, L. A., and D. C. Montgomery (1974). *Operations Research in Production Planning, Scheduling and Inventory Control*. New York: Wiley.

Lawler, E. L., J. K. Lenstra, A. H. G. Rinnooy Kan, and D. B. Shmoys (1985). *The Traveling Salesman Problem*. New York: Wiley.

Liao, C., and R. Huang (1991). "An Algorithm for Minimizing the Range of Lateness on a Single Machine," *Journal of the Operational Research Society* 42, 183–186.

Chapter 4

Baker, K. R., and J. B. Martin (1974). "An Experimental Comparison of Solution Algorithms for the Single-Machine Tardiness Problem," *Naval Research Logistics Quarterly* 21, 187–199.

Bector, C. R., Y. P. Gupta, and M. C. Gupta (1989). "V-Shape Property of Optimal Sequence of Jobs About a Common Due Date on a Single Machine," *Computers and Operations Research* 16, 583–588.

Cheng, T. C. E. (1990). "Common Due-Date Assignment and Scheduling for a Single Processor to Minimize the Number of Tardy Jobs," *Engineering Optimization* 16, 129–136.

Cheng, T. C. E. (1991). "Optimal Assignment of Total-Work-Content Due-Dates and Sequencing in a Single-machine Shop," *Journal of the Operational Research Society* 42, 177–181.

Emmons, H. (1969). "One-Machine Sequencing to Minimize Certain Functions of Job Tardiness," *Operations Research* 17, 701–715.

Gupta, S. K., and J. Kyparisis (1987). "Single Machine Scheduling Research," *Omega* 15, 207–227.

Heller, J. (1960). "Some Numerical Experiments for an $M * J$ Flow Shop and Its Decision-Theoretical Aspects," *Operations Research* 8, 178–184.

Kiran, A. S., and A. T. Unal (1991). "A Single-Machine Problem with Multiple Criteria," *Naval Research Logistics* 38, 721–727.

Maxwell, W. L. (1970). "On Sequencing n Jobs on One Machine to Minimize the Number of Late Jobs," *Management Science* 16, 295–297.

Moore, J. M. (1968). "Sequencing n Jobs on One Machine to Minimize the Number of Tardy Jobs," *Management Science* 15, 102–109.

Rau, J. G. (1973). "Selected Comments Concerning Optimization Theory for Functions of Permutations," *Symposium on the Theory of Scheduling and Its Applications,* S. E. Elmaghraby (ed.). New York: Springer-Verlag, 167–200.

Sen, T., and S. K. Gupta (1984). "A State-of-Art Survey of Static Scheduling Research Involving Due Dates," *Omega* 12, 63–76.

Sen, T., F. M. E. Raiszadeh, and P. Dileepan (1988). "A Branch-and-Bound Approach to the Bicriterion Scheduling Problem Involving Total Flowtime and Range of Lateness," *Management Science* 34, 254–260.

"Sequencing and Scheduling: Algorithms and Complexity," *Handbooks in Operations Research and Management Science, Vol. 4: Logistics of Production and Inventory,* S. C. Graves, A. H. G. Rinnooy Kan, and P. Zipkin (eds.). New York: North-Holland.

Sidney, J. B. (1972). "A Comment on a Paper of Maxwell," *Management Science* 18, 716–717.

Smith, W. E. (1956). "Various Optimizations for Single State Production," *Naval Research Logistics Quarterly* 3, 59–66.

Sturm, L. J. M. (1970). "A Simple Optimality Proof of Moore's Sequencing Algorithm," *Management Science* 17, 116–118.

Vig, M. M., and K. J. Dooley (1991). "Dynamic Rules for Due-Date Assignment," *International Journal of Production Research* 29, 1361–1377.

Wilkerson, L. J., and J. D. Irwin (1971). "An Improved Algorithm for Scheduling Independent Tasks," *AIIE Transactions* 3, 239–245.

Woeginger, G. (1991). "On Minimizing the Sum of k Tardiness," *Information Processing Letters* 38, 253–256.

Chapter 5

Agin, N. (1966). "Optimum Seeking with Branch and Bound," *Management Science* 13, B176–B185.

Bagchi, U., Y. L. Chang, and R. S. Sullivan (1987). "Minimizing Absolute and Squared Deviations for Completion Times with Different Earliness and Tardiness Penalties and a Uniform Due Date," *Naval Research Logistics Quarterly* 34, 739–751.

Baker, K. R. (1974). *Introduction to Sequencing and Scheduling.* New York: Wiley.

Baker, K. R., and J. B. Martin (1974). "An Experimental Comparison of Solution Algorithms to the Single Machine Tardiness Problem," *Naval Research Logistics Quarterly* 21, 187–199.

Baker, K. R., and A. G. Merten (1973). "Scheduling with Parallel Processors and Linear Delay Costs," *Naval Research Logistics Quarterly* 20, 793–804.

Baker, K. R., and Z. Su (1974). "Sequencing with Due-Dates and Early Start Times to Minimize Maximum Tardiness," *Naval Research Logistics Quarterly* 21, 171–176.

Browne, S., and U. Yechiali (1990). "Scheduling Deteriorating Jobs on a Single Processor," *Operations Research* 38, 495–498.

Elmaghraby, S. E. (1967). "On the Expected Duration of PERT Type Networks," *Management Science* 13, 299–306.

Elmaghraby, S. E. (1971). "A Graph-Theoretic Interpretation of the Sufficiency Conditions for the Contiguous-Binary-Switching (CBS) Rule," *Naval Research Logistics Quarterly* 18, 339–344.

Gupta, S. K., A. S. Kunnathur, and K. Dandapani (1987). "Optimal Repayment Policies for Multiple Loans," *Omega* 15, 323–330.

Heller, J. (1960). "Some Numerical Experiments for an $M * J$ Flow Shop and Its Decision-Theoretical Aspects," *Operations Research* 8, 178–184.

Lawler, E. L. (1979). "Efficient Implementation of Dynamic Programming Algorithms for Sequencing Problems," MC Report No., BW106/79, Amsterdam.

Lawler, E. L., and D. E. Wood (1966). "Branch and Bound Methods: A Survey," *Operations Research* 14, 699–719.

Mitten, L. G. (1970). "Branch and Bound Methods: General Formulation and Properties," *Operations Research* 18, 24–34.

Moore, J. M. (1968). "Sequencing n jobs on One Machine to Minimize the Number of Tardy Jobs," *Management Science* 15, 102–109.

Morton, T. E., and P. Ramnath (1992). "Guided Forward Search in Tardiness Scheduling of Large One Machine Problems," GSIA, Carnegie Mellon University, Pittsburgh, PA.

Rachamadugu, R. V., and T. E. Morton (1982). "Myopic Heuristics for the Single Machine Weighted Tardiness Problem," Working Paper 30-82-83, GSIA, Carnegie Mellon University, Pittsburgh, PA.

Randolph, P. H., G. H. Swinson, and C. Ellingsen (1973). "Stopping Rules for Sequencing Problems," *Operations Research* 21, 1309–1315.

Rau, J. G. (1971). "Minimizing a Function of Permutations of n Integers," *Operations Research* 19, 237–240.

Rau, J. G. (1973). "Selected Comments Concerning Optimization Theory for Functions of Permutations," *Symposium on the Theory of Scheduling and Its Applications,* S. E. Elmaghraby (ed.). New York: Springer-Verlag.

Reiter, S., and G. Sherman (1965). "Discrete Optimization," *SIAM Journal* 13, 864–889.

Schwimer, J. (1972). "On the n-Job, One-Machine, Sequence-Independent Scheduling Problem with Tardiness Penalties: A Branch and Bound Solution," *Management Science* 18, B301–B313.

Sen, T., and B. Borah (1991). "On the Single-Machine Scheduling Problem with Tardiness Penalties," *Journal of the Operational Research Society* 42, 695–702.

Srinivasan, V. (1971). "A Hybrid Algorithm for the One-Machine Sequencing Problem to Minimize Total Tardiness," *Naval Research Logistics Quarterly* 18, 317–327.

Sundararaghavan, P. S., and M. U. Ahmed (1984). "Minimizing the Sum of Absolute Lateness in Single-Machine and Multi-Machine Scheduling," *Naval Research Logistics Quarterly* 31, 325–333.

Chapter 6

Aarts, E. H. L., and P. J. M. Van Laarhoven (1985a). "A New Polynomial Time Cooling Schedule," *Proceedings of the IEEE International Conference on Computer-Aided Design,* Santa Clara, CA, 206–208.

Aarts, E. H. L., and P. J. M. Van Laarhoven (1985b). "Statistical Cooling: A General Approach to Combinatorial Optimization Problems," *Philips Journal of Research* 40, 193–226.

Carroll, D. C. (1965). "Heuristic Sequencing of Single and Multiple Component Jobs," Sloan School of Management, M.I.T., Cambridge, MA.

Cerny, V. (1985). "Thermodynamical Approach to the Traveling Salesman Problem: An Efficient Simulation Algorithm," *Journal of Optimal Theory Applications* 45, 41–51.

Coroyer, C., and Z. Liu (1991). "Effectiveness of Heuristics and Simulated Annealing for the Scheduling of Concurrent Tasks: An Empirical Comparison," Institute Nationale de Recherche en Informatique et en Automatique, Le Chesnay, France.

DARPA Neural Network Study (1987–88). Fairfax, VA: AFCEA International.

Dell'Amico, M., and M. Trubian (1991). "Applying Tabu-Search to the Job-Shop Scheduling Problem," Politecnico di Milano, Italy.

Della Croce, F., R. Tadei, and G. Volta (1992). "A Genetic Algorithm for the Job Shop Problem," D.A.I., Politecnico di Torino, Italy.

Dobson, G., and U. S. Karmarkar (1989). "Simultaneous Resource Scheduling to Minimize Weighted Flow Times," *Operations Research* 37, 592–600.

Eglese, R. W. (1990). "Simulated Annealing: A Tool for Operational Research," *European Journal of Operational Research* 46, 271–281.

Erlenkotter, D. (1978). "A Dual-Based Procedure for Uncapacitated Facility Location," *Operations Research* 26, 992–1009.

Fisher, H., and G. L. Thompson (1963). "Probabilistic Learning Combinations of Local Job-Shop Scheduling Rules," *Industrial Scheduling,* J. F. Muth and G. L. Thompson (eds.). Englewood Cliffs, NJ: Prentice-Hall, 225–251.

Fisher, M. L. (1973). "Optimal Solution of Scheduling Problems Using Lagrange Multipliers: Part I," *Operations Research* 21, 1114–1121.

Fisher, M. L. (1981). "The Lagrangian Relaxation Method for Solving Integer Programming Problems," *Management Science* 27, 1–18.

Fisher, M. L. (1985). "An Applications Oriented Guide to Lagrangian Relaxation," *Interfaces* 15(2), 10–21.

Fisher, M. L, A. J. Greenfield, R. Jaikumar, and J. T. Lester (1982). "A Computerized Vehicle Routing Application," *Interfaces* 12, 42–52.

Geoffrion, A. M. (1974). "Lagrangian Relaxation and Its Uses in Integer Programming," *Mathematical Programming Study* 2, 82–114.

Glover, F. (1990). "Tabu Search: A Tutorial," *Interfaces* 20(4), 74–94.

Johnson, D. S., C. R. Aragon, L. A. McGeoch, and C. Schevon (1989). "Optimization by Simulated Annealing: An Experimental Evaluation (Part I)," *Operations Research* 37, 865–892.

Laarhoven, P. J. M., E. H. L. Aarts, and J. K. Lenstra (1992). "Job Shop Scheduling by Simulated Annealing," *Operations Research* 40, 113–124.

Lo, Z. P., and B. Bavarian (1991). "Scheduling with Neural Networks for Flexible Manufacturing Systems," *Proceedings of the 1991 IEEE International Conference on Robotics and Automation* (Cat. No. 91CH2969-4), Sacramento, CA, 9–11.

Lundy, M., and A. Mees (1986). "Convergence of an Annealing Algorithm," *Mathematical Programming* 34, 111–124.

Marquez, L., T. Hill, M. O'Conner, and W. Remus (1992). "Neural Network Models for Forecast: A Review," *Proceedings of the Hawaii International Conference on Systems Science,* IEEE.

Matsuo, H., C. J. Suh, and R. S. Sullivan (1988). "A Controlled Search Simulated Annealing Method for the General Jobshop Scheduling Problem," Working Paper 03-04-88, Department of Management, The University of Texas, Austin.

Morton, T. E. (1981). "Forward Algorithms for Forward Thinking Managers," *Applications of Management Science* I, 1–55.

Morton, T. E. and P. Ramnath (1992). "Guided Forward Search in Tardiness Scheduling of Large One Machine Problems," GSIA, Carnegie Mellon University.

Ow, P. S., and T. Morton (1988). "Filtered Beam Search in Scheduling," *International Journal of Production Research* 26, 35–62.

Ow, P. S., and T. E. Morton (1989). "The Single Machine Early/Tardy Problem," *Management Science* 35, 177–191.

Panwalker, S. S., M. L. Smith, and A. Seidmann (1982). "Common Due-Date Assignment to Minimize Total Penalty for the One-Machine Scheduling Problem," *Operations Research* 30, 391–399.

Potts, C. N., and L. N. Van Wassenhove (1985). "A Branch and Bound Algorithm for the Total Weighted Tardiness Problem," *Operations Research* 33, 363–377.

Rachamadugu, R. V., and T. E. Morton (1982). "Myopic Heuristics for the Single Machine Weighted Tardiness Problem," Working Paper 30-82-83, GSIA, Carnegie Mellon University, Pittsburgh, PA.

Ragavachari, M. (1986). "A V-shape Property of Optimal Schedule of Jobs About a Common Due Date," *European Journal of Operational Research* 23, 401–402.

Raghavachari, M. (1988). "Scheduling Problems with Non-Regular Penalty Functions—A Review," *Opsearch* 25, 144–164.

Romeo, F., and A. L. Sangiovanni-Vincentelli (1985). "Probabilistic Hill Climbing Algorithms: Properties and Applications," *Proceedings of the 1985 Chapel Hill Conference on VLSI,* Chapel Hill, NC, 393–417.

Seidmann, A., S. S. Panwalker, and M. L. Smith (1981). "Optimal Assignment of Due-Date for a Single Processor Scheduling Problem," *International Journal of Production Research* 19, 393–399.

Sundararaghavan, P. S., and M. U. Ahmed (1984). "Minimizing the Sum of Absolute Lateness in Single-Machine and Multi-Machine Scheduling," *Naval Research Logistics Quarterly* 31, 325–333.

Szwarc, W. (1988). "Minimizing Absolute Lateness in Single Machine Scheduling with Different Due Dates," Working Paper, University of Wisconsin, Milwaukee.

Szwarc, W. (1989). "Single-Machine Scheduling to Minimize Absolute Deviation of Completion Times from a Common Due Date," *Naval Research Logistics Quarterly* 36, 663–673.

Vaithyanathan, S., and J. P. Ignizio (1992). "A Stochastic Neural Network for Resource Constrained Scheduling," *Computers and Operations Research* 19, 241–254.

Van Laarhoven, P. J. M. (1988). *Theoretical and Computational Aspects of Simulated Annealing*. Ph.D. Thesis, Erasmus University, Rotterdam.

Van Laarhoven, P. J. M., and E. H. L. Aarts (1987). *Simulated Annealing: Theory and Applications*. Dordrecht, The Netherlands: Reidel.

Van Laarhoven, P. J. M., E. H. L. Aarts, and J. K. Lenstra (1992). "Job Shop Scheduling by Simulated Annealing," *Operations Research* 40, 112–129.

Vani, V., and M. Ragahvachari (1987). "Deterministic and Random Single Machine Sequencing with Variance Minimization," *Operations Research* 35, 111–120.

Weiss, S. M., and I. Kapouleas (1989). "An Empirical Comparison of Pattern Recognition, Neural Nets and Machine Learning Classification Methods," *Proceedings of the 11th International Joint Conference on AI—IJCAI '89,* Detroit.

Yano, C., and Y. Kim (1986). "Algorithms for Single Machine Scheduling Problems Minimizing Tardiness and Earliness," Technical Report #86-40, Department of Industrial Engineering, University of Michigan, Ann Arbor.

Zhou, D. N., V. Cherkassky, T. R. Baldwin, and D. W. Hong (1990). "Scaling Neural Network for Job Shop Scheduling," *IJCNN International Joint Conference on Neural Networks* (Cat. No. 90CH2879-5), San Diego, CA, 17–21.

Zhou, D. N., V. Cherkassky, T. R. Baldwin, and T. R. Olson (1991). "A Neural Network Approach to Job Shop Scheduling," *IEEE Transactions on Neural Networks* 2, 175–179.

Chapter 7

Adams, J., E. Balas, and D. Zawack (1988). "The Shifting Bottleneck Procedure for Job Shop Scheduling," *Management Science* 34, 391–401.

Adiri, I., E. Frostig, and A. H. G. Rinnooy Kan (1991). "Scheduling on a Single Machine with a Single Breakdown to Minimize Stochastically the Number of Tardy Jobs," *Naval Research Logistics* 38, 261–271.

Ahmadi, R., and U. Bagchi (1986a). "Single Machine Scheduling to Minimize Earliness Subject to Deadlines," Working Paper 85/86-4-17, Department of Management, University of Texas, Austin.

Ahmadi, R., and U. Bagchi (1986b). "Just-in-Time Scheduling in Single Machine Systems," Working Paper 85/86-4-21, Department of Management, University of Texas, Austin.

Ahmed, M. U., and P. S. Sundararaghavan (1990). "Minimizing the Sum of Absolute Deviation of Job Completion Times from Their Due Dates in a Single Machine Scheduling Problem," *IIE Transactions* 22, 288–290.

Alidaee, B. (1990). "A Heuristic Solution Procedure to Minimize Makespan on a Single Machine with Non-linear Cost Functions," *Journal of the Operational Research Society* 41, 1065–1068.

Alidaee, B. (1991). "Single Machine Scheduling with Nonlinear Cost Functions," *Computers and Operations Research* 18, 317–322.

Bagchi, U. (1987a). "Due Date or Deadline Assignment to Multi-Job Orders to Minimize Total Penalty in One Machine Scheduling Problem," Presented at the ORSA/TIMS Joint National Conference, St. Louis.

Bagchi, U. (1987b). "Scheduling to Minimize Earliness and Tardiness Penalties with a Common Due Date," Working Paper, Department of Management, University of Texas, Austin.

Bagchi, U. (1989). "Simultaneous Minimization of Mean and Variation of Flow Time and Waiting Time in Single Machine Systems," *Operations Research* 37, 118–125.

Bagchi, U., Y. L. Chang, and R. S. Sullivan (1987). "Minimizing Absolute and Squared Deviations for Completion Times with Different Earliness and Tardiness Penalties and a Common Due Date," *Naval Research Logistics Quarterly* 34, 739–751.

Bagchi, U., R. S. Sullivan, and Y. L. Chang (1986). "Minimizing Mean Absolute Deviation of Completion Times About a Common Due Date," *Naval Research Logistics Quarterly* 33, 227–240.

Bagchi, U., R. S. Sullivan, and Y. L. Chang (1987). "Minimizing Mean Squared Deviation of Completion Times About a Common Due Date," *Management Science* 33, 894–906.

Bagga, P. C., and N. K. Chakravati (1968). "Optimal *m*-Stage Production Schedule," *Canadian Operations Research Journal* 6, 71–78.

Baker, K. R. (1974). *Introduction to Sequencing and Scheduling.* New York: Wiley.

Baker, K., and A. Chadowitz (1989). "Algorithms for Minimizing Earliness and Tardiness Penalties with a Common Due Date," Working Paper No. 240, Amos Tuck School of Business Administration, Dartmouth College, Hanover, NH.

Baker, K. R., and G. D. Scudder (1990). "Sequencing with Earliness and Tardiness Penalties: A Review," *Operations Research* 38, 22–36.

Baker, K. R., and Z. Su (1974). "Sequencing with Due-Dates and Early Start Times to Minimize Maximum Tardiness," *Naval Research Logistics Quarterly* 21, 171–176.

Banerjee, B. P. (1965). "Single Facility Sequencing with Random Execution Times," *Operations Research* 13, 358–364.

Bector, C. R., Y. P. Gupta, and M. C. Gupta (1990). "Minimizing the Flow-time Variance in Single-machine Systems," *Journal of the Operational Research Society* 41, 767–779.

Birge, J., J. B. G. Frenk, J. Mittenthal, and A. H. G. Rinnooy Kan (1990). "Single-Machine Scheduling Subject to Stochastic Breakdowns," *Naval Research Logistics Quarterly* 37, 661–677.

Brown, A., and Z. Lomnicki (1966). "Some Applications of the Branch and Bound Algorithm to the Machine Scheduling Problem," *Operational Research Quarterly* 17, 173–186.

Browne, S., and U. Yechiali (1990). "Scheduling Deteriorating Jobs on a Single Processor, *Operations Research* 38, 495–498.

Burns, R. N. (1976). "Scheduling to Minimize the Weighted Sum of Completion Times with Secondary Criteria," *Naval Research Logistics Quarterly* 23, 125–129.

Campbell, H. G., R. A. Dudek, and M. L. Smith (1970). "A Heuristic Algorithm for the *n*-Job *m*-Machine Sequencing Problem," *Management Science* 16, B630–B637.

Carlier, J. (1982). "The One Machine Sequencing Problem," *European Journal of Operational Research* 11, 42–47.

Carlier, J., and E. Pinson (1989). "An Algorithm for Solving the Job Shop Problem, *Management Science* 35, 164–176.

Carroll, D. C. (1965). *Heuristic Sequencing of Single and Multiple Component Jobs.* Ph.D. Thesis, Sloan School of Management, M.I.T., Cambridge, MA.

Chambers, R. J., R. L. Carraway, T. J. Lowe, and T. L. Morin (1991). "Dominance and Decomposition Heuristics for Single Machine Scheduling," *Operations Research* 39, 639–647.

Chand, S., and H. Schneeberger (1986). "A Note on the Single-Machine Scheduling Problem with Weighted Completion Time and Maximum Allowable Tardiness," *Naval Research Logistics Quarterly* 33, 551–557.

Chand, S., and H. Schneeberger (1988). "Single Machine Scheduling to Minimize Weighted Earliness Subject to No Tardy Jobs," *European Journal of Operational Research* 34, 221–230.

Chen, C., and R. L. Bulfin (1990). "Scheduling Unit Processing Time Jobs on a Single Machine with Multiple Criteria," *Computers and Operations Research* 17, 1–7.

Cheng, T. C. E. (1990). "A Note on a Partial Search Algorithm for the Single-Machine Optimal Common Due-date Assignment and Sequencing Problem," *Computers and Operations Research* 17, 321–324.

Cheng, T. C. E., and H. G. Kahlbacher (1991). "A Proof for the Longest-Job-First Policy in One-Machine Scheduling," *Naval Research Logistics* 38, 715–720.

Chu, C., and M.-C. Portmann (1992). "Some New Efficient Methods to Solve the $n/1/r_i/\Sigma T_i$ Scheduling Problem," *European Journal of Operational Research* 58, 404–413.

Cobham, A. (1954). "Priority Assignment in Waiting Line Problems," *Operations Research* 2, 70–76.

Coffman, E. G., A. Nozari, and M. Yannakis (1989). "Optimal Scheduling of Products with Two Subassemblies on a Single Machine," *Operations Research* 37, 426–436.

Crabill, T. B., and W. L. Maxwell (1969). "Single Machine Sequencing with Random Processing Times and Random Due-Dates," *Naval Research Logistics Quarterly* 16, 549–554.

Davis, J., and J. Kanet (1988). "Single Machine Scheduling with a Nonregular Convex Performance Measure," Working Paper, Department of Management, Clemson University, Clemson, SC.

De, P., J. B. Ghosh, and C. E. Wells (1989). "A Note on the Minimization of Mean Squared Deviation of Completion Times About a Common Due Date," *Management Science* 35, 1143–1147.

De, P., J. B. Ghosh, and C. E. Wells (1990). "Scheduling About a Common Due Date with Earliness and Tardiness Penalties," *Computers and Operations Research* 17, 231–241.

De, P., J. B. Ghosh, and C. E. Wells (1991a). "On the Minimization of the Weighted Number of Tardy Jobs with Random Processing Times and Deadline," *Computers and Operations Research* 18, 457–463.

De, P., J. B. Ghosh, and C. E. Wells (1991b). "Scheduling to Minimize Weighted Earliness and Tardiness About a Common Due-Date," *Computers and Operations Research* 18, 465–475.

Della Croce, F., R. Tadei, and G. Volta (1992). "A Genetic Algorithm for the Job Shop Problem," D.A.I., Politecnico di Torino, Italy.

Dileepan, P., and T. Sen (1988). "Bicriterion Static Scheduling Research for a Single Machine," *Omega* 16, 53–59.

Eilon, S., and I. Chowdhury (1977). "Minimizing Waiting Time Variance in the Single Machine Problem," *Management Science* 23, 567–575.

Emmons, H. (1969). "One-Machine Sequencing to Minimize Certain Functions of Job Tardiness," *Operations Research* 17, 701–715.

Emmons, H. (1975a). "One Machine Sequencing to Minimize Mean Flow Time with Minimum Number Tardy," *Naval Research Logistics Quarterly* 22, 585–592.

Emmons, H. (1975b). "A Note on a Scheduling Problem with Dual Criteria," *Naval Research Logistics Quarterly* 22, 615–616.

Emmons, H. (1987). "Scheduling to a Common Due Date on Parallel Common Processors," *Naval Research Logistics Quarterly* 34, 803–810.

Fathi, Y., and H. W. L. Nuttle (1990). "Heuristics for the Common Due Date Weighted Tardiness Problem," *IIE Transactions* 22, 215–225.

Fife, D. W. (1965). "Scheduling with Random Arrivals and Linear Loss Functions," *Management Science* 11, 429–437.

Fisher, M. L. (1976). "A Dual Algorithm for the One-Machine Scheduling Problem," *Mathematical Programming* 11, 229–251.

Fry, T., and G. K. Leong (1987). "A Bi-criterion Approach to Minimizing Inventory Costs on a Single Machine When Early Shipments Are Forbidden," *Computers and Operations Research* 14, 363–368.

Fry, T., G. Leong, and T. Rakes (1987). "Single Machine Scheduling: A Comparison of Two Solution Procedures," *Omega* 15, 277–282.

Fry, T. D., R. D. Armstrong, and J. H. Blackstone (1987). "Minimizing Weighted Absolute Deviation in Single Machine Scheduling," *IEEE Transactions* 19, 445–449.

Fry, T., K. Darby-Dowman, and R. Armstrong (1988). "Single Machine Scheduling to Minimize Mean Absolute Lateness," Working Paper, College of Business Administration, University of South Carolina, Columbia.

Fry, T. D., R. D. Armstrong, and H. Lewis (1989). "A Framework for Single Machine Multiple Objective Sequencing Research," *Omega* 17, 595–607.

Fry, T. D., R. D. Armstrong, and L. D. Rosen (1990). "Single Machine Scheduling to Minimize Mean Absolute Lateness: A Heuristic Solution," *Computers and Operations Research* 17, 105–112.

Fry, T. D., and J. H. Blackstone (1988). "Planning for Idle Time: A Rationale for Underutilization of Capacity," *International Journal of Production Research* 26, 1853–1859.

Gapp, W., D. S. Mankekar, and L. G. Mitten (1965). "Sequencing Operations to Minimize In-Process Inventory Costs," *Management Science* 11, 476–484.

Garey, M., R. Trajan, and G. Wilfong (1988). "One-Processor Scheduling with Symmetric Earliness and Tardiness Penalties," *Mathematics of Operations Research* 13, 330–348.

Glazebrook, K. D. (1984). "Scheduling Stochastic Jobs on a Single Machine Subject to Breakdowns," *Naval Research Logistics Quarterly* 31, 251–264.

Glazebrook, K. D. (1987). "Evaluating the Effects of Machine Breakdowns in Stochastic Scheduling Problems," *Naval Research Logistics Quarterly* 34, 319–355.

Gupta, J. N. D. (1972). "Heuristic Algorithms for Multistage Flow Shop Problem," *AIIE Transactions* 4, 11–18.

Gupta, S. K., and T. Sen (1983). "Minimizing a Quadratic Function of Job Lateness on a Single Machine," *Engineering Costs and Production Economics* 7, 187–194.

Gupta, S. K., and T. Sen (1984). "Minimizing the Range of Lateness on a Single Machine," *Journal of Operational Research* 35, 853–587.

Gupta, S. K., and J. Kyparisis (1987). "Single Machine Scheduling Research," *Omega* 15, 207–227.

Hall, N. G. (1986). "Single- and Multiple-Processor Models for Minimizing Completion Time Variance," *Naval Research Logistics Quarterly* 33, 49–54.

Hall, N., W. Kubiak, and S. Sethi (1989). "Deviation of Completion Times About a Restrictive Common Due Date," Working Paper 89-19, College of Business, The Ohio State University, Columbus.

Hall, N. G., W. Kubiak, and S. P. Sethi (1991). "Earliness–Tardiness Scheduling Problems, II: Deviation of Completion Times About a Restrictive Common Due Date," *Operations Research* 39, 847–856.

Hall, N. G., and M. E. Posner (1989). "Weighted Deviation of Completion Times About a Common Due-Date," Working Paper, College of Business, Ohio State University, Columbus.

Hall, N. G., and M. E. Posner (1991). "Earliness–Tardiness Scheduling Problems, I: Weighted Deviation of Completion Times About a Common Due Date," *Operations Research* 39, 836–846.

Hall, N. G., S. P. Sethi, and C. Sriskandarajah (1991). "On the Complexity of Generalized Due Date Scheduling Problems," *European Journal of Operational Research* 51, 100–109.

Heck, H., and S. Roberts (1972). "A Note on the Extension of a Result on Scheduling with Secondary Criteria," *Naval Research Logistics Quarterly* 19, 403–405.

Hochbaum, D. S., and R. Shamir (1989). "An $O(n \log^2 n)$ Algorithm for the Maximum Weighted Tardiness Problem," *Information Processing Letters* 31, 215–219.

John, T. C. (1989). "Tradeoff Solutions in Single Machine Production Scheduling for Minimizing Flow-Time and Maximum Penalty," *Computers and Operations Research* 16, 471–479.

John, T. C., and Y. Wu (1987). "Minimum Number of Tardy Jobs in Single Machine Scheduling with Release Dates—An Improved Algorithm," *Computers and Industrial Engineering* 12, 223–230.

Kanet, J. J. (1981a). "Minimizing the Average Deviation of Job Completion Times About a Common Due Date," *Naval Research Logistics Quarterly* 28, 643–651.

Kanet, J. J. (1981b). "Minimizing Variation of Flow Time in Single-Machine Systems," *Management Science* 27, 1453–1459.

Lakshminarayan, S., R. Lakshminarayan, R. L. Papineau, and R. Rochette (1978). "Optimal Single-Machine Scheduling with Earliness and Tardiness Penalties," *Operations Research* 26, 1079–1082.

Lawrence, S. R. (1991). "Scheduling a Single Machine to Maximize Net Present Value," *International Journal of Production Research* 29, 1141–1160.

Lawrence, S. R., and T. E. Morton (1989). "Resource-Constrained Multi-project Scheduling with Tardy Costs: Comparing Myopic, Bottleneck and Resource Pricing Heuristics", Working Paper #88-89-8, GSIA, Carnegie Mellon University, Pittsburgh, PA. (To appear in the *European Journal of Operational Research.*)

Lee, I. (1991). "A Worst-Case Performance of the Shortest-Processing-Time Heuristic for Single Machine Scheduling," *Journal of the Operational Research Society* 42, 895–901.

Liao, C., and R. Huang (1991). "An Algorithm for Minimizing the Range of Lateness on a Single Machine," *Journal of the Operational Research Society* 42, 183–186.

Liu, J., and B. L. MacCarthy (1991). "Effective Heuristics for the Single Machine Sequencing Problem with Ready Times," *International Journal of Production Research* 29, 1521–1533.

McMahon, G. B. (1969). "Optimal Production Schedules for Flow Shops," *Canadian Operations Research Journal* 7, 141–151.

Mitten, L. G. (1959). "Sequencing *n* Jobs on Two Machines with Arbitrary Time Lags," *Management Science* 5, 293–303.

Montagne, E. R. Jr. (1969). "Sequencing with Time Delay Costs," *Industrial Engineering Research Bulletin* 5.

Morton, T. E. and D. Pentico (1993). "Scheduling in Dynamic Flow Shops with Weighted Average Flow Time," *Proceedings, Northeast Decision Sciences Institute,* 321–323.

Morton, T. E., and P. Ramnath (1992), "Guided Forward Search in Tardiness Scheduling of Large One Machine Problems, GSIA, Carnegie Mellon University, Pittsburgh, PA.

Ow, P. S., and T. E. Morton (1989). "The Single Machine Early/Tardy Problem," *Management Science* 35, 177–191.

Potts, C. N., and L. N. Van Wassenhove (1982). "A Decomposition Algorithm for the Single Machine Total Tardiness Problem," *Operations Research Letters* 1, 177–181.

Potts, C. N., and L. N. Van Wassenhove (1983). "An Algorithm for Single Machine Sequencing with Deadlines to Minimize Total Weighted Completion Time," *European Journal of Operational Research* 12, 379–383.

Potts, C. N., and L. N. Van Wassenhove (1985). "A Branch and Bound Algorithm for the Total Weighted Tardiness Problem," *Operations Research* 33, 363–377.

Potts, C. N., and L. N. Van Wassenhove (1988). "Algorithms for Scheduling a Single Machine to Minimize the Weighted Number of Late Jobs," *Management Science* 34, 843–858.

Rachamadugu, R. V., and T. E. Morton (1982). "Myopic Heuristics for the Single Machine Weighted Tardiness Problem," Working Paper 30-82-83, GSIA, Carnegie Mellon University, Pittsburgh, PA.

Rothkopf, M. H. (1966). "Scheduling with Random Service Times," *Management Science* 12, 707–713.

Sarin, S., G. Steiner, and E. Erel (1991). "Sequencing Jobs on a Single Machine with a Common Due Date and Stochastic Processing Times," *European Journal of Operational Research* 51, 188–198.

Schrage, L. E. (1968). "A Proof of the Optimality of the Shortest Remaining Processing Time Discipline," *Operations Research* 16, 687–690.

Schultz, C. R. (1989). "An Expediting Heuristic for the Shortest Processing Time Dispatching Rule," *International Journal of Production Research* 27, 31–41.

Sen, T., and S. K. Gupta (1984). "A State-of-Art Survey of Static Scheduling Research Involving Due Dates," *Omega* 12, 63–76.

Smith, W. E. (1956). "Various Optimizations for Single State Production" *Naval Research Logistics Quarterly* 3, 59–66.

Szwarc, W. (1989). "Single Machine Scheduling to Minimize Absolute Deviation of Completion Times from a Common Due Date," *Naval Research Logistics Quarterly* 36, 663–673.

Szwarc, W. (1990). "Parametric Precedence Relations in Single Machine Scheduling," *Operations Research Letters* 9, 133–140.

Szwarc, W., M. E. Posner, and J. J. Liu (1988). "The Single Machine Problem with Quadratic Cost Function of Completion Times," *Management Science* 34, 1480–1499.

Uzsoy, R., and L. A. Martin-Vega (1990). "Scheduling Semiconductor Test Operations: Minimizing Maximum Lateness and Number of Tardy Jobs on a Single Machine," Research Report No. 90-4, Department of Industrial and Systems Engineering, University of Florida, Gainesville.

Vepsalainen A., and T. E. Morton (1987). "Priority Rules and Leadtime Estimation for Job Shop Scheduling with Weighted Tardiness Costs," *Management Science* 33, 1036–1047.

Yano, C., and Y. Kim (1991). "Algorithms for Single Machine Scheduling Problems Minimizing Tardiness and Earliness," *European Journal of Operational Research* 52, 167–178.

Zdrzalka, S. (1989). "Scheduling Jobs on a Single Machine with Periodic Release Date/Deadline Intervals," *European Journal of Operational Research* 40, 243–251.

Chapter 8

Afentakis, P. (1985). "Simultaneous Lot Sizing and Sequencing for Multistage Production Systems," *IIE Transactions* 17, 327–331.

Ahmadi, J. J., R. H. Ahmadi, S. Dasu, and C. S. Tang (1989). "Batching and Scheduling Jobs on Batch and Discrete Processors," Anderson School of Management, University of California, Los Angeles.

Ahn, B., and J. Hyun (1990). "Single Facility Multi-class Job Scheduling," *Computers and Operations Research* 17, 265–272.

Baker, K. R. (1974). *Introduction to Sequencing and Scheduling.* New York: Wiley.

Baker, K. R., and A. G. Merten (1974). "Scheduling with Parallel Processors and Linear Delay Costs," *Naval Research Logistics Quarterly* 20, 793–804.

Baker, K. R., and Z. Su (1974). "Sequencing with Due-Dates and Early Start Times to Minimize Maximum Tardiness," *Naval Research Logistics Quarterly* 21, 171–176.

Bianco, L., S. Ricciardelli, G. Rinaldi, and A. Sassano (1988). "Scheduling Tasks with Sequence-Dependent Processing Times," *Naval Research Logistics Quarterly* 35, 177–184.

Deb, R. K., and R. F. Serfozo (1984). "Optimal Control of Batch Service Queues," *Advances in Applied Probability* 5, 340–361.

Ding, F. (1990). "A Pairwise Interchange Solution Procedure for a Scheduling Problem with Production of Components at a Single Facility," *Computers and Industrial Engineering* 18, 325–331.

Dobson, G., and U. S. Karmarkar (1986). "Large-Scale Shop Scheduling: Formulations and Decomposition," Graduate School of Management, University of Rochester, Rochester, NY.

Dobson, G., U. Karmarkar, and J. Rummel (1987). "Batching to Minimize Flow Times on One Machine," *Management Science* 33, 784–799.

Eilon, S. (1985). "Multi-Product Batch Production on a Single Machine—Problem Revisited," *Omega* 13, 453–468.

Elmaghraby, S. E. (1968). "The Machine Scheduling Problem—Review and Extensions," *Naval Research Logistics Quarterly* 15, 587–598.

Flood, M. M. (1955). "The Traveling Salesman Problem," *Operations Research* 4, 61–75.

Gavett, J. W. (1965). "Three Heuristic Rules for Sequencing Jobs to a Single Production Facility," *Management Science* 2, B166–B176.

Glazebrook, K. D. (1981). "On Nonpreemptive Strategies in Stochastic Scheduling," *Naval Research Logistics Quarterly* 28, 289–300.

Goyal, S. K. (1975). "Scheduling a Multi-product Single-Machine System," *Operational Research Quarterly* 24, 261–266.

Gupta, S. K., and J. Kyparisis (1987). "Single Machine Scheduling Research," *Omega* 15, 207–227.

Horn, W. A. (1972). "Single Machine Job Sequencing with Treelike Precedence Ordering and Linear Delay Penalties," *SIAM Journal of Applied Mathematics* 23, 189–202.

Jacobs, F. R., and D. J. Bragg (1988). "Repetitive Lots: Flow-Time Reductions Through Sequencing and Dynamic Batch Sizing," *Decision Sciences* 19, 281–294.

Karg, R., and G. L. Thompson (1964). "A Heuristic Approach to Solving Travelling Salesman Problems," *Management Science* 10, 225–248.

Karmarkar, U. S. (1983). "Lot Sizes, Manufacturing Lead Times and Utilization," Graduate School of Management, University of Rochester, Rochester, NY.

Karmarkar, U. S. (1987). "Lot-Sizing and Sequencing Delays," *Management Science* 33, 419–423.

Kono, H., and Z. Nakamura (1990). "The Balanced Lot Size for a Single-Machine Multiproduct Lot Scheduling Problem," *Journal of Operations Research Society of Japan* 33, 119–138.

Kunnnathur, A. S., and S. K. Gupta (1990). "Minimizing the Makespan with Late Start Penalties Added to Processing Times in a Single Facility Scheduling Problem," *European Journal of Operational Research* 47, 56–64.

Lawler, E. L. (1973). "Optimal Sequencing of a Single Machine Subject to Precedence Constraints," *Management Science* 19, 544–546.

Lawler, E., J. Lenstra, A. Rinooy Kan, and D. Shmoys (1985). *The Traveling Salesman Problem*. New York: Wiley.

Lee, C.-Y., R. Uzsoy, and L. A. Martin-Vega (1990). "Efficient Algorithms for Scheduling Batch Processing Machines," School of Industrial Engineering, Purdue University, West Lafayette, IN.

Lee, C.-Y., R. Uzsoy, and L. A. Martin-Vega (1991). "Efficient Algorithms for Scheduling Semiconductor Burn-In Operations," School of Industrial Engineering, Purdue University, West Lafayette, IN.

Little, J. D. C., K. G. Murty, D. W. Sweeny, and C. Karel (1963). "An Algorithm for the Traveling Salesman Problem," *Operations Research* 11, 972–989.

Mason, A. J., and E. J. Anderson (1991). "Minimizing Flow Time on a Single Machine with Job Classes and Setup Times," *Naval Research Logistics Quarterly* 38, 333–350.

Morton, T. E., and B. G. Dharan (1978). "Algoristics for Sequencing with Precedence Constraints," *Management Science* 24, 1011–1020.

O'Grady, P. J., and C. Harrison (1988). "Search Based Job Scheduling and Sequencing with Setup times," *Omega* 16, 547–552.

Oliff, M. D., and E. E. Burch (1985). "Multiproduct Production Scheduling at Owens–Corning Fiberglass," *Interfaces* 15(5), 25–34.

Plenert, G. (1990). "Bottleneck Scheduling for an Unlimited Number of Products," *Journal of Manufacturing Systems* 9, 324–331.

Potts, C. N. (1991). "Scheduling Two Job Classes on a Single Machine," *Computers and Operations Research* 18, 411–415.

Sidney, J. B. (1972). "One Machine Sequencing with Precedence Relations and Deferral Costs," Faculty of Commerce and Business Administration, University of British Columbia, Vancouver.

Sidney, J. B. (1975). "Decomposition Algorithms for Single-Machine Sequencing with Precedence Relations and Deferral Costs," *Operations Research* 23, 283–298.

Silver, E. A., and H. C. Meal (1973). "A Heuristic Selecting Lotsize Requirements for the Case of a Deterministic Time-Varying Demand Rate and Discrete Opportunities for Replenishment," *Production and Inventory Management* 14 (second quarter), 64–77.

Szwarc, W. (1990). "Parametric Precedence Relations in Single Machine Scheduling," *Operations Research Letters* 9, 133–140.

Woolsey, G. (1982). "An Essay on the Setup, Tear-down Problem or Being Clean, Being Profitable, or Both," *Interfaces* 12(4), 11–13.

Yano, C. A., and H. L. Lee (1989). "Lot-Sizing with Random Yields: A Review," Technical Report No. 89-16, Department of Industrial and Systems Engineering, The University of Michigan, Ann Arbor.

Zdrzalka, S. (1991). "Approximation Algorithms for Single-Machine Sequencing with Delivery Times and Unit Batch Set-Up Times," *European Journal of Operational Research* 51, 199–209.

Chapter 9

Adams J., E. Balas, and D. Zawack (1988). "The Shifting Bottleneck Procedure for Job Shop Scheduling," *Management Science* 34, 391–401.

Carlier, J. (1982). "The One Machine Sequencing Problem," *European Journal of Operational Research* 11, 42–47.

Chapter 10

Adams, J., E. Balas, and D. Zawack (1988). "The Shifting Bottleneck Procedure for Job Shop Scheduling," *Management Science* 34, 391–401.

Alidaee, B. (1990). "A Heuristic Solution Procedure to Minimize Makespan on a Single Machine with Non-linear Cost Functions," *Journal of Operational Research Society* 41, 1065–1068.

Carlier, J., and E. Pinson (1989). "An Algorithm for Solving the Job Shop Problem," *Management Science* 35, 164–176.

Daniels, R. L., and R. K. Sarin (1989). "Single Machine Scheduling with Controllable Processing Times and Number of Jobs Tardy," *Operations Research* 37, 981–989.

Fawcett, S. E., and J. N. Pearson (1991). "Understanding and Applying Constraint Management in Today's Manufacturing Environment," *Production and Inventory Management* 32, 46–55.

Glassey, C. R., and R. G. Petrakian (1989). "The Use of Bottleneck Starvation Avoidance with Queue Predictions in Shop Floor Control," Research Report ESRC 89-23, University of California, Berkeley.

Goldratt, E. M. (1990). *What's This Thing Called Theory of Constraints?* Milford, CT: North River Press.

Gupta, J. N. D., and S. K. Gupta (1988). "Single Facility Scheduling with Nonlinear Processing Times," *Computers and Industrial Engineering* 14, 387–393.

Jacobs, F. R. (1984). "OPT Uncovered: Many Production Planning and Scheduling Concepts Can Be Applied With or Without the Software," *Industrial Engineering* 16(10), 32–41.

Kunnathur, A. S., and S. K. Gupta (1990). "Minimizing the Makespan with Late Start Penalties Added to Processing Times in a Single Facility Scheduling Problem," *European Journal of Operational Research* 47, 56–64.

Lundrigan, R. (1986). "What Is This Thing Called OPT?" *Production and Inventory Management* 27 (second quarter), 2–12.

Meleton, M. P. (1986). "OPT: Fantasy or Breakthrough?" *Production and Inventory Management* 27 (second quarter), 13–21.

Morton, T. E., and D. Pentico (1993). "Scheduling in Dynamic Flow Shops with Weighted Average Flow Time," *Proceedings, Northeast Decision Sciences Institute,* 321–323.

Pence, N. E., J. D. Megeath, and J. S. Morrell (1990). "Coping with Temporary Bottlenecks in a Several-Stage Process with Multiple Products," *Production and Inventory Management* 31(3), 5–6.

Chapter 11

Alt, H., T., Hagerup, K. Mehlhorn, and F. P. Preparata. (1987). "Deterministic Simulation of Idealized Parallel Computers on More Realisitic Ones," *SIAM Journal of Computing* 16, 808–835.

Balakrishnan, A. (1989). "Preemptive Scheduling of Hybrid Parallel Machines," *Operations Research* 37, 301–313.

Brumelle, S. (1971). "Some Inequalities for Parallel-Server Queues," *Operations Research* 19, 402–413.

Cheng, T. C. E., and C. C. S. Chin (1990). "A State-of-the-Art Review of Parallel-Machine Scheduling Research," *European Journal of Operational Research* 47, 271–292.

Cho, Y., and S. Sahni (1980). "Scheduling Independent Tasks with Due Times on a Uniform Processor System," *Journal of the Association of Computing Machinery* 27, 550–563.

Coffman, E. G. Jr., G. S. Lueker, and A. H. G. Rinnooy Kan (1988). "Asymptotic Methods in the Probabilistic Analysis of Sequencing and Packing Heuristics," *Management Science* 34, 266–290.

DeBruin, A., A. H. G. Rinnooy Kan, and H. W. J. M. Trienekens (1988). "A Simulation Tool for the Performance Evaluation of Parallel Branch and Bound Algorithms," *Mathematical Programming Series B* 42, 245–271.

Eastman, W. L., S. Even, and I. M. Isaacs (1964). "Bounds for the Optimal Scheduling of n Jobs on m Processors," *Management Science* 11, 268–279.

Emmons, H. (1987). "Scheduling to a Common Due Date on Parallel Common Processors," *Naval Research Logistics Quarterly* 34, 803–810.

Epstein, S., Y. Walderman, and B. Dickman (1992). "Deterministic Multiprocessor Scheduling with Multiple Objectives," *Computers and Operations Research* 19, 743–749.

Federgruen, A., and H. Groenvelt (1986). "Preemptive Scheduling of Uniform Machines by Ordinary Network Flow Techniques," *Management Science* 32, 341–349.

Fischetti, M., S. Martello, and P. Toth (1989). "The Fixed Job Schedule Problem with Working-Time Constraints," *Operations Research* 37, 395–403.

Fischetti, M., S. Martello, and P. Toth (1992). "Approximation Algorithms for Fixed Job Schedule Problems," *Operations Research* 40, S96–S108.

Gonsalez, T., and S. Sahni (1978). "Preemptive Scheduling of Uniform Processor Systems," *Journal of the ACM* 25, 92–101.

Graham, R. L., E. L. Lawler, J. K. Lenstra, and A. H. G. Rinnooy Kan (1979). "Optimization and Approximation in Deterministic Sequencing and Scheduling: A Survey," *Annals of Discrete Mathematics* 5, 287–326.

Hall, N. G. (1986). "Single- and Multiple-Processor Models for Minimizing Completion Time Variance," *Naval Research Logistics Quarterly* 33, 49–54.

Herrbach, L. A., and J. Y.-T. Leung (1990). "Preemptive Scheduling of Equal Length Jobs on Two Machines to Minimize Mean Flow Time," *Operations Research* 38, 487–494.

Horn, W. A. (1974). "Some Simple Scheduling Algorithms," *Naval Research Logistics Quarterly* 21, 177–185.

Hu, T. C. (1961). "Parallel Sequencing and Assembly Line Problems," *Operations Research* 9, 841–848.

Ikura, Y., and M. Gimple (1986). "Scheduling Algorithms for a Single Batch Processing Machine," *Operations Research Letters* 5, 61–65.

Janiak, A., and J. Grabowski (1987). "Job-shop Scheduling with Resource–Time Models of Operations," *European Journal of Operational Research* 28, 58–73.

Kämpke, T. (1989). "Optimal Scheduling of Jobs with Exponential Service Times in Identical Parallel Processors," *Operations Research* 37, 126–133.

Kumar A., and T. E. Morton (1991). "Dynamic Job Routing in a Heavily Loaded Shop," GSIA, Carnegie Mellon University, Pittsburgh, PA.

Lawler, E. L., and C. U. Martel (1989). "Preemptive Scheduling of Two Uniform Machines to Minimize the Number of Late Jobs," *Operations Research* 37, 314–318.

Lee, C.-Y., R. Uzsoy, and L. A. Martin-Vega (1991). "Efficient Algorithms for Scheduling Semiconductor Burn-In Operations," School of Industrial Engineering, Purdue University, West Lafayette, IN.

Liu, Z., and E. G. Coffman (1992). "On the Optimal Stochastic Scheduling of Out-Forests," *Operations Research* 40, S67–S75.

Lloyd, E. L. (1981). "Concurrent Task Systems," *Operations Research* 29, 189–201.

Martel, C. (1982). "Scheduling Uniform Machines with Release Times, Deadlines and Due Dates," *Journal of the ACM* 29, 812–829.

Martello, S., and P. Toth (1986). "A Heuristic Approach to the Bus Driver Scheduling Problem," *European Journal of Operational Research* 24, 106–117.

McNaughton, R. (1959). "Scheduling with Deadlines and Loss Functions," *Management Science* 6, 1–12.

Muntz, R. R., and E. G. Coffman (1969). "Optimal Preemptive Scheduling on Two-Processor Systems," *IEEE Transactions on Computers* 18, 1014–1020.

Muntz, R. R., and E. G. Coffman (1970). "Preemptive Scheduling of Real-Time Tasks on Multiprocessor Systems," *Journal of the ACM* 17, 324–338.

Murata, T. (1980). "Synthesis of Decision-Free Concurrent Systems for Prescribed Resources and Performance," *IEEE Transactions on Software Engineering* 6, 525–530.

Pinedo, M., and G. Weiss (1984). "Scheduling Jobs with Exponentially Distributed Processing Times and Intree Precedence Constraints on Two Parallel Machines," *Operations Research* 33, 1381–1388.

Root, J. G. (1965). "Scheduling with Deadlines and Loss Functions on k Parallel Machines," *Management Science* 11, 460–475.

Rothkopf, M. H. (1966). "Scheduling Independent Tasks on Parallel Processors," *Management Science* 12, 437–447.

Sen, T., and S. K. Gupta (1984). "A State-of-Art Survey of Static Scheduling Research Involving Due Dates," *Omega* 12, 63–76.

Sundararaghavan, P. S., and M. U. Ahmed (1984). "Minimizing the Sum of Absolute Lateness in Single-Machine and Multi-Machine Scheduling," *Naval Research Logistics Quarterly* 31, 325–333.

Tang, C. S. (1990). "Scheduling Batches on Parallel Machines with Major and Minor Setups," *European Journal of Operational Research* 46, 28–37.

Weber, R. R. (1982). "Scheduling Jobs with Stochastic Processing Requirements on Parallel Machines to Minimize Makespan or Flowtime," *Journal of Applied Probability* 19, 167–182.

Weber, R. R. (1986). "Stochastic Scheduling on Parallel Processors and Minimization of Concave Functions of Completion Times" (preprint).

Weber, R. R., P. Varaiya, and J. Warland (1986). "Scheduling Jobs with Stochastically Ordered Processing Times on Parallel Machines to Minimize Expected Flowtime," *Journal of Applied Probability* 23, 841–847.

Wittrock, R. J. (1990). "Scheduling Parallel Machines with Major and Minor Setup Times," *International Journal of Flexible Manufacturing Systems* 2, 329–341.

Xu, S. J., S. P. R. Kumar, and P. B. Mirchandani (1992). "Scheduling Stochastic Jobs with Increasing Hazard Rate on Identical Parallel Machines," *Computers and Operations Research* 19, 535–543.

Yamazaki, G., and H. Sakasegawa (1984). "An Optimal Design Problem for Limited Processor-Sharing Systems," *SEP Discussion Paper Series* 259, University of Tsukuba, Japan.

Yang, C. I., J. Wang, and R. C. T. Lee (1989). "A Branch-and-Bound Algorithm to Solve the Equal-Execution-Time Job Scheduling Problem with Precedence Constraint and Profile," *Computers and Operations Research* 16, 257–269.

Young, J. W. Jr. (1982). "Prevention Techniques Ease Concurrent Processing Snags," *Computerworld* 16, 27–28.

Chapter 12

Agnetis, A., C. Arbib, M. Lucertini, and F. Nicolo (1990). "Part Routing in Flexible Assembly Systems," *IEEE Transactions on Robotics and Automation* 6, 697–705.

Arbib, C., M. Lucertini, and F. Nicolo (1991). "Workload Balance and Part-Transfer Minimization in Flexible Manufacturing Systems," *International Journal of Flexible Manufacturing Systems* 3, 5–15.

Arbib, C., and L. Perugin (1989). "Optimal Part Routing in Flexible Manufacturing Systems," *Proceedings of the Tenth International Conference in Production Research,* Nottingham, UK.

Balakrishnan, A. (1989). "Preemptive Scheduling of Hybrid Parallel Machines," *Operations Research* 37, 301–313.

Bartholdi, J. J., and K. L. McCroan (1990). "Scheduling Interviews for a Job Fair," *Operations Research* 38, 951–960.

Blazewicz, J., H. A. Eislet, G. Finke, G. Laporte, and J. Welgarz (1992). "Scheduling Tasks and Vehicles in a Flexible Manufacturing System," *International Journal of Flexible Manufacturing Systems* 4, 5–16.

Brown, G. G., C. E. Goodman, and R. K. Wood (1990). "Annual Scheduling of Atlantic Fleet Naval Combatants," *Operations Research* 38, 249–259.

Brown, G. G., G. W. Graves, and D. Ronen (1987). "Scheduling Ocean Transportation of Crude Oil," *Management Science* 33, 335–346.

Browne, S., and U. Yechiali (1991). "Dynamic Scheduling in Single-Server Multiclass Service Systems with Unit Buffers," *Naval Research Logistics* 38, 383–396.

Calabrese, J. M., and W. H. Hausman (1991). "Simultaneous Determination of Lot Sizes and Routing Mix in Job Shops," *Management Science* 37, 1043–1057.

Carter, M. W., and C. A. Tovey (1992). "When Is the Classroom Assignment Problem Hard?" *Operations Research* 40, S28–S39.

Cheng, T. C. E., and C. C. S. Chin (1990). "A State-of-the-Art Review of Parallel-Machine Scheduling Research," *European Journal of Operational Research* 47, 271–292.

Crawford, J. L., and G. B. Sinclair (1977). "Computer Scheduling of Beer Tanker Deliveries," *International Journal of Physical Distribution* 7, 294–304.

Dantzig, D. B., and D. R. Fulkerson (1954). "Minimizing the Number of Tankers to Meet a Fixed Schedule," *Naval Research Logistics Quarterly* 1, 217–222.

De, Prabhu, and T. E. Morton (1980). "Scheduling to Minimize Makespan on Unequal Parallel Processors," *Decision Sciences* 11, 586–602.

Dobson, G., U. Karmarkar, and J. Rummel (1987). "Batching to Minimize Flow Times on Parallel Heterogenous Machines," Revised Working Paper No. QM8535, University of Rochester, Rochester, NY.

Dondeti, V. R., and H. Emmons (1992). "Fixed Job Scheduling with Two Types of Processors," *Operations Research* 40, S76–S85.

Gavett, J. W. (1968). Chapter 6 of *Production and Operations Management*. New York: Harcourt, Brace and World.

Ibarra, O. H., and C. E. Kim (1977). "Heuristic Algorithms for Scheduling Independent Tasks on Non-identical Processors," *Journal of the ACM* 24, 280–289.

Jafari, M. A. (1992). "An Architecture for Shop-floor Controller Using Colored Petri Nets," *International Journal of Flexible Manufacturing Systems* 4, 159–181.

Kellerer, H., and G. J. Woeginger (1992). "UET-Scheduling with Constrained Processor Allocations," *Computers and Operations Research* 19, 1–8.

Kolen, A., J. K. Lenstra, and C. H. Papadimitriou (1986). "Interval Scheduling Problems," Working Paper, Centre for Mathematics and Computer Science, Amsterdam.

Kumar, A., and T. E. Morton (1991). "Dynamic Job Routing in a Heavily Loaded Shop," GSIA, Carnegie Mellon University, Pittsburgh, PA.

Lawler, E. L., and J. Labetoulle (1978). "On Preemptive Scheduling of Unrelated Parallel Processors by Linear Programming," *Journal of the ACM* 25, 612–619.

Lin, G. Y.-J., and J. J. Solberg (1991). "Effectiveness of Flexible Routing Control," *International Journal of Flexible Manufacturing Systems* 3, 189–211.

Morton, T. E., and Prabhu De (1982). "Scheduling to Minimize Maximum Lateness on Unequal Parallel Processors," *Computers and Operations Research* 9, 221–232.

Schweitzer, P. J., A. Seidmann, and P. B. Goes (1991). "Performance Management in a Flexible Manufacturing System," *International Journal of Flexible Manufacturing Systems* 4, 17–50.

Sengupta, S., and R. P. Davis (1992). "Quality Implications of Machine Assignment Decisions in a Flexible Manufacturing System," *Applied Mathematical Modelling* 16, 86–93.

Wang, L., and W. E. Wilhelm (1992). "A Recursion Model for Cellular Production/Assembly Systems," *International Journal of Flexible Manufacturing Systems* 4, 129–158.

Weiss, G., and M. Pinedo (1980). "Scheduling Tasks with Exponential Service Times on Non-identical Processors to Minimize Various Cost Functions," *Journal of Applied Probability* 17, 187–202.

Younis, M. A., and M. S. Mahmoud (1992). "An Algorithm for Dynamic Routing in Flexible Manufacturing Systems Under an Unpredicted Failure," *Applied Mathematical Modelling* 16, 141–147.

Chapter 13

Agnetis, A., C. Arbib, and K. E. Stecke (1990). "Optimal Two Machine Scheduling in a Flexible Flow System," *Proceedings of the Second International Conference on Computer Integrated Manufacturing,* Renssalaer Polytechnic Institute, Renssalaer, NY.

Ashour, S. (1970). "An Experimental Investigation and Comparative Evaluation of Flowshop Scheduling Techniques," *Operations Research* 18, 541–548.

Baker, K. R. (1974). *Introduction to Sequencing and Scheduling.* New York: Wiley.

Baker, K. R. (1975). "A Comparative Survey of Flowshop Algorithms," *Operations Research* 23, 62–73.

Brown, A., and Z. Lomnicki (1966). "Some Applications of the Branch and Bound Algorithm to the Machine Scheduling Problem," *Operational Research Quarterly* 17, 173–186.

Bruno, J. P., P. Downey, and G. N. Frederickson (1981). "Sequencing Tasks with Exponential Service Times to Minimize the Expected Flow Time or Makespan," *Journal of the ACM* 28, 100–113.

Campbell, H. G., R. A. Dudek, and M. L. Smith (1970). "A Heuristic Algorithm for the *n* Job *m* Machine Sequencing Problem," *Management Science* 16, 630–637.

Dannenbring, D. G. (1977). "An Evaluation of Flowshop Sequencing Heuristics," *Management Science* 23, 1174–1182.

Dudek, R. A., M. L. Smith, and S. S. Panwalkar (1974). "Use of a Case Study in Sequencing/Scheduling Research," *Omega* 2, 253–261.

Dudek, R. A., S. S. Panwalkar, and M. L. Smith (1992). "The Lessons of Flowshop Scheduling Research," *Operations Research* 40, 7–13.

Dutta, S. K., and A. A. Cunningham (1975). "Sequencing Two Machine Flowshops with Finite Intermediate Storage," *Management Science* 21, 989–996.

Ewacha, K., I. Rival, and G. Steiner (1990). "Permutation Schedules for Flow Shops with Precedence Constraints," *Operations Research* 38, 1135–1139.

Foley, R. D., and S. Suresh (1984). "Stochastically Minimizing the Makespan in Flow Shops," *Naval Research Logistics Quarterly* 31, 551–557.

Foley, R. D., and S. Suresh (1986). "Scheduling n Non-overlapping Jobs and Two Stochastic Jobs in a Flow Shop," *Naval Research Logistics Quarterly* 33, 123–128.

Glazebrook, K. D. (1979). "Scheduling Tasks with Exponential Service Times on Parallel Processors," *Journal of Applied Probability* 16, 685–689.

Gupta, J. N. D. (1971). "A Functional Heuristic Algorithm for the Flowshop Scheduling Problem," *Operational Research Quarterly* 22, 39–48.

Gupta, J. N. D. (1972). "Heuristic Algorithms for Multistage Flow Shop Problem," *AIIE Transactions* 4, 11–18.

Gupta, J. N. D. (1976). "Optimal Flowshop with No Intermediate Storage Space," *Naval Research Logistics Quarterly* 23, 235–243.

Gupta, J. N. D., and R. A. Dudek (1971). "Optimality Criteria for Flow Shop Schedules," *AIIE Transactions* 3, 199–205.

Heller, J. (1960). "Some Numerical Experiments for an $M * J$ Flow Shop and Its Decision-Theoretical Aspects," *Operations Research* 8, 178–184.

Ignall, E., and L. E. Schrage (1965). "Application of the Branch and Bound Technique to Some Flow Shop Scheduling Problems," *Operations Research* 13, 400–412.

Johnson, S. M. (1954). "Optimal Two- and Three-Stage Production Scheduling with Setup Times Included," *Naval Research Logistics Quarterly* 1, 61–68.

Kämpke, T. (1987). "On the Optimality of Static Priority Policies in Stochastic Scheduling on Parallel Machines," *Journal of Applied Probability* 24, 430–448.

King, J. R., and A. S. Spachis (1980). "Heuristics for Flowshop Scheduling," *International Journal of Production Research* 18, 347–357.

Lomnicki, Z. (1965). "A Branch-and-Bound Algorithm for the Exact Solution of the Three-Machine Scheduling Problem," *Operational Research Quarterly* 16, 89–100.

Maggu, P. L., G. Das, and R. Kumar (1981). "On Equivalent-Job for Job-Block in $2 \times n$ Sequencing Problem with Transportation Times," *Journal of the Operations Research Society of Japan* 24, 136–146.

Maggu, P. L., M. L. Singhal, N. Mohammad, and S. K. Yadav (1982). "On N-Job 2-Machine Flowshop Scheduling Problem with Arbitrary Time Lags and Transportation Times of Jobs," *Journal of the Operations Research Society of Japan* 25, 219–227.

McMahon, G. B. (1969). "Optimal Production Schedules for Flowshops," *Canadian Operations Research Journal* 7, 141–151.

McMahon, G. B., and P. G. Burton (1967). "Flowshop Scheduling with the Branch and Bound Method," *Operations Research* 15, 473–481.

Mitten, L. G. (1959). "Sequencing n Jobs on Two Machines with Arbitrary Time Lags," *Management Science* 5, 293–303.

Muth, E. J. (1979). "The Reversibility Property of Production Line," *Management Science* 25, 152–158.

Nagasawa, H., K. Nango, N. Hirabayashi, and N. Nishiyama (1989). "Method for Distributing Tools in 2-Machine Flowshop Type FMS," *Transactions of the Japanese Society of Mechanichal Engineers Part C* 55, 1133–1146.

Palmer, D. S. (1965). "Sequencing Jobs Through a Multi-Stage Process in the Minimum Total Time—A Quick Method of Obtaining a Near Optimum," *Operational Research Quarterly* 16, 101–107.

Panwalkar, S. S., and A. W. Khan (1975). "An Improved Branch and Bound Procedure for *n* * *m* Flowshop Problems," *Naval Research Logistics Quarterly* 22, 787–790.

Panwalkar, S. S., and A. W. Khan (1976). "An Ordered Flowshop Sequencing Problem with Mean Completion Time Criterion," *International Journal of Production Research* 14, 631–635.

Panwalkar, S. S., and A. W. Khan (1977). "A Convex Property of the Ordered Flowshop Sequencing Problem," *Naval Research Logistics Quarterly* 24, 159–162.

Panwalkar, S. S., R. A. Dudek, and M. L. Smith (1973). "Sequencing Research and Industrial Sequencing Problem," *Proceedings of Symposium on Theory of Scheduling and Its Application,* S. Elmaghraby (ed.). New York: Springer-Verlag, 29–38.

Pinedo, M. (1982). "Minimizing the Expected Makespan in Stochastic Flow Shops," *Operations Research* 30, 148–162.

Smith, M. L., S. S. Panwalkar, and R. A. Dudek (1975). "Flowshop Sequencing with Ordered Processing Time and Matrices," *Management Science* 21, 544–549.

Smith, M. L., S. S. Panwalkar, and R. A. Dudek (1976). "Flowshop Sequencing Problem with Ordered Processing Time Matrices: A General Case," *Naval Research Logistics Quarterly* 23, 481–486.

Smith, R. D., and R. A. Dudek (1967). "A General Algorithm for Solution of the *n*-Job *m*-Machine Sequencing Problem of the Flow Shop," *Operations Research* 15, 71–82 and Errata 17, 756 (1969).

Srikandarajah, C., and S. P. Sethi (1989). "Scheduling Algorithms for Flexible Flowshops: Worst and Average Case Performance," *European Journal of Operational Research* 43, 143–160.

Stern, H. I., and G. Vitner (1990). "Scheduling Parts in a Combined Production-Transportation Work Cell," *Journal of the Operational Research Society* 41, 625–632.

Szwarc, W. (1971). "Elimination Methods in the *m* * *n* Sequencing Problem," *Naval Research Logistics Quarterly* 18, 295–305.

Szwarc, W. (1973). "Optimal Elimination Methods in the *m* * *n* Flow Shop Scheduling Problem," *Operations Research* 21, 1250–1259.

Widmer, M., and A. Hertz (1989). "A New Heuristic Method for the Flowshop Sequencing Problem," *European Journal of Operational Research* 41, 186–193.

Wismer, D. A. (1972). "Solution of Flowshop Scheduling Problem with No Intermediate Queues," *Operations Research* 20, 689–697.

Chapter 14

Baker, K. R. (1974). *Introduction to Sequencing and Scheduling.* New York: Wiley.

Brooks, G. H., and C. R. White (1965). "An Algorithm for Finding Optimal or Near-Optimal Solutions to the Production Scheduling Problem," *Journal of Industrial Engineering* 16, 34–40.

Conway, R. W. (1965). "Priority Dispatching and Job Lateness in a Job Shop," *Journal of Industrial Engineering* 16, 228–237.

Graves, S. C., H. C. Meal, D. Stefek, and A. H. Zeghmi (1983). "Scheduling of Re-entrant Flow Shops," *Journal of Operations Management* 3, 197–207.

Gupta, J. N. D. (1972). "Heuristic Algorithms for Multistage Flow Shop Problem," *AIIE Transactions* 4, 11–18.

Ignall, E., and L. E. Schrage (1965). "Application of the Branch and Bound Technique to Some Flow Shop Scheduling Problems," *Operations Research* 13, 400–412.

Lashine, S., B. Foote, and A. Ravindran (1991). "A Nonlinear Mixed Integer Goal Programming Model for the Two-Machine Closed Flow Shop," *European Journal of Operational Research* 55, 57–70.

Morton, T. E., and D. Pentico (work in process). "Comparing Myopic, OPT, and Bottleneck Dynamics Heuristics in a Generalized Flowshop," GSIA, Carnegie Mellon University, Pittsburgh, PA.

Ow, P. S. (1985). "Focused Scheduling in Proportionate Flowshops," *Management Science* 31, 852–869.

Park, T., and H. J. Steudel (1991). "A Model for Determining Job Throughput Times for Manufacturing Flow Line Workcells with Finite Buffers," *International Journal of Production Research* 29, 2025–2041.

Taillard, E. (1990). "Some Efficient Heuristic Methods for the Flow Shop Sequencing Problem," *European Journal of Operational Research* 47, 65–74.

Vepsalainen, A., and T. E. Morton (1987). "Priority Rules and Leadtime Estimation for Job Shop Scheduling with Weighted Tardiness Costs," *Management Science* 33, 1036–1047.

Widmer, Marino, and A. Hertz (1989). "A New Heuristic Method for the Flow Shop Sequencing Problem," *European Journal of Operational Research* 41, 186–193.

Wismer, D. A. (1972). "Solution of the Flow Shop Scheduling Problem with No Intermediate Queues," *Operations Research* 20, 689–697.

Chapter 15

Bai, S., N. Srivatsan, and S. B. Gershwin (1990). "Scheduling Manufacturing Systems with Work-in-Progress Inventory: Single Part Type Systems," VLSI Memo No. 90-604, MIT Microsystems Research Center.

Baker, K. R. (1968). "Priority Dispatching in the Single Channel Queue with Sequence-Dependent Setups," *Journal of Industrial Engineering* 19, 203–206.

Baker, K. R. (1974). *Introduction to Sequencing and Scheduling,* New York: Wiley.

Bakhru, A. N., and M. R. Rao (1964). "An Experimental Investigation of Job-Shop Scheduling," Research Report, Department of Industrial Engineering, Cornell University, Ithaca, NY.

Bowman, E. H. (1959). "The Schedule-Sequencing Problem," *Operations Research* 7, 621–624.

Brooks, G. H., and C. R. White (1965)."An Algorithm for Finding Optimal or Near Optimal Solutions to the Production Scheduling Problem," *Journal of Industrial Engineering* 16, 34–40.

Dondeti, V. R., and J. Emmons (1992). "Fixed Job Shop Scheduling with Two Types of Processors," *Operations Research* 40, Supplement 1, S76–S85.

Giffler, B., and G. L. Thompson (1960). "Algorithms for Solving Production Scheduling Problems," *Operations Research* 8, 487–503.

Greenberg, H. (1968). "A Branch-and-Bound Solution to the General Scheduling Problem," *Operations Research* 16, 353–361.

Jerimiah, B., A. Lalchandani, and L. Schrage (1964). "Heuristic Rules Toward Optimal Scheduling," Research Report, Department of Industrial Engineering, Cornell University, Ithaca, NY.

Lasserre, J. B. (1992). "An Integrated Model for Job-Shop Planning and Scheduling," *Management Science* 38, 1201–1211.

Lozinski, C. and C. R. Glassey (1988). "Bottleneck Starvation Indicators for Shop-Floor Control," *IEEE Transactions on Semiconductor Manufacturing* 1, 147–153.

Manne, A. S. (1960). "On the Job-Shop Sequencing Problem," *Operations Research* 8, 219–223.

Morton, T. E., S. Kekre, S. Lawrence, and S. Rajagopalan (1988). "SCHED-STAR: A Price Based Shop Scheduling Module," *Journal of Manufacturing and Operations Management* 1, 131–181

Story, A. E., and H. M. Wagner (1963). "Computational Experience with Integer Programming for Job-Shop Scheduling," *Industrial Scheduling*, J. F. Muth and G. L. Thompson (eds.), Englewood Cliffs, NJ: Prentice-Hall.

Uzsoy, R., and C.-Y. Lee (1991). "A Review of Production Planning and Scheduling Models in the Semiconductor Industry," Unpublished paper.

Uzsoy, R., L. A. Martin-Vega, C. Y. Lee, and P. A. Leonard (1991). "Production Scheduling Algorithms for a Semiconductor Test Facility," *IEEE Transactions on Semiconductor Manufacturing* 4, 270–280.

Vepsalainen, A., and T. E. Morton (1987). "Priority Rules and Leadtime Estimation for Job Shop Scheduling with Weighted Tardiness Costs," *Management Science* 33, 1036–1047.

Wagner, H. M. (1959). "An Integer Linear-Programming Model for Machine Scheduling," *Naval Research Logistics Quarterly* 6, 131–140.

Wein, L. M. (1988). "Scheduling Semiconductor Wafer Fabrication," *IEEE Transactions on Semiconductor Manufacturing* 1, 115–129.

Wein, L. M., and P. B. Chevelier (1992). "A Broader View of the Job-Shop Scheduling Problem," *Management Science* 38, 1018–1033.

Wein, L. M., and J. Ou (1991). "The Impact of Processing Time Knowledge on Dynamic Job-Shop Scheduling," *Management Science* 37, 1002–1014.

Chapter 16

Adams, J., E. Balas, and D. Zawack (1988). "The Shifting Bottleneck Procedure for Job Shop Scheduling," *Management Science* 34, 391–401.

Aneja, Y. P., and N. Singh (1990). "Scheduling Production of Common Components at a Single Facility," *IIE Transactions* 22, 234–237.

Baker, C. T., and B. P. Dzielinski (1960). "Simulation of a Simplified Job Shop," *Management Science* 16, 311–323.

Baker, K. R. (1968). "Priority Dispatching in the Single Channel Queue with Sequence-Dependent Setups," *Journal of Industrial Engineering* 19, 203–206.

Baker, K. R. (1969). *Control Policies for an Integrated Production Inventory System*. Ph.D. Thesis, Department of Operations Research, Cornell University, Ithaca, NY.

Barnes, W. J., and J. B. Chambers (1992). "Solving the Job Shop Problem Using Tabu Search," Graduate Program in Operations Research, Technical Report Series, The University of Texas, Austin.

Berry, W. L. (1972). "Priority Scheduling and Inventory Control in Job Lot Manufacturing Systems," *AIIE Transactions* 4, 267–276.

Bitran, G. R., and D. Tirupati (1988a). "Planning and Scheduling for Epitaxial Wafer Production," *Operations Research* 36, 34–49.

Bitran, G. R., and D. Tirupati (1988b). "Multiproduct Queueing Networks with Deterministic Routing; Decomposition Approach and the Notion of Interference," *Management Science* 34, 75–100.

Buffa, E. S., and W. H. Taubert (1972). *Production–Inventory Systems*. Homewood, IL: Richard D. Irwin, Inc.

Carlier, J. (1982). "The One Machine Sequencing Problem," *European Journal of Operational Research* 11, 42–47.

Carroll, D. C. (1965). *Heuristic Sequencing of Single and Multiple Component Jobs*. Ph.D. Thesis, Sloan School of Management, M.I.T., Cambridge, MA.

Chu, C., M. C. Portmann, and J. M. Proth (1992). "A Splitting-Up Approach to Simplify Job-Shop Scheduling Problems," *International Journal of Production Research* 30, 859–870.

Conway, R. W. (1965a). "Priority Dispatching and Work In Process Inventory in a Job Shop," *Journal of Industrial Engineering* 16, 123–130.

Conway, R. W. (1965b). "Priority Dispatching and Job Lateness in a Job Shop," *Journal of Industrial Engineering* 16, 228–237.

Conway, R. W., B. M. Johnson, and W. L. Maxwell (1960). "An Experimental Investigation of Priority Dispatching," *Journal of Industrial Engineering* 11, 221–229.

Conway, R. W., W. L. Maxwell, and L. W. Miller (1967). *Theory of Scheduling*. Reading, MA: Addison-Wesley.

Day, J. E., and M. H. Hottenstein (1970). "Review of Sequencing Research," *Naval Research Logistics Quarterly* 17, 11–39.

Dayhoff, J. E., and R. W. Atherton (1986). "Signature Analysis of Dispatch Schemes in Wafer Fabrication," *IEEE Transactions on Components, Hybrids and Manufacturing Technology* 9, 498–507.

Dell'Amico, M., and M. Trubian (1991). "Applying Tabu-Search to the Job Shop Scheduling Problem," Politecnico di Milano, Italy.

Della Croce, F., R. Tadei, and G. Volta (1992). "A Genetic Algorithm for the Job Shop Problem," D.A.I., Politecnico di Torino, Italy.

Emery, J. C. (1969). "Job Shop Scheduling by Means of Simulation and an Optimum Seeking Search," *Proceedings of the Conference on Simulation*, Los Angeles.

Giffler, B., G. L. Thompson, and V. Van Ness (1963). "Numerical Experience with the Linear and Monte Carlo Algorithms for Solving Production Scheduling Problems," *Industrial Scheduling*, J. F. Muth and G. L. Thompson (eds.). Englewood Cliffs, NJ: Prentice-Hall.

Hottenstein, M. P. (1970). "Expediting in Job-Order-Control Systems: A Simulation Study," *AIIE Transctions* 2, 46–54.

Kubiak, W., S. X. C. Lou, and Y.-M. Wang (1990). "Mean Flow Time Minimization in Reentrant Job Shops with Hub," Faculty of Management, University of Toronto.

Laarhoven, P. J. M., E. H. L. Aarts, and J. K. Lenstra (1992). "Job Shop Scheduling by Simulated Annealing," *Operations Research* 40, 113–124.

Lawrence, S., and T. E. Morton (work in process). "An Integrative Dispatch Heuristic Comparison Study for Job Shops," GSIA, Carnegie Mellon University, Pittsburgh, PA.

LeGrande, E. (1963). "The Development of a Factory Simulation System Using Actual Operating Data," *Management Technology* 3.

Lozinski, C., and C. R. Glassey (1988). "Bottleneck Starvation Indicators for Shop-Floor Control," *IEEE Transactions on Semiconductor Manufacturing* 1, 147–153.

Matsuo, H., C. J. Suh, and R. S. Sullivan (1989). "A Controlled Search Simulated Annealing Method for the General Jobshop Scheduling Problem," Working Paper 03-04-88, Department of Management, The University of Texas, Austin.

Maxwell, W. L., and M. Mehra (1968). "Multiple-Factor Rules for Sequencing with Assembly Constraints," *Naval Research Logistics Quarterly* 15, 241–254.

Moore, J. M., and R. C. Wilson (1967). "A Review of Simulation Research in Job Shop Scheduling," *Production and Inventory Management* 8 (first quarter), 1–10.

Morton, T. E., S. Kekre, S. Lawrence, and S. Rajagopalan (1988). "SCHED-STAR: A Price Based Shop Scheduling Module,' *Journal of Manufacturing and Operations Management* 1, 131–181.

Morton, T. E., and S. Lawrence (work in process). GSIA, Carnegie Mellon University, Pittsburgh, PA.

Morton, T., and P. Ramnath (1992). "Guided Forward Search In Tardiness Scheduling of Large One Machine Problem," GSIA, Carnegie Mellon University, Pittsburgh, PA.

Muth, J. F., and G. L. Thompson (eds.) (1963). *Industrial Scheduling.* Englewood Cliffs, NJ: Prentice-Hall.

Nanot, Y. R. (1963). *An Experimental Investigation and Comparative Evaluation of Priority Disciplines in Job Shop-Like Queueing Networks.* Ph.D. Thesis, UCLA, Los Angeles, CA.

Nelson, R. T. (1970). "A Simulation Study of Labor Efficiency and Centralized Labor Assignment in a Production System Model," *Management Science* 17, B97–B106.

Nugent, C. N. (1964). *On Sampling Approaches to the Solution of the n-by-m Static Sequencing Problem.* Ph.D. Thesis, Cornell University, Ithaca, NY.

Pai, A. R., and K. L. McRoberts (1971). "Simulation Research in Interchangeable Part Manufacturing," *Management Science* 17, B732–B743.

Philipoom, P. R., and T. D. Fry (1990). "The Robustness of Selected Job-shop Dispatching Rules with Respect to Load Balance and Work-flow Structure," *Journal of the Operational Research Society* 41, 897–906.

Raman, N., F. B. Talbot, and R. V. Rachamadugu (1989). "Due Date Based Scheduling in a General Flexible Manufacturing System," *Journal of Operations Management* 8, 115–132.

Rowe, A. J. (1958). "Sequential Decision Rules in Production Scheduling," Ph.D. Thesis, UCLA, Los Angeles, CA.

Russo, F. J. (1958). *A Heuristic Approach to Alternate Routing in a Job Shop.* Master's Thesis, M.I.T., Cambridge, MA.

Taillard, E. (1990). "Some Efficient Heuristic Methods for the Flow Shop Sequencing Problem," *European Journal of Operational Research* 47, 65–74.

Uzsoy, R., and C.-Y. Lee (1991). "A Review of Production Planning and Scheduling Models in the Semiconductor Industry," Work-In-Progress, Department of Industrial Engineering, Purdue University, West Lafayette, IN.

Van Laarhoven, P. J. M., E. H. L. Aarts, and J. K. Lenstra (1992). "Job Shop Scheduling by Simulated Annealing," *Operations Research* 40, 112–129.

Vepsalainen, A., and T. E. Morton (1987). "Priority Rules and Leadtime Estimation for Job Shop Scheduling with Weighted Tardiness Costs," *Management Science* 33, 1036–1047.

Vepsalainen, A., and T. E. Morton (1988). "Improving Local Priority Rules with Global Leadtime Estimates," *Journal of Manufacturing and Operations Management* 1, 102–118.

Wayson, R. D. (1965). *The Effect of Alternate Machines on Two Priority Dispatching Disciplines in the General Job Shop.* M.S. Thesis, Department of Operations Research, Cornell University, Ithaca, NY.

Chapter 17

Archibald, R. D. and R. L. Villoria (1967). *Network-Based Management Systems.* New York: Wiley.

Baker, K. R. (1974). *Introduction to Sequencing and Scheduling.* New York: Wiley.

Battersby, A. (1967). *Network Analysis for Planning and Scheduling,* London: Macmillan.

Cleland, D. I. (1969). "Project Management," *Systems, Organizations, Analyses, Management: A Book of Readings,* D. I. Cleland and W. R. King (eds.). New York: McGraw-Hill, 281–290.

Cleland, D. I., and W. R. King (1975). *Systems Analysis and Project Management.* New York: McGraw-Hill.

Elmaghraby, S. E. (1967). "On the Expected Duration of PERT Type Networks," *Management Science* 13, 299–306.

Fulkerson, D. R. (1961). "A Network Flow Computation for Project Cost Curves," *Management Science* 7, 167–178.

Fulkerson, D. R. (1962). "Expected Critical Path Lengths in PERT Networks," *Operations Research* 10, 808–817.

Garman, M. B. (1972). "More on Conditional Sampling in the Simulation of Stochastic Networks," *Management Science* 19, 90–95.

Hartley, H. O., and A. W. Worthman (1966). "A Statistical Theory for PERT Critical Path Analysis," *Management Science* 12, B469–B481.

Kerzner, H. (1982). *Project Management: A Systems Approach to Planning, Scheduling, and Controlling.* New York: Van Nostrand Reinhold.

Klingel, A. R. (1966). "Bias in PERT Project Completion Time Calculations for a Real Network," *Management Science* 13, B194–B201.

MacCrimmon, K. R., and C. A. Ryavec (1964). "An Analytical Study of the PERT Assumptions," *Operations Research* 12, 16–37.

Martin, J. J. (1965). "Distribution of the Time Through a Directed Acyclic Network," *Operations Research* 13, 46–66.

Miller, R. W. (1963). *Schedule, Cost and Project Control with PERT.* New York: McGraw-Hill.

Moder, J. G., and C. R. Phillips (1970). *Project Management with CPM and PERT.* New York: Van Nostrand Reinhold.

Parks, W. H., and K. L. Ramsig (1969). "The Use of the Compound Poisson in PERT," *Management Science* 15, B397–B402.

Ringer, L. J. (1971). "A Statistical Theory for PERT in Which Completion Times of Activities Are Interdependent," *Management Science* 17, 717–723.

Sidney, J. B. (1977). "Optimal Single-Machine Scheduling with Earliness and Tardiness Penalties," *Operations Research* 25, 62–69.

Steiner, G. A., and W. G. Ryan (1968). *Industrial Project Management.* New York: Macmillan.

Swanson, L. A., and H. L. Pazer (1971). "Implication of the Underlying Assumptions of PERT," *Decision Sciences* 2, 461–480.

Van Slyke, R. M. (1965). "Monte Carlo Methods and the PERT Problem," U.S. Government Research Reports, Document Number AD-412, 731.

Wiest, J. D., and F. K. Levy (1977). *A Management Guide to PERT/CPM.* Englewood Cliffs, NJ: Prentice-Hall.

Chapter 18

Archibald, R. D., and R. L. Villoria (1967). *Network-Based Management Systems.* New York: Wiley.

Baker, K. R. (1974). *Introduction to Sequencing and Scheduling.* New York: Wiley.

Balas, E. (1970). "Project Scheduling with Resource Constraints." *Applications of Mathematical Programming Techniques,* E. M. L. Beale (ed.). London: English University Press, 187–200.

Balas, E., and A. Vazacoupolos (work in process). GSIA, Carnegie Mellon University, Pittsburgh, PA.

Bennington, G. E., and L. F. McGinnis (1973). "A Critique of Project Planning with Constrained Resources," *Symposium on the Theory of Scheduling and Its Applications,* S. E. Elmaghraby (ed.). New York: Springer-Verlag, 1–28.

Calica, A. (1965). "Fabrication and Assembly Operations," *IBM Systems Journal* 4, 94–104.

Cleland, D. I. (1969). "Project Management," *Systems, Organizations, Analyses, Management: A Book of Readings,* D. I. Cleland and W. R. King (eds.). New York: McGraw-Hill, 281–290.

Cleland, D. I., and W. R. King (1975). *Systems Analysis and Project Management.* New York: McGraw-Hill.

Cooper, D. F. (1976). "Heuristics for Scheduling Resource-Constrained Projects: An Experimental Investigation," *Management Science* 22, 1186–1194.

Davis, E. W. (1973). "Project Scheduling Under Resource Constraints—Historical Review and Categorization of Procedures," *AIIE Transactions* 5, 297–313.

Davis, E. W., and G. E. Heidhorn (1971). "An Algorithm for Optimal Project Scheduling Under Multiple Resource Constraints," *Management Science* 17, B803–B816.

Davis, E. W., and J. H. Patterson (1975). "A Comparison of Heuristic and Optimum Solutions in Resource-Constrained Project Scheduling," *Management Science* 21, 944–955.

Doersch, R. H., and J. H. Patterson (1977). "Scheduling a Project to Maximize Its Present Value: A Zero–One Programming Approach," *Management Science* 23, 882–889.

Dumond, J. (1992). "In a Multi-resource Environment, How Much Is Enough?" *International Journal of Production Research* 30, 395–410.

Fendley, L. (1968). "Toward the Development of a Complete Multi-project Scheduling System," *Journal of Industrial Engineering* 19, 505–515.

Gorenstein, S. (1972). "An Algorithm for Project (Job) Sequencing with Resource Constraints," *Operations Research* 20, 835–850.

Janiak, A., and J. Grabowski (1987). "Job-shop Scheduling with Resource–Time Models of Operations," *European Journal of Operational Research* 28, 58–73.

Johnson, T. J. R. (1967). *An Algorithm for the Resource-Constrained Project Scheduling Problem.*" Ph.D. Thesis, M.I.T., Cambridge, MA.

Kelly, J. E. (1963). "The Critical Path Method: Resources Planning and Scheduling," *Industrial Scheduling*, J. Muth and G. Thompson (Eds.). Englewood Cliffs, NJ: Prentice-Hall.

Kerzner, H. (1982). *Project Management: A Systems Approach to Planning, Scheduling, and Controlling.* New York: Van Nostrand Reinhold.

Kurtulus, I., and E. W. Davis (1982). "Multi-project Scheduling: Categorization of Heuristic Rules of Performance," *Management Science* 18, 161–172.

Lawrence, S., and T. E. Morton (1989). "Resource-Constrained Multiproject Scheduling with Tardy Costs: Comparing Myopic, Bottleneck, and Resource Pricing Heuristics," GSIA, Carnegie Mellon University, Pittsburgh, PA. (To appear in the *European Journal of Operational Research.*

Lawrence, S., and T. E. Morton (work in process). GSIA, Carnegie Mellon University, Pittsburgh, PA.

Li, R. K.-Y., and R. J. Willis (1991). "Alternative Resources in Project Scheduling," *Computers and Operations Research* 18, 663–668.

Mason, A. T., and C. L. Moodie (1971). "A Branch and Bound Algorithm for Minimizing Cost in Project Scheduling," *Management Science* 18, B158–B173.

Morris, P. W. G. (1982). "Managing Project Interfaces—Key Points for Project Success," *Project Management: A Systems Approach to Planning, Scheduling, and Controlling,* D. I. Cleland and W. R. King (eds.). New York: Van Nostrand Reinhold, 16–55.

Morton, T. E., S. Kekre, S. Lawrence, and S. Rajagopalan (1988). "SCHED-STAR: A Price-Based Shop Scheduling Module," *Journal of Manufacturing and Operations Management* 1, 131–181.

Patterson, J. H. (1973). "Alternative Methods of Project Scheduling with Limited Resources," *Naval Research Logistics Quarterly* 20, 767–784.

Patterson, J. H. (1982). "Exact and Heuristic Solution Procedures for the Constrained Resource Project Scheduling Problem," Vols. I–IV, monograph, Department of Operations Management, Indiana University, Bloomington.

Patterson, J. H. (1984). "A Comparison of Exact Approaches for Solving the Multiple Constrained Resources, Project Scheduling Problem," *Management Science* 30, 854–867.

Patterson, J. H., and W. D. Huber (1974). "A Horizon-Varying, Zero–One Approach to Project Scheduling," *Management Science* 20, 990–998.

Patterson, J. H., and G. W. Roth (1976). "Scheduling a Project Under Multiple Resource Constraints: A Zero–One Programming Approach," *AIIE Transactions* 8, 449–455.

Pritsker, A. A. B., L. J. Watters, and P. M. Wolfe (1969). "Multiproject Scheduling with Limited Resources: A Zero–One Programming Approach," *Management Science* 16, 93–108.

Schrage, L. E. (1971). "Obtaining Optimal Solutions to Resource Constrained Network Scheduling Problems," *Proceedings of the Systems Engineering Conference,* Phoenix, AZ.

Stinson, J., E. W. Davis, and B. M. Khumawala (1978). "Multiple Resource-Constrained Scheduling Using Branch and Bound," *AIIE Transactions* 10, 252–259.

Talbot, F. B. (1982). "Resource-Constrained Project Scheduling with Time–Resource Trade-offs: The Nonpreemptive Case," *Management Science* 28, 1197–1210.

Talbot, F. B., and J. H. Patterson (1978). "An Efficient Integer Programming Algorithm with Network Cuts for Solving Resource-Constrained Scheduling Problems," *Management Science* 24, 1163–1174.

Thesen, A. (1976). "Heuristic Scheduling of Activities Under Precedence Restrictions," *Management Science* 23, 412–422.

Thomas, E., and D. Coveleski (1973). "Planning Nuclear Equipment Manufacturing," *Interfaces* 3(3), 18–29.

Vepsalainen, A., and T. E. Morton (1987). "Priority Rules and Leadtime Estimation for Job Shop Scheduling with Weighted Tardiness Costs," *Management Science* 33, 1036–1047.

Wiest, J. D. (1967). "A Heuristic Model for Scheduling Large Projects with Limited Resources," *Management Science* 13, B359–B377.

Ye, M., and G. B. Williams (1991). "An Expediting Heuristic for the Shortest Processing Time Dispatching Rule," *International Journal of Production Research* 29, 209–213.

Zaloom, V. (1971). "On the Resource Constrained Project Scheduling Problem," *AIIE Transactions* 3, 302–305.

Chapter 19

Agnetis, A., C. Arbib, M. Lucertini, and F. Nicolo (1990). "Part Routing in Flexible Assembly Systems," *IEEE Transactions on Robotics and Automation* 6, 697–705.

Arbel, A., and A. Seidmann (1984). "Performance Evaluation of Flexible Manufacturing Systems," *IEEE Transactions on Systems, Man and Cybernetics* SMC-14, 132–140.

Arbib, C., M. Lucertini, and F. Nicolo (1991). "Workload Balance and Part-Transfer Minimization in Flexible Manufacturing Systems," *International Journal of Flexible Manufacturing Systems* 3, 5–15.

Arbib, C., and L. Perugin (1989). "Optimal Part Routing in Flexible Manufacturing Systems," *Proceedings of the Tenth International Conference in Production Research,* Nottingham, UK.

Blazewicz, J., H. A. Eislet, G. Finke, G. Laporte, and J. Welgarz (1992). "Scheduling Tasks and Vehicles in a Flexible Manufacturing System," *International Journal of Flexible Manufacturing Systems* 4, 5–16.

Buzacott, J. A., and J. G. Shantikumar (1980). "Models for Understanding Flexible Manufacturing Systems," *AIIE Transactions* 12, 339–350.

Buzacott, J. A., and D. D. Yao (1986). "Flexible Manufacturing Systems: A Review of Analytical Models," *Management Science* 32, 890–905.

Chan, B. W. M. (1992). "Tool Management for Flexible Manufacturing," *Intelligent Manufacturing* 5, 255–265.

Chandra, J., and J. Talavage (1991). "Intelligent Dispatching for Flexible Manufacturing," *International Journal of Production Research* 29, 2259–2278.

Daniels, R. L. (1990). "A Multi-objective Approach to Resource Allocation in Single Machine Scheduling," *European Journal of Operational Research* 48, 226–241.

Das, S. R., and B. M. Khumawala (1991). "An Efficient Heuristic for Scheduling Batches of Parts in a Flexible Flow System," *International Journal of Flexible Manufacturing Systems* 3, 121–147.

De Souza, R. B. R., and R. Bell (1991). "A Tool Cluster Based Strategy for the Management of Cutting Tools in Flexible Manufacturing Systems," *Journal of Operations Management* 10, 73–91.

De Warra, D., and M. Widmer (1991). "Loading Problems with Tool Management in Flexible Manufacturing Systems: A Few Integer Programming Models," *International Journal of Flexible Manufacturing Systems* 3, 71–82.

Dobson, G., and U. S. Karmarkar (1989). "Simultaneous Resource Scheduling to Minimize Weighted Flow Times," *Operations Research* 37, 592–600.

Dupont-Gatelmand, C. (1982). "A Survey of Flexible Manufacturing Systems," *Journal of Manufacturing Systems* 1, 1–16.

Eaves, B. C. (1986). "A Flexible Manufacturing and Operator Scheduling Model Solved by Deconvexification Over Time," Technical Report SOL 86-14, Stanford University, Stanford, CA.

Ghosh, S., S. A. Melnyk, and B. L. Ragatz (1992). "Tooling Constraints and Shop Floor Scheduling: Evaluating the Impact of Sequence Dependency," *International Journal of Production Research* 30, 1237–1253.

Hutchison, J., and B. Khumawala (1990). "Scheduling Random Flexible Manufacturing Systems with Dynamic Environments," *Journal of Operations Management* 9, 335–351.

Iwata, K., Y. Murotsu, F. Oba, and K. Yasuda (1982). "Production Scheduling of Flexible Manufacturing Systems," *Journal of Manufacturing Systems* 1, 1–16.

Janiak, A., and J. Grabowski (1987). "Job-shop Scheduling with Resource–Time Models of Operations," *European Journal of Operational Research* 28, 58–73.

Kimemia, J., and S. B. Gershwin (1983). "An Algorithm for the Computer Control of a Flexible Manufacturing System," *IIE Transactions* 15, 353–362.

Kimemia, J., and S. B. Gershwin (1985). "Flow Optimization in Flexible Manufacturing Systems," *International Journal of Production Research* 23, 81–96.

Klahorst, H. T. (1981). "Flexible Manufacturing Systems: Combining Elements to Lower Costs, Add Flexibility," *Industrial Engineering* 13, 112–117.

Leachman, R. C., and V. S. Sohoni (1990). "Automated Shift Scheduling as a Tool for Problem Identification and People Management in Semiconductor Factories," Department of Industrial Engineering and Operations Research, University of California, Berkeley.

Lin, G. Y.-J., and J. J. Solberg (1992). "Integrated Shop Floor Control Using Autonomous Agents," *IEE Transactions on Design and Manufacturing* 24, 57–71.

Mukhopadhyay, S. K., B. Maithi, and S. Garg (1991). "Heuristic Solution to the Scheduling Problems in Flexible Manufacturing System," *International Journal of Production Research* 29, 2003–2024.

Nagasawa, H., K. Nango, N. Hirabayashi, and N. Nishiyama (1989). "Method for Distributing Tools in 2-Machine Flowshop Type FMS," *Transactions of the Japanese Society of Mechanical Engineering Part C* 55, 1133–1146.

O'Keefe, R. M., and T. Kasirajan (1992). "Interaction Between Dispatching and Next Station Selection Rules in a Dedicated Flexible Manufacturing System," *International Journal of Production Research* 30, 1753–1772.

Park, T., and H. J. Steudel (1991). "A Model for Determining Job Throughput Times for Manufacturing Flow Line Workcells with Finite Buffers," *International Journal of Production Research* 29, 2025–2041.

Rabinowitz, G., A. Mehrez, and S. Samaddar (1991). "A Scheduling Model for Multirobot Assembly Cells," *International Journal of Flexible Manufacturing Systems* 3, 149–180.

Raman, N., F. B. Talbot, and R. V. Rachamadugu (1989). "Due Date Based Scheduling in a General Flexible Manufacturing System," *Journal of Operations Management* 8, 115–132.

Sabuncuoglu, I., and D. L. Hommertzheim (1992). "Dynamic Dispatching Algorithm for Scheduling Machines and Automated Guided Vehicles in a Flexible Manufacturing System," *International Journal of Production Research* 30, 1059–1079.

Schweitzer, P. J., A. Seidmann, and P. B. Goes (1991). "Performance Management in a Flexible Manufacturing System," *International Journal of Flexible Manufacturing Systems* 4, 17–50.

Sengupta, S., and R. P. Davis (1992). "Quality Implications of Machine Assignment Decisions in a Flexible Manufacturing System," *Applied Mathematical Modelling* 16, 86–93.

Sethi, S. P., C. Sriskandarajah, G. Sorger, J. Blazewicz, and W. Kubiak (1992). "Sequencing of Parts and Robot Moves in a Robotic Cell," *International Journal of Flexible Manufacturing Systems* 4, 331–358.

Shanthikumar, J. G. (1982). "On the Superiority of Balanced Load in a Flexible Manufacturing System," Department of Industrial Engineering and Operations Research, Syracuse University, Syracuse, NY.

Shanthikumar, J. G., and K. E. Stecke (1986). "Reducing Work-In-Progress Inventory in Certain Classes of Flexible Manufacturing Systems," *European Journal of Operational Research* 26, 266–271.

Sriskandarajah, C., P. Labet, and R. Germain (1986). "Scheduling Methods for a Manufacturing System," *Flexible Manufacturing Systems: Methods and Studies*, A. Kusiak (ed.). Amsterdam: Elsevier, 173–189.

Stecke, K. E. (1989). "Algorithms for Efficient Planning and Operation of a Particular FMS," *International Journal of Flexible Manufacturing Systems* 1, 287–324.

Suri, R., and W. Dille (1984). "On Line Optimization of Flexible Manufacturing Systems Using Perturbation Analysis," First ORSA/TIMS Conference on Flexible Manufacturing Systems, Ann Arbor, MI.

Suri, R., and R. R. Hildebrant (1984). "Modelling Flexible Manufacturing Systems and Using Mean-Value Analysis," *Journal of Manufacturing Systems* 3, 27–38

Veermani, D., D. M. Upton, and M. M. Barash (1992). "Cutting-Tool Management in Computer-Integrated Manufacturing," *International Journal of Flexible Manufacturing Systems* 4, 237–265.

Wang, L., and W. E. Wilhelm (1992). "A Recursion Model for Cellular Production/Assembly Systems," *International Journal of Flexible Manufacturing Systems* 4, 129–158.

Wittrock, R. J. (1989). "The 'Orchard' Scheduler for Manufacturing Systems," IBM Research Report RC 15275, IBM T. J. Watson Research Center, Yorktown Heights, NY.

Younis, M. A., and M. S. Mahmoud (1992). "An Algorithm for Dynamic Routing in Flexible Manufacturing Systems Under an Unpredicted Failure," *Applied Mathematical Modelling* 16, 141–147.

Chapter 20

Ahmed, I., and W. W. Fisher (1992). "Due-Date Assignment, Job Order Release and Sequencing Interaction in Job Shop Scheduling," *Decision Sciences* 23, 633–647.

Baker, K. R. (1984). "Sequencing Rules and Due-Date Assignments in a Job Shop," *Management Science* 30, 1093–1104.

Baker, K. R., and J. W. Bertrand (1981). "An Investigation of Due-Date Assignment Rules with Constrained Tightness," *Journal of Operations Management* 1, 109–120.

Baker, K. R., and G. D. Scudder (1989). "On the Assignment of Optimal Due Dates," *Journal of the Operational Research Society* 40, 93–95.

Bector, C. R., Y. P. Gupta, and M. C. Gupta (1988). "Determination of an Optimal Common Due Date and Optimal Sequence in a Single Machine Job Shop," *International Journal of Production Research* 26, 613–628.

Bector, C. R., Y. P. Gupta, and M. C. Gupta (1990). "Optimal Schedule on a Single Machine Using Various Due Date Determination Methods," *Computers in Industry* 15, 245–253.

Berkley, B. J. (1991). "Tandem Queues and Kanban-Controlled Lines," *International Journal of Production Research* 29, 2057–2081.

Berkley, B. J., and A. S. Kiran (1991). "A Simulation Study of Sequencing Rules in a Kanban-controlled Flow Shop," *Decision Sciences* 22, 559–582.

Bieleck, T., and P. R. Kumar (1991). "Optimality of Zero-Inventory Policies for Unreliable Manufacturing Systems," *Operations Research* 36, 532–541.

Bitran, G. R., and L. Chang (1987). "A Mathematical Programming Approach to a Deterministic Kanban System," *Management Science* 33, 427–441.

Bolander, S. F., and S. G. Taylor (1990a). "Process-Flow Scheduling: Basic Cases," *Production and Inventory Management* 31(3), 1–4.

Bolander, S. F., and S. G. Taylor (1990b). "Process-Flow Scheduling: Mixed-Flow Cases," *Production and Inventory Management* 31(4), 1–6.

Box, R. E., and D. G. Herbe, Jr. (1988). "A Scheduling Model for LTV Steel's Cleveland Works' Twin Strand Continuous Slab Caster," *Interfaces* 18(1), 42–56.

Chan, W., A. Li, and H. C. Co (1989). "Machining Parameter Selection for a Stochastic Flow System," *IIE Transactions* 21, 241–249.

Cheng, T. C. E. (1984). "Optimal Due-Date Determination and Sequencing of *n* Jobs on a Single Machine," *Journal of the Operational Research Society* 35, 433–437.

Cheng, T. C. E. (1987). "An Algorithm for the CON Due Date Determination and Sequencing Problem," *Computers and Operations Research* 14, 537–542.

Cheng, T. C. E. (1988). "Optimal Common Due-Date with Limited Completion Time Deviation," *Computers and Operations Research* 15, 91–96.

Cheng, T. C. E. (1989). "A Heuristic for Common Due-Date Assignment and Job Scheduling on Parallel Machines," *Journal of the Operational Research Society* 40, 1129–1135.

Cheng, T. C. E., and M. C. Gupta (1989). "Survey of Scheduling Research Involving Due-Date Determination Decisions," *European Journal of Operational Research* 38, 156–166.

Deleersnyder, J. L., T. J. Hodgson, R. E. King, P. J. O'Grady, and A. Savva (1992). "Integrating Kanban-type Pull Systems and MRP-type Push Systems: Insights from a Markovian Model," *IEE Transactions on Design and Manufacturing* 24, 43–56.

Diaz, A., L. Sancho, R. Garcia, and J. Larraneta (1991). "A Dynamic Scheduling and Control System in an ENSIDESA Steel Plant," *Interfaces* 21(5), 53–62.

DiMascolo, M., Y. Frein, Y. Dallery, and R. David (1991). "A Unified Modelling of Kanban Systems Using Petri Nets," *International Journal of Flexible Manufacturing Systems* 3, 275–307.

Dumond, J. (1985). *Evaluation of Due Date Assignment Rules for Dynamically Arriving Projects.* Ph.D. Thesis, Indiana University, Bloomington.

Dumond, J., and V. Mabert (1988). "Evaluating Project Scheduling and Due Date Assignment Procedures: An Experimental Analysis," *Management Science* 34, 101–118.

Eilon, S., and I. J. Chowdhury (1976). "Due-Date in Jobshop Scheduling," *International Journal of Production Research* 14, 223–237.

Eilon, S., and R. M. Hodgson (1967). "Job Shop Scheduling with Due-Dates," *International Journal of Production Research* 6, 1–13.

Elvers, D. (1973). "Job Shop Dispatching Rules Using Various Delivery Date Setting Criteria," *Production and Inventory Management* 14 (fourth quarter), 62–69.

Fry, T. (1990). "Controlling Input: The Real Key to Shorter Lead-Times," *International Journal of Logistics Management* 1, 7–12.

Glassey, C. R., and R. G. Petrakian (1989). "The Use of Bottleneck Starvation Avoidance with Queue Predictions in Shop Floor Control," Research Report ESRC 89-23, University of California, Berkeley.

Glassey, C. R., and M. G. C. Resende (1988a). "A Scheduling Rule for Job Release in Semiconductor Fabrication," *Operations Research Letters* 7, 213–217.

Glassey, C. R., and M. G. C. Resende (1988b). "Closed-Loop Job Release Control for VLSI Circuit Manufacturing," *IEEE Transactions on Semiconductor Manufacturing* 1, 36–46.

Goyal, S. K., and S. G. Deshmukh (1992). "A Critique of the Literature on Just-In-Time Manufacturing," *International Journal of Operations and Production Management* 12, 18–28.

Gravel, M., and W. L. Price (1988). "Using the Kanban in a Job Shop Environment," *International Journal of Production Research* 26, 1105–1118.

Inman, R. R., and R. L. Bulfin (1991). "Sequencing JIT Mixed-Model Assembly Lines," *Management Science* 37, 901–904.

Karmakar, U. S. (1989). "Capacity Loading and Release Planning with Work-In-Progress (WIP) and Leadtimes," *Journal of Manufacturing and Operations Management* 2, 105–123.

Kimura, O., and H. Terada (1981). "Design and Analysis of Pull System, a Method of Multi-Stage Production Control," *International Journal of Production Research* 19, 241–253.

Krajewksi, L. J., B. E. King, L. P. Ritzman, and D. S. Wong (1987). "Kanban, MRP and Shaping the Manufacturing Environment," *Management Science* 33, 39–57.

Kubiak, W., and S. Sethi (1991). "A Note on 'Level Schedules for Mixed-Model Assembly Lines in Just-In-Time Production Systems,'" *Management Science* 37, 121–122.

Leachman, R. C. (1989). "A Queue Management Policy for the Release of Factory Work Orders," Presented at ORSA/TIMS Conference, Vancouver.

Lefrancois, P., and M.-C. Roy (1990). "Estimation of Mean Flow Time in a Rolling-Mill Facility," *Journal of Manufacturing and Operations Management* 3, 134–152.

Mahmoodi, F., K. J. Dooley, and P. J. Star (1990). "An Evaluation of Order Releasing and Due-Date Assignment Heuristics in a Cellular Manufacturing System," *Journal of Operations Management* 9, 548–573.

Mejabi, O., and G. S. Wasserman (1992). "Basic Concepts of JIT Scheduling," *International Journal of Production Research* 30, 141–149.

Miltenburg, J. (1989). "Level Schedules for Mixed-Model Assembly Lines in Just-In-Time Production Systems," *Management Science* 35, 192–207.

Miltenburg, J., and G. Sinnamon (1992). "Algorithms for Scheduling Multilevel Just-In-Time Production Systems," *IEE Transactions on Design and Manufacturing* 24, 121–130.

Narayan, V., and T. E. Morton (1992). "Processing Speed vs. Capacity, Cost and Congestion," GSIA, Carnegie Mellon University, Pittsburgh, PA.

Park, T., and H. J. Steudel (1991). "A Model for Determining Job Throughput Times for Manufacturing Flow Line Workcells with Finite Buffers," *International Journal of Production Research* 29, 2025–2041.

Quaddus, M. A. (1987). "A Generalized Model of Optimal Due-Date Assignment by Linear Programming," *Journal of the Operational Research Society* 38, 353–359.

Ragatz, G. L., and V. A. Mabert (1984). "A Simulation Analysis of Due Date Assignment Rules," *Journal of Operations Management* 5, 27–39.

Roderick, L. M., D. T. Phillips, and G. L. Hogg (1992). "A Comparison of Order Release Strategies in Production Control Systems," *International Journal of Production Research* 30, 683–694.

Sarker, B. R., and J. A. Fitzsimmons (1989). "The Performance of Push and Pull Systems: A Simulation and Comparative Study," *International Journal of Production Research* 27, 1715–1731.

Seidmann, A., S. S. Panwalker, and M. L. Smith (1981). "Optimal Assignment of Due-Date for a Single Processor Scheduling Problem," *International Journal of Production Research* 19, 393–399.

Sen, T., and S. K. Gupta (1984). "A State-of-the-Art Survey of Static Scheduling Research Involving Due Dates," *Omega* 12, 62–76.

Spearman, J. L., W. J. Hopp, and D. L. Woodruff (1989). "A Hierarchical Control Architecture for Constant Work-In-Progress (CONWIP) System," *Journal of Manufacturing and Operations Management* 3, 147–171.

Sugimori, Y., K. Kusunoki, F. Cho, and S. Uchikawa (1977). "Toyota Production System and Kanban System Materialization of Just-In-Time and Respect-For-Human System," *International Journal of Production Research* 15, 553–564.

Sumichrist, R. T., and R. S. Russell (1990). "Evaluating Mixed-Model Assembly-line Sequencing Heuristics for Just-in-Time Production Systems," *Journal of Operations Management* 9, 371–390.

Takahashi, K., and T. Kateeshock (1992). "Expert System for Refinery Off-Site Facility Management," *ISA Transactions* 31, 67–75.

Taylor, F. W. (1907). "On the Art of Cutting Metals," *ASME Transactions* 28, 310–350.

Una, A. T., A. S. Kiran, and R. Uzsoy (1991). "Due-Date Determination on a Single Machine with Part Type Dependent Setup Times," USC Working Paper, University of Southern California, Los Angeles.

Weeks, J. K. (1979). "A Simulation Study of Predictable Due-Dates," *Management Science* 25, 363–373.

Wein, L. M. (1991). "Due-Date Setting and Priority Sequencing in a Multiclass M/G/1 Queue," *Management Science* 37, 834–850.

Chapter 21

Adachi, T., C. L. Moodie, and J. J. Talvadge (1988). "A Pattern Recognition-Based Method for Controlling a Multi-Loop Production System," *International Journal of Production Research* 26, 1943–1957.

Adachi, T., C. L. Moodie, and J. J. Talvadge (1989). "A Rule-Based Control Method for a Multi-Loop Production System," *Artificial Intelligence in Engineering* 4, 115–125.

Adelsberger, H. H., and J. J. Kanet (1991). "The LEITSTAND—A New Tool for Computer-Integrated Manufacturing," *Production and Inventory Management* 32 (1), 43–48.

Bai, X., N. Srivatsan, and S. B. Gershwin (1990). "Hierarchical Real-Time Scheduling of a Semiconductor Fabrication Facility," *Proceedings of the Ninth IEEE International Electronics Manufacturing Technology Symposium,* Washington, DC.

Bector, C. R., Y. P. Gupta, and M. C. Gupta (1988). "Determination of an Optimal Common Due Date and Optimal Sequence in a Single Machine Job Shop," *International Journal of Production Research* 26, 613–628.

Bensana, E., G. Bel, and D. Dubois (1988). "OPAL: A Multi-Knowledge-Based System for Industrial Job Shop Scheduling," *International Journal of Production Research* 26, 795–819.

Bitran, C. R., E. A. Haas, and A. C. Hax (1980). "Hierarchical Production Planning: A Multi-stage System," Technical Report 179, Operations Research Center, M.I.T., Cambridge, MA.

Bitran, C. R., E. A. Haas, and A. C. Hax (1981). "Hierarchical Production Planning: A Single Stage System," *Operations Research* 29, 717–743.

Cheng, T. C. E. (1990). "Common Due-Date Assignment and Scheduling for a Single Processor to Minimize the Number of Tardy Jobs," *Engineering Optimization* 16, 129–136.

Cheng, T. C. E. (1991). "Optimal Assignment of Total-Work-Content Due-Dates and Sequencing in a Single-Machine Shop," *Journal of the Operational Research Society* 42, 177–181.

Cheng, T. C. E., and M. C. Gupta (1989). "Survey of Scheduling Research Involving Due-Date Determination Decisions," *European Journal of Operational Research* 38, 156–166.

Daniels, R. L., and R. K. Sarin (1989). "Single Machine Scheduling with Controllable Processing Times and Number of Jobs Tardy," *Operations Research* 37, 981–984.

Fox, Mark S., B. P. Allen, S. F. Smith, and G. A. Strom (1983). *ISIS: A Constraint-Directed Reasoning Approach to Job Shop Scheduling. A System Summary.* The Robotics Institute, Carnegie Mellon University, Pittsburgh, PA.

Fry, T. D., J. F. Cox, and J. H. Blackstone, Jr. (1992). "An Analysis and Discussion of the Optimized Production Technology Software and Its Use," *Production and Operations Management* 1, 229–242.

Gershwin, S. B., R. Akella, and Y. F. Choong (1985). "Short-Term Production Scheduling of an Automated Manufacturing Facility," *IBM Journal of Research and Development* 29, 392–400.

Glover, F., G. Jones, D. Karney, D. Klingman, and J. Mote (1979). "An Integrated Production, Distribution and Inventory Planning System," *Interfaces* 9(5), 21–35.

Godin, Victor B. (1978). "Interactive Scheduling: Historical Survey and State of the Art," *AIIE Transactions* 10, 331–337.

Graves, S. (1982). "Using Lagrangian Techniques to Solve Hierarchical Production Planning Problems" *Management Science* 28, 260–275.

Hadavi, K., and K. Voight (1987). "An Integrated Planning and Scheduling Environment," *Proceedings of the Simulation and Artificial Intelligence Society in Manufacturing*, Long Beach, CA: Society of Engineers.

Hax, A. C., and H. C. Meal (1975). *Hierarchical Integration of Production Planning and Scheduling, TIMS Studies in Management Sciences* 1, 53–69.

Jaikumar, R. (1974). "An Operational Optimization Procedure for Production Scheduling," *Computers and Operations Research* 1, 191–200.

Klingman, D. (1977). "Finding Equivalent Network Formulations for Constrained Network Problems," *Management Science* 23, 737–744.

Kusiak, A., and M. Chen (1988). Expert Systems for Planning and Scheduling Manufacturing Systems," *European Journal of Operational Research* 34, 113–130.

Leachman, R. C. (1986). "Preliminary Design and Development of a Corporate-Level Production Planning System for the Semiconductor Industry," OR Center, University of California, Berkeley.

Leachman, R. C., and A. Gascon (1988). "A Heuristic Scheduling Policy for Multi-item Single-Machine Production Systems with Time-Varying, Stochastic Demands," *Management Science* 34, 377–390.

Lecocq, P., and T. Guiot (1988). "Expert System for Production Planning and Scheduling," *Knowledge Based Production Management Systems. Proceedings of the IFIP*, WG 5.7 Working Conference, Galway, Ireland.

Lehoczky, J. P., and L. Sha (1986). "Performance of Real-Time Bus Scheduling Algorithms," *ACM Performance Evaluation Review*, Special Issue Vol. 14, No. 1.

Lehoczky, J. P., L. Sha, and J. Strosnider (1987). "Aperiodic Scheduling in a Hard Real-Time Environment," *Proceedings of the IEEE Real-Time Systems Symposium* 261–270.

Liu, J. W. S., K. J. Lin, and S. Natarajan (1987). "Scheduling Real-Time, Periodic Jobs Using Imprecise Results," *Proceedings of the IEEE Real-Time Systems Symposium* 252–260.

Lundrigan, R. (1986). "What Is This Thing Called OPT?" *Production and Inventory Management* 27 (second quarter), 2–12.

Marsh, C. A. (1985). "MARS—An Expert System Using the Automated Reasoning Tool to

Schedule Resources," *Robotics and Expert Systems,* Proceedings of Robex's 85–Instrument Society of America, 123–125.

Meleton, A. S. (1986). "OPT—Fantasy or Breakthrough," *Production and Inventory Management* 27 (second quarter), 13–21.

Morton, T. E., S. Kekre, S. Lawrence, and S. Rajagopalan (1988). "SCHED-STAR: A Price Based Shop Scheduling Module," *Journal of Manufacturing and Operations Management* 1, 131–181.

Morton, T. E., S. Lawrence, and G. L. Thompson (1986). "MRP-STAR: PATRIARCH's Planning Module," GSIA, Carnegie Mellon University, Pittsburgh, PA.

Newson, E. F. P. (1975). "Multi-item Lot Size Scheduling by Heuristic Part I: With Fixed Resources" and "Part II: With Variable Resources," *Management Science* 21, 1186–1203.

Osleeb, J. P., and A. G. Cromley (1977). "Location-Throughput-Allocation Problem with Nonlinear Throughput Costs," *Geographical Analysis* 9, 142–159.

Ow, P. S., and S. F. Smith (1987). "Viewing Scheduling as an Opportunistic Problem Solving Process," *Annals of OR: Approaches to Intelligent Decision Support.* Basel, Switzerland: Baltzer Scientific Publishing Company.

Prietula, M. J., and G. L. Thompson (work in process). GSIA, Carnegie Mellon University, Pittsburgh, PA.

Rajkumar, R. (1989). *Task Synchronization in Real-Time Systems.* Ph.D. Thesis, Carnegie Mellon University, Pittsburgh, PA.

Rao, Raghav H., and B. P. Lingaraj (1988). "Expert Systems in Production and Operations Management: Classification and Prospects," *Interfaces* 18 (6), 80–91.

Roundy, R., W. Maxwell, Y. Herer, S. Tayur, and A. Getzler (1988). "A Price-Directed Approach to Real-Time Scheduling of Production Operations," Technical Report No. 823, School of Operations Research and Industrial Engineering, College of Engineering, Cornell University, Ithaca, NY.

Sathi, Arvind, and Contributors: Mark S. Fox and Mike Greenberg (1985). *Representation of Activity Knowledge for Project Management.* Pittsburgh, PA: Carnegie Mellon University, The Robotics Institute.

Savell, D., R. Perez, and S. Koh (1989). "Scheduling Semiconductor Wafer Production: An Expert System Implementation," *IEEE Expert* 4, 9–15.

Shaw, M. J. (1988). "Knowledge-Based Scheduling in Flexible Manufacturing Systems: An Integration of Pattern-Directed Inference and Heuristic Search," *International Journal of Production Research* 26, 821–844.

Shaw, M. J., S. Park, and N. Raman (1992). "Intelligent Scheduling with Machine Learning Capabilities: The Induction of Scheduling Knowledge," *IEE Transactions on Design and Manufacturing* 24, 156–168.

Shaw, M. J., and A. B. Whinston (1989). "An Artificial Intelligence Approach to the Scheduling of Flexible Manufacturing Systems," *IIE Transactions* 21, 170–183.

Smith, S. F., N. Muscettola, D. C. Matthys, Peng Si Ow, and J. Y. Potvin (1990). "OPIS: An Opportunistic Factory Scheduling System," *Proceedings of the Third International Conference on Industrial and Expert Systems (IEA/AIE 90),* Charleston, SC.

Spence, A. M., and D. J. Welter (1987). "Capacity Planning of a Photolithography Work Cell in a Wafer Manufacturing Line," *Proceedings of the IEEE Conference on Robotics and Automation* 702–708.

Sprunt, B., L. Sha, and J. Lehoczky (1989). "Aperiodic Task Scheduling for Hard-Real-Time Systems," *Real-Time Systems*, 27–60.

Srinivasan, A., and T. E. Morton (work in process). "LOGJAM—A Proposed System for Strategic Management of Bottleneck Congestion in Logistic System Networks," GSIA, Carnegie Mellon University, Pittsburgh, PA.

Srivatsan, N., and S. B. Gershwin (1990). "Selection of Setup Times in a Hierarchically Controlled Manufacturing System," *Proceedings of the 29th IEEE Conference on Decision and Control,* Hawaii.

Sullivan, G., and K. Fordyce (1990). "IBM Burlington's Logistics Management System," *Interfaces* 20(1), 43–64.

APPENDIX B
GLOSSARIES

B.1 ACRONYM GLOSSARY

This glossary contains primarily acronyms, heuristic names, and software names frequently appearing in the book away from their initial defining material.

AC	Average cost
AI	Artificial intelligence
AOA	Activity on arc project network
AON	Activity on node project network
AWINQ	Anticipated work in next queue heuristic
B&B	Branch-and-bound
BD	Bottleneck dynamics
BOTFLOW	Bottleneck heuristics for the flow shop
BOTJOB	Bottleneck heuristics for the job shop
CAE	Cost-at-end preemption
CALLISTO	Expert system project management software
CDS	Heuristic, makespan flow shop
CIM	Computer integrated manufacturing
CONWIP	American system to approximate JIT
COVERT	A heuristic for weighted tardiness
CPM	Critical path method
CRIT. RATIO	A heuristic for weighted tardiness
DEDD	Dynamic EDD heuristic
DP	Dynamic programming
DSS	Decision support system

EDD	Earliest due date heuristic
EEDD	Earliest expected due date heuristic
EFT	Early finish time heuristic
EORP	Embedded one-resource problem
ERORP	Embedded relaxed one-resource problem
EST	Early start time heuristic
EXPET	A heuristic for the early–tardy problem
FASFS	First shop arrival first served
FCFS	First come first served
FMC	Flexible manufacturing center
FMS	Flexible manufacturing system
FOFO	First on first off heuristic
GA	Genetics algorithm heuristic method
GUPTA	Heuristic, makespan flow shop
HODGSON	Procedure to minimize number of tardy jobs
IBM	International Business Machines
ISIS	An expert system scheduling software
JIT	Just-in-time
JOHNSON	Algorithm, two-machine makespan flow shop
KANBAN	Japanese card system to implement JIT
LEITSTAND	Popular DSS scheduling systems (German)
LFT	Late finish time heuristic
LINET	Heuristic for early–tardiness problem
LOGJAM	A bottleneck methodology for logistics
LP	Linear programming
LPT	Longest processing time heuristic
LST	Late start time heuristic
LTI	Lead time iteration
LWKR	Least work remaining
MERLE	An expert system scheduling software
MIS	Management information systems
MLP	Multiple large project problem
MONTAGNE	A heuristic for weighted tardiness
MRP	Material requirements planning
MRP-II	Corporate integrated version of MRP
MRP-STAR	MRP updated with bottleneck dynamics
MST	Minimum slack time heuristic
NC	Numerically controlled
NLP	Nonlinear programming
NPV	Net present value
OPAL	An expert system scheduling software
OPIS	An expert system scheduling software
OPNDD	EDD but with operation (local) due dates
OPT	A proprietary bottleneck method software
OR	Operations research

PALMER	Heuristic, makespan flow shop
PATRIARCH	A bottleneck dynamics scheduling software
PDEDD	Preemptive dynamic EDD
PDWSPT	Preemptive dynamic WSPT
PERT	Program evaluation and review technique
PRODUCE	Early interactive scheduling system
R&M	Rachamadugu and Morton tardiness heuristic
SA	Simulated annealing heuristic method
SBI	Simpler version of SBM
SBII	More complex version of SBM
SBM	Shifting bottleneck method
SC	Spread-cost preemption
SCHED-STAR	NPV, BD, scheduling software
S/OP	Slack per remaining operation heuristic
SPT	Shortest processing time heuristic
TS	Tabu search heuristic method
TSC	Tight-spread-cost preemption
TSPT	Truncated SPT heuristic
TWORK	Least work in the shop heuristic
WIP	Work in process inventory
WSEPT	WSPT using expected processing time
WSPT	Weighted shortest processing time heuristic
WTWORK	Weight per TWORK heuristic

B.2 SYMBOL GLOSSARY

Chapter 4

j	Job index
k	Machine index
m	Number of machines
n	Number of jobs
t	Current time
t_s	Scheduling start time
t_e	Scheduling end time
p_j	Job j processing time on single machine
p_{jk}	Job j processing time on machine k
w_j	Job flow (lateness) weight
w_{Tj}	Job tardiness weight
w_{Nj}	Job "drop dead" weight
w_{Ej}	Job earliness weight
r_j	Job availability time
ρ_k	Machine utilization factor
a_k	Machine availability time

d_j	Job completion due date
dd_{jk}	Job estimated due date off machine k
C_j	Job completion time
F_j	Job flowtime
L_j	Job lateness
T_j	Job tardiness
E_j	Job earliness
C_{max}	Makespan objective
F_{wt}	Weighted flowtime objective
L_{wt}	Weighted lateness objective
T_{wt}	Weighted tardiness objective
F_{max}	Maximum flowtime objective
L_{max}	Maximum lateness objective
T_{max}	Maximum tardiness objective
N_{wt}	Weighted number of tardy jobs
ET_{wt}	Weighted earliness plus weighted tardiness
$\delta(x)$	1 for positive x, 0 otherwise
$E[\ ]$	Expectation operator (statistics)
D	Dollar value of job
A	Customer priority factor
z	Objective function value
f	Objective function

Chapters 5 and 6

$iR(t)j$	Job i is preferred to come before job j
$\pi(i, t)$	Priority of job i at time t
π_{jk}	Priority of job j on machine k
P_6	Collection of all job sequences ending in job 6
σ	General subsequence constrained to be last
$R(t)$	Heuristic price of resource at time t
$R_k(t)$	Heuristic price of machine k at time t
I	Firm's cost of capital
$S_j(t)$	Slack of job j at time t
$U_j(t)$	Marginal forecast cost of delaying job j at t
Δ	A small delay in a sequencing decision

Chapters 7–9

$\pi_{j\ idle}$	Corrected priority for inserted idletime
T_{ij}	Setup time if job j comes immediately after job i
Δ_{kji}	Net savings of sequence kji over kij with setups
π_{kji}	Priority for ji preceded by k with setups
π_{kj}	Priority for j if preceded by k with setups
J, I	Indices for lots in lot-sizing model

QJ	Actual lot size
dd_j	Local machine, derived due date
π_{pre}	Priority of a given preemption
ji	Operation i of job j
HT_{ji}	Head time of ji
TT_{ji}	Tail time of ji
MT_j	Make time (total processing time) of j
ST	Span time (relaxed makespan)
HLT_{ji}	Head lead time of ji
TLT_{ji}	Tail lead time of ji (also called lead time)
r_{ji}	Arrival time at ji's machine
d_{ji}	Derived due date off ji's machine

Chapter 10–12

t_{kr}	Time reaching machine k if route r
ρ_k	Resource utilization factor
m'	Effective number of parallel machines
$p(N, k)$	Processing time of Nth job on machine k
R_{agg}	Aggregate machine price
I_j	Idleness caused by j
N_j	Normalized regret of not assigning j to its best place
K_{jk}	Nonnormalized regret of assigning j to k
τ_{jk1}, τ_{jk2}	Arrival, departure transport times, j to k

Chapters 13–16

E	Economic makespan
l_{jk}	Lead time from machine k for job j
τ	Tardiness factor
s_j	Actual start time of j

Chapters 17–21

$A < B$	A is a predecessor of B
$j \in A(t)$	j is an active activity at time t
RA_k	Amount of resource k available
$P(j)$	Immediate predecessors of activity j
r_{jk}	Rate that activity j uses resource k
i	Designation of a project among several
ij	Activity j with project i
P_{ij}	Predecessors in i to ij
l_{ij}	Late finish time of j in project i
e_{ij}	Early finish time of j in project i
$(qf)_{ij}^n$	Queuing time forecast for ij in nth iteration

α	Iteration smoothing factor
$(lf)_{ij}^{n}$	Resource adjusted ij late start times in nth iteration
$(RU)_{ij}$	Total earlier resource usage if ij scheduled early
$A_{ijk}(P_{ij})$	Usage of resource k to do ij if processing time is P_{ij}
$(AM)_{ijq}$	Cost of doing ij by method q
$v(x)$	Fraction of a batch wasted, continuous processes

APPENDIX C

GETTING STARTED WITH THE PARSIFAL™ SOFTWARE

This section includes instructions for setting up the *Parsifal* software on your computer. The instructions are written with the assumption that your floppy disk drive is called **A:** and your hard disk drive is called **C:**. If you are using the package on a computer with different names for the drives, merely substitute your floppy or hard disk drive letter for the ones used below.

C.1   HARDWARE REQUIREMENTS

To use the *Parsifal* software package that is included with this book, you must have and IBM or compatible computer with the minimum equipment listed below.

- Hard disk drive
- DOS 3.0 or higher
- 640K memory
- 80286 processor

In addition, to print reports as discussed in section C.8, you must have a printer attached.

C.2   MAKING A BACKUP COPY

Before you install the enclosed disk, it is strongly recommended that you make a backup copy. Making a backup copy will allow you to have a safe, clean copy of the program for archival purposes. Remember, however, that making more than one backup copy for archival purposes is in violation of the copyright law.

To make a backup of the *Parsifal* disk:

1. Place your DOS disk in the drive A of your computer.
2. At the **A:\>** type **DISKCOPY A: A:** and press ↵.
3. When prompted for the source disk, place the original *Parsifal* disk into drive A and press ↵.
4. When prompted for the target disk, place a blank disk into drive A and press ↵.
5. Continue to follow the directions on screen to complete the copy.

When the disk copy is completed, remove the target disk and label it **Parsifal Backup Disk**. Put the original *Parsifal* disk in a safe place and use the backup for installation.

C.3   INSTALLING PARSIFAL

To install the program, place the *Parsifal Backup Disk* into drive A of your computer and follow these steps:

1. Type **A:** and press ↵.
2. At the **A:\>**, type **INSTALL** and press ↵.

The *Parsifal* Installation Program will appear on screen. You will be prompted for the name of the hard disk drive and subdirectory where you want to install the program. When prompted, type in the appropriate drive letter and subdirectory name. The default is to install the program on a hard disk drive named **C:** in a subdirectory named **PARSIFAL**. After installation is completed, take out the *Parsifal Backup Disk*. The program is now loaded on your hard disk.

C.4   STARTING PARSIFAL

To start the *Parsifal* program, go to the subdirectory on your hard disk where you have installed the program. If you used the default installation setup, you can start the program by following these instructions:

1. At the DOS prompt, type **C:** and press ↵.
2. At the **C:\>**, type **CD\PARSIFAL** and press ↵.
3. At the **C:\PARSIFAL>** prompt, type **PARSIFAL** and press ↵.

The *Parsifal* program will load automatically. For instructions on using the package, please refer to the User's Guide in Appendix D. It provides detailed information on the program.

C.5   TECHNICAL NOTES AND TROUBLESHOOTING

If you have difficulty running *Parsifal*, you can consult this section for technical notes and troubleshooting advice. If you are not familiar with the commands mentioned in DOS, then consult your lab coordinator or systems adminstrator for additional help. In addition, you can refer to your DOS manual.

Check the file called **CONFIG.SYS** in the root directory of your hard disk. It should have at least the following parameters for FILES and BUFFERS:

FILES=40
BUFFERS=40

In addition, you may have a statement in your **CONFIG.SYS** file that reads as follows:

SHELL=C:\DOS\COMMAND.COM /E:1024

This command allocates memory to DOS. The first part of the statement, C:\DOS, refers to the location of the file **COMMAND.COM** on your hard disk. If you have **COMMAND.COM** in the root directory, this statement will look different. The second part of the statement allocates 1024 bytes of memory for DOS. This number is conservative; DOS can operate with as little as 170 bytes. You can edit this statement to lower the number, therefore allocating more memory to run the *Parsifal* program.

C.6   USER DOCUMENTATION

In addition to Appendix D in this book, there is a file on the *Parsifal* disk called **README.TEX** which is installed on your computer. This file contains a detailed tutorial on using the program. To read this file, follow the steps below:

1. At the **C:\PARSIFAL>** prompt, type **README** and press ↵.
2. Browse through the file on screen by using the ↑ and ↓ cursor keys.
3. To exit the **README** file, press ↵.

You will be returned to DOS. To print this information on your printer, type the following at the **C:\PARSIFAL>** prompt:

PRINT C:\PARSIFAL\README.TEX and press ↵.

This file will print on any type of printer that is capable of printing straight ASCII text from DOS. The full **README.TEX** file is over 50 pages long, so make sure to have plenty of paper in your printer.

C.7 STORING USER FILES

If you manually define problems using the *Parsifal* software, then the files are stored in a subdirectory called **USER** which can be found under the subdirecories for each program. For your information, below is a listing of the location of any user defined problems:

Single Machines	**C:\PARSIFAL\SINGLE\USER**
Parallel Machines	**C:\PARSIFAL\SINGLE\USER**
Flow Shops	**C:\PARSIFAL\JOBSHOP\FLOWSHP\USER**
Job Shops	**C:\PARSIFAL\JOBSHOP\JOBSHP\USER**

Parallel machines problems are essentially input sets for a single machine with speed factors correcting for the multiple machine case. Hence, the problems for parallel machines are stored in the same place as those for single machines.

All job input sets are stored in files with the extension **SET**. In case of Flow Shops and Job Shops, there is an additional file which contains the Shop input set and is saved with the extension **SHP**. To erase user defined files, use the following command at the DOS prompt:

DEL LOCATION\FILENAME.SET

For example, to delete a single machines problem called **ASSIGN1** from the hard disk, type the following at the DOS prompt:

DEL C:\PARSIFAL\SINGLE\USER\ASSIGN1.SET

However, to delete a flow shop problem called **HOMEWORK** from the hard disk, you would need to type the following two statements at the DOS prompt:

DEL C:\PARSIFAL\JOBSHOP\FLOWSHP\USER\HOMEWORK.SET
DEL C:\PARSIFAL\JOBSHOP\FLOWSHP\USER\HOMEWORK.SHP

The first statement deletes the job input file, and the second deletes the shop input file.

C.8 PRINTING REPORTS

The software allows you to save the output report from each run. If you do so, then you can find the output reports in the following directories:

Single Machines	**C:\PARSIFAL\SINGLE**
Parallel Machines	**C:\PARSIFAL\PAR**
Flow Shops	**C:\PARSIFAL\JOBSHOP\FLOWSHP**
Job Shops	**C:\PARSIFAL\JOBSHOP\JOBSHP**

The output reports are saved in ASCII format under the name that you specify. You can give both a name and extension to the report file. In addition, you can view, edit or print the file using any word processing or editing program. To print output reports through DOS, use the following command at the DOS prompt:

PRINT LOCATION\FILENAME.EXT

For example, to print a single machines problem called**ASSIGN1.TXT** from the hard disk, type the following at the DOS prompt:

PRINT C:\PARSIFAL\SINGLE\ASSIGN1.TXT

The filename and extension that you specify must conform to DOS standards, so the filename cannot exceed 8 characters and the extension can be no longer than 3 characters. Most word processing programs look for ASCII text files with the extension TXT, so you may want to use that extension to make your files easier to find.

C.9 EXECUTING CUSTOMIZED OBJECTIVE FUNCTIONS AND HEURISTICS

Whenever the program is called upon to run a user-defined objective function, it calls the program **CUSTOMOBJ.BAT**. When it runs user-defined heuristics it calls the program **CUSTHEU.BAT**. You can add whatever objective functions and heuristics by writing these programs in almost any programming language you like. In the subdirectory **C:\PARSIFAL\SAMPCODE** there are sample programs in BASIC, FORTRAN, PASCAL, and C which illustrate how to generate custom objective functions and/or heuristics. The programs each have comment lines which give detailed information for generating user-defined routines.

For example, if you compiled the user defined objective program**SAMPOBJ.C**, which is found in the subdirectory**C:\PARSIFAL\SAMPCODE**, you can create an executable file called **SAMPOBJ.EXE** in the same directory. Then you would create a DOS batch file called **CUSTOBJ.BAT** with the following line:

C:\PARSIFAL\SAMPCODE\SAMPOBJ

If you change subdirectories in the batch file then you have to return to the same subdirectory at the end of your routine. The software writes the necessary data into a file and reads the output from your program from a pre-defined file. The names of these files can be found in the example programs. The software will write data only in the subdirectory **C:\PARSIFAL\SINGLE** when using Single Machines and in the **C:\PARSIFAL\PAR** subdirectory when using Parallel Machines. In addition, it will read the data files from the same subdirectories. So take care that you read and write your own files in the same subdirectories.

PARSIFAL™ SOFTWARE
USER GUIDE

D.1 OVERVIEW

Parsifal makes it easy for the user to learn scheduling, whether at the beginning, intermediate, or advanced level. In addition, *Parsifal*:

- Includes all sorts of shop environments, from very simple to complex: one-machine shops, parallel-machine shops, flow shops, job shops.
- Allows the choice of a number of possible objective functions including makespan, lateness, tardiness, number of tardy jobs, maximum lateness, etc. In addition, the knowledgeable user can specify customized objectives.
- Avoids almost all the usual input drudgery by providing a large number of pre-specified problem input sets from a read-only library.
- Includes features for creating new problems from scratch as desired, for immediate use and/or to save as a new input set.
- Allows more advanced users to create larger sets of similar problems to obtain average results and/or do separate statistical studies using a Monte Carlo input facility.
- Provides a number of different solution methods and approximations, including:

manual move	dispatch heuristics
bottleneck dynamics	neighborhood search
tabu search	simulated annealing
beam search	breadth-first branch-and-bound
depth-first branch-and-bound	

- Lets the user save both input and output. Newly created problems can be saved in a user problem file for future easy input. Output can be saved in a file for later editing or printing.

For simplicity, the instructions in this Appendix use the following conventions:

- Commands, filenames and drive names are printed in a simple text font (e.g., **CD\PARSIFAL**)

- References to the chapters in this book are noted with the following symbol: ➲

D.2 GETTING STARTED

The following section assumes that *Parsifal* has previously been installed on your computer according to the installation instructions in Appendix C.

To start the program, follow these steps:

1. At the DOS prompt, type **CD\PARSIFAL** and press ↵.
2. Type **PARSIFAL** and press ↵.

The title screen will appear, which includes the startup options as listed in Figure D.1.

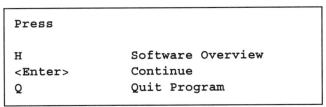

```
Press

H                Software Overview
<Enter>          Continue
Q                Quit Program
```

Figure D.1. Startup Screen

- Press **H** for **Software Overview** to view a help file called **README.TEX**. This file is also available on disk and can be printed by typing **PRINT README.TEX** at the DOS prompt. Once you have viewed the help file, press **ESC** to return to the startup screen.
- Press ↵ to move forward into the program.
- Press **Q** for **Quit Program** to leave *Parsifal*.

While there are cursor choices at the bottom of each screen, for the sake of brevity this documentation does not show the bottom line menu for every screen unless it contains new cursor choices or if the cursor choices are used differently.

Throughout the program, the following keys will always be available to perform the functions listed:

H	Provides you with on-screen help using the Software Overview.
↵	Moves you to the next screen in the program.
Q	Lets you quit the program and return to DOS.

In addition, **ESC** is always used to go back a level in the menu structure. However, it is not possible to move backward from a screen that shows computations or outputs from a run, so **ESC** is not shown as a bottom line menu option on this type of screen. At computation or output screens, you can always press ↵ to go forward to the next non-computational screen and then go back from there. To exit a help screen, press ↵.

To move past the startup screen, press ↵. You will be brought to the Main Menu, which gives you the choice of shop types as shown in Figure D.2.

```
                CHOOSE THE SHOP TYPE
     Single Machine
     Parallel Machines
     Flow Shop
     Job Shop
     <Esc> Previous Menu   Q Quit Program   H Help
```

Figure D.2. Shop Type Options

To select the desired option, simply type the appropriate first letter. It is necessary to understand the single-machine problem thoroughly before attempting the other options. Press **S** to view the Single Machine problem.

D.3   THE SINGLE-MACHINE PROBLEM

D.3.1   Input

When you have selected the Single-Machine option from the Main Menu, you are brought to the Job Input Menu as shown in Figure D.3.

This menu gives you the option to use data from an input file or to enter in new data. For the sake of simplicity, you can start by using data that has already been created in an input file. Manual input is not discussed until section D.3.9 of this Appendix. Press **F** to select File Input. You will be brought to the Single Machine Job Input Set Menu as shown in Figure D.4.

```
       SINGLE MACHINE
       JOB INPUT
File Input - Choose a problem from the input file
Manual Input - Create a new problem from scratch
<Esc> Previous Menu   Q   Quit   H   Help
```

Figure D.3. Single Machine Job Input Options

You can move around this menu by using the ↑ and ↓ arrow keys to highlight the different options. Below is a brief summary of the options:

Tutorial Problem	Uses the Tutorial problem from this User Guide
Previous Problem	Loads the last problem used by the program. This makes it easy to solve the same problem using a number of methods.
Static Problems	Includes twenty static problems called S1 to S20, each with a brief description of problem job characteristics, such as number of jobs, tardiness factor, and so on. Problems generated statistically may actually consist of perhaps 5 similar problems. You can highlight any one of the twenty problems and select it for use.
Dynamic Problems	Includes twenty dynamic problems called D1 to D20.
Larger Dynamic Problems	Includes twenty larger dynamic problems called L1 to L20.
User Problems	Displays problems which have been previously saved by the user. You can define the filenames; if there are many user-defined problems, the filenames can be scrolled on screen.

```
       SINGLE MACHINE
       JOB INPUT SETS
Tutorial Problem
Previous Problem
Static Problems
Dynamic Problems
Larger Dynamic Problems
User Problems
```

Figure D.4. Single Machine Job Input Set Menu

For this User Guide, we'll step through the tutorial problem. The tutorial problem has 12 jobs and tardiness factor = 0.7. To select the Tutorial Problem from the Single Machine Job Input Set Menu, highlight the menu option and press ↵.

The problem data set will appear on screen, as shown in Figure D.5, and will include one line for each job. Each line gives the job identification number, arrival time, processing time, due date, and weight. You can re-display this job input screen at other points during the program by using the V key to View it. The View option is only available on screens that include the **V   View** option in the bottom line menu. To return to the program while viewing the job input screen, press ↵.

```
SINGLE MACHINE
PROBLEM INPUT DISPLAY--1 MACHINE--Tutorial
    JOB      ARRIVAL   PROCESS  DUE DATE   WEIGHT
     1         0         1         7         2
     2         0         3        18         8
     3         0         1         8         1
     4         0         1         4         1
     5         0         3        13         7
     6         0         1         1         1
     7         0         2         0         4
     8         0         2         6         6
     9         0         3         8         7
    10         0         1        11         2
    11         0         1        11         4
    12         0         1         7         4
<Cursor Keys>  Scroll Table   <Enter> Exit
```

Figure D.5. Job Input for Tutorial Problem

➲ Job input issues are discussed in Chapters 4.1 and 4.2 of the text. Static one-machine problems are also discussed in Chapter 4.

You cannot back up from this screen by using the **ESC** key since it is not accessible at a numerical screen. To back up to the previous screen to select another data set, you must press ↵ to go forward to a non-numerical screen, and then press **ESC** to return to the first non-numerical screen before this, and then forward again. This is much easier than it sounds at first.

For now, press ↵ to move to the next screen. The Single Machine Solution Method Menu will appear, as shown in Figure D.6. This screen allows you to select the solution method for the data set.

```
SINGLE MACHINE
  SOLUTION METHOD (Choose One)
Manual Move and/or Dispatch Heuristics
Neighborhood Search--Tutor
Neighborhood/Tabu/Annealing Search
Beam Search/Branch & Bound
```

Figure D.6. Single Machine Solution Method Menu

The remainder of this section will discuss each of these solution methods in turn.

3.2 Manual Move

To select **Manual Move and/or Dispatch Heuristics** on the Single Machine Solution Method Menu in Figure D.6, highlight it and press⏎. The Single Machine Objectives Choices Menu in Figure D.7 will appear with a list of possible objectives to minimize, such as makespan, weighted lateness, or weighted tardiness.

➲ Objectives are discussed in Chapter 4.2.

You can select either one objective or multiple objectives. To choose only one objective, highlight it by using the ↑ and ↓ keys. Press the **SPACEBAR** to mark it. Then press ⏎ to move to the next screen. Similarly, to choose several objectives, highlight one objective by using the ↑ and ↓ keys and press the **SPACEBAR**. This will mark the first objective. Use the same procedure to mark other objectives. When you have finished marking objectives, press ⏎ to move to the next screen.

```
SINGLE MACHINE
  OBJECTIVE CHOICES (One or More)
Makespan
Maximum Lateness
Maximum Tardiness
Average Flowtime
Average Lateness
Weighted Flowtime
Weighted Lateness
Average Tardiness
Weighted Tardiness
Number of Tardy Jobs
Weighted Number of Tardy Jobs
Custom User-Defined Objective
```

Figure D.7. Single Machine Objective Choices Menu

To learn more about the manual mode in a simple way, we will step through a tutorial of just one objective, that of weighted tardiness. To select Weighted Tardiness, highlight it and press the **SPACEBAR** to mark it, then press ↵ to move on. The Single Machine Heuristics Choices Menu will appear, as shown in Figure D.8.

```
        SINGLE MACHINE
        HEURISTIC CHOICES   (One or More)
MANUAL MOVE      Interactive Creation of Solution
RANDOM           Equi-Probability Selection
FCFS             First Come First Served
SPT              Shortest Process Time
WSPT             Weighted Shortest Process Time
LPT              Longest Process Time
WLPT             Weighted Longest Process Time
EDD              Earliest Due Date
SLACK            Least Slack
HODGSON          Hodgson's Algorithm
WTD. HODGSON     Hodgson's Alg. Modified for Weights
R&M              Rachamadugu & Morton
CUSTOM           User's Choice (Requires Subroutine)
```

Figure D.8. Single Machine Heuristic Choices Menu

Like the Objective Choices Menu, you can select more than one option on this menu by using the ↑ and ↓ keys to highlight each choice and pressing the **SPACEBAR** to select each one. After making all your selections, you can press ↵ to move on.

The options on the Heuristic Choices Menu will be explained in detail later in this guide. This section of the tutorial will only review Manual Move. To select Manual Move in the Heuristic Choices Menu, highlight it and press the **SPACEBAR**. Then press ↵ to move to the next screen.

```
        SINGLE MACHINE
        OUTPUT FORMAT (One Choice)
Summary: Values of Heuristic(s) vs. Benchmark
Detailed: Input, Heuristic Values, Schedules
```

Figure D.9. Single Machine Output Format Menu

You are now in manual mode. The next screen is the Single Machine Output Format Menu, as shown in Figure D.9. This menu gives options for the output report, either Summary or Detailed. A Summary report merely includes the values of the heuristics vs. the benchmark value. A Detailed report shows the input, the heuristic values and the schedules.

For this tutorial, the summary is enough. To select the Summary option, highlight it using the ↑ and ↓ keys and press ↵. You will be moved to the next screen, which allows you to manually manipulate the order of jobs and see the results of changes on the objective function.

The manual mode initially schedules the jobs in the order given from the problem input, and gives an evaluation of how good the solution is. The manual mode is limited to job sets with no more than 16 jobs. The user can then move individual jobs one at a time forward or back in the schedule to try to improve the solution. Remember that you are using the Tutorial problem with the weighted tardiness objective function. The initial manual screen is shown in Figure D.10.

```
SINGLE MACHINE
MANUAL MOVE--1 MACHINE--Tutorial--WEIGHTED TARDINESS
  JOB  ARRIV  PROC  DUE D   WT   COMPL  LATE  TARDY   W/P
   1     0     1     7     2     1     -6     0     2.00
   2     0     3    18     8     4    -14     0     2.67
   3     0     1     8     1     5     -3     0     1.00
   4     0     1     4     1     6      2     2     1.00
   5     0     3    13     7     9     -4     0     2.33
   6     0     1     1     1    10      9     9     1.00
   7     0     2     0     4    12     12    12     2.00
   8     0     2     6     6    14      8     8     3.00
   9     0     3     8     7    17      9     9     2.33
  10     0     1    11     2    18      7     7     2.00
  11     0     1    11     4    19      8     8     4.00
  12     0     1     7     4    20     13    13     4.00
  (1)   (2)   (3)   (4)   (5)   (6)    (7)   (8)    (9)
              WEIGHTED TARDINESS   268.0
     <Space> Lock/Unlock Job Position <X> Clear* (etc.)
```

Figure D.10. Initial Manual Solution for Tutorial, Weighted Tardiness

This screen contains columns of data marked at the bottom with the column number in parentheses. The first five columns repeat the input data: (1) job number, (2) arrival time, (3) process time, (4) due date, and (5) weight. The remaining columns are as follows:

Column (6) The completion time of the job, which is just the maximum of the completion time of the previous job and the current job's arrival time which gives the new start time, plus the process time to give its finish time.

Column (7)   The lateness of the job in this schedule which is the completion time minus the due date.

Column (8)   The tardiness of the job, which is just the lateness, except that negative lateness is replaced by zero.

Column (9)   The weight of the job divided by the processing time of the job, sometimes called the "benefit/cost" ratio.

➲ See Chapter 4.4, Propositions 10, 11, 12 for more information on the benefit/cost ratio.

Since problem input has been defined to be in integer format, all of these columns are in integers for the one-machine problem, except the last column.

For the weighted lateness objective, or for the weighted tardiness objective for jobs known to be tardy, the size of the benefit/cost ratio is a good indicator of the priority of the job. On the other hand, using earliest due date as the priority is a good way to minimize maximum tardiness (➲ Chapter 4.4, Propositions 13 and 14). The Figure D.at the bottom of the table presents the weighted tardiness for the current schedule, which is the sum over all jobs of weight times tardiness for each. (➲ Chapter 4.5 and 7.2). As we try to move jobs around in the schedule to improve it, we can watch the objective to see whether we are really improving the solution.

As you are solving a problem with Manual Move, you may wish to start over with the original data. Since you cannot back up using **ESC** (remember, it does not work in the middle of computations), you must move forward by pressing ↵ to get to the Run Summary screen. Since this is also a computational screen, go forward again by pressing ↵ to get to the End of Run screen. From there, you can press **ESC** twice to return to the Output Format Menu. At that menu, you can move forward again through the Heuristic Choices Menu to view the Manual Move screen with the original data.

At the Manual Move screen, you can manually change the order of the jobs. To move around a job, follow these steps:

- Using the ↑ or ↓ key, move to the job to highlight it.
- When the job is highlighted, press the **SPACEBAR** to mark it with a + sign.
- Once the job has been marked, you can press the ↑ or ↓ key to move it up or down in the order.
- When you have moved the job to the new position, press the **SPACEBAR** again. The job will be placed into the new position and marked with an asterisk (*) to show that the job has been moved.

As you move around jobs, you can watch the objective change to reflect the effect of each job in its new position. The change in the objective can help you decide where to leave the job. Since an asterisk (*) is used to mark all moved jobs, you can easily see what has been changed. To remove the asterisks, you press **X**.

```
SINGLE MACHINE
MANUAL MOVE--1 MACHINE--Tutorial--WEIGHTED TARDINESS
  JOB   ARRIV  PROC  DUE D   WT   COMPL  LATE  TARDY   W/P
   7      0     2      0      4     2      2      2     2.00
   6      0     1      1      1     3      2      2     1.00
   4      0     1      4      1     4      0      0     1.00
   8      0     2      6      6     6      0      0     3.00
   1      0     1      7      2     7      0      0     2.00
  12      0     1      7      4     8      1      1     4.00
   3      0     1      8      1     9      1      1     1.00
   9      0     3      8      7    12      4      4     2.33
  10      0     1     11      2    13      2      2     2.00
  11      0     1     11      4    14      3      3     4.00
   5      0     3     13      7    17      4      4     2.33
   2      0     3     18      8    20      2      2     2.67
  (1)    (2)   (3)    (4)    (5)   (6)    (7)    (8)    (9)
              WEIGHTED TARDINESS   103.0
```

Figure D.11. Dispatch EDD Heuristic Table for Tutorial, Weighted Tardiness

Try to find a manual solution to this problem in systematic fashion. To do this, it is useful to first construct a fairly good solution in a standard way. You know that the dispatch EDD heuristic (➲ Chapter 4.4) will minimize the maximum tardiness to a good approximation. To construct this solution, determine the job with earliest due date which is available at time zero (in this case all jobs are), and move it to the first position. Then find the earliest due date job which can start at the first job's finish, and so forth. Once this procedure is finished, you should have the schedule shown in Figure D.11. Notice that the maximum tardiness has indeed been reduced considerably, although there are now more tardy jobs. The objective has been improved from 268.0 to 103.0.

Now we can apply a different idea: moving low benefit/cost jobs which are tardy anyway to a later position in the schedule, but being careful not to move enough past a given job that it would try to start before its arrival time, and cause idle time on the machine. The software warns of this by showing an **I** next to job numbers which are currently causing such inserted idleness. This will not happen in the static case.

First move job 3 to the end of the list, which improves the objective from 103.0 to 86.0. Then move job 10 just before job 3, which improves the objective from 86.0

to 81.0. Note that it is somewhat risky to move past a job which then becomes early, since the benefit/cost interchange assumes credit is being given for being early, while there is no such credit for weighted tardiness. Finally, move job 1 down just before job 2, but do not move it past it, since job 2 would have negative lateness. (You may want to see what happens to the solution value if you do.) This improves the tardiness from 81.0 to 75.0. After some trials, you'll learn that further improvement is too much work, and so you can stop. You should see the screen from Figure D.12.

```
SINGLE MACHINE
MANUAL MOVE--1 MACHINE--Tutorial--WEIGHTED TARDINESS
  JOB   ARRIV  PROC  DUE D   WT    COMPL  LATE  TARDY   W/P
   7      0     2      0      4      2     2      2     2.00
   6      0     1      1      1      3     2      2     1.00
   4      0     1      4      1      4     0      0     1.00
   8      0     2      6      6      6     0      0     3.00
  12      0     1      7      4      7     0      0     4.00
   9      0     3      8      7     10     2      2     2.33
  11      0     1     11      4     11     0      0     4.00
   5      0     3     13      7     14     1      1     2.33
   1      0     1      7      2     15     8      8     2.00
   2      0     3     18      8     18     0      0     2.67
  10      0     1     11      2     19     8      8     2.00
   3      0     1      8      1     20    12     12     1.00
  (1)    (2)   (3)    (4)    (5)    (6)   (7)    (8)   (9)
              WEIGHTED  TARDINESS   75.0
```

Figure D.12. Benefit/Cost Improvement to Schedule Table

Although this is a vast improvement over the original weighted tardiness of 268.0, it is not optimal, as you shall soon see by comparing the result with a benchmark. Press ⏎ to indicate that the current solution represents the final version of the manual heuristic.

The output for the run will appear next on a Run Summary screen. It shows the objective value of your solution as 75.0 and the benchmark soluation at 59.0, with the percentage deviation from benchmark as 27%.

Press ⏎ to move to the next screen, which gives you following options for the End of the Run:

Return to Solution Methods   Choose this option to return to the Solution Methods Menu as shown in Figure D.6.

Quit the Program   Choose this option to exit from *Parsifal.*

Save Output   Choose this option to save the results of your run. You will be asked for a filename for the output file. If you enter a filename which has been previously used, the program will over-write the old file of the same name. The program saves the information displayed according to the OUTPUT FORMAT chosen earlier. See Appendix C for more details on saving, printing and deleting files.

Choose Return to Solution Methods now to try various heuristics on this sample problem.

D.3.3   Dispatch Heuristics

From the Solutions Methods Menu, choose `Manual Move and/or Dispatch Heuristics` to use the dispatch heuristics. At the next screen, which is the Objective Choices Menu in Figure D.7, select `Weighted Tardiness` as before. The Heuristics Choices Menu as shown in Figure D.8 will appear.

At this point, you should try a number of dispatch heuristics and see how well they solve the problem. (➲ Chapter 7). You may as well try several heuristics at once. Try the following: Random, Weighted Shortest Processing Time, Earliest Due Date and Rachamadugu & Morton. Highlight each of these five heuristics in turn by pressing the **SPACEBAR**. As each heuristic is marked, a + will appear next to it. When you have finished marking the heuristics, press ↵ to finish making selections. Then press ↵ again to move to the next screen, which contains the Output Format Menu as shown in Figure D.9. Choose **Summary** and press ↵ to calculate the solution for each heuristic. The program will display a summary screen as shown in Figure D.13.

```
SINGLE MACHINE
RUN SUMMARY
HEURISTICS--1 MACHINE--Tutorial--WEIGHTED
TARDINESS
Heuristic    Objective Value    % Deviation
RANDOM            136.              131.0
WSPT              148.              151.0
EDD                95.               61.0
R&M                64.                8.5
BENCHMARK          59.                0.0
```

Figure D.13. Comparison of Heuristics vs. Benchmark

Compared to the detailed listing used in the last run, the summary screen does not show the problem input and solution sequences for each heuristic.

Remember that the manual solution was 75.0, with a deviation of 27%. Thus of these seven one pass heuristics, R&M performed the best, at about 9% above the benchmark, and our manual construction was second, at 27% above the benchmark. The benchmark used was R&M as a starter heuristic, followed by general pairwise interchange. Note also that the values on your screen for the Random heuristic might not be the same as those shown here, since it is random by nature. Press ↵ and again choose **Return to Solution Methods** to try more elaborate heuristics.

D.3.4   Neighborhood Search--Tutor

Recall that classical neighborhood search methods begin with a starter heuristic, and make small changes looking for improvements a little at a time, terminating eventually in a local optimum for the problem. Tabu Search and Simulated Annealing are modern variations which we discuss later in section 3.5.

Note that for problems with dynamic arrivals, manual move, neighborhood methods and tree methods may call for a job to be scheduled before it has actually arrived. In this case idleness will be inserted into the schedule until the job does arrive.

A mediocre starter heuristic here would be Random; a good one would be R&M. There are three types of repeated small changes that are often made: Adjacent Pairwise Interchange, General Pairwise Interchange, and K-Move (➲ Chapter 5.3). All possible changes from a given solution are called the "neighborhood". More generally we speak of adjacent pairwise interchange neighborhoods, general pairwise interchange neighborhoods, k-move neighborhoods, and so on.

A good way to get started with neighborhood search methods is to select **Neighborhood Search -- TUTOR** from the Solution Methods Menu as shown in Figure D.14.

```
SINGLE MACHINE
SOLUTION METHODS   (Any One)
Manual Move and/or Dispatch Heuristics
Neighborhood Search--TUTOR
Neighborhood/Tabu/Annealing Search
Beam Search/Branch & Bound
```

Figure D.14. Single Machine Solution Methods Menu

After choosing **Neighborhood Search--TUTOR**, you move to the Objective Choices Menu, where you can select **Weighted Tardiness** as before. At the following screen, which displays the Heuristic Choices Menu, choose **WSPT** as an example of a poor starting heuristic. Later we will try EDD as an example of a fair starting heuristic, and finally R&M as an example of a good starting heuristic. The Neighborhood Style Menu will appear, as shown in Figure D.15.

```
   SINGLE MACHINE
   NEIGHBORHOOD STYLE--TUTOR    (Choose One or More)
 Adjacent Pairwise Interchange
 General Pairwise Interchange
 K-Move
```

Figure D.15. Single Machine Neighborhood Style Menu

To keep things simple for this example, select **Adjacent Pairwise Interchange** (you can select several at a time). When prompted, choose **100** for the number of moves. At the Output Format Menu, select **Summary Output** and press ⏎. The program will compute the starting heuristic part of the solution.

The screen in Figure D.16 shows us the start of the problem for adjacent pairwise interchange with weighted tardiness as the objective and WSPT as the starter heuristic. The screen is basically the same as the Manual Move Screen, except for different headings.

```
SINGLE MACHINE NEIGHBORHOOD TUTOR TRACE--ADJACENT-WSPT SEED
--Tutorial--WEIGHTED TARDINESS
```

JOB	ARRIV	PROC	DUE D	WT	COMPL	LATE	TARDY	W/P
12	0	1	7	4	1	-6	0	4.00
11	0	1	11	4	2	-9	0	4.00
8	0	2	6	6	4	-2	0	3.00
2	0	3	18	8	7	11	0	2.67
5	0	3	13	7	10	-3	0	2.33
9	0	3	8	7	13	5	5	2.33
1	0	1	7	2	14	7	7	2.00
7	0	2	0	4	16	16	16	2.00
10	0	1	11	2	17	6	6	2.00
3	0	1	8	1	18	10	10	1.00
4	0	1	4	1	19	15	15	1.00
6	0	1	1	1	20	19	19	1.00

```
ITERATION   1              WEIGHTED TARDINESS 148.0
```

Figure D.16. Neighborhood TUTOR Trace Screen

At this point, you have the following options:

- Press ↵ to complete the run segment finish automatically, with the table showing how the value of the objective function improves with each change.
- Press **s** to have only the current interchange carried out, and the objective value will decrease appropriately. This can be repeated until you press ↵ to request that computations be completed automatically. In this way, pressing **s** "steps" you through the run.
- Press **A** to abort the run. In this case, the output will indicate that the run was aborted, and show the objective value at termination.

When you complete the run, the summary output will appear on screen. You can use **ESC** to go back to the Heuristic Choices Menu, and execute a run for EDD the same way. After you have completed the run for EDD, you can execute a similar run for R&M. You could also have chosen several heuristics to begin with, and would then see an output screen like the one shown in Figure D.17.

```
SUMMARY NEIGHBORHOOD TUTOR--ADJACENT--Tutorial--WEIGHTED
TARDINESS
Start Heur.  Before     After     % Before    % After     Steps
   WSPT      148.0      134.0      151.0       127.0         2
   EDD        95.0       75.0       61.0        27.0        10
   R&M        64.0       59.0        8.5         0.0         3
- - - - - - - - - - - - - - - - - - - - - - - - - - - - - - - - - -
BENCHMARK          59.0
```

Figure D.17. Run Summary -- Neighborhood TUTOR,
Adjacent Pairwise Interchange

Some interesting facts are apparent from this summary:

- Good starter heuristics are better than poor starting heuristics;
- Adjacent pairwise interchange is not a very powerful search.

```
SUMMARY NEIGHBORHOOD TUTOR--GENERAL--Tutorial--WEIGHTED
TARDINESS
Start Heur.  Before     After     % Before    % After     Steps
   WSPT      148.0       62.0      151.0         5.1         8
   EDD        95.0       66.0       61.0        12.0         8
   R&M        64.0       59.0        8.5         0.0         3
- - - - - - - - - - - - - - - - - - - - - - - - - - - - - - - - - -
BENCHMARK          59.0
```

Figure D.18. Run Summary -- Neighborhood TUTOR,
General Pairwise Interchange

Press ↵ and then **ESC** to go back to the Neighborhood Style Menu as shown in Figure D.15 and try the same analysis for general pairwise interchange. General interchange gets better answers starting with lower quality heuristics, but requires much much more time per step. The summary report for the run is shown in Figure D.18.

Press ↵ and then **ESC** to go back to the Neighborhood Style Menu as shown in Figure D.15 and try the same analysis for for k-move. After you choose **k-move**, you are asked to choose a value for k. For small problems a value of k about 50% of the number of jobs is a reasonable choice, while for very large problems a fixed value of k=20 is usually sufficient. Choose k=6 and perform the same analysis. The summary is shown in Figure D.19.

We have deliberately chosen a value of k here that will spend about the same amount of time per iteration on a 12-job problem as does the general pairwise interchange. Notice for this example that it seems to require somewhat fewer iterations for problems with bad start heuristics, and to achieve much better results for such problems, than does the general pairwise interchange method.

```
SUMMARY NEIGHBORHOOD  TUTOR--6 MOVE--Tutorial--WEIGHTED
TARDINESS
Start Heur. Before      After     % Before   % After     Steps
   WSPT      148.0       59.0      151.0       0.0          5
   EDD        95.0       59.0       61.0       0.0          8
   R&M        64.0       59.0        8.5       0.0          3
- - - - - - - - - - - - - - - - - - - - - - - - - - - - - - - - -
   BENCHMARK              59.0
```

Figure D.19. Run Summary -- Neighborhood TUTOR, k-move using k=6

In any of these more complicated search methods, especially if the input are small integers, ties in making choices are quite common. Due to round off errors, this can cause occasional answers differing from those given here, mostly in the sequence. The objective value itself should not be affected.

D.3.5 Neighborhood/Tabu/Annealing Search

Now that you have studied classical neighborhood search, you can look at some more recent and extended neighborhood search methods such as Tabu Search and Simulated Annealing. (➲ Chapter 6.3)

- **Neighborhood search** only allows you to make small changes that are clear improvements at each step, and so must stop at a local optimum.
- **Tabu search** is a more sophisticated version that lets you make the best new move available, even if it is not as good as the current position. You always

save the best move to date. This is like walking up out of a valley in order to get down into a deeper valley later. The main problem is not to allow yourself to repeat recent moves, so you don't go right back to the same valley you've just been in. In non-linear programming one speaks of "hill-climbing". Since you are minimizing costs, you need to think of "valley-descending"! You simply make a "tabu" list to prevent repeating these moves. Hence the name of the technique. This will be explained in more detail below when you actually do tabu search.

- **Simulated annealing** works similarly by randomly choosing the small improvement and occasionally allowing a move that temporarily makes things worse. This is something like jumping around in the valley on a pogo stick somewhat at random, and sooner or later jumping out of this valley into the better one.

It is important to notice that by its nature simulated annealing will get a different answer every time due to randomness. Thus your answers will not agree with those given here. It also makes it useful to repeat a run a number of times to get the best answer.

```
SINGLE MACHINE
NEIGHBORHOOD SEARCH METHOD
(Choose One or More)

    Neighborhood -- Adjacent Pairwise
    Neighborhood -- General Pairwise
    Neighborhood -- K-move

    Tabu Search -- Adjacent Pairwise
    Tabu Search -- General Pairwise
    Tabu Search -- K-move

    Annealing -- Adjacent Pairwise
    Annealing -- General Pairwise
    Annealing -- K-move
```

Figure D.20. Neighborhood Search Method Menu

To use these methods, go back to the Solution Method screen as shown in Figure D.6 and choose **Neighbor-hood/Tabu/Annealing Search** as the next method. At the Objective Choices Menu, pick **Weighted Tardiness** as the objective function. At the Heuristic Choices Menu, choose **WSPT** as the initial heuristic or "seed". It is better to use a "bad" starting heuristic so you can watch the procedure make lots of improvements.

After choosing **WSPT** at the Heuristic Choices Menu, you will be brought to the Neighborhood Search Method Screen as shown in Figure D.20. You can choose one or more of these options. For this exercise, choose **Neighborhood-Adjacent Pairwise**, **Tabu Search-Adjacent Pairwise** and **Annealing-Adjacent Pairwise**. Again, it is better to choose a poor starting seed (WSPT) and a poor search neighborhood (Adjacent Pairwise) to make the problem harder. Like in previous menu screens, press ↵ to move to confirm your choices and move to the next screen.

A line will appear on the bottom of the screen which asks for the frequency of the trace steps. Type **5**, which which means that at every fifth improvement, the screen will show the current objective function and sequence. You could type **1** to make every step visible, or **500** not to trace at all. When prompted, type in **100** for the maximum number of moves in Neighborhood Search.

Next a line will appear on the bottom of the screen asking for the maximum number of moves for tabu search; type in **100**. The next line asks for the length of the tabu list. It is not costly to use the tabu list, to avoid repeating moves type **1000**. This idea is sometimes called using an "infinite" tabu list.

When prompted for the total number of iterations for simulated annealing, type **100**. A line appears on the bottom of the screen asking for a starting value of (normalized) temperature. Enter **2**. Next you are asked at what iteration to lower the temperature. Type **50**. Finally, you need to indicate by what factor between 0 and 1 to lower the temperature at that point; try **0.25**. Now computations start. Remember that you have asked for Neighborhood--Adjacent, Tabu--Adjacent, and Annealing-- Adjacent. Remember also that for this problem the benchmark value was 59.0.

```
SINGLE MACHINE
NEIGHBOR TRACE--ADJACENT--WSPT SEED--
Tutorial--WEIGHTED TARDINESS


This step 1.
This value 148.


This sequence 12-11-8-2-9-5-1-7-10-3-6-4.
```

Figure D.21. Adjacent Neighborhood Search Trace Screen

Parsifal starts with the Neighborhood-Adjacent run. First, the problem after the starting heuristic is displayed, as shown in Figure D.21. After one iteration, neighborhood search finds a local minimum, and terminates as seen in Figure D.22.

```
This step 2.
This value 134. Local minimum, terminate.

This sequence 12-11-8-9-2-5-1-7-10-3-6-4
```

Figure D.22. Adjacent Neighborhood Search Trace Screen
After 1 Iteration

Press ↵ to continue, and the program switches to tabu search. *Parsifal* again displays the problem after the starting heuristic, as shown in Figure D.23. Press **s** a number of times to move 5 steps at a time. A composite of the resulting screens are seen in Figure D.24.

```
SINGLE MACHINE
TABU TRACE--ADJACENT--WSPT SEED--Tutorial--
WEIGHTED TARDINESS

This step 1
Best step 1  This value 148.0

Best value 148.0
Best sequence 12-11-8-2-9-5-1-7-10-3-6-4.
```

Figure D.23. Adjacent Tabu Search Trace Screen

You can see that tabu search quickly got down to the value of 134.0 found by neighborhood search. There were a lot of alternate solutions with value 134.0 (no repeats are allowed) which tabu searched before breaking down quickly to a new local minimum of 104.0. Notice that after reaching a best value of 59.0, the program continues to search for a hoped for way out of this (local?) minimum until time runs out.

This step 6	This value134.0	Best step 2	Best value134.0
11	134.0	2	134.0
26	134.0	2	134.0
41	104.0	41	104.0
51	68.0	48	68.0
76	63.0	54	59.0
91	64.0	54	59.0
101	59.0	54	59.0

Figure D.24. Adjacent Tabu Search Trace Iterations

If you press ↵ to finish the run automatically, the bottom line will read **Computing..** for a long time! Then a trace for the final iteration will summarize the problem. Press ↵ one more time to switch to annealing search.

Since a simulated annealing is random, two different runs with different starting seeds may give entirely different answers. For this reason, in practice simulated annealing is often run a number of times, and the best result is chosen. The annealing search results are displayed in Figure D.25. As you can see, the results suggest that we may have chosen our annealing parameters well. To move to the summary of all three runs, press ↵.

```
SINGLE MACHINE
ANNEAL TRACE--ADJACENT--WSPT SEED--Tutorial--WEIGHTED
TARDINESS

This step  1        Best step  1
This value  148.0 Best value  148.0
Best sequence 4-6-7-11-10-9-12-1-3-8-5-2


This step11  This value130.0  Best step10  Best value130.0
        51              68.0          50           68.0
        61              59.0          59           59.0
       101              81.0          59           59.0
```

Figure D.25. Adjacent Simulated Annealing Trace Screen

A summary run will appear on screen, as shown in Figure D.26. Note that it does not print out all of the input parameters from all three runs. You have to make a note of them if necessary for future reference.

```
SINGLE MACHINE
RUN SUMMARY
NEIGHBORHOOD SEARCH-WSPT SEED-Tutorial-WEIGHTED TARDINESS

Method        Before    After    % Before  % After
Neighbor-Adj  148.0     134.0    151.0     127.0
Tabu-Adj.     148.0      59.0    151.0       0.0
Anneal-Adj.   148.0      59.0    151.0       0.0
-----------------------------------------------------------
BENCHMARK    59.0
```

Figure D.26. Run Summary Screen of Comparative Search Methods

D.3.6 Beam Search/Branch-and-Bound

Beam Search and **Branch-and-Bound** are based on the idea of a decision tree. Start at a node representing the state in which no sequencing choices have yet been made. Create a branch for every job which might be chosen first. At each such choice, make a branch for every job which might be chosen second, and so forth. These methods search over all these choices, while not bothering to search parts of the tree which are unlikely. Beam search does this in an approximate fashion, while depth-first branch-and-bound and breadth-first branch-and-bound do this in a way guaranteed to give the optimal solution given enough time. The running time for beam search is polynomial in the size of the problem, while the times to solve the latter problems exactly is exponential in the size of the problem; they are thus not really practical for large problems. A heuristic can be created from these procedures simply by stopping after a given amount of computation and taking the best solution to date. Such a procedure is not really designed to provide good answers quickly, and hence must be tested against the kinds of heuristics presented here.

```
SINGLE MACHINE
TREE SEARCH METHOD (Choose One or More)


Beam Search
Best-First Branch-and-Bound
Depth-First Branch-and-Bound
```

Figure D.27. Single Machine Tree Search Method Menu

Now go back up the menu again using **ESC** until you get to Solution Method Menu in Figure D.6. This time select **Beam Search/Branch-and-Bound** and press ↵. At the Heuristic Choices Menu, select a guide heuristic for beam search of **R&M**. At the Tree Search Menu shown in Figure D.27, select **Beam Search**.

D.3.7 Beam Search (⌾ Chapter 6.2)

At this point, you are asked for some specifics about the beam search. In response to the prompts, choose the following:

Beam width	**8**
Objective function	**Weighted Tardiness**

You cannot choose the trace frequency; you get one trace for each of the twelve levels. The resulting trace is shown in Figure D.28. Like with the Tabu Search screen, you press **S** to trace each level. The second trace screen is shown in Figure D.29 with a composite of further trace screens.

```
SINGLE MACHINE
BEAM TRACE--WIDTH 8--R&M GUIDE--Tutorial--
WEIGHTED TARDY

Level  1
Best value  64.0


Best sequence 7-X-X-X-X-X-X-X-X-X-X-X
```

Figure D.28. Beam Search Trace Screen

The 7-8 under Best sequence in Figure D.29 just means that the best sequence found to date starts with 7-8 for the first two levels.

```
Level  2
Best value  64.0
Best sequence   7-8-X-X-X-X-X-X-X-X-X-X

Level      3              Best value      64.0
Level      4              Best value      64.0
Level      5              Best value      62.0
Level      6              Best value      62.0
Level      7              Best value      62.0
Level      8              Best value      60.0
Level      9              Best value      59.0
Level     10              Best value      59.0
Level     11              Best value      59.0
Level     12              Best value      59.0
```

Figure D.29. Beam Search Trace Results

Thus beam search with beam width 8 using the guide heuristic of R&M achieves a final value of 59.0 by level 9, which is the best benchmark thusfar (and turns out to be optimal as you can see below). By contrast, if this same problem were run using the poor guide heuristic of WSPT the result obtained would have been 69, or about 17% higher. Press ↵ and to see the output screen, as shown in Figure D.30.

While this is a good result, notice that as for simulated annealing we must guess at a parameter, in this case beam width. If several different beam widths need to be tried, execution time goes up. Of course, you could simply start with a large beam width. The tradeoffs here are somewhat difficult, and experience will help you to evaluate them.

```
SINGLE MACHINE
12-R&M-SEED-Tutorial-WEIGHTED TARDINESS

Starting Value       64.0
Ending Value  59.0
Starting Sequence   7-8-9-12-11-1-5-10-2-3-4-6
Ending Sequence     7-8-9-12-1-10-5-11-3-2-4-6
```

Figure D.30. Beam Search Run Summary Screen

D.3.8   Branch-and-Bound

Branch-and-bound is an exact method, which is only practical for relatively small problems (➲ Chapter 5.5). You will, however, be able to verify that the benchmark of 59.0 for the 12-job problem is, in fact, optimal.

At the Solution Methods Menu in Figure D.6, choose **Beam Search/Branch-and-Bound**. At the Heuristic Choices Menu, select **R&M**. This brings us back to the Tree Search Method Menu as shown in Figure D.27. At that menu, choose **Best-First Branch-and-Bound**.

```
SINGLE MACHINE
BEST-FIRST TRACE--R&M START--Tutorial--WEIGHTED TARDINESS

THIS STEP 1 LOWER BOUND    0.0
BEST STEP 1 BEST VALUE     64.0
BEST SEQ 7-8-12-9-11-1-5-10-2-3-4-6

THIS STEP50   BEST STEP    1   LOWER BOUND50.0   BEST VALUE64.0
        1001              687                53.0              59.0
        1501              687                54.0              59.0
        2001              687                55.0              59.0
        2501              687                56.0              59.0
        3001              687                56.0              59.0
        3501              687                56.0              59.0
        4001              687                57.0              59.0
        4501              687                57.0              59.0
        5001              687                58.0              59.0
```

Figure D.31. Best-First Branch-and-Bound Trace Results

Note that branch-and- bound routines tend to be complex and rather specific. Therefore, it is important to note that the two branch-and- bound routines are only available for use with static problems and the weighted tardiness objective.

At this point, you are asked for the frequency of trace. Type **500**. Press ↵ to perform the search. The starting trace is shown in Figure D.31 with successive traces.

When the optimum is achieved, the final trace is displayed, as shown in Figure D.32. Note that it's finally been proven that the benchmark of 59.0 is optimal. The routine spent only 10% of its time achieving an optimal solution, and the other 90% of the time verifying optimality.

Best-first search often has problems in that the storage of nodes to expand may exceed the computer's available memory. In this case, eliminating nodes with the highest lower bound is a good heuristic. In fact, you may often be able to see that the nodes eliminated in this way were not needed. This trick helped the current problem achieve optimality. It is interesting to note that with a weaker upper bound such as WSPT, many more iterations would have been required, and optimality of the final answer could not be proved.

```
SINGLE MACHINE
BEST FIRST TRACE--R&M START--Tutorial--
WEIGHTED TARDINESS

This Step   6322
Lower Bound  59.0
Best Step     687   Best Value    59.0
Best Seq 7-12-8-9-1-11-5-10-6-2-4-3

              OPTIMUM ACHIEVED!
```

Figure D.32. Best-First Branch-and-Bound Trace Final Results

Now go back and select **depth-first search** with **R&M** as the upper-bound heuristic, and a trace frequency of **1000**. The resulting screen is shown in Figure D.33. Press S to go the next trace. After 1000 steps, a compiled version of the output is shown in Figure D.34.

```
SINGLE MACHINE
DEPTH-FIRST TRACE--R&M START--Tutorial--
WEIGHTED TARDY

This Step    1      Lower Bound 12.0
Best Step    1      Best Value   64.0
Best Seq 7-8-12-9-11-1-5-10-2-3-4-6
```

Figure D.33. Depth-First Branch-and-Bound Trace Screen

Note that best-first search found the optimal solution 10% of the way through its computation, while depth-first search found it 25% of the way through. This seems to support the idea of using truncated branch-and-bound as a heuristic, at least for smaller problems.

```
This Step2001  Best Step1876  Lower Bound16  Best Value60.0
        3001            2197              16            59.0
        4001            2197              18            59.0
        5001            2197              18            59.0
        6001            2197              26            59.0
        7001            2197              26            59.0

Final trace:
This Step 7711  Best Step 2197
Lower Bound 59.0  Best Value 59.0
Best Seq 7-12-8-9-1-11-5-10-6-2-4-3

OPTIMUM ACHIEVED!
```

Figure D.34. Depth-First Branch-and-Bound Trace Results

D.3.9   Manual Input

Up to this point, you have been working entirely with the Tutorial problem provided with the program. Now go back to the Single Machine Job Input Set Menu in Figure D.4 and choose **Manual Input** to experiment with using your own data sets.

After selecting Manual Input, you will be brought to the Manual Input Menu, as shown in Figure D.35. This menu gives you options for creating the following:

- A single, individual problem;
- A number of similar problems (statistically generated) with a common name.

```
SINGLE MACHINE
SINGLE VS. MULTIPLE INPUT (Pick One)

Single Problem
Multiple Problems (Statistical)
```

Figure D.35. Manual Input Menu

Whichever you choose, the next screen as shown in Figure D.36 gives you two options for storing the set:

- Save the set as a new job set under an unique filename.
  If you ask to save it, the program asks you for a name for the user file. (➲ Appendix C Section 7 and 8 for specifications for naming, storing and printing user files.) You are also allowed to attach a short description of up to 65 characters to the file.
- Do not save the new job as a file. Rather, treat it as a temporary job set.

In both cases, the problem is also available for your immediate use and is stored in the temporary file called **Previous Problem**. The next time you use the program, you may come back to this problem by selecting **Previous Problem** at the Job Input Set Menu in Figure D.4. Unless you have subsequently overwritten this problem, you can continue working on it.

Note that job sets may not be deleted directly within the software. Appendix C Section 7 explains how this may be accomplished using standard DOS file management commands.

```
SINGLE MACHINE
SAVE OPTION   (Pick One)

Save New Job Set
Temporary Job Set
```

Figure D.36. Manual Input Save Options Menu

Choose the single-problem option and the program will prompt you to provide the number of jobs. Then, you are prompted for each job by number and the following integer information (separated by spaces):

- Arrival time
- Process time
- Due date
- Weight

Your actual input could look like the sample in Figure D.37. Input must be in integer form. Press ↵ to display the problem input in table form; press it again to move to the Solution Methods screen. If you are not sure that you have generated a reasonable problem, you may view it or play with it a bit in manual mode at this point.

The statistical mode is more advanced, and may be skipped initially. In case you have chosen the Multiple Problems (Statistical) mode, the input might look like the sample in Figure D.38, where SD = standard deviation.

```
SINGLE MACHINE
SINGLE MANUAL INPUT

Number of jobs:   5
Job number 1>   0   6   17   4
Job number 2>   0   8   8    1
Job number 3>   0   1   2    2
Job number 4>   0   4   0    3
Job number 5>   0   6   4    3
```

Figure D.37. Sample Single Manual Input

The mean due date is automatically set to achieve the tardiness factor (rough percentage of jobs which will be tardy). Difficult problems very often have tardiness factors between about 0.5 and 0.8. The mean weight is automatically calculated as mean weight/process multiplied by mean process time. For each replication, the arrival date, process time, due dates, and weights are chosen from the appropriate normal distribution. Results are rounded to the nearest integer in all cases. Negative results are truncated to 0, or to 1 in the case of process time. Note that the resulting data may have biased parameters.

```
SINGLE MACHINE
MULTIPLE MANUAL (STATISTICAL) INPUT

Number of jobs:   5
Number replications:   10
Mean arrival date:   0
SD of arrival date:   0
Mean process time:   5.1
SD of process time:   2.3
SD of due date:   4.4
Mean weight/process:   3.6
SD weights:   1.3
Tardiness factor:   0.7
Integer random seed:   4672341
```

Figure D.38. Sample Multiple Manual Input

In this case, **Previous Problem** will now contain these generated 10 similar problems. If the Manual mode is called, the user will have to solve all ten problems manually before proceeding! The Output screen will show the results for each of the 10 problems, as well as results averaged over all 10 runs. The same idea holds for all the heuristic methods. Clearly, you should be careful about entering 10 problems at once into Branch-and-Bound! Detailed discussion of formats of job data files, the master files, custom heuristic subroutine files, and objective files are given in Appendix C.

D.4 THE PARALLEL-MACHINES PROBLEM (➲ CHAPTER 11.2 AND 11.3)

You are now ready to leave the single-machine problem and go on to the parallel machines problem. Chapter 11 tells us that in many ways the parallel machines problem can be treated in an aggregate fashion as a single machine problem. Consider a single-machine problem with processing times and due dates. Define speed factors, and assume that the sum of the speed factors is 1.0, so that the m machines have total capacity equal to that in the single-machine problem. Thus the processing time of job j on machine k is its one-machine processing time divided by the speed of that machine, and the average speed factor is 1/m or an average process time over the machines. Thus for 3 equal machines, the speed factor is 1/3, so that all processing times from the single-machine problem are simply multiplied by 3.

This means that if we have one-machine job input and speed factors, we can construct all desired data for the parallel machine problem in terms of these formulas, but leave process times unchanged (except for slack, see below) in heuristic formulas. Chapter 11.3.2 in the text summarizes the situation:

(1) Proposition 1 says that there is a single optimal sequence for the jobs such that each can be scheduled in turn on the first available machine, for the case of equal machines.

(1') For the general case, instead schedule each job in turn on the machine that would finish it first. (We often approximate (1)' by (1)).

(2) Several parallel machines can roughly be aggregated in terms of capacity into a single machine whose speed m' is the sum of the individual speeds.

(3) Similar heuristics may be constructed as for the one- machine problem, simply by leaving process time unaffected in the formulas except, in the calculation of slack time, where the actual processing time must be used. We approximate this as for equal machines.

D.4.1 Input

In the parallel machines option, the first screen after the input screen is the Shop Data Menu, as shown in Figure D.39. For this exercise, choose **Equal Machines**. When prompted for the number of machines, enter **2**. At this point, the program will move on to the Problem Input Display. If you choose Parallel Machines, the screen prompts you for the speed factors which will be normalized. Enter the following numbers, separated by spaces: **0.5** and **0.5**.

```
PARALLEL MACHINES
SHOP DATA

Equal Machines
Parallel Machines
```

Figure D.39. Parallel Machines Shop Data Menu

Note that for unequal machines, the calculated process times of job j on machine k will not generally be an integer. The program does not round these, but treats them as exact decimals. At this point the program will also move on to display Job Input, which is identical in detailed design and purpose to the single-machine case, since job input is in fact identical to the one machine case for equal and proportional machines.

```
PARALLEL MACHINES
MACHINE SELECTION FOR JOB (Pick One)

1. "Start First" Scheduling
2. "Finish First" Scheduling
```

Figure D.40. Machine Selections Menu

Parsifal next allows you to decide whether to schedule on the machine which can start it first, or on the machine which can finish it first, using the menu in Figure D.40. For equal machines the two options are the same. The second option is much more expensive computationally, and should only be chosen when machines differ greatly in speeds. For this exercise, choose **Start First**.

```
PARALLEL MACHINES
SOLUTION METHOD

1. Manual Move and/or Dispatch Heuristics
2. Neighborhood/Tabu/Annealing Search
```

Figure D.41. Parallel Machines Solution Methods Manual

Next, the Solution Methods Menu will appear, as shown in Figure D.41. The Solution Methods Menu for Parallel Machines differs from that for Single Machines, as discussed below:

- Neighborhood Search--Tutor has been omitted from the parallel machines section since you should now be comfortable using these routines directly.

- Beam Search/Branch-and-Bound have been omitted due to their complexity, storage, and computational requirements.

D.4.2 Manual Move

Select to use the **Tutorial** problem and two equal machines, with **Weighted Tardiness** as the objective and **Manual Move** as the heuristic, using **Summary** Output. You will be brought directly into the manual mode.

You will get the screen shown in Figure D.42, which is almost the same as for the one-machine case since you use exactly the same arrival times, aggregate process times, due dates, weights, and w/p. In calculating the completion time, you need to divide the processing time by the speed of the appropriate machine. Thus for job 7, aggregate process time is 2 but process time on either machine 1 or 2 is 4 (since two jobs can be done at once). The column displaying tardiness has been omitted because of space constraints; you can always calculate that value from the lateness column. This display keeps track of which machine was first available for the job, so that you can see the status of the machines at all times.

```
PARALLEL MACHINES
MANUAL MOVE--2 MACHINES 0.5   0.5 --Tutorial--WEIGHTED TARDY

JOB   ARRIV  PROC  DUE D   WT    COMPL  LATE MACHINE  W/P
 1      0     1     7      2     2.0   -5.0    1       2.0
 2      0     3    18      8     6.0  -12.0    2       2.7
 3      0     1     8      1     4.0   -4.0    1       1.0
 4      0     1     4      1     6.0   +2.0    1       1.0
 5      0     3    13      7    12.0   -1.0    1       2.3
 6      0     1     1      1     8.0   +7.0    2       1.0
 7      0     2     0      4    12.0  +12.0    2       2.0
 8      0     2     6      6    16.0  +10.0    1       3.0
 9      0     3     8      7    18.0  +10.0    2       2.3
10      0     1    11      2    18.0   +7.0    1       2.0
11      0     1    11      4    20.0   +9.0    1       4.0
12      0     1     7      4    20.0  +13.0    2       4.0
     WEIGHTED TARD.        289.0
```

Figure D.42. Initial Manual Solution Screen

The weighted tardiness of 289.0 is only about 8% greater than the weighted tardiness of 268.0 for the same schedule and problem in the one-machine case.

On many screens within the parallel machine, flow shop, job shop analyses, you may press **G** to see a visual representation of the current schedule using a Gantt

Chart. As usual, pressing ⏎ returns you to the previous screen. As you did in the one-machine case, apply the EDD heuristic to this problem to get the screen shown in Figure D.43.

```
MANUAL MOVE--2 MACHINES 0.5 0.5 --Tutorial--WEIGHTED TARDY

 JOB   ARRIV  PROC  DUE D   WT    COMPL  LATE MACHINE  W/P
  7      0     2      0      4     4.0   +4.0    1      2.0
  6      0     1      1      1     2.0   +1.0    2      1.0
  4      0     1      4      1     4.0    0.0    2      1.0
  8      0     2      6      6     8.0   +2.0    1      3.0
  1      0     1      7      2     6.0   -1.0    2      2.0
 12      0     1      7      4     8.0   +1.0    2      4.0
  3      0     1      8      1    10.0   +2.0    1      1.0
  9      0     3      8      7    14.0   +6.0    2      2.3
 10      0     1     11      2    12.0   +1.0    1      2.0
 11      0     1     11      4    14.0   +3.0    1      4.0
  5      0     3     13      7    20.0   +7.0    1      2.3
  2      0     3     18      8    20.0   +2.0    2      2.7

       WEIGHTED TARD.      156.0
```

Figure D.43. Manual Move Screen, EDD Solution

This is indeed a vast improvement over our original objective value of 289. However, it is different from that in the one-machine value as shown in Figure D.11, which had an objective of 103.0. You can also explore the effects of benefit/cost improvement as shown in Figure D.12 for the one-machine case. Again, we can move job 3 to the end of the list, since it will be tardy anyway, and has a low benefit/cost ratio. Then as before, job 10 can similarly be moved next to the last for further improvement. Finally, job 1 can be moved down, but only part of the way. What is the best value that you can achieve for the objective?

```
PARALLEL MACHINES
RUN SUMMARY HEURISTICS --2 MACHINE 0.5 0.5 -- Tutorial --
WEIGHTED TARDY

     Heuristic    Objective Value   % Deviation

     MANUAL MOVE       156.0            44.0
     BENCHMARK         108.0             0.0
```

Figure D.44. Manual Move, Run Summary Screen

Suppose, however, that we are currently satisfied with the EDD solution. We may get an idea of its quality by comparing it with the benchmark. The benchmark is still the R&M heuristic followed by general pairwise interchange. Press ⏎ to indicate that the current solution represents the final version of your manual heuristic. This takes you to the output screen as shown in Figure D.44. Note that in the detailed version of the output, the Gantt Charts appear automatically on screen.

You can see that the value of the benchmark for this problem is 108.0. What was the percentage error for your solution improved over EDD? You can then move to the next screen to save the output, return to solution methods, or quit the program. Select to return to the solution methods, and now you can try various heuristics on the problem.

D.4.3   Dispatch Heuristics

After returning to solution methods, select to solve the same problem using **Weighted Tardiness** objective and a number of dispatch heuristics: **Random**, **WSPT**, **EDD** and **R&M**. This combination results in the table shown in Figure D.45.

```
PARALLEL MACHINES
RUN SUMMARY HEURISTICS -- 2 MACHINE 0.5   0.5 -- Tutorial--
WEIGHTED TARDY

Heuristic Objective Value      % Deviation

RANDOM          188.0                 74.0
WSPT            183.0                 69.0
EDD             156.0                 44.0
R&M             114.0                  5.6
--------------------------------------------------------------
BENCHMARK       108.0                  0.0
```

Figure D.45. Run Summary Screen Comparison of Dispatch Heuristics

Comparing these results with those of the one-machine problem, we see that they are quite similar. R&M is only about 6% worse than the benchmark, compared with about 9% in the one machine case. The benchmark here is still R&M followed by general pairwise interchange. Notice that the percentage deviation of each heuristic from the benchmark is smaller than for the one-machine case. This is interesting, but hard to evaluate without more evidence.

D.4.4   Neighborhood/Tabu/Annealing Search

Go back to the Solution Methods menu and choose **Neighborhood/Tabu/ Annealing Search**. On the usual preliminary screens, select **Weighted**

Tardiness as the objective function and **WSPT** as the starting heuristic. Then you get the same nine choices as for the one-machine problem. Select the same three to investigate: **Neighborhood-Adjacent, Tabu Search-Adjacent** and **Annealing-Adjacent**. When prompted, specify the following parameters for the search:

Frequency of trace step:	20
Maximum number of moves in neighborhood search	100
Maximum moves in tabu search	100
Length of the tabu list	1000
Maximum moves in simulated annealing	100
Initial annealing temperature	5

```
PARALLEL MACHINES
NEIGHBOR TRACE--ADJACENT--WSPT SEED 2 MACHINES
0.5 0.5--Tutorial--WEIGHTED TARDY

This step  1
This value 183.0
This sequence 11-12-8-2-1-5-9-10-7-3-4-6
 .  .  .  .  .
This step  3
This value 178.0
Local minimum, terminate
This sequence 11-12-8-2-1-5-9-10-7-3-4-6
```

Figure D.46. Neighborhood Search Trace Results

In addition, set the temperature to drop after 60 steps to 0.5 of the original. For brevity, give just the first and last step for each of the three methods.

```
PARALLEL MACHINES
TABU TRACE--ADJACENT--WSPT SEED 2 MACHINES
0.5 0.5--Tutorial--WEIGHTED TARDY

This step  1 Best step  1
This value 183.0    Best value 183.0
Best sequence 11-12-8-2-1-5-9-10-3-7-4-6
 .  .  .  .  .  .
This step  101      Best step   23
This value  117.0   Best value 117.0
Best sequence  12-11-8-9-5-1-10-2-7-4-3-6
MAXIMUM MOVES EXCEEDED: TERMINATING RUN
```

Figure D.47. Tabu Search Trace Results

Parsifal first displays the neighborhood search, as shown in composite in Figure D.46. Note that ordinary adjacent pairwise interchange after a poor starting seed of WSPT does not do as well as the simple R&M dispatch heuristic without searching. The program then performs the tabu trace search, as shown in Figure D.47.

```
PARALLEL MACHINES
ANNEAL TRACE -- ADJACENT -- WSPT SEED 2
machines 0.5 0.5 -- Tutorial -- WEIGHTED
TARDY

This step    1     Best step    1
This value   189.0 Best value   189.0
This sequence   11-12-8-2-1-5-10-3-9-4-6-7
. . . . . .
This step    101   Best step    92
This value   130.0 Best value   116.0
This sequence   12-8-9-11-1-10-7-5-2-6-3-4
```

Figure D.48. Simulated Annealing Search Trace Results

Finally, the program performs the simulated annealing run, as shown in Figure D.48. After the runs are completed, you can explore other search neighborhoods such as general neighborhood search and k-move for this problem.

D.5   THE FLOW SHOP

Thus far, you have considered only single-machine and parallel-machine problems, which may be considered single-resource problems in the sense that there is a single queue of jobs, and hence a single sequence to be optimized. You can turn now to multiple-queue multiple-resource problems. Actually both flow shops and job shops could be treated at once, since the only difference is that for flow shops all jobs follow the same route through the shop, while for job shops each group of jobs will follow a different route. However, this one difference makes a significant practical difference for you in visualizing and gaining insight into problems, so it's better to develop the flow and job shop separately. Go all the way back to the Shop Type Menu in Figure D.2 and choose **Flow Shop**.

D.5.1   Input

After selecting the flow shop option, the next screen asking for Job Input Options will appear, similar to Figure D.3. Select **File Input** first. Since the files cannot usefully separate job data and shop data, your choices will be complete sets of problem data. You will see the format later when veiwing Manual Input. The next screen contains the Problem Data Menu, as shown in Figure D.49.

```
FLOW SHOP
PROBLEM DATA

Previous Problem
Small Problems
Larger Problems
User Problems
```

Figure D.49. Flow Shop Problem Data Menu

Problems that you create will be stored by the names that you provide. As before, first choose one of these groups, and then a problem within the group, using the short descriptions given of each problem. The next screens lets you choose an objective function and a heuristic for the problem.

Now, go back to Manual Input. Because you are now dealing with much larger problems, it becomes tedious to specify the entire input by hand. Thus you are given two options, as shown in Figure D.50.

```
FLOW SHOP
TYPE OF MANUAL INPUT (Pick One)

Single Problem--Detailed Input
Single/Multiple Problem(s)--Statistical Input
```

Figure D.50. Flow Shop Manual Input Menu

Whichever you choose, the next screen asks you whether to save the new set, or just treat it as temporary, as seen in Figure D.51.

```
FLOW SHOP
SAVE OPTION (Pick One)

Save New Job Set
Temporary Job Set
```

Figure D.51. Flow Shop Manual Input Save Menu

Choose the single-problem detailed-input option. The program first prompts you to give the number of jobs and number of machines. Then for each job, by number, you must type in the arrival time, due date, and weight, followed by an operation processing time for each operation of the job. You can separate the initial input from the process-time input by a / to make things easier to read. Actual input might look like the sample shown in Figure D.52.

```
FLOW SHOP SINGLE
DETAILED INPUT

Number of jobs:  6
Number of machines: 4
Job number 1>    0   17   4.6 /  2    4    7   1
Job number 2>    0    8   1.0 /  4    7    8   2
Job number 3>    0    2   4.5 /  0    6    9   4
Job number 4>    0    6   1.0 /  5    6    9   0
Job number 5>    0    0   0.5 /  1    1    4   3
Job number 6>    0   22   1.5 /  2   20   17   4
```

Figure D.52. Sample Flow Shop Detailed Input

Note that floating-point numbers are only allowed for weights. All other values must be in integer form. Press ↵ to bring up the problem input display screen. As for the one-machine and parallel-machine cases, you can press v from an appropriate screen to redisplay this input. The table can be scrolled to the right if there are a larger number of machines. We do not discuss the statistical generation option here, since it is identical to that for the job shop, and will be presented in that section.

D.5.2   Objectives and Heuristics

An Objectives Choices Menu appears as shown in Figure D.53, which lets you evaluate several objectives during a single simulation run. For this example, again select **Weighted Tardiness** and move on to chooose the solutions method.

```
FLOW SHOP
OBJECTIVE CHOICES     (One or More)

Makespan
Maximum Lateness
Maximum Tardiness
Weighted Flow
Weighted Lateness
Weighted Tardiness
Weighted Number of Tardy Jobs
```

Figure D.53. Flow Shop Objective Choices Menu

There are many differences between the solution methods available for flow/job shops and those available in the single-machine and parallel-machine menus. Here manual move is not particularly practical, due both to display and computational complexity, so it is not included. And while neighborhood search, tabu, annealing, beam search and branch-and-bound are feasible, they are quite computationally intensive for an educational package, and have not been implemented.

However, there are a rich variety of dispatch heuristic methods available for the flow shop and the job shop (➲ especially Chapters 10, 13-16), using the modern techniques of heuristic pricing and leadtime iteration. The software allows a thorough exploration of these options for flow/job shops as shown in Figure D.54. For additional details, see the help screens in the program and Chapter 16.

```
FLOW SHOP
HEURISTICS  (One or More)

RANDOM               Equi-Probability Selection
FAMFS                First at Machine First Served
FASFS                First in Shop First Served
WSPT                 Weighted Shortest Processing Time
WLWKR                Weighted Least Work Remaining
WTWORK               Weighted Total Work
WLPT                 Weighted Longest Processing Time
EGDD                 Earliest Global Due Date
EODD                 Earliest Local Due Date
MST                  Minimum Slack Time
S/OP                 Minimum Slack/Remaining Operations
WTD HODGSON          Hodgson Modified for Weights
CRIT                 Critical Ratio
WCOVERT              Weighted Covert
R&M                  Rachamadugu & Morton
```

Figure D.54. Flow Shop Heuristics Menu

Some of these methods depend on accurate estimation of operation due dates, and/or slack times. These heuristics can be improved by running the problem, using the improved estimates of lead times to generate better due dates and slacks, and indeed repeating this procedure several times. The heuristics affected by this are EODD, MST, S/OP, WTD HODGSON, WCOVERT, R&M. If any of these are among the heuristics you select for the run, the screen shown in Figure D.55 will come up.

If you choose leadtime iteration, you are asked for the initial leadtime factor k. Leadtime is initially estimated for any operation as k times the remaining process time of the job. The default leadtime is 3. After you input leadtime, you are then asked for the maximum number of leadtime iterations. Again, the default is 3, but more than 5 to 10 iterations is rarely useful.

```
FLOW SHOP
LEADTIME ESTIMATION

No Leadtime Iteration
Leadtime Iteration
```

Figure D.55. Flow Shop Leadtime Estimation Menu

Some of these heuristics can benefit from consideration of the differing load on different resources, and estimation of a "price" for using that resource. These heuristics include: WSPT, WCOVERT, WTD HODGSONS and R&M. If you choose any of these, the menu in Figure D.56 will appear.

```
FLOW SHOP
RESOURCE PRICING OPTIONS (Pick One)

Myopic
Bottleneck
User Prices--No Price Iteration
User Prices--Price Iteration
Static Queuing--No Price Iteration
Static Queuing--Price Iteration
Dynamic Queuing--No Price Iteration
Dynamic Queuing--Price Iteration
```

Figure D.56. Flow Shop Resource Pricing Options Menu

For each operation without price iteration, the option prices shown are used throughout the run. But for an option such as "Static Queuing--Price Iteration", static queuing prices are used in the first iteration, and the standard pricing iteration procedure is used thereafter.

If you select price iteration, you will be asked for the maximum number of iterations, with a default of 5. If you select user prices, you will asked to enter a price for each machine in order (separated by spaces). Prices entered only need to reflect the relative importance of the machines. For example, you could input the relative load on each machine.

In addition, note the following about the pricing menu:

● **Myopic** chooses a price of 1.0 for the current machine for any job, and 0 for other remaining machines.

- **Bottleneck** chooses a price of 1.0 for the highest-utilization machine that the job has not yet processed on, and 0 for other remaining machines.
- **Static Queuing** chooses a price for each machine increasing non-linearly in its historical utilization.
- **Dynamic Queuing** chooses a price for each machine based both on its historical utilization and its current line length.

```
FLOW SHOP
OUTPUT REPORT (One Choice)

1. Summary: Values of Heuristic(s) Vs. Benchmark

2. Detailed:  Input, Heuristic Values, Gantt Charts
```

Figure D.57. Flow Shop Output Report Menu

The next screen asks for the Output Report desired, as shown in Figure D.57. A sample summary report is shown in Figure D.58.

```
FLOW SHOP
RUN SUMMARY HEURISTICS--SOME PROBLEM--FLOWSHOP--Weighted
Tardiness

5 LEADTIME ITERATIONS; 5 PRICE ITERATIONS
STATIC QUEUING PRICING

Heuristic          Objective Value   % Deviation
Random                  174             132.0
WSPT                    142              89.0
R&M                      75               0.0
EODD                     95              27.0
BENCHMARK                75
```

Figure D.58. Run Summary Screen Comparing Heuristics

If you choose detailed output, the report consists of a repetition of the input problem for the run, followed by the above comparison of the heuristic performances, followed by a final Gantt Chart for each heuristic. You can press ↵ again to get to the End of Run options, as shown in Figure D.59.

If you choose to make another run of the flow shop, however, you are returned to Objective Choices rather than Solution Methods, as in the one-machine case.

```
FLOW SHOP
END OF RUN (Choose at Least One)

Save Output
Return to Objective Choices
Quit the Program
```

Figure D.59. Flow Shop End of Run Menu

D.6   THE JOB SHOP

As already noted, the only real difference between the flow shop and the job shop is that for the flow shop, all jobs follow the same routes, while in the job shop, jobs may be grouped into clusters each of which follow the same routes. Thus almost everything in the software has been set up the same for both, except for manual input. Manual single problem input is treated fully here, since it differs from the Flow Shop version somewhat. The multiple problem statistical input case is also treated, which was not discuss in the flow shop section.

D.6.1   Single Problem--Detailed Input

The program first prompts you to provide the number of jobs, number of machines, and number of routes. Then, for each route, the machine prompts for the route length (number of operations), and the machine numbers visited in sequence.

```
JOB SHOP--SINGLE PROBLEM
DETAILED INPUT

#Jobs:  6        #Machines:  4       #Routes 3

Route # 1 >    2 / 1   4
Route # 2 >    6 / 1   4   2   4   3   4
Route # 3 >    2 / 2   4

Job #1 >   1   0   17   4.6 / 2 4
Job #2 >   1   0    8   1.0 / 4 7
Job #3 >   2   0    2   4.5 / 1 6 6 4 4 5
Job #4 >   1   0    6   1.0 / 5 6
Job #5 >   2   0   10   4.5 / 0 6 4 4 9 1
Job #6 >   3   0   22   1.3 / 2 2
```

Figure D.60. Job Shop Detailed Input Sample

Note re-entrant machines are permitted; that is, a job may visit a machine more than once. Separate the entries by spaces rather than commas, and put a slash after route length. Then, for each job, enter the route number, arrival time in shop, global due date, weight, slash, followed by job process time on each machine in the route.

Actual input could look like the screen in Figure D.60. In this example, machine 4 was set up to be a bottleneck. Once again, floating-point numbers are only allowed for weights. All other values must be in integer form. This input format is practical for small problems, but for larger problems we probably prefer to use statistical input.

D.6.2   Job Shop--Single/Multiple Problem(s)--Statistical Input

If you select **Statistical Input**, the problem input might look something like the sample in Figure D.61. All random variables have the same standard deviation divided by the mean, here 40%. Relative machine speeds together with average time on route allow determination of average time at each machine for each route. Job routes are determined at random with the probabilities given. The mean (weight/total process time) is also randomized.

```
JOB SHOP--SINGLE/MULTIPLE STATISTICAL INPUT

#Jobs: 30        #Machines: 8        #Routes: 3

Number of Replications: 2
(St. Dev.)/(Mean) of Random Factors:   0.3
Relative Machine Speeds>  1  4  3  5  2  2  4  3
Route #1>   5 / 1 3 4 7 3
Route #2>   8 / 1 2 3 4 5 6 7 8
Route #3>   9 / 1 2 3 4 5 3 6 7 6
Proportion on Route>  0.25 0.35 0.40
Average Time on Route>   30.5   60.1   70.2
Mean (weight)/(total process time) :    1.1
Utilization of Bottleneck Machine:   0.85
Tardiness Factor:   0.7
Random Seed:   654231
```

Figure D.61. Job Shop Statistical Input Sample

Overall process times are then multiplied by a constant to give the right utilization of the bottleneck machine. Finally, due dates are tightened or loosened to give the right tardiness factor. Note that the bottleneck machine is determined by adding up total expected processing time for the problem on each possible machine, and selecting the largest.

INDEX